11th Hungarian Conference on Materials Science 2017

IOP Conference Series: Materials Science and Engineering
Volume 426

Balatonkenese, Hungary
15 – 17 October 2017

ISBN: 978-1-5108-7365-0
ISSN: 1757-8981

Printed from e-media with permission by:

Curran Associates, Inc.
57 Morehouse Lane
Red Hook, NY 12571

Some format issues inherent in the e-media version may also appear in this print version.

Printed by Curran Associates, Inc. (2018)

For permission requests, please contact the Institute of Physics
at the address below.

Institute of Physics
Dirac House, Temple Back
Bristol BS1 6BE UK

Phone: 44 1 17 929 7481
Fax: 44 1 17 920 0979

techtracking@iop.org

Additional copies of this publication are available from:

Curran Associates, Inc.
57 Morehouse Lane
Red Hook, NY 12571 USA
Phone: 845-758-0400
Fax: 845-758-2633
Email: curran@proceedings.com
Web: www.proceedings.com

Table of contents

Volume 426

11th Hungarian Conference on Materials Science

15–17 October 2017, Balatonkenese, Hungary

Accepted papers received: 3 September 2018
Published online: 18 October 2018

Preface

Papers

IOP Conf. Series: Materials Science and Engineering **426** (2018) 011001 doi:10.1088/1757-899X/426/1/011001

Preface

The series "Hungarian Conference and Exhibition on Materials Science, Testing and Informatics" (later simply "Hungarian Conference on Materials Science") was intended to provide a forum where Hungarian and foreign scientists and research groups interested in metals and alloys, silicates, polymers and composites get an opportunity to exchange and publish ideas and establish new and integrated partnerships. The 11th Hungarian Conference on Materials Science was held at the Balaton lakeshore in Balatonkenese, October 15-17, 2017.

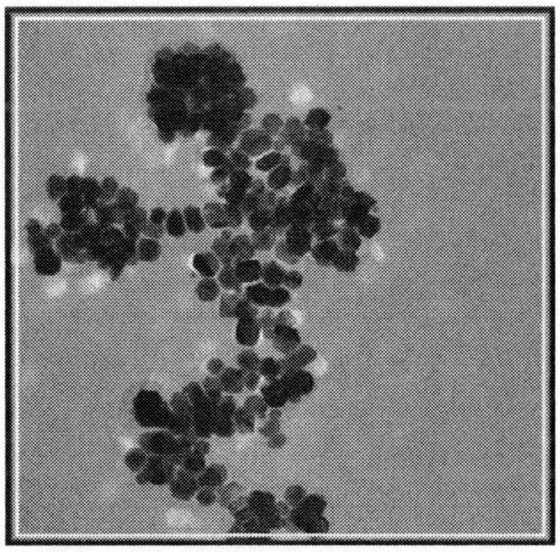

The winner of the photo competition of the 11th Hungarian Conference on Materials Science was Mr. Koppány Juhász, PhD student from the Department of Applied and Environmental Chemistry, University of Szeged.
The title of the photo is "Astronaut"
There can be seen platinum nanoparticles, forming a human shape.

The organisers present this volume as a peer-reviewed selection of papers submitted to publication by participants

The organisers express their gratitude to all sponsors. Thanks are due to organising societies, as the Hungarian Materials Science Society and to other Hungarian scientific and engineering societies for their support.

Budapest, July, 2018.

Peter J. SZABÓ
Chairman of Organising Committee

József GYULAI, András ROÓSZ
Co-chairmen of Scientific Committee

IOP Conf. Series: Materials Science and Engineering **426** (2018) 011002 doi:10.1088/1757-899X/426/1/011002

Peer review statement

All papers published in this volume of *IOP Conference Series: Materials Science and Engineering* have been peer reviewed through processes administered by the proceedings Editors. Reviews were conducted by expert referees to the professional and scientific standards expected of a proceedings journal published by IOP Publishing.

Characterization of grain structure and crystallographic texture variation during the production of a Properzi semi-product

D A Angel[1], A Hlavacs[1], D Petho[1], V Mertinger[1], P Barkoczy[1] and M Benke[1]

[1] University of Miskolc, Institute of Physical Metallurgy, Metalforming and Nanotechnology, Miskolc-Egyetemváros, Hungary

E-mail: femangel@uni-miskolc.hu

Abstract. The solidification on the casting wheel is the first step of the Properzi process which is followed by rolling in several steps, in this case 11 rolling steps. The uniqueness of the technology is that the rolling occurs from 3 directions. Therefore, a completely unique texture develops in the product compared to classic rolling. This aluminium semi-product is the base product for wire drawing. The characteristic of the strand greatly influences the success of the wire drawing therefore, knowing the crystallographic texture is demanded. In this paper, the grain structure and crystallographic texture variation during the full production technology of a Properzi semi-product is characterised. The examined strand includes all the technological steps of the process (casting and the 11 rolling steps). To perform such investigation, strand the complete production must be stopped for sampling because of the closed and semi-continuous type of the process, Therefore, the presented work is unique. To characterize the semi- product, cross-sectional and longitudinal-sectional samples were prepared. Grain structure and X-ray diffraction (XRD) texture examinations were carried out on the two sample series. Based on the results it can be stated, that despite of the grain refinement, initially a columnar structure formed and a strong <100> texture developed. During rolling, the <111> fibre texture developed. After the last step, the initial <100> texture is still present besides the <111> fibre texture. Consequently, the initial conditions of the crystallization highly determine the developed texture of the semi- product.

1. Introduction

The mechanical properties and formability of the products and semi-products are highly determined by the crystallographic texture, therefore knowing the texture development during the technology steps helps to understand the inter process behaviour and final texture of the product.

The Properzi technology is a unique and different process compared to classic rolling. While during classic rolling the rolling occurs from 2 directions, for the Properzi process, the rolling occurs from 3 directions, where the angle between the neighboring rolls are is 120°. Due to this configuration, a unique and different crystallographic texture compared to classic rolling develops within the product. The process itself is a half-continuous, closed system. Because of that, sampling is very complicated and requires the stop of the manufacturing line. In this paper, a strand produced by Properzi process was examined after every technological step. The aim of this paper is to characterize the crystallographic texture and grain structure variation during the full Properzi production technology. To achieve this aim, XRD texture and grain structure examinations were carried out.

2. Experimental

The examined aluminium strand was produced by Properzi process. For sampling, the whole process was stopped and a section of the strand including all technological steps (casting and 11 rolling steps) was taken out. The obtained strand is shown in Fig. 1.

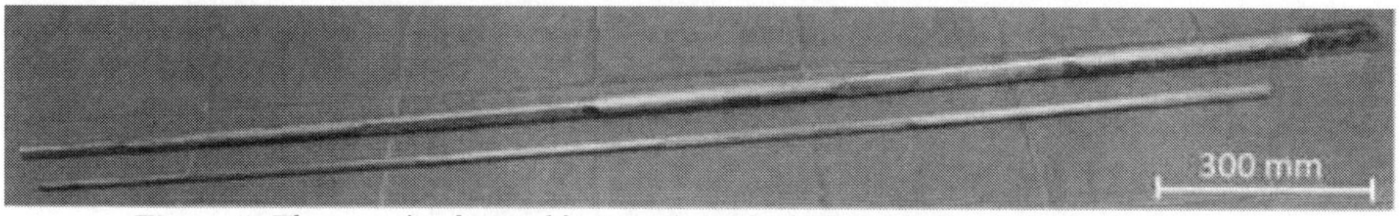

Figure 1. The examined strand in two pieces including all the technological steps

XRD texture measurements were carried out with a Bruker D8 Advance X-ray diffractometer equipped with an Eulerian cradle. For the measurements, circular cross-sectional samples were machined from each technological step (Fig. 2.). The first step of the texture measurement was to record the full interference function of the samples, followed by incomplete (up to 75° tilting) pole figure measurements. Afterwards, the Orientation Distribution Function (ODF) and the volume fractions of the favoured unit cell orientations were calculated. The Eulerian angles of ODF are connected to the coordinate system of classic rolling. Furthermore, the rotation axes of the Eulerian angles are bonded to the sample holder of the X-ray diffractometer. Therefore, to properly define the unit cell orientations in the Properzi coordinate system, the rotation axes and directions of the Eulerian angles had to be implemented into the coordinate system of the Properzi samples [1-4].

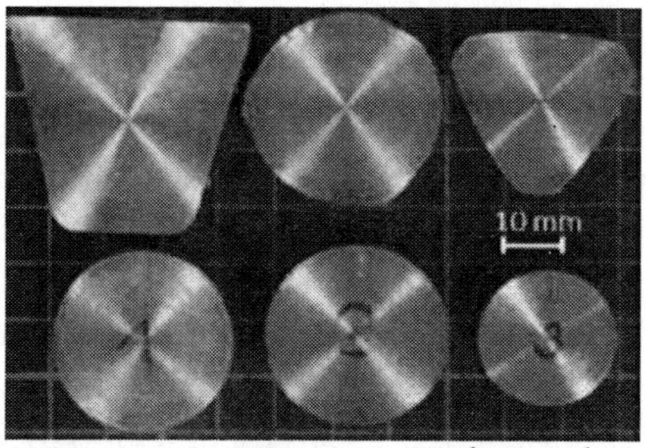

Figure 2. Cross-section samples after the 1st, 2nd and 3rd manufacturing steps prepared for XRD measurements

The grain structure examinations were carried out on two sample series, cross-sectional (Fig.3a) and longitudinal-sectional (Fig. 3b). The cross-sectional samples were turned out from each technological step, meanwhile the longitudinal-sectional samples were machined out from the transitions between two subsequent steps. called transitional zones. After sample' preparations (grinding, polishing), two different etching techniques were used for both sample series. First, macro etching with caustic containing HNO_3, HCL, HF and subsequently, re-preparation of the same sample with Barker etchant with caustic containing distilled water, HBF_4 were performed. For the examination of the grain structure for macro etching, a Canon EOS 700D + EFS 60 mm SLR camera and for Barker etching Zeiss AxioVert 100 optical microscope with polarized light were used [5-6].

IOP Conf. Series: Materials Science and Engineering **426** (2018) 012001 doi:10.1088/1757-899X/426/1/012001

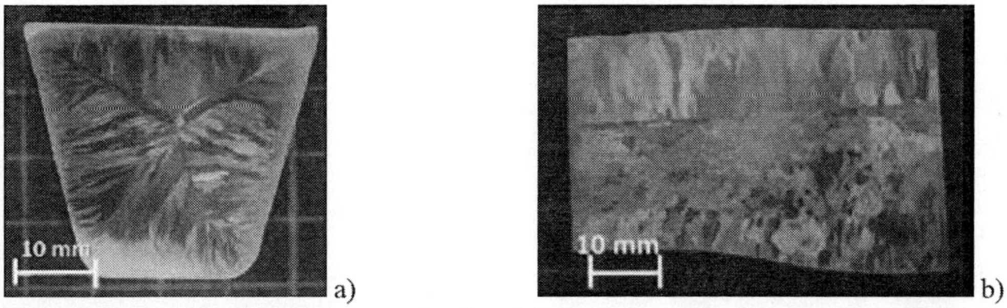

Figure 3. The 1st cross-sectional sample a) and the 1st longitudinal-sectional sample b)

3. Results and Discussion

Fig. 4 shows the results of the macro and barker etched cross-sectional samples after the 1st, 5th and 12th technology steps. Due to the macro etching, the grain structure is visible for the naked eye. After each step, it can be observed that 4 zones formed according to the 4 directions of heat transfer during solidification. It can be also seen that the zones meet in the centre of the strand, furthermore, the grains formed a columnar structure within the zones despite grain refinement. As the technology proceeds the grains are more severely formed.

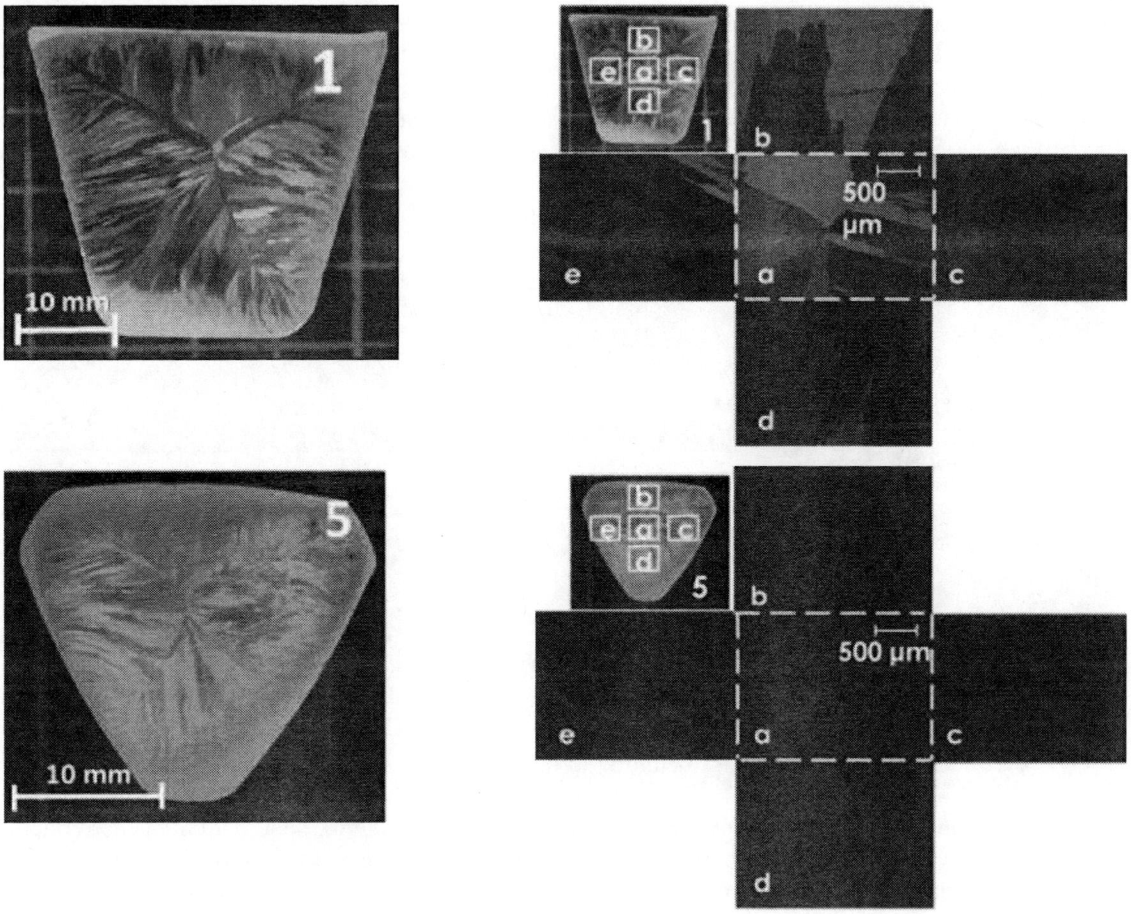

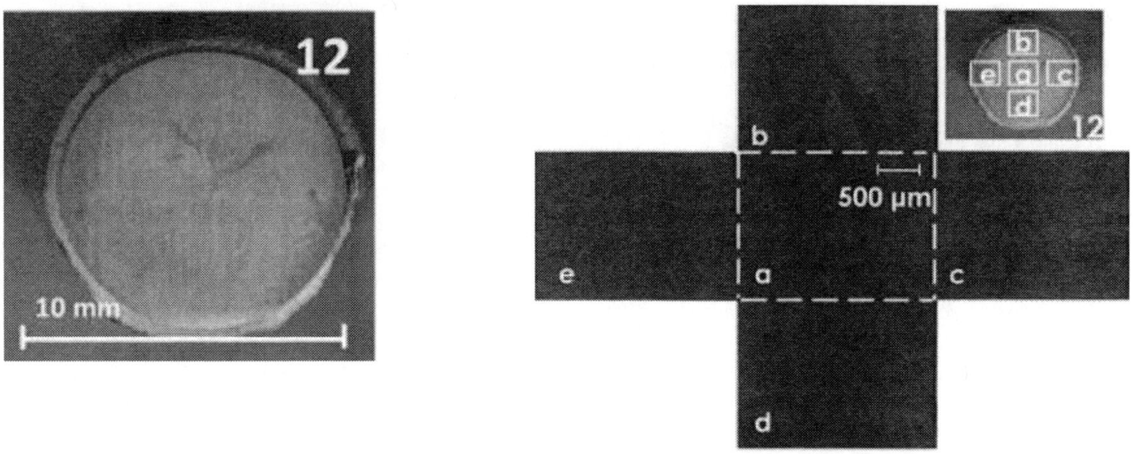

Figure 4. The grain structure of the cross-sectional samples after the 1st, 5th and 12th technological steps by macro and Barker etching

As shown in longitudinal section series of Fig. 5, moving from the initial (as-cast) state towards to the last rolling step, the grain structure changes. It is confirmed that despite grain refinement was applied, initially a columnar structure was formed. In the middle of the columnar structure, a nearly equiaxed area (NEA) is visible which is parallel with the longitudinal direction of the strand and is surrounded by perpendicular columnar grains. During rolling, the grains in the NEA and the columnar grains are elongated and rotated into the rolling direction, creating a fiber-like grain structure. After the 7th - 8th technology steps, the two areas cannot be separated any longer and etching the end of the technology the fiber structure is becoming finer.

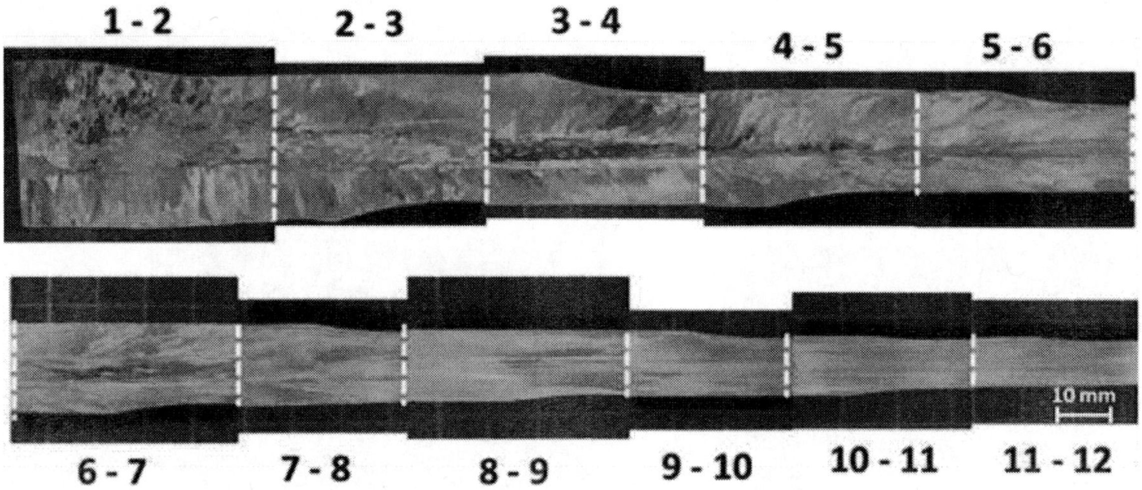

Figure 5. The grain structure of the longitudinal-sectional samples by macro etching in all technology steps

In the longitudinal-sectional sample series the effect of rolling is the most evident at the reduction zones (transition zones) Because of that, Barker etching was performed at these zones Fig. 6 shows the results of the Barker etching in the areas of the 1st – 2nd, 5th – 6th and 11th – 12th transitional zones. The above mentioned same phenomena - the initial columnar structure turns into a fiber structure - can be observed as the technology proceeds. Furthermore, due to the Barker etching, the change in the orientation of the grains can be seen.

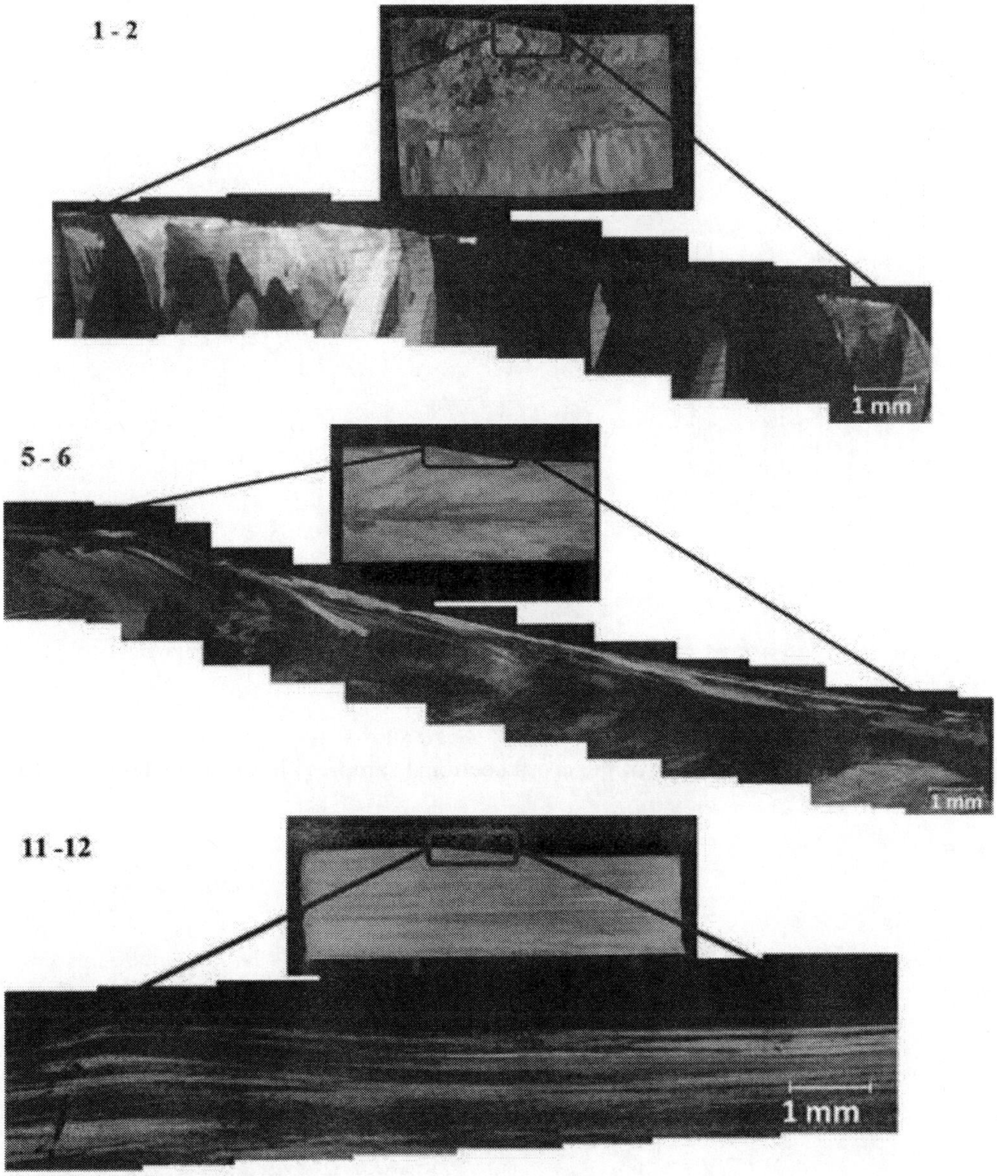

Figure 6. The grain structure of the longitudinal-sectional samples between the $1^{st}-2^{nd}$, $5^{th}-6^{th}$ and $11^{th}-12^{th}$ transitional zones

The full XRD spectra (interference functions) of the circular cross-sectional samples are shown in the Fig. 7. To eliminate the decreasing intensity of the peaks resulting from the decreasing examined volume, the intensity values of the peaks on each XRD spectra have been normalized to the {111} peak of the 12th sample. It is shown in Fig. 7 that after the initial technological step, the casting, a strong {200} (equivalent to {100}) texture developed. As the technology proceeds, a {111} texture is formed. In the final step, a strong {111} texture rules besides the initial {200} texture.

After transforming the calculated ODF into the Properzi coordinate system, the favored orientations of the unit cells and their volume fractions were calculated which presented in Fig. 8. The favored unit cell orientations are in a good agreement with the results obtained from the interference functions.

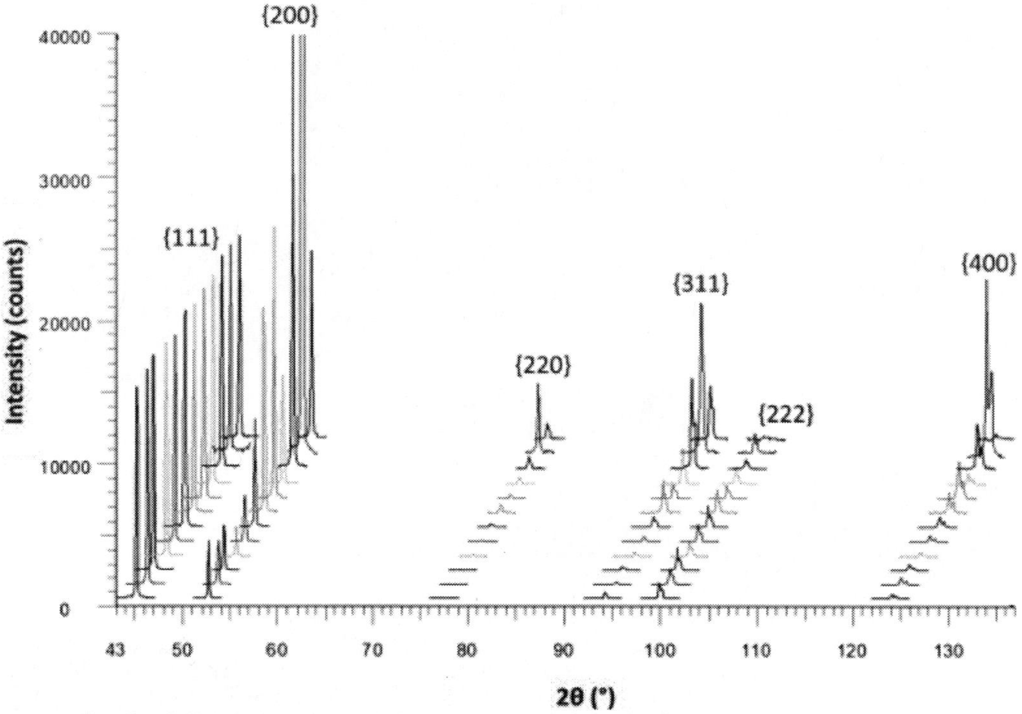

Figure 7. The full XRD spectra of the cross-sectional samples after each technological step

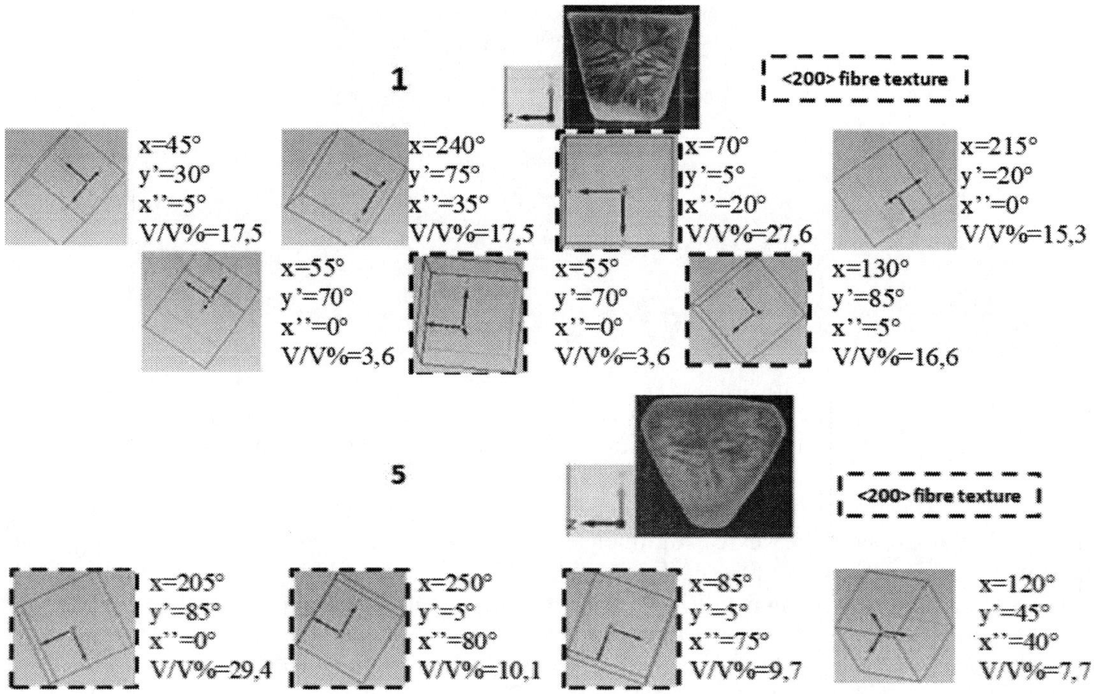

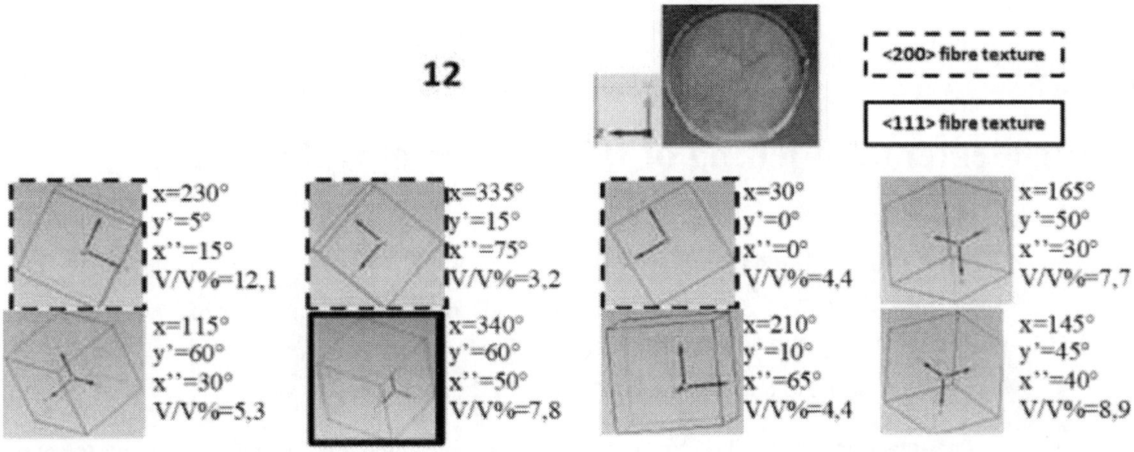

Figure 8. The favored orientations of the unit cells after the 1st, 5th and 12th technological steps

4. Summary

Based on the results of grain structure examinations, it can be stated that a strong columnar structure formed in the as-cast state despite of the applied grain refinement. The columnar structure turned into a fibre structure due to the rolling steps and as the technology proceeded, the fibre structure became finer.

Due to the formed columnar structure, a strong {200} texture developed after solidification, which remained in the material as the technology proceeded, although a strong {111} fibre texture developed due to the rolling. After the final step, besides the forming {111} fibre texture, the initially developed {200} fibre texture still can be observed.

It can be stated that the initially developed grain structure and crystallographic texture have a major influence on the final texture of the semi- product, thus, the properties of the semi- product.

Acknowledgment

The described research work was supported through the National Research, Development and Innovation Office – NKFIH K119566 project.

The described article was carried out as part of the GINOP-2.2.1-15-2016-00018 project implemented in the framework of the Szechenyi 2020 program. The realization of this project is supported by the European Union.

References

[1] U.F. Kocks, C.N. Tomé, H.R. Wenk, Texture and Anisotropy, Cambridge University Press, 1998
[2] Olaf Engler, Valerie Randle, Introduction to texture analysis, SecondEdition, 2010
[3] Satyam Suwas, Ranjit Kumar Ray, Crystallographic Texture of Materials, 2014
[4] ASTM E81: 1977 (1982) Preparing Quantitative Pole Figures of Metals
[5] Hlavács Adrienn, A mikroszerkezet változása Properzi öntvehengerlés során, Diploma thesis, Miskolc, Hungary 2014 (in Hungarian)
[6] Dr. Bárczy Pál, Dr. Fusch Erik, Metallográfia I., Tankönyvkiadó, 1981 (in Hungarian)

IOP Conf. Series: Materials Science and Engineering **426** (2018) 012002 doi:10.1088/1757-899X/426/1/012002

Finite element modelling of thin chip removal process

B Z Balázs[1] and M Takács[1]

[1] Budapest University of Technology and Economics, Department of Manufacturing Science and Engineering, Budapest, Hungary

E-mail: balazs@manuf.bme.hu, tm@manuf.bme.hu

Abstract. Micro milling is a continuously spreading, and therefore heavily researched technology. In this subject plenty of publications are highlighting the fact that there are many issues, which still has to be solved. Micro milling realizes thin chip removal process. In this paper 2D finite element simulation was performed to investigate the characteristics of small sized chip formation. Results of micro milling experiments were analysed too. This paper presents the results regarding effects of the cutting parameters (cutting speed, feed per tooth) on cutting temperature and machining forces at 2D finite element. Results based on different material models were compared with each other too.

1. Introduction

Nowadays, the role of miniaturization is becoming higher in many areas of our lives, but the production of these components requires the application of advanced technologies. One of the most important ones is micro milling, which typically produces structures between 1-999 μm on a wide range of materials. Such components are required for example in military industry, space technology, production of electrical components and in medical sciences.

Micro milling can be considered as transformation of conventional sized milling into the size range, which is less than 1 mm. Although the tool geometry and the kinematic system at micro end milling are actually the same, as that of conventional milling, there are many differences, which are originating from the size reduction. All these phenomena can be summarized as size effect.

Due to the size reduction, the effect of the tool edge radius is not negligible, because its size is comparable with the chip thickness. According to Aramcharoen and Mativenga [1], the ratio between the uncut chip thickness and the cutting tool edge radius is an important parameter in micro milling. Machining of hardened tool steel AISI H13 was investigated, and it was shown that the role of size effect is higher, if this ratio is less than one.

Other critical parameter considering the size effect is the minimum chip thickness (h_{min}). If the chip thickness is less than a critical value, there will be no material separation. The minimum value was considered by many researchers as a function of the cutting tool edge radius (r_β). De Oliveira et al. [2] investigated the ratio of h_{min}/r_β on AISI 1045, and it was found to be between 0.22 and 0.36. Cuba Ramos et al. [3] determined this ratio on AISI 1045 as 0.29, and Kang et al. [4] found a value of 0.3. Wu et al. [5] conducted finite element modelling in Deform FEM program and the results were validated micro turning experiments. The effect of the tool edge radius and the grain size were studied at a cutting speed of 314 mm/sec, the uncut chip thickness was 3 and 11 μm, and the cutting edge radius were 0, 2, 5, and 8 μm. Oxygen free copper with 20 and 60 μm grain size was investigated as workpiece material. It was found that the larger the cutting edge radius the larger the cutting force. Considering both factors, the effect of the grain size is more significant. The authors stated that the decreased grain size increases the forces and the specific energy. The chip formation requires the movement of the dislocations, but the grain boundaries block them. Smaller grain size will result in more grain boundaries, which have a more intensive blocking effect. It was also found that in the case of a 5 μm cutting edge radius, the shearing force will have a ratio of 45% of the cutting force, while the ploughing force will be 55% of the cutting force. Therefore the larger cutting tool edge radius generates larger differences.

Zhou et al. [6] predicted the cutting forces based on finite element modeling of orthogonal cutting on mirror die made of NAK 80, and the results were validated experimentally. The model has taken into account the tool trajectory, run-out, cutting edge radius, rake angle, mechanical and physical properties of the material, and the interaction between the cutting tool and the workpiece. The tool diameter was 0.4 mm, and cutting velocity was set in the range of v_c=12-36 m/min, and the uncut chip thickness was found to be between 0.1-1.2 μm. The largest von Mises stress (2000 MPa) raised in the secondary shear zone, the highest temperature was 210 °C. Based on the simulations the effect of the cutting speed on the forces is negligible in the investigated parameters range. Predicted and measured forces approached each other well.

Yang et al. [7] investigated the temperature distribution, effective stress and the cutting forces at various cutting edge radii in case of micro milling on Al2024-T6 material by means of numerical and experimental approaches. The tool diameter was 0.5 mm with a rake angle of 5° and with a clearance angle of 12°. The set parameters were as follows: n=22 282 rotation/min, a_p=10 μm, f_z=0.1 μm, and the cutting edge radius were 0, 3.2, 5 and 7 μm. The temperature decreased with the increasing of the cutting edge radius, the values for the above listed radii were 57.5, 51.5, 45.4 and 40.4 °C, respectively. The generated heat is concentrated to a small area at the contact point of the tool and the workpiece. In addition, effective stress has decreased, but the cutting forces have increased. The temperatures were slightly higher in the simulations than that of the experimental results, which can be explained by the fact that the tool edge radius increases during machining.

Pratap et al. [8] studied the cutting forces, the temperature and the stress distribution on Ti-6Al-4V material by finite element method, taking into account the effect of cutting speed, feed rate, uncut chip thickness, and cutting edge radius. The simulated results were validated by micro milling experiments. The diameter of the tool was 0.4 mm, the spindle rotational speed was chosen in the range of 15 000–35 000 rpm, the feed per tooth was set the range of 1 to 5 μm, and the depth of cut was 30 μm. At v_c=31,425 m/min, f_z=1 μm and a_p=30 μm the maximum stress (2467 MPa) was generated in the primary shear zone, which can be explained by the size effect. The maximum calculated temperature was as high as 845.3 °C. The cutting forces increased with the intensity of the cutting process, which could be caused by the growth of the cutting edge radius caused by high cutting temperature and accelerated wear accordingly.

Afazov et al. [9] investigated the micro milling process on AISI 4340 by experimental and finite element methods. According to the suggestion, novel force model factors should have been taken into account, which directly affect the chip formation, such as tool run-out, cutting edge radius, cutting speed, cutting tool geometry, and the workpiece preheating. It has been found that preheating of the workpiece material to 600 °C resulted in a higher stability limit due to the softening and the decrease of the cutting forces. Considering the tool geometry, the authors stated that an increase of the rake angle from 0° to 8° creates larger force components, while the stability limits will be lower.

2. Aims and circumstances

The main aim of this research is the better understanding of the micro milling process by studying the generated temperatures and forces during the cutting process. 2D finite element simulations were performed by Deform software. The chip formation phenomenon has been studied very close to the shear zone. The practical investigation of real material flow and temperature distribution is very complicated because of the limited size of the process. Finite element simulations were used by many researchers to investigate the cutting processes [8, 10].

The geometry applied in the simulations was based on a two flutes carbide end milling cutter produced by Magafor. The diameter of the tool is 0.3 mm. The geometric characteristics are as follows:

rake angle (γ) is 35°, clearance angle (α) is 20°, and helix angle (λ) is 30°. Based on the scanning electron microscopic images the cutting edge radius in the simulation was set to 3 μm. Machining of AISI 1045 was analyzed by means of finite element simulation. Based on the material model available in Deform software the material properties at room temperature are as follows: E=212 000 MPa, thermal conductivity is 41.7 W/(m·K), the thermal expansion is $1.19 \cdot 10^{-5}$ m/°C, which are changing with the temperature. The Poisson ratio is 0.3. The tool material is WC and the properties are as follows: E=650 000 MPa, Poisson ratio is 0.25, thermal conductivity is 59 W/(m·K) and the thermal expansion is $5 \cdot 10^{-6}$ m/°C. In case of carbide material the values are not changing with the temperature. Special material behaviors of chip removal process, such as intensive deformation and high strain rate require the application of visco-plastic models. Three different material models were applied and evaluated to make a comparison: Oxley's equation [11], Johnson-Cook constitutive equation [12], and a material model based on the flow stress data originated from the researchers of Engineering Research Centre (ERC) of Ohio State University [13]. Considering our previous investigation, the Oxley's equation provides the best approximation of the practical values in the case of normalized AISI 1045 steel [14]. Shear friction coefficient was fixed at 0,6 based on other studies [5, 15]. Two additional meshing windows were applied on the workpiece with even smaller element sizes (e.g. 0.5 μm), to ensure the appropriate mesh density in the near of the chip formation. The effect of the mesh size was investigated too.

In this study, the effects of two factors were investigated, such as cutting speed and feed per tooth. The levels of the experimental design are shown in the Table 1. The minimum chip thickness was taken into consideration at determination of the experimental values of feed per tooth. Systematic series of finite element simulation was carried out based on the full factorial design of experiments (DoE) method, which is often used during the scientific researches [16, 17, 18].

Table 1. Experimental design

Factors	Levels
v_c (m/min)	30, 60, 90, 120, 150
f_z (μm/tooth)	1, 3, 5, 8

At 2D simulations orthogonal cutting process is examined actually, where there are basically two important directions, as shown in the Figure 1. Direction of the cutting force is parallel to the machined surface. The perpendicular direction to the surface of the machined workpiece is the direction of the

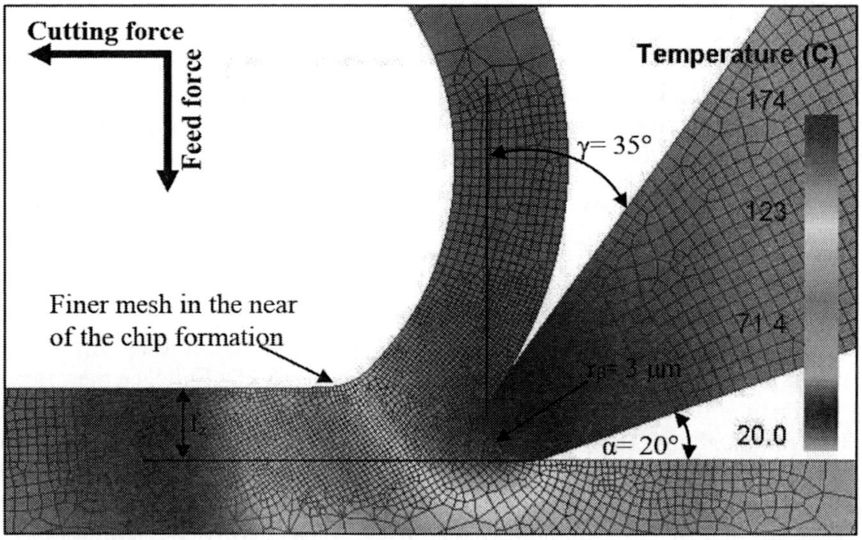

Figure 1. 2D finite element model of thin chip removal process

passive force, which can be considered as feed force in the case of milling process. There are some simplifications in the 2D simulations compared to the real milling process. E.g. the chip thickness was constant during the simulation, and it corresponds to the maximum thickness of the material layer removed by one edge of a micro end milling cutter. Other important difference lays in the width of the chip, which is influenced by the setting of the depth of cut. It is considered to have a unit thickness. In the paper the feed force was analysed and discussed too, because it has a great importance if the chip thickness is less than the critical value. In that case, the tool can pass through the material several times without chip removal, which can cause local hardening. Although, no parameter was chosen below the minimal chip thickness in this paper, but a worn tool can change the ratio of cutting edge radius and chip thickness in real life, so a given feed per tooth value can be shifted to the non-preferable range.

3. Results and discussion

3.1. Tool temperature

The temperature has a great importance, as it affects the tool wear and the cutting force [9]. The temperature of the tool was investigated in detail by means of finite element simulation of thin chip removal process. Figure 2. shows the characteristic change of the temperature value in any point of the tool, which is in physical contact with the machined workpiece. It can be seen that the temperature changing rate is much higher at the starting point of the cutting than in the later position of the tool tip. The cutting length of 0.45 mm corresponds approximately to a half rotation of the tool in the case of an

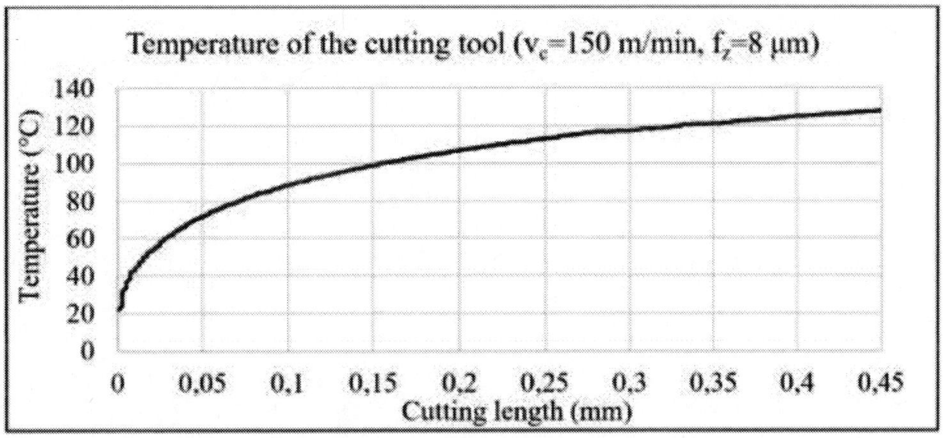

Figure 2. Finite element simulated temperature of the cutting tool

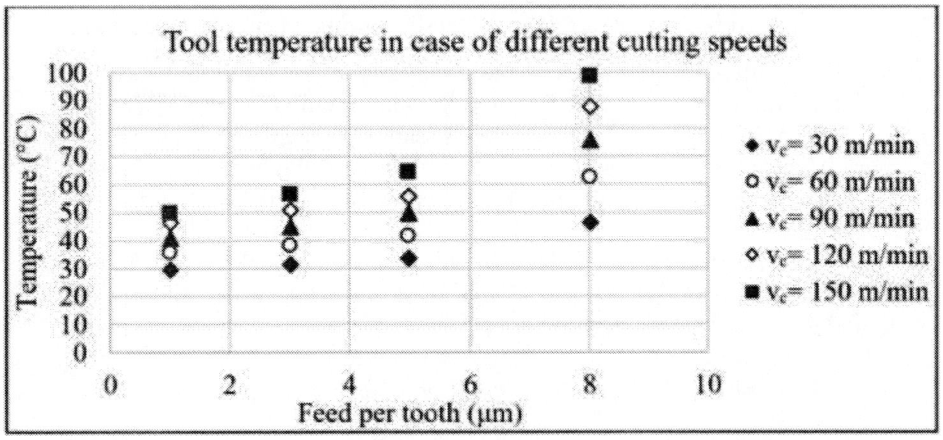

Figure 3. Effect of the cutting parameters on the simulated tool temperature

IOP Conf. Series: Materials Science and Engineering **426** (2018) 012002 doi:10.1088/1757-899X/426/1/012002

end milling cutter with a diameter of 0.3 mm, which means the full impact of one edge. The tool temperature at thin chip removal process can be evaluated as quite low, similar to the results of other researchers [7]. It must be taken into account that there is no continuous chip formation during the end milling process. In the case of full groove milling, there is a half rotation for the cooling of one edge of the tool.

Large number of simulations were performed regarding the tool temperature. Figure 3. shows the temperature values as a function of feed per tooth at different cutting velocities after a cutting length of 0.15 mm. It can be seen that the larger the feed per tooth, and the larger the cutting velocity the larger the temperature.

If mesh size of the model is decreased, higher temperature values were calculated.

3.2. Machining forces

The cutting force and passive force were examined by means of finite element simulation too. Because of the simplification of the 2D model the passive force component corresponds to the feed force in the case of milling process. It was found that a larger cutting speed slightly reduces the cutting force in the investigated parameter range, which is consistent with the conventional size milling, where the cutting velocity has a weak effect on the main cutting force [19]. However, the effect of the feed per tooth is more significant (see Figure 4.). The simulated cutting forces are relatively high, nearly one order of magnitude higher than our experimental results (e.g. ~1 N vs. ~10 N). The reason of that may be the difference between the 2D simulation and the actual kinematic of milling process. E.g. the chip cross section is not taken into consideration by the 2D. The width of the chip (depth of cut) can not be defined in this case, although it has great importance on the forces.

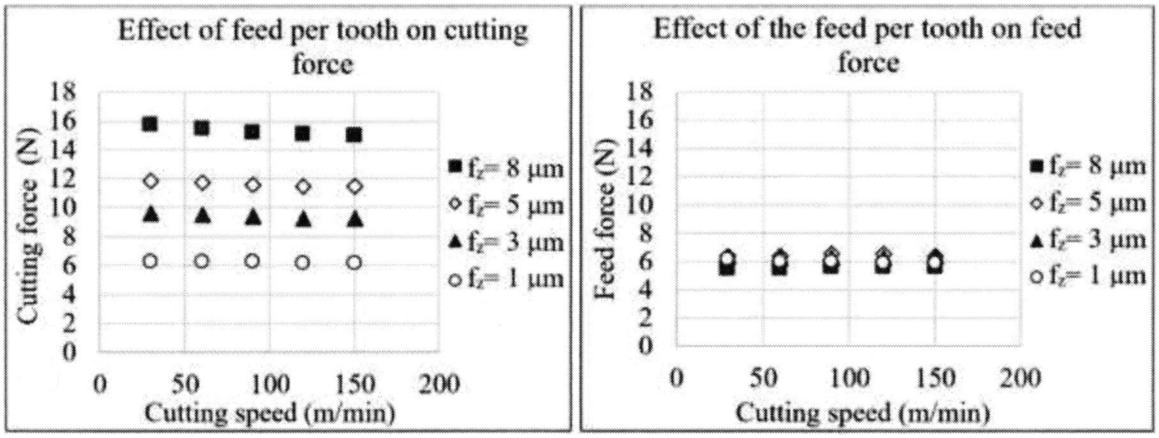

Figure 4. Effect of the cutting parameters on the simulated cutting force and feed force

As a result of the simulations the values of the passive force varied within a relatively small range (see Figure 4.), the minimum value was 5.48 N, and the maximum value was 6.5 N at f_z=8 μm and f_z=5 μm, respectively. The ratio of F_f/F_c is also examined. It can be stated that this ratio decreases, if feed per tooth is set larger. In the case of f_z=1 μm the ratio is as high as 0.95, at f_z=3 μm it is 0.67, if f_z=5 μm is set, it will be 0.55, and in the case of f_z=8 μm the ratio is found to be 0.36.

Three different material models were applied and evaluated. At v_c=120 m/min and f_z=3 μm the Oxley's model provided F_c=9.31 N, F_f=6.3 N, the Johnson-Cook model resulted in F_c=9.36 N, F_f=6.36 N, and the ERC model led to F_c=10.33 N, F_f=5.58 N. It can be seen that the results of Oxley's and Johnson-Cook models are very close to each other.

It was observed that in the case of finer mesh size the cutting force (in X direction) is increased, and the feed force (in Y direction) is decreased. The deviation of the results decreased significantly at smaller mesh size.

IOP Conf. Series: Materials Science and Engineering **426** (2018) 012002 doi:10.1088/1757-899X/426/1/012002

4. Conclusions

The demand for better understanding of the micro milling process initiated performing 2D finite element simulation, which was carried out by means of Deform software. In this phase of research AISI 1045 material was investigated. The cutting temperatures and forces close to the cutting zone were studied in detail. Full groove milling experiments were also performed to make a comparison between the theoretical and practical results.

Based on the finite element calculation the temperature of the tool during the cutting process is slightly low, which increases by both the increasing feed per tooth and the cutting speed.

The cutting speed has a small effect on the cutting force in the examined parameter range. There is a slight reduction of the values, if cutting speed increases. There is a general observation considering the cutting force values determined by FEM simulation. The results are located above the measured values by an order of magnitude. The reason can be the 2D simplification of the model, where the width of the chip is considered as unity.

The passive force (which is the feed force in case of milling) change is small in the investigated parameter range. Considering the four feed per tooth values (1, 3, 5 and 8 μm), the highest passive force value belongs to f_z=5 μm, and the lowest one to f_z=8 μm. The F_f/F_c ratio decreased from ~0.95 to ~0.36 with increasing feed per tooth.

Acknowledgement

Current research work is part of the research project K 84177 “Modelling and dynamical investigation of high precision cutting of hard surfaces”. The author gratefully acknowledges the financial support of the Hungarian Scientific Research Fund (OTKA).

References

[1] Aramcharoen A and Mativenga P T 2009 Size effect and tool geometry in micromilling of tool steel *Precis. Eng.* **33** no. 4 pp 402–407

[2] Oliveira F B, Rodrigues A R, Coelho R T and Souza A F 2015 Size effect and minimum chip thickness in micromilling *Int. J. Mach. Tools Manuf.* **89** pp 39–54

[3] Ramos A C, Autenrieth H, Strauß T, Deuchert M, Hoffmeister J and Schulze V 2012 Characterization of the transition from ploughing to cutting in micro machining and evaluation of the minimum thickness of cut *J. Mater. Process. Technol.* **212**, no. 3, pp 594–600

[4] Kang I S, Kim J S and Seo Y W 2010 Investigation of cutting force behaviour considering the effect of cutting edge radius in the micro-scale milling of AISI 1045 steel *Proc. Inst. Mech. Eng. Part B J. Eng. Manuf.* **225** pp 163–171

[5] Wu X, Li L, He N, Yao C and Zhao M 2016 Influence of the cutting edge radius and the material grain size on the cutting force in micro cutting *Precis. Eng.* **45** pp 359–364

[6] Zhou L, Peng F, Yan R, Dong Q and Yang C 2015 Prediction and Experimental Validation of Micro End-Milling Forces with Finite Element Method *Intelligent Robotics and Applications* pp 664–675

[7] Yang K, Liang Y, Zheng K, Bai Q and Chen W 2011 Tool edge radius effect on cutting temperature in micro-end-milling process *Int. J. Adv. Manuf. Technol.* **52**, no. 9–12 pp 905–912

[8] Pratap T, Patra K and Dyakonov A A 2015 Modeling Cutting Force in Micro-Milling of Ti-6Al-4V Titanium Alloy *Procedia Eng.* **129** pp 134–139

[9] Afazov S M, Zdebski D, Ratchev S M, Segal J and Liu S 2013 Effects of micro-milling conditions on the cutting forces and process stability *J. Mater. Process. Technol.* **213** no. 5 pp 671–684

[10] Farkas B Zs 2014 Finite Element Analysis of Orthogonal Hard Turning with Different Tool Geometries *Key Eng. Mater.* **581** pp 163–168

[11] Oxley P L B 1989 Mechanics of Machining, An Analytical Approach to Assessing Machinability *Halsted Press* New York

IOP Conf. Series: Materials Science and Engineering **426** (2018) 012002 doi:10.1088/1757-899X/426/1/012002

[12] Jaspers S P F C and Dautzenberg J H 2002 Material behavior in conditions similar to metal cutting: flow stress in the primary shear zone *J. Mater. Process. Technol.* **122** pp 322-330
[13] Sartkulvanich P, Koppka F and Altan T 2004 Determination of flow stress for metal cutting simulation – a progress report *J. Mater. Process. Technol.* **146** pp 61-71
[14] Takacs M 2013 Numerical Simulation of Chip Removal Process Performed by Cutting Insert with Unique Edge Geometry *Proc. 2nd Int. Conf. Mech. Eng. Mechatron.* Paper No. 202
[15] Wu X, Li L and He N 2017 Investigation on the burr formation mechanism in micro cutting *Precis. Eng.* **47** pp 191–196
[16] Swain N, Venkatesh V, Kumar P, Srinivas G, Ravishankar S and Barshilia H C 2017 An experimental investigation on the machining characteristics of Nimonic 75 using uncoated and TiAlN coated tungsten carbide micro-end mills *CIRP J. Manuf. Sci. Technol.* **16** pp 34–42.
[17] Srinivasa Y V and Shunmugam M S 2013 Mechanistic model for prediction of cutting forces in micro end-milling and experimental comparison *Int. J. Mach. Tools Manuf.* **67** pp 18–27
[18] Geier N and Szalay T 2017 Optimisation of process parameters for the orbital and conventional drilling of uni-directional carbon fibre-reinforced polymers (UD-CFRP) *Measurement* **110** pp 319–334
[19] Biró I, Czampa M and Szalay T 2014 Experimental Model for the Main Cutting Force in Face Milling of a High Strength Structural Steel *Period. Polytech. Mech. Eng.* **59** pp 16–22

IOP Conf. Series: Materials Science and Engineering **426** (2018) 012003 doi:10.1088/1757-899X/426/1/012003

A simple correlation between texture and earing

M Benke[1], A Hlavacs[1], D Petho[1], D A Angel[1], M Sepsi[1], E Nagy[2], V Mertinger[1]

[1] University of Miskolc, Institute of Physical Metallurgy, Metalforming and Nanotechnology, Miskolc-Egyetemváros, Hungary
[2] MTA-ME Materials Research Group, Miskolc-Egyetemváros, Hungary

E-mail: fembenke@uni-miskolc.hu

Abstract. The phenomenon of earing has been known since the rise of deep drawing process. It has been early recognized, that earing is caused by the plastic anisotropy of textured materials. Over the past decades, there have been many methods proposed to find a way to predict earing through either mechanical tests, the combination of texture data evaluation and strain calculations, or finite element analysis. The objective of the present study is to propose a simple, yet suitable method, with which the average earing value of a blank can be estimated. The method solely requires measured texture (ODF) data and is based on defining every individual texture component pole as the vectoral combination of their (+) and (-) earing direction components. The method was tested on a sample series with different average earing values taken from an industrial rolled and annealed, 3003 type Al semi-product coil. Satisfactory correlation was found between estimated and measured earing numbers.

1. Introduction

During deep drawing of a cylindrical cup, different height forms along the circumference if the blank is anisotropic. The phenomenon is called earing and is caused by the plastic anisotropy of textured polycrystalline sheets. The demand to connect the observed earing to texture exists since the 1940s. Over decades, there have been quite many attempts to find the correlation between earing and texture. Fukui and Kudo [1], later Grewen [2] found that earing is predictable from the Lankford number (Δr). Tucker's method [3] was based on single slip and Schmid's law and was able to predict earing of a single crystalline material. His concept was applied to polycrystalline cases by Kanetake et al. [4], where tensile radial stress, compressive tangential stress and zero normal stress (to blank surface) was assumed during deep drawing and the bulk material was simplified to be an agglomerate of individual single crystals with certain orientations whose volume fractions were obtained from the orientation distribution function (ODF). Using empirical values in the work-hardening rule of crystals on earing, they accurately calculated not only the positions, but also the height of peaks within a blank. The model proposed by Da Costa Viana et al. [5] was based on combining ODF data with yield locus calculation. They assumed that the radial strains within sheets are inversely proportional to yield stresses. The method was able to predict the shape of earing, but it was not clarified whether their method predicted ear height. Pochinetto et al. [6] proposed a more sophisticated calculation, in which stress equilibrium equations were used. They obtained very good correlation between predicted and measured earing. Rodrugues and Bate [7] improved the previous concept, but their solution was suitable only for the four-ear cases. Moreover, they assumed that during deep drawing, tangential stress is compressive, normal stress is zero and radial stress is also zero, which is only valid for the outside edge of the blank. This was refined by Van Houtte [8]. All the above described methods predict the locations of peaks and troughs within a deep draw print

IOP Conf. Series: Materials Science and Engineering **426** (2018) 012003 doi:10.1088/1757-899X/426/1/012003

through additional mechanical tests, or evaluating the ODF and calculating the strain field or yield locus. Recently, earing is usually analyzed through finite element (FE) simulations. For those, the function describing the material's plastic anisotropy, which is implemented as user subroutines, is in the focus during development [9-11].

The aim of the present manuscript is to present a simple, yet suitable method to estimate earing directly and solely from the texture measurement (ODF) data. At first, only the value of average earing of a blank is dealt with.

2. Experimental

2.1. Sampling

The examined samples were taken from an industrial size, 3003 type aluminum coil manufactured by ARCONIC-Köfém Ltd. where the main production steps include closed direct chill casting, homogenization, hot-rolling, cold-rolling, annealing and finishing operations. In the annealed state, full-width sections were cut from the front and middle (along the length) of the coil. Samples were taken from five locations across the cross sections, namely: center, (C) side 1 quarter (Q1), side 1 edge (E1), side 2 quarter (Q2), side 2 edge (E2) for XRD and deep draw tests, according to Fig. 1. In the figure, it can also be seen that the XRD and deep draw test samples were taken out from as close as possible.

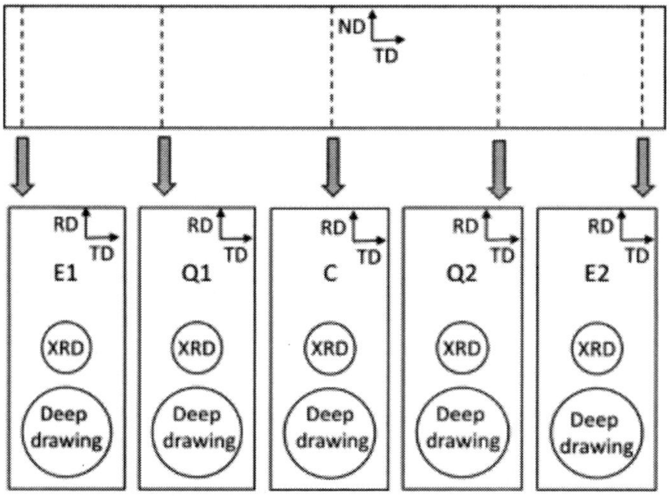

Figure 1. Sampling locations for XRD texture and deep drawing tests along the cross sections of the industrial coil. The reference rolling coordinate system is also marked.

2.2. Texture measurements

X-Ray diffraction measurements were performed with a Bruker D8 Advance type diffractometer equipped with an Eulerian cradle and CoKαtube (40 kV, 40 mA). The partial (0°-75°) {111}, {200} and {220} pole figures were measured and corrected for defocusing. After recalculation, the ODFs were obtained using the harmonic method with TexEval software. The main texture components were calculated with 10° scatter.

2.3. Determination of earing

It is commonly observed, that after cold rolling, the ears appear in the 45°(+n*90°) directions relative to rolling direction (RD), while after annealing in the RD and transverse direction (TD) (+n*90°) directions (Fig. 2.a). (Cold rolling earing is often called "-" earing, while annealed is called "+" earing. However, the signs are only labels, they have no mathematical sense.)

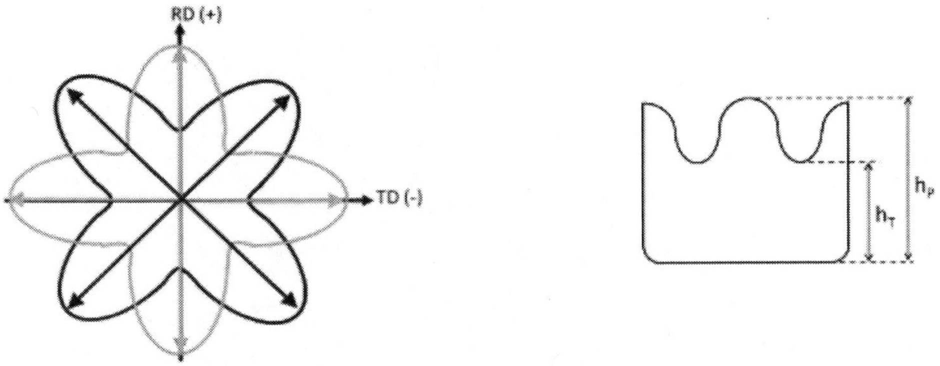

a) b)

Figure 2. a) Peak and valley heights in case of earing, b) earing directions after cold rolling (black) and annealing (grey).

In general, earing is characterized by deep-draw tests, which, in this case, were carried out by ARCONIC-Köfém Ltd. After deep drawing, the average height of the peaks (h_P^{AV}) and troughs (h_T^{AV}) were measured according to (1) and (2), respectively (Fig. 2. b).

$$h_P{}^{AV} = \frac{\sum_1^n h_P}{n_P} \tag{1}$$

$$h_T{}^{AV} = \frac{\sum_1^n h_T}{n_P} \tag{2}$$

Afterwards, the difference (h_D) between average peak height (h_P^{AV}) and troughs (h_T^{AV}) was calculated according to (3), and the mean height (h^{AV}) was calculated by (4).

$$h_D = h_P{}^{AV} - h_T{}^{AV} \tag{3}$$

$$h^{AV} = \frac{h_P{}^{AV} + h_T{}^{AV}}{2} \tag{4}$$

Finally, earing (Z) was calculated according to (5).

$$Z = \frac{h_D}{h^{AV}} * 100 \tag{5}$$

2.4. Calculation of the Vectoral Earing Ratio

In Fig. 3, poles of the main texture components (C, S, B, Cube, G) [12, 13] are presented on the {h00} pole figure. It can be clearly seen, that in the recrystallized state, the Cube and G components can be found along the RD and TD (+n*90°) directions. Comparing to Fig. 2. these are the directions, where the ears (peaks) appear after annealing. On the other hand, poles of the C, S and B components can be found close to the 45° (+n*90°) directions relative to RD. Again, these are the directions, where the ears appear after cold rolling (Fig. 2.). This purely empirical observation had already been recognized by many experts previously [8, 14-19]. Hence, the correlation between {h00} poles of the main texture components and the macroscopic formability is recognized.

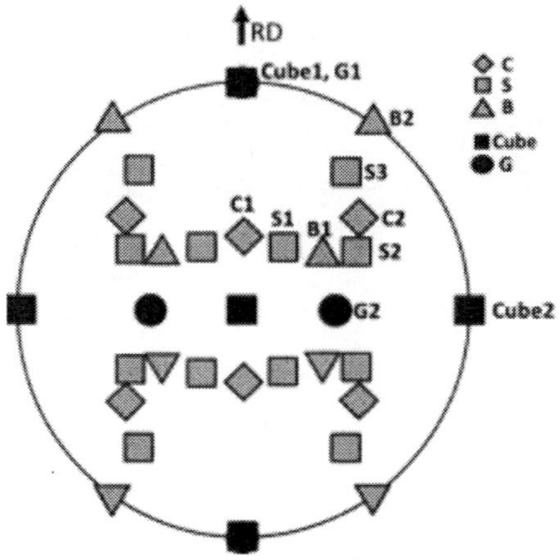

Figure 3. Poles of the main texture components (C, S, B, Cube, G) on the {200} pole figure [20].

This correlation can be more precisely described if it is examined, what is the contribution of an individual texture component pole P to the "+" and "-" earing. This is shown in Fig. 4, as follows. The direction of "+" earing overlaps with the RD and TD, while the direction of "-" earing is marked with the line 45° relative to RD. The location vector L (pointing to P from the origin O) can be resolved as the sum of vectors OD+DP, where OD is the contribution of P to (-) earing, and DP to (+) earing. If the pole coordinates, α (the altitude angle of P) and β (the azimuth angle of P) are known, OD and DP can be calculated. Due to the stereographic projection, the length of L linearly increases with α and reaches 1 if α=90°. Thus, L can be calculated according to (6).

$$L = \frac{\alpha}{90} \tag{6}$$

The projection of L to TD (ProjTD) can be calculated within the triangle OPT as follows:

$$Proj_{TD} = |L| \cos(90 - \beta) \tag{7}$$

Similarly, the projection of L to RD (ProjRD) can be obtained from the triangle OPR:

$$Proj_{RD} = |L| cos\beta \tag{8}$$

If the pole P is located between RD and (-) (Fig. 4 a.), OD can be obtained within the ODT triangle as:

$$OD = \frac{Proj_{TD}}{cos45°} \tag{9}$$

Afterwards, the projection of OD to RD (r) within the ODR' triangle is:

$$r = OD * sin45° \tag{10}$$

Finally, DP can be achieved by (11):

$$DP = Proj_{RD} - r \tag{11}$$

If the pole P is located between (-) and TD (Fig. 4 b.), OD is as determined from the triangle ODR as follows:

$$OD = \frac{Proj_{RD}}{cos45°} \quad (12)$$

Hence, the projection of OD to TD (t) within the ODT' triangle is:

$$t = OD * cos45° \quad (13)$$

At the end, DP can written as:

$$DP = Proj_{TD} - t \quad (14)$$

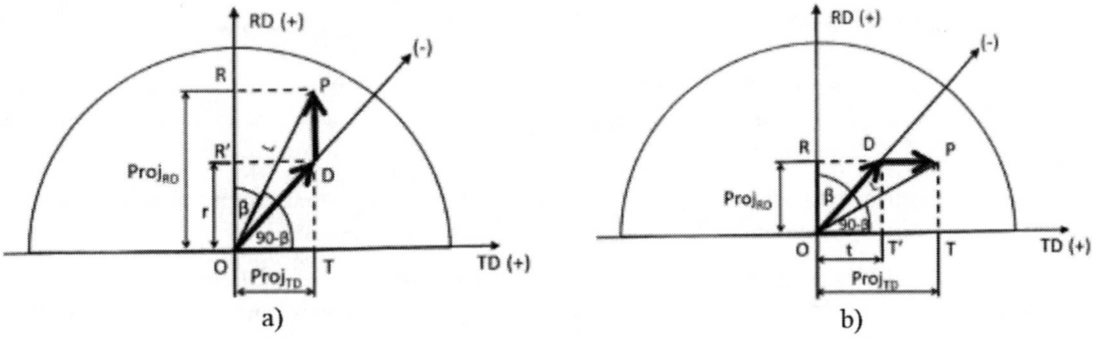

Figure 4. The (+) and (-) vectoral components of an individual pole P. a) P is between RD and (-) b) P is between (-) and TD

In Fig. 3, it was seen that one texture component has multiple poles on the β=0°-90° quarter pole figure (namely, C1, C2, S1, S2, S3, B1, B2, Cube1, Cube2, G1, G2). The (+) and (-) vectoral components ($COMP_{i,j}^{+} \equiv DP$ and $COMP_{i,j}^{-} \equiv OD$, respectively) of all main texture component poles were calculated and multiplied with the volume fractions of the texture components (V_i). This previous calculation was performed only within the β=0°-90° quarter pole figure. However, since the calculation of earing is based on the average of all peak heights and not only the maximal peak height of the ears, the full pole figure (the β=0°-360°) must be considered. Therefore, the vectoral components were multiplied with the total number of the poles ($n_{i,j}$) (15) and (16) to obtain the vectoral sum of all texture component poles in the (+) and (-) directions (V_+ and V_-, respectively). Finally, the larger one of V_+ and V_- was divided with the smaller one (17) to get the Vectoral Earing Ratio (VER).

$$V_+ = \sum_{i=1}^{n} n_{i,j}(V_i * COMP_{i,j}^{+}) \quad (15)$$

$$V_- = \sum_{i=1}^{n} n_{i,j}(V_i * COMP_{i,j}^{-}) \quad (16)$$

$$VER = \frac{MAX(V_+,V_-)}{MIN(V_+,V_-)} \quad (17)$$

3. Results

Figure 5. shows the variation of measured (+) type earing together with the calculated texture component volume fractions across the front cross section (a) and the middle cross section (b) of the examined coil. It is apparent that earing strongly varies across both sample series. It is high at the edges (~5-6%), while low in the center regions (~1-2%). The texture component volume fractions also show significant deviations. The strongest texture component is the Cube, which, similarly to earing, is highest at the edges. However, no straight correlation can be seen between either texture component volume fraction and earing.

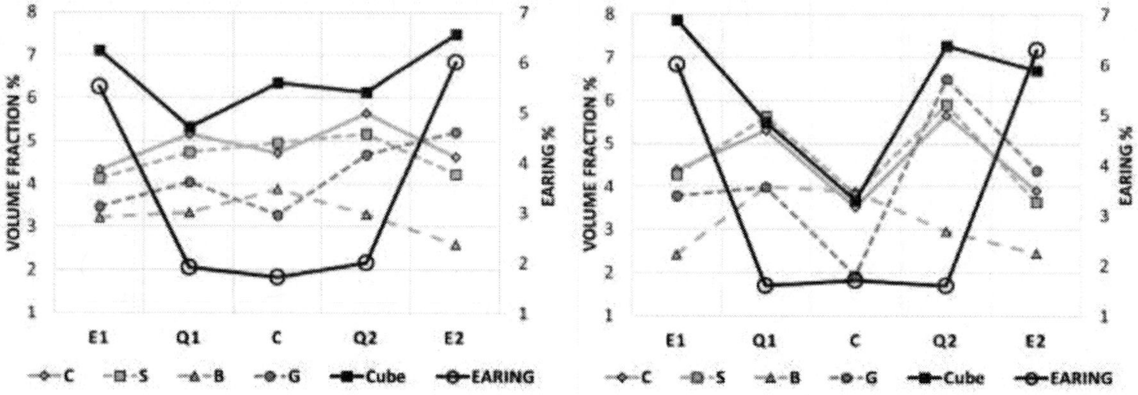

a) b)

Figure 5. Variation of measured (+) type earing and texture component volume fractions a) across the front cross section b) across the middle cross section.

Figure 6. shows the measured (+) type earing and the calculated Vectoral Earing Ratio variation across the front (a) and middle (b) cross sections. In the sample series taken from the front cross section, the tendency of Vectoral Earing Ratio has a good correlation with that of the measured earing. In case of series from the middle cross section, the two variations are in a good agreement, except the sample Q2, where Vectoral Earing Ratio is somewhat high compared to the measured earing. However, this slight deviation can originate from the fact that the samples for deep drawing and XRD examinations were not taken from the same exact location, and, as seen, the texture of the sample series shows a quite strong texture inhomogeneity along the coil length and across the cross section.

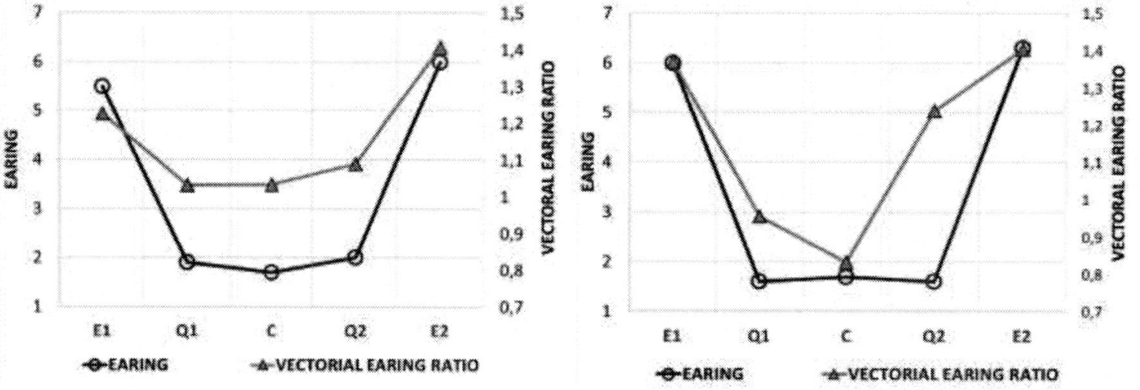

a) b)

Figure 6. Variation of measured (+) type earing and calculated Vectoral Earing Ratios a) across the front cross section b) across the middle cross section.

4. Conclusions

It was seen that the main texture component volume fractions, as single quantities, did not exhibit a straightforward correlation with the measured earing across the cross sections of 3003 type aluminum sheet being in rolled+annealed state. This is understandable, since all texture components have an effect (smaller or larger) on formability, thus, the observed earing is governed by the combination of all texture components that are present. A method was presented in this manuscript which proposes a geometrical connection between the texture components and the magnitude of earing (both (+) and (-) type). The proposed calculation is based on defining all texture component poles as the vectoral combination of their (+) and (-) earing direction components. The method supposes that unless the pole is located exactly on the (+) or (-) directions, it has a contribution to both (+) and (-) type earing. It follows that from the calculation point of view, the method does not differentiate between rolling and recrystallization texture components. The calculated Vectoral Earing Ratio is the ratio of the sum of (+) and (-) earing contribution of all poles considered. The obtained results show that a good correlation was achieved between the calculated Vectoral Earing Ratio and earing.

Acknowledgments
The described research work was supported through the National Research, Development and Innovation Office – NKFIH K119566 project.

References
[1] Fukui S and Kudo H 1950 *Rep. Inst. Sci. Tech. Univ. Tokyo 4* p 33
[2] Grewen J 1971 Quantitative Analysis of Textures *Proc. Int. Seminar (Cracow)* p 195
[3] Tucker G E G 1961 *Acta Metall.* **9** 275-286
[4] Kanetake N, Tozawa Y and Otani T 1983 *Int. J. Mech. Sci.* **25** 337-345
[5] Da Costa Viana C S, Davies G J and Kallend J S 1978 *Textures of Materials. Proc. ICOTOM 5* **2** (Germany: Springer) p 447
[6] Pochinetto A, Pernot M and Penelle R 1981 *Deformation of Polycrystals* (Denmark: Riso National Laboratory) p 205
[7] Rodrigues P M B and Bate P S 1985 *Texture in Non-Ferrous Metals and Alloys* AIME Warrendale PA p 173
[8] Van Houtte P, Cauwenberg G and Aernoudt E 1987 *Mat. Sci Eng.* 95 115-124
[9] Aretz H, Aegerter J and Engler O 2010 Proc. *NUMIFORM* American Institute of Physics **1252** 417-424
[10] Engler O and Kalz S 2004 *Mat. Sci. Eng. A* **373** 350-362
[11] Engler O and Hirsch J 2007 *Mat. Sci. Eng.* A **452-453** 640-651
[12] Kocks U F, Tomé C N and Wenk H-R 1998 Texture and Anisotropy (UK: Cambridge University Press)
[13] Engler O and Randle V 2010 *Introdusction to Texture Analysis* Second Edition (USA: CRC Press)
[14] Baldwin W M, Howald T S and Ross A W 1945 *Met. Tech. Tech. Pub.* **1808** p 86
[15] Roberts W T 1966 *Texture Control in Sheet Metal Ind.* **43** p 237
[16] Hug H 1965 *Aluminium und Aluminiumlegierungen* (Germany: Springer) p 412
[17] Blade J C and Morris P L 1975 *Texture and the Properties of Materials Proc ICOTOM 4* (UK: Metals Society, Cambridge) p 171
[18] Sheppard T and Zaidi M A 1982 Met. Tech **9** p 368
[19] Kao P-W 1985 *Mat. Sci. Eng.* **74** p 147
[20] Ma Q, Mao W, Feng H and Yu Y 2006 *Script. Mat.* **54** 1901-1905

IOP Conf. Series: Materials Science and Engineering **426** (2018) 012004 doi:10.1088/1757-899X/426/1/012004

Determination of dislocation density in an Inconel 600 nickel alloy by XLPA and automated EBSD

T Berecz[1], B Jóni[2] and J Ginsztler[3]

[1] Assistant professor, Department of Materials Science and Technology, Budapest University of Technology and Economics, Budapest, Hungary
[2] PhD student, Loránd Eötvös University, Department of Materials Physics, Budapest, Hungary
[3] Professor emeritus, Department of Materials Science and Technology, Budapest University of Technology and Economics, Budapest, Hungary

E-mail: berecz@eik.bme.hu

Abstract. There are two widely used examination methods to study microstructure of polycrystalline materials, namely the automated electron backscatter diffraction (EBSD) and the X-ray diffraction line profile analysis (XLPA). Both methods are suitable to determine quality and quantity of crystallographic defects or irregularities within crystalline structures. However, the EBSD method can be used to estimate the density of only geometrically necessary dislocations (GND), while the XLPA method can give the total dislocation density. A novel software was developed for the determination of GND density from misorientations measured between the neighbouring pixels in EBSD images. Therefore EBSD method is a local, while XLPA method is a non-local procedure to determine dislocation densities. In fact, the calculated GND density depends on the applied step size of EBSD measurements, namely GND densities increase with decreasing the step size. So this leads to a question of how large applied step size of EBSD measurements gives a good approximation for the real GND density in a given polycrystalline material.

1. Introduction

In a previous paper [1] dislocation densities were determined through by using both EBSD and XLPA measurements in ferrous lath martensite. However, the ferrous lath martensite morphology differs from the microstructure of usual polycrystalline materials: it has a characteristic multilevel microstructure. A parent austenite grain contains several packets (the group of laths with the same habit plane). Each packet is divided into parallel blocks and a block is further subdivided into laths. The martensite lath is a single crystal of martensite which contains high density of lattice defects, such as dislocations. The size of single martensite laths is very small (typically few micrometers), therefore they cannot be seen clearly by optical microscope. The blocks and packets are separated by high-angle grain boundaries with the misorientation larger than 15°, while between the laths the misorientation angle is smaller (~10°) [2-5].

It was found that the smaller the step size the higher the dislocation density. As an example, for the smallest (35 nm) and largest (500 nm) applied step sizes the dislocation densities were about 6 and 30×10^{14} $1/m^2$, respectively. This tendency was explained by the fine subgrain structure in the laths, revealed by transmission electron microscopy (TEM). TEM pictures demonstrated that the laths consist

IOP Conf. Series: Materials Science and Engineering **426** (2018) 012004 doi:10.1088/1757-899X/426/1/012004

of subgrains in the size range between 50 and 100 nm. This was in good agreement with the average value of crystallite size (71 nm) determined by XLPA. Since the subgrain size was under 100 nm, the more reliable dislocation density could be acquired with the smallest step size (35 nm).

Nevertheless, in common polycrystalline materials, the microstructure does not contain any smaller units within crystallites and sizes of the crystallites may vary from tens to hundreds of micrometers. Therefore determination of a reliable GND density through EBSD measurements seems to be difficult. In order to study the effect of step size on GND density, EBSD measurements were performed with different step sizes and these results of GND densities were compared the total dislocation density determined by XLPA.

2. Experimental material and methods

EBSD and XLPA measurements were performed on an EBSD reference sample given and prepared for EBSD measurements by EDAX-TSL. This sample is a 6.35 mm diameter rod made from Inconel 600 nickel alloy with a recrystallized microstructure. Its composition according to our measurements is as follows: Ni ~70%, Cr ~20% and Fe ~10%.

EBSD measurement were carried out on a Philips XL30 scanning electronmicroscope equipped with an EDAX-TSL type EBSD system and an OIM Analysis 6.2 software configuration. X-ray diffraction measurements were performed by a special high resolution double crystal diffractometer using CoKα1 radiation (λ = 0.1789 nm).

3. Results and discussion

3.1. Determination of total dislocation density by XLPA

The total dislocation density was determined by XLPA using the CMWP fitting procedure. The parameters of the microstructure obtained by XLPA are shown in Table 1.

Table 1. In this table q is describing the edge/screw character of dislocations, ρ is the dislocation density, m and σ are the median and variance of coherently scattering domain size, and <x> is area-weighted mean crystallite size.

ρ [10^{13} m^{-2}]	q [-]	M* [-]	m [nm]	σ [-]	<x> [nm]
3(±1)	2.1(±0.1)	3.0(±0.3)	97	0.28	120(±20)

The area-weighted mean crystallite size is about 120 nm for the sample. XLPA is sensitive to the coherently scattering domain size, so not only the high angle grain boundaries are detected, but the low-angle boundaries as well [6, 7]. Subgrains with low angle domain boundaries, as well as dislocation walls are the structures the XLPA is sensitive to, in contrast with the EBSD technique which cannot resolve these low angle grain boundaries. This is why the sub-grain size determined by diffraction methods like XLPA is always smaller than that determined by EBSD.

XLPA yields a total dislocation density of 3×10^{13} 1/m^2 in this sample. As it will be seen later this value is higher than GND values obtained by EBSD. However, it should be noted that the method based on EBSD technique gives only the density of the geometrically necessary dislocations, while XLPA provides the total dislocation density (including both geometrically necessary and statistically stored dislocations).

XLPA also gives a parameter describing the edge/screw character of dislocations. The experimental value of parameter q is 2.1. Within the experimental error limits it is in good agreement with the arithmetic average of the theoretical values calculated for pure edge and screw dislocations, 1.6 and 2.3, respectively. This observation suggests that the character of the dislocation structure is mixed.

3.2. Methodology of GND density determination from EBSD misorientations

Determination of density of GNDs is possible via mapping the misorientation relations in a certain sample. Orientation of a "point" on the sample being investigated can be measured by EBSD technique. The advantage of this method is that arbitrary areas (e.g. a single grain) can be selected for the measurements thus local dislocation density can be earned. On the other hand, only superficial investigation can be carried out. In this section the main steps of derivation of dislocation density from misorientation relations is described.

Orientation of a lattice point can be characterized by the Euler-angles (ϕ_1; ϕ_2; φ). As EBSD measurements carried out in separate points on the surface of the sample, Euler-angles can be assigned to each measured point. In fact, scanning the surface with the primer electron beam means that orientation (and other properties) are measured in thousands of points on the surface of the sample. In this way, all neighbouring points can be characterized by their orientation deviation, i.e. the misorientation.

Following the guidelines of Pantleon et al. misorientations of adjacent measuring points are calculated in quaternionic representation [8, 9, 10]. From these mathematical objects, Nye's dislocation density tensor can be determined as the follows.

$$\boldsymbol{\alpha} = \begin{bmatrix} \alpha_{11} & \alpha_{12} & \alpha_{13} \\ \alpha_{21} & \alpha_{22} & \alpha_{23} \\ \alpha_{31} & \alpha_{32} & \alpha_{33} \end{bmatrix} \tag{1}$$

In Eq. 1, α_{ij} are unambiguous functions of the misorientation and the distance between two measuring points, i.e. the resolution of the measurement. From Nye's tensor GND density can be directly determined via Eq. 2:

$$\rho^* = \frac{1}{b}(|\alpha_{12}| + |\alpha_{13}| + |\alpha_{21}| + |\alpha_{23}| + |\alpha_{33}|), \tag{2}$$

where ρ^* is the dislocation density, b is the length of Burgers-vector (in this case $|b| = 2{,}49$ Å). Since the scanning is two-dimensional measuring points do not have neighbours in the z-direction (perpendicular to the plane of the sample). Hence, Nye-tensor is an incomplete matrix with five elements, and the dislocation density obtained in this way called – more precisely – quasi-dislocation density. After calculating the local ρ^* values, the global dislocation density for the measured area can be obtained by averaging.

3.3. Determination of GND density by EBSD

In order to calculate GND densities through EBSD measurements, at first a misorientation range should be determined. This range with a higher and a lower limit contains the misorientations considered for calculations. A higher misorientation limit should be avoided for taking into account grain boundaries (as high angle misorientations) and misidentified points (tiny grains or single pixels with high angle boundaries inside a larger grain mostly caused by structural faults). Fig. 1 shows a misorientation distribution for the sample obtained by EBSD with 1.8 μm step size. It reveals that the majority of misorientations what can be taken in account to calculate GND densities are smaller than 5°.

Fig. 2 shows a misorientation map of the same EBSD measurement with low (<5°) and high (>5°) angle boundaries. (This kind of map shows misorientation of each point with different colours.) It reveals that low angle boundaries can be seen only within the individual grains, while the majority of high angle boundaries can be seen between the adjacent grains (as high angle boundaries) and inside the grains as twin boundaries or misidentified points. Therefore, only misorientations smaller than 5° can be used for determination of GND density.

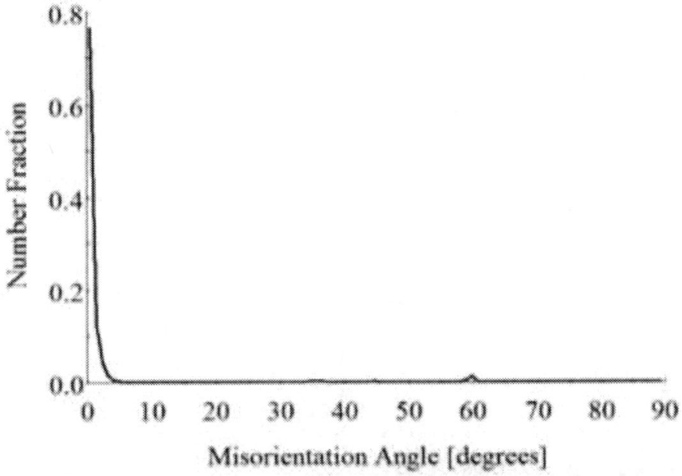

Figure 1. Misorientation distribution of the sample obtained by EBSD with 1.8 μm step size.

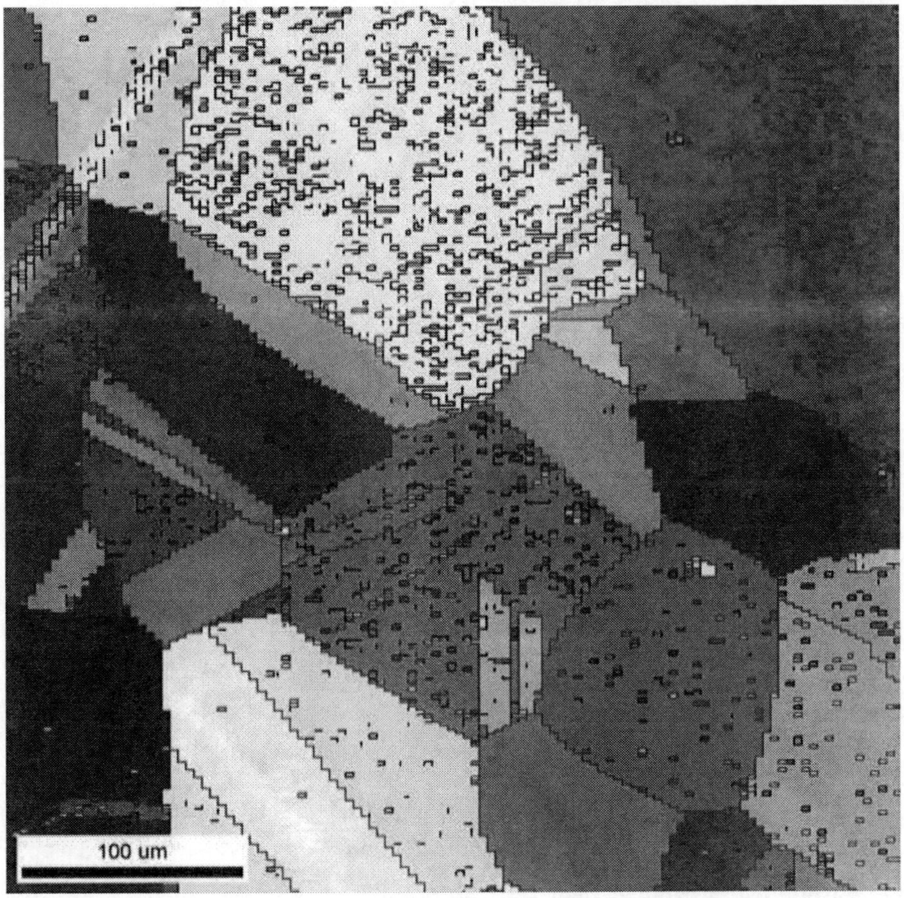

Figure 2. Misorientation map of the sample obtained by EBSD with 1.8 μm step size with low (<5°, black lines) and high (>5°, red lines) angle boundaries.

However, there is an uncertainty in the determination of the Euler angles from the Kikuchi patterns obtained by EBSD which can result an error in the calculation of the dislocation density from the misorientations. In order to estimate the error in the misorientation angles the same measurements were carried out on an undeformed single crystalline Si wafer. In this single crystal orientation does not change, therefore differences between the crystallographic orientations of the neighbouring EBSD image pixels should be practically zero.

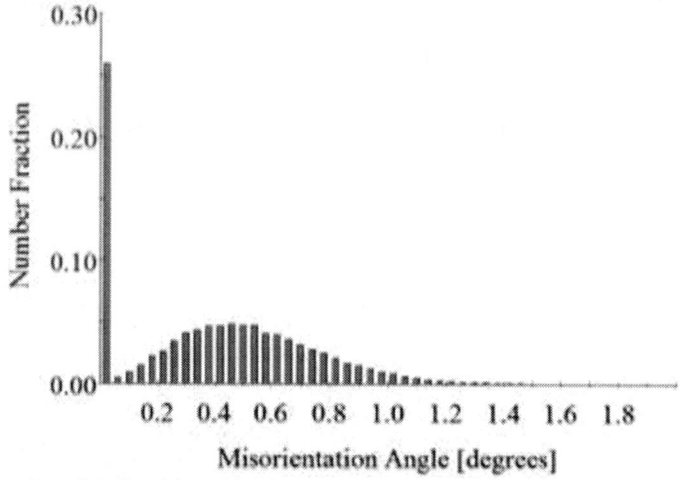

Figure 3. Misorientation distribution of a Si single crystal obtained by EBSD with 2.0 μm step size.

Fig. 3 shows the misorientation distribution for this Si single crystal obtained with an applied step size of 2.0 μm. This distribution reveals that certain misorientations can be observed under 1.5° even if the material is practically defect-free.

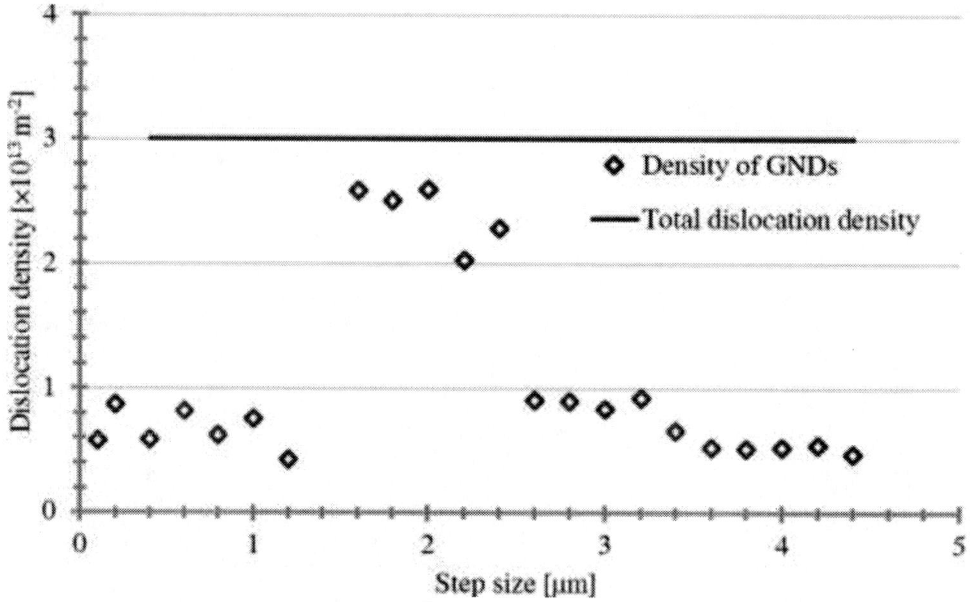

Figure 4 Density of GNDs (and total dislocation density) versus applied step size.

Therefore, misorientations higher than 1.5° can only be used in the calculation of the dislocation density. The other thing why a lower limit should be applied is the high fractions of small misorientations under ~0.1° (see Fig. 5) what would increase unreasonably the calculated dislocation densities.

Namely, if both misorientations between a pixel and its two neighbours (at the right side and below the studied pixel) are smaller than 1.5°, the dislocation density is taken as zero in that pixel. If any of the two misorientations is higher than 5°, the pixel is ignored from the dislocation density calculation. For all other pixels the dislocation density was calculated from the misorientations, as described in the section above. Fig. 4 shows the calculated dislocation densities determined for the whole studied area as a function of applied step size calculated from EBSD data.

It shows that calculated densities of GNDs are between 0.5 and 1×10^{13} 1/m^2 up to ~1.2 μm step size, then grow up to ~2.5×10^{13} 1/m^2 and decreases to ~0.5×10^{13} 1/m^2. The small calculated dislocation densities at <1.5 μm step sizes can be understood from the Fig. 5. It reveals that while the majority of misorientations are under 1.5° at 4.4 μm step size, until then almost all the misorientations are smaller than 1.5°at 0.4 μm step size The smaller the step size, the more number fraction of misorientations will be smaller than 1.5°, therefore they are ignored during calculation of GND density, thus it will also be small.

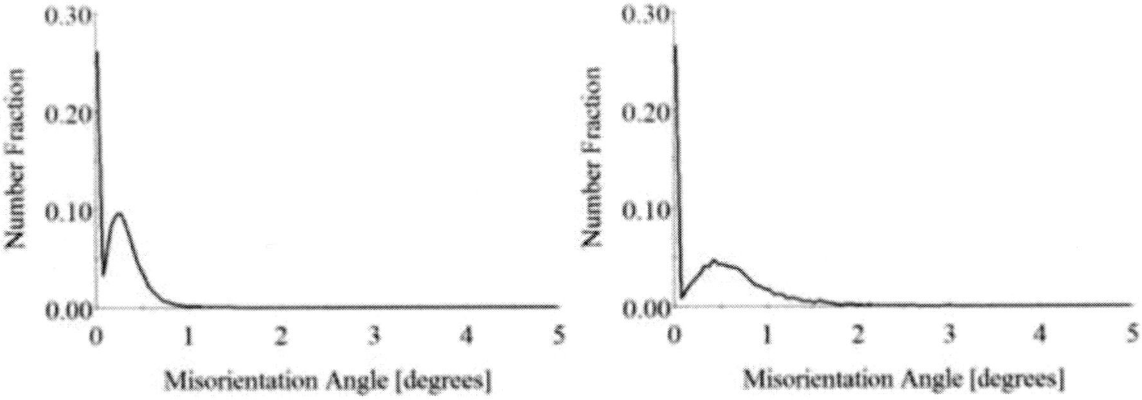

Figure 5. Misorientation distribution for the Ni sample obtained by EBSD with 0.4 (left) and 4.4 μm (right) step size.

Fig. 4 also reveals that the calculated dislocation densities are smaller (~0.5×10^{13} 1/m^2) at larger (2.4-4.4 μm) step sizes. This is because the larger the step size the more misorientations are omitted for calculating GND densities during scanning. Accordingly, it seems that step sizes between 1.6 and 2.4 μm would give reliable results for GND density in a common polycrystalline material. These GND densities are smaller (70-85%) than the total dislocation density and their values are close to it.

4. Conclusions

Regarding our experimental results contradictions are to be observed. On the one hand, they are appropriate from the point of view that calculated GND densities are smaller than the total dislocation density at every step sizes. On the other hand, however, based on these measurements it is difficult to specify an optimal step size for calculating correct GND density in common polycrystalline materials. Two main reasons can be considered as explanations for that:

a. In fact, the total dislocation density (3×10^{13} 1/m^2) in the sample examined by XLPA is very low. A sample with much more total dislocation density (due to cold working) should be more suitable for this kind of measurements. However, it would make sample preparation for EBSD to be difficult.
b. It seems to be another problem that the uncertainty in the determination of the Euler angles from the Kikuchi patterns (1.5°, determined by EBSD on Si single crystal) is too large. Especially, when it is compared to the upper limit of calculation (5°). It can be as high as 30% of that. It means that a major proportion of the misorientations smaller than 5° should be ignored during determination of GND density.

Acknowledgement
Authors express their thanks to András Csóré PhD student for his help.

IOP Conf. Series: Materials Science and Engineering **426** (2018) 012004 doi:10.1088/1757-899X/426/1/012004

References

[1] Berecz T *et al* 2016 *Mat Char* **113** 117
[2] Kitahara H *et al* 2006 *Acta Mater* **54** 1279
[3] Furuhara T *et al* 2003 *J. Phys. IV France* **112** 255
[4] Morito S *et al* 2003 *Acta Mater* **51** 1789
[5] Morito S *et al* 2006 *Mat Sci Eng A-Struct* **438–440** 237
[6] Ungár T *et al* 2005 *Powder Diff* **20(4)** 366
[7] Varbai B *et al* 2017 *Period Polytech Mech* **61(1)** 68
[8] He W *et al* 2008 *Mat Sci Eng A-Struct* **494** 21
[9] Pantleon W 2008 *Scripta Mater* **58** 994
[10] Szabó P J and Csóré A 2017 *Mater Sci Forum* **885** 275

IOP Conf. Series: Materials Science and Engineering **426** (2018) 012005 doi:10.1088/1757-899X/426/1/012005

Comparative study on theoretical and experimental evaluation of forming limit diagrams

G Béres[1], Z Weltsch[1] and M Tisza[2]

[1] John von Neumann University, Faculty of GAMF Engineering and Computer Science, Department of Materials Technology, Kecskemét, Hungary
[2] University of Miskolc, Faculty of Mechanical Engineering and Informatics, Institute of Materials Sciences & Technology, Miskolc-Egyetemváros, Hungary

E-mail: beres.gabor@gamf.uni-neumann.hu

Abstract. Forming limit diagrams (FLDs) are generally used to reflect the strain limits of sheet metal materials. These diagrams contain essential information for planning of high series stamping processes, however the experimental definition is still a difficult and time-consuming procedure. Five or more samples are needed to cover the different strain paths sufficiently, from pure shear ($\varepsilon_1 = -2\varepsilon_2$) up to biaxial stretch forming ($\varepsilon_1 = \varepsilon_2$), moreover the experimental detection of the onset of local necking depends on the capability of the measuring equipment and person. The determination of the measurement results is further complicated by the material and other influencing conditions, like strain path (deformation history), rolling direction, sheet thickness, temperature, measuring grid size, tool surfaces, friction and vibrations. All these things supports the development of clearly theoretical and semi-empirical models for defining FLDs. This paper presents a comparative study of some kind of well-known theoretical models and experimental FLDs of automotive dual phase (DP) thin sheets. The experimental determination was carried out by Nakajima tests on five different sample geometries. We used follow-up strain analysis on the cracked samples. The results highlighted that most of the theoretical and semi-empirical models can be appropriately used in the negative quadrant of the FLD despite of the effect of the friction. However, in the positive quadrant, there are some more visible deviations due to outer conditions like lubricant.

1. Introduction

Due to the importance of forming limit diagrams, many researchers - mostly from Europe and the USA - have dealt with the failure modes of sheet metals since the 50s. Lankford [1] and then more researchers [2-11] showed that the anisotropy coefficient (r) and the work-hardening coefficient (n) are strongly related to sheet formability. However, regarding a large number of press-shop measurements carried out by the NADDRG (North American Deep-drawing Group), the correlations do not enable us to answer Pierce's [12] questions: "what mechanical properties are required satisfactory to produce a certain pressing trouble-free? Furthermore, when a pressing is running successfully, how close is it to failure?" The answers are in Keeler's publications from the 60s [13, 14], that with the investigation of major and minor principal surface strains in a blank, a limit strain graph could be plotted in biaxial stretching, i.e. from plane strain tension (ε_2=0) up to equibiaxial tension (ε_2=ε_1). Referring the negative strain ratio, Goodwin published a paper in 1968 [15]. Based on these three papers, the Keeler-Goodwin diagram i.e. the forming limit diagram was born. Keeler [16, 17] still proposed the circle grid design as the method of strain measurement, even gave some application examples in the press-shop also. Pearce [18],

IOP Conf. Series: Materials Science and Engineering **426** (2018) 012005 doi:10.1088/1757-899X/426/1/012005

Kleemola [19], Hecker [20] and Ayres and Brewer [21] worked out the concept that which deformed ellipse (or square) and how should be measured.

Tisza has been also comprehensively studying sheet metal formability and forming limits in the last two decades. His research group carried out the so-called "star specimen" to monitor the FLDs' sensitivity for the effect of the rolling direction [22]. They found more intense sensitivity perpendicular and 45° to the rolling direction. In other paper [23], an automatic evaluation system for determining forming limit diagrams was studied. Finite element simulation results based on automatically defined forming limit curve (FLC) showed good agreement with experimental strain fields in case of stainless steel samples. Furthermore, forming limit prediction at single point incremental sheet forming (SPIF) process was also investigated [24]. They performed a systematic experiment series to find the FLC for both the negative and positive minor strain quadrants. The experimental results showed that the forming limit curve for SPIF represents two straight lines with a negative slope in the positive quadrant and with a positive slope in the negative minor strain range, but both are above the conventional forming limit curve.

The experimental technique worked out by Nakajima [25, 26] is the most popular in FLCs' determinations due to its practicalities. In Hungarian research laboratories, other test methods like Marciniak, hydraulic bulge test, etc. are less widespread. Marciniak's name is rather known because of his physical based theoretical FLC definitions.

Marciniak and Kuczinski [27] proposed that the failure starts with local inhomogeneity, and thus gave the theoretical solution for both sides of the FLCs. Similarly, with the investigation of the physical occurrence of a local necking in the sheet, Swift [28], Hill [29] and Stören and Rice [30] also determined the failure in plane strain under different loading conditions. Swift used the Mises-Hencky yield criterion, and the major stress components to predict the major and minor limit strains with the consideration of the strain-hardening capacity of the metal. Hill predicted the failure in plane strain state (FLC_0 point) equal with the onset of local necking, i.e. with the Nádai strain-hardening exponent. In his solution, a 45° slope straight line predicts the description of the negative quadrant ($-\varepsilon_2$). Stören and Rice deduced FLDs from continuum mechanic equations. Although the FLC_0 point was predicted the same way as Hill did, their new bifurcation method led to new relationships, called vertex theory, in both negative and positive ranges between the major and minor strains. Chow et al. [31] extended later this theory for anisotropic materials.

Among the semi-empirical methods, which main advantage is the easier definition of the necessary material parameters, Keeler and Brazier [32] were one of the firsts who made a comprehensive study about the strain limits from simple properties of materials. They experienced similarity in the geometry refers to FLDs of more material grades. They specified a linear correlation between the FLC_0 point and material parameters – like sheet thickness and strain hardening exponent – and then they assumed a curve shifting along the major strain axis. Similar context can be found in the paper of Cayssilas [33] who already took into consideration the strain rate-hardening exponent, too. He further developed his predicting model more times [34] and validated the results with many experiments at Arcelor Research S. A. Abspoel et al. [35] from Tata Steel Research Development also proposed a method for FLC evaluation from mechanical properties originated from tensile test results. With the investigation of approximately fifty steel grades, they defined a relationship statistically between the total elongation and the major and minor strains at four different strain paths in the FLC. The equations consider the anisotropic behavior and the sheet thickness too. Similarly, from tensile test results, Levy and Tyne [36] deduced the major strains to define stress based forming limit curve, subsequently. They used Hill anisotropic yield criterion to evaluate the FLC_0 i.e. the major strain corresponding to the zero minor strain – and thus identified a linear approximation for both sides of the FLD.

This paper is a review about the observation of different limit strains both theoretically and experimentally for automotive DP steels.

2. Experimental analysing of formability

2.1. Applied materials

Three types of conventional DP steels as DP600, DP800 and DP1000 with the basic mechanical properties according to MSZ EN ISO 6892-1:2010 tensile tests' results, with 30 mm/min displacement velocity were applied in this study. The tensile specimens with gauge length of 80 mm and 1mm thickness were manufactured in three directions: parallel, perpendicular and 45° to the rolling direction (RD). The stress-strain curves in the coordinate system of true (logarithmic) and equivalent quantities and the Lankford coefficients in different directions as the tensile tests' results can be seen in figure 1.

It can be seen that DP steels show nearly isotropic behaviour. Figure 1.b presents the Lankford coefficients in the different directions, which also highlights that the transverse-thickness strain ratios are close to unity. The main mechanical properties are summarized in Table 1, where the $\bar{r}$-value is calculated as:

$$\bar{r} = \frac{r_0 + r_{90} + 2r_{45}}{4} \quad (1)$$

in which r_0, r_{90} and r_{45} denotes the measured values in 0°, 90°and 45° to the rolling direction. In Table 1, Ag belongs to the engineering uniform elongation. The values of strain hardening exponent (n) and the strength coefficient (K) in the Nádai expression (2) are fitted by the least-squares method.

$$\sigma_{eq} = K \cdot \varepsilon_{eq}^{n} \quad (2)$$

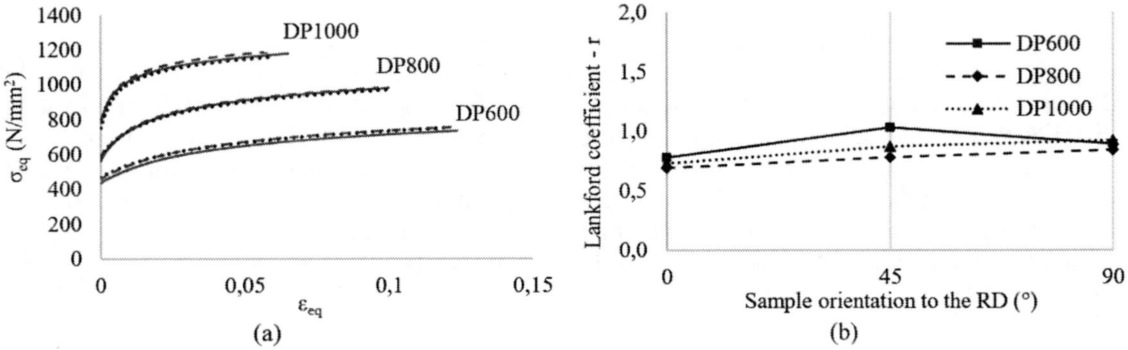

Figure 1. Flow curves of DP600, DP800 and DP1000 materials: continuous line parallel, dashed line perpendicular, dotted line 45° to the RD (a); and Lankford coefficients in 0°, 45° and 90° to the RD (b)

Table 1. Main mechanical properties of the investigated DP steels

	UTS N/mm²	YS N/mm²	A_{80} (%)	Ag (%)	K	n	$\bar{r}$
DP600	656	445	20.6	13.6	918	0.112	0.92
DP800	879	571	16.0	10.8	1217	0.104	0.75
DP1000	1099	767	10.6	7.0	1481	0.083	0.76

2.2. Forming Limit Diagrams

Five different geometries embodied the minimum necessary numbers of samples (figure 2.a) to cover the FLCs from pure shear up to equi-biaxial tension. We used rectangular grids on the sheet surfaces to determine the major (ε_1) and minor (ε_2) logarithmic strains belonging to the onset of local necking. An Erichsen 142 type universal sheet formability tester equipment was applied to form the prepared samples without any lubricant (figure 2.b). It has a practical significance, since the sheet panels carry various oils, such as mill oil, washer oil, and press lubricant in the automotive industry. As a result, the lubrication system become more complicated, and lubrication performance become poorer in the press-shop compared to the laboratory conditions [32]. The stroke speed of the punch was 30 mm/min. The machine automatically stopped itself at the load drop caused by the occurrence of crack.

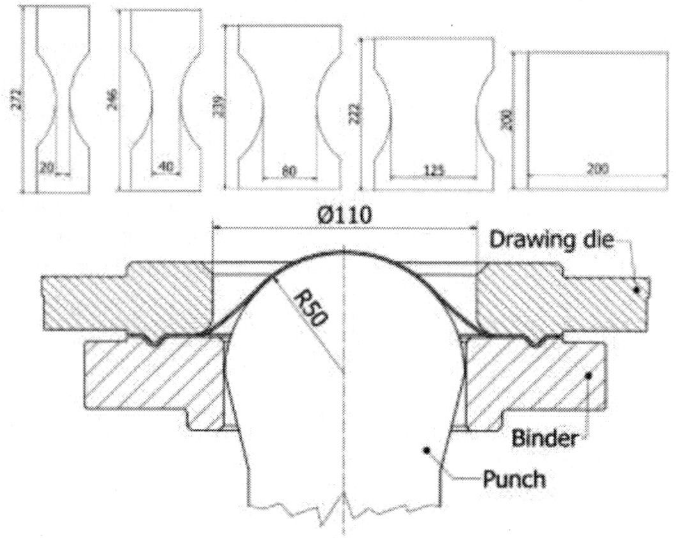

(a)

(b)

Figure 2. Sample geometries with different cross-widths: 20, 40, 80, 125 and 200 mm respectively (a); and the arrangement of tools at the Nakajima test with 100 mm punch diameter (b)

We used Vialux Autogrid® strain-analyzer system in agreement with Mercedes-Benz Manufacturing Hungary Ltd. for the detection of local strains. We adhered to the description of "ISO 12004-2:2008 Determination of forming-limit curves in the laboratory", during the calculation of the limit values. It specifies that the necking strain is equal with the maximum point of the best fit second order inverse parabola curve (3), which is superposed for the strain values in the relevant section containing the neck, see figure 3.a, which illustrates a DP600 sample with 20 mm cross-width. The inner boundaries of the curve fitting are found there, where the second derivative of the measured strains gives the positive maximum at each side of the crack. Figure 3.b shows the major strains as well as the derivative along the section's length.

$$f(x) = \frac{1}{ax^2 + bx + c}, \tag{3}$$

here x denotes the curvilinear length of the sample's cross section, while a, b, c are constants.

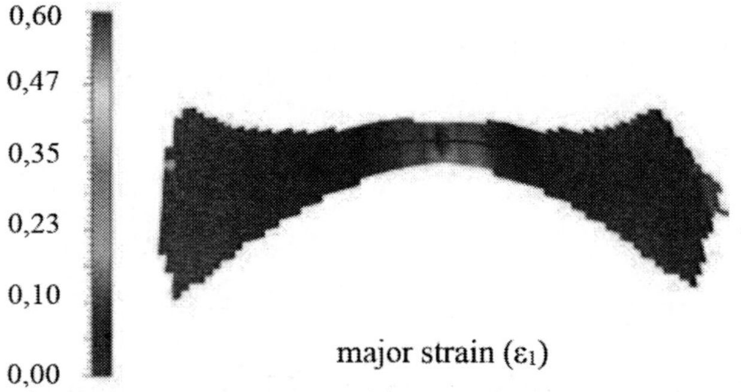

(a)

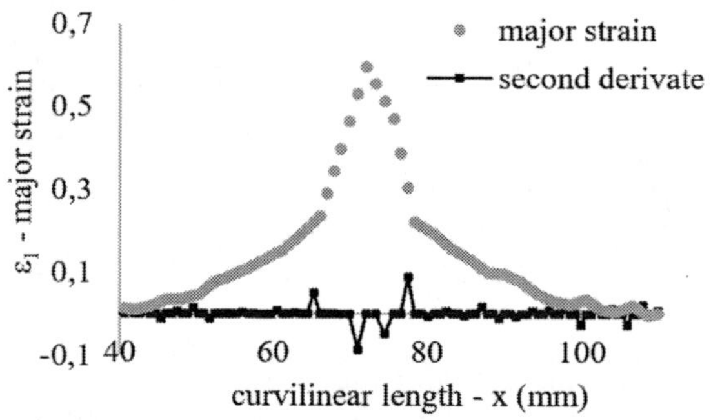

(b)

Figure 3. Strain distribution in a DP600 sample highlighting the relevant section with black line (a); and the measured major strains and those second derivatives (b)

Ignoring the cracked points, the fitted inverse parabola curve can be received, which first derivative is identical with the strain at the onset of local necking: figure 4.a. The measured limit values obtained with this method for the investigated steels and approximated linearly in the coordinate system of the major and minor strains are illustrated in figure 4.b.

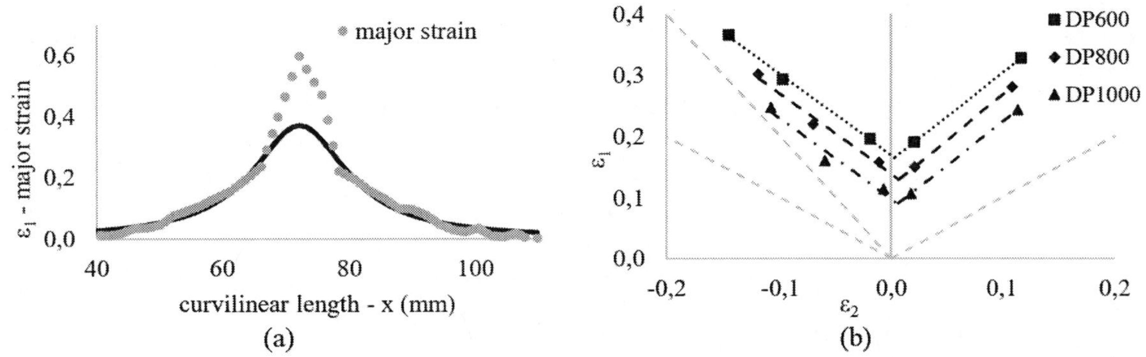

Figure 4. Major strain points with the fitted inverse parabola curve (a); experimentally defined limit strains and the forming limit curves of the investigated steels (b)

3. Theoretical FLC prediction

Though the theoretical FLC prediction models disregard the effect of the friction, some of them are in good agreement with the experimental results in the negative quadrant of the FLC and in plane strain tension.

Hill assumed that the necessary criterion of local necking is the occurrence of plane strain state in the sheet ($d\varepsilon_2 = 0$). Consequently, his solution does not allow the localized necking for biaxial stretching, i.e. when $\varepsilon_2 > 0$. According to his equations (4), the FLC_0 point is equivalent with the strain-hardening exponent (n) and thus the limit major strain (ε_1*) increases linearly with a slope of -1 in the negative quadrant. Comparing the left hand side of the FLC points given by his method with the experiments, it can be observed that the experimental curves have more powerful slope than the theoretical ones (figure 5). It is visible in figure (b), that the slopes exceed one third the -1 value for all steels. Besides, the FLC_0 points are obviously under predicted by Hill.

$$\varepsilon_1{}^* = \frac{n}{1+\rho} \qquad -1 < \rho \leq 0 \tag{4}$$

where ρ means the strain ratio

$$\rho = \frac{\varepsilon_2}{\varepsilon_1} \tag{5}$$

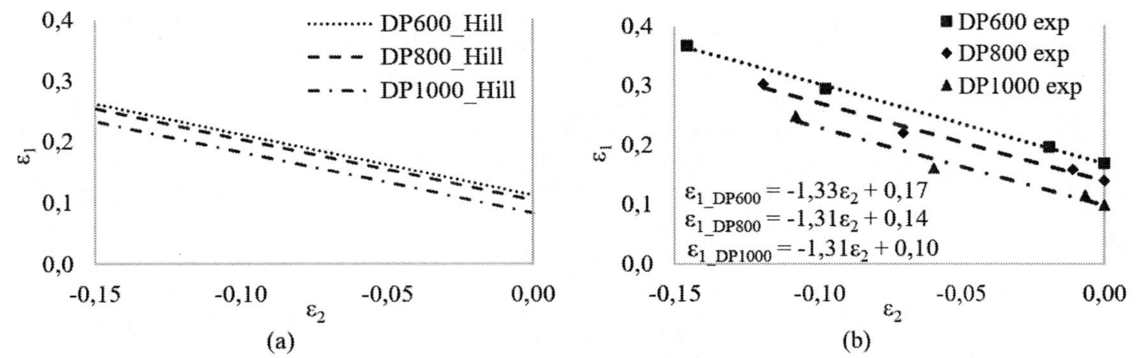

Figure 5. Major and minor limit strains belonging to the localized necking in Hill's theory (a); and to the experiments (b)

It is generally noticed that the investigated evaluation models somewhat predict under the plane strain point and the left hand side in the FLCs, however are relatively close to the experiments. Meanwhile, the calculated values spectacularly under predict the right hand side of the experimental FLCs due to the friction. The curves in both sides have steeper rising in ε_1 direction caused by the poor lubricant and produce earlier failure considering the minor strain values. It can be seen in figure 6 also, where the evaluated FLC by Swift and the experimental strain data are listed respectively.

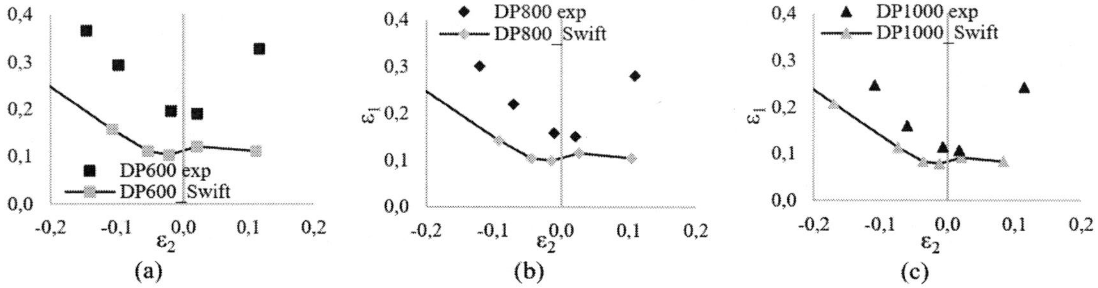

Figure 6. Major and minor limit strains belonging to the localized necking in Swift's theory (continuous line) and to the experimental strain points for DP600 (a), DP800 (b) and DP1000 (c) shown by the signs of measured points

Based on Swift's and Hill's works, Stören and Rice developed a new bifurcation method, which approximated the FLC with second order and root functions. Based on their description, when $\rho \geq 0$ ($\varepsilon_2 \geq 0$) the FLC curves can be more precisely predicted as the neck forms along the direction of the minimum principal strain. For $\rho < 0$ ($\varepsilon_2 < 0$), the neck inception along the direction of zero extension results their calculation close to Hill's and Swift's theories (figure 7).

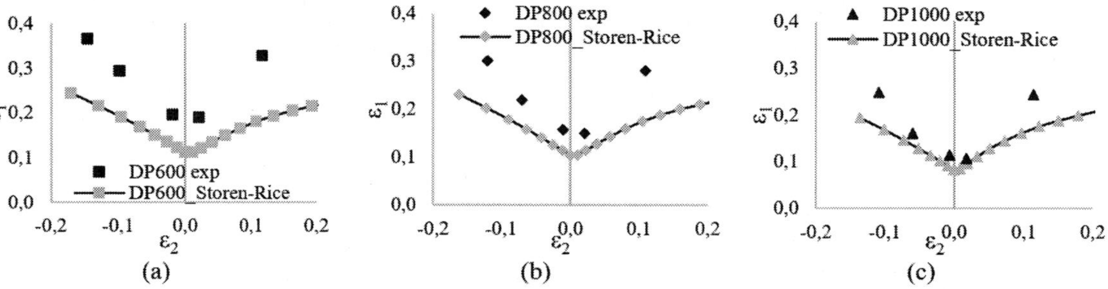

Figure 7. Major and minor limit strains belonging to the localized necking in Stören and Rice's theory (continuous line) and to the experimental strain points for DP600 (a), DP800 (b) and DP1000 (c)

In North America, FLCs for mild steels, high strength steels (HSS) and advanced high strength steels (AHSS) have been studied and approximated by the Keeler-Brazier method based on the initial sheet thickness (t) and the strain-hardening exponent (n) of the material. The Keeler-Brazier equations (6, 7, 8) are strongly recommended for the prediction of AHSS materials, and as it is visible in figure 8, the evaluated curves are very close to the measured points in the negative range of ε_2, and in plane strain, where the effect of the friction is less significant due to the sample geometries. At the right to ε_1, this model also under predict under the poor lubricant experiments naturally.

$$\mathrm{FLD_0} = \ln\left[1 + \frac{(23.3 + 14.13t)\cdot n}{0.21}\right] \qquad n \leq 0.21 \tag{6}$$

$$\varepsilon_1{}^* = \ln\left|0.6(e^{\varepsilon_2{}^*} - 1)\right| + e^{FLD_0} \qquad 0 < \varepsilon_2{}^* \tag{7}$$

$$\varepsilon_1{}^* = FLD_0 - \varepsilon_2{}^* \qquad \varepsilon_2{}^* < 0 \tag{8}$$

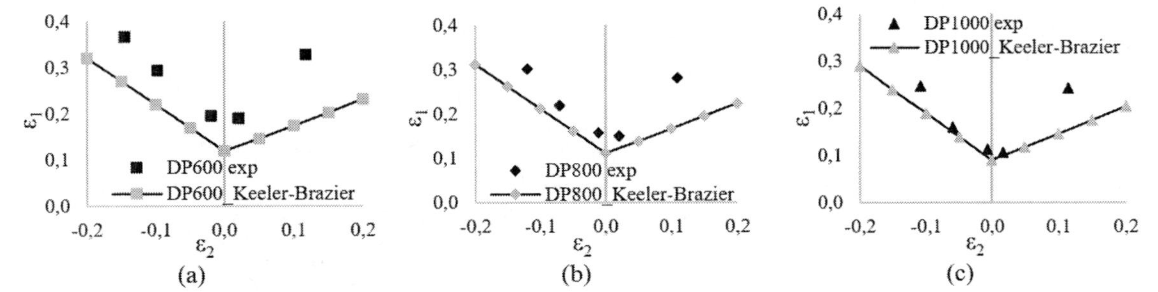

Figure 8. Major and minor limit strains belonging to the localized necking in Swift's theory (continuous line) and to the experimental strain points for DP600 (a), DP800 (b) and DP1000 (c)

Levy and Tyne got the strain data in plane strain from tensile tests' results with using the Hill '48 plasticity assumption (9) and the fact that the logarithmic, axial uniform strain (ε_{crit}) is identical with ε_1* if ε_2 is equal to zero. According to their theory, the negative side of the FLCs can be predicted by a linear assumption with a slope -1 (45°) line based on Hill's solution. Experimental FLCs available from Levy and Green [37] showed that the average slope of the positive side is close to 0.53 for HSLA and DP steels. If ε_{crit} is determined from a given hardening law (2), FLD_0 is obtained by (9, 10), and for any values of ρ, ε_1 and ε_2 can be calculated from the slope values (figure 9).

$$\varepsilon_{eq} = \varepsilon_{crit} = \frac{1+\bar{r}}{\sqrt{(1+2\bar{r})}} \sqrt{\varepsilon_1^2 + \varepsilon_2^2 + \frac{2\bar{r}}{1+\bar{r}} \varepsilon_1 \cdot \varepsilon_2} \tag{9}$$

$$\varepsilon_1{}^* = \varepsilon_{crit} \quad if \quad \varepsilon_2 = 0 \tag{10}$$

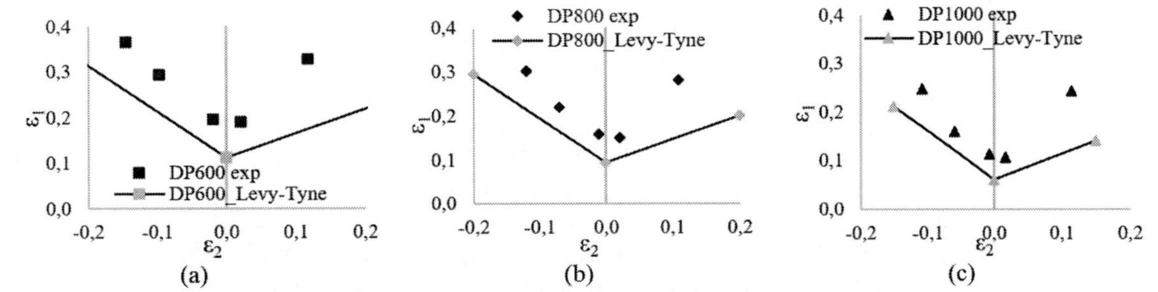

Figure 9. Major and minor limit strains belonging to the localized necking in Swift's theory (continuous line) and to the experimental strain points for DP600 (a), DP800 (b) and DP1000 (c)

It is worth mentioning that the predicted curves according to the Levy-Tyne theory are also closer to the measured data at the left hand side than at the right.

4. Conclusion

Results of different theoretical methods for predicting the FLC were compared to the experimental curves for DP600, DP800 and DP1000 steels. Although friction has important effect, the experiments were carried out without any lubricant and the effect of the friction have not been also examined.

Despite the enhanced friction, many theories are relatively close to the left hand side of the experimental FLC. This is due to the fact, that most evaluations in the negative minor strain range or accept Hill's solution either be close to it, which slope is only differs with one-third to the experiments. Furthermore, at the DP800 and DP1000 steels, the predicted curves almost totally overlap the measured strains in plane strain tension. The effect of the friction can be rather observed in the positive quadrant of the FLDs, where the contact areas are larger between the samples and the punch.

Acknowledgement
This research is supported by EFOP-3.6.1-16-2016-00014 project. The Project is supported by the Hungarian Government and co-financed by the European Social Fund.

References

[1] Lankford W I, Snyder S C, Bauscher J A. 1950. *Trans. ASM* **42** pp 1196-1232
[2] Sowerby R, Johnson W 1975. *Mat. Sci. Eng.* **20** pp 101-111
[3] Mellor P B, Parmar A 1978. *Mechanics of sheet metal forming. Material behavior and deformation analysis* (New York/London, Plenum Press) pp 53-57
[4] Mellor P B 1982. *Mechanics of materials* ed Hopkins G (Oxford, Pergamon Press) pp 383-415
[5] Chatfield D A, Keeler S P 1971. *Technology for using sheet steel designing for formability*, (Metal Progress), pp 60-63
[6] Banabic D, Dorr R I 1992. *Formability of thin sheet metals* (Romania), Bucarest, O. I. D. I. C. M.
[7] Whiteley R L 1960. *Trans. ASM* **52** pp 154-157
[8] Hill R 1950. *The mathematical theory of plasticity* (Oxford, Clarendon Press)
[9] Hosford W F, Backofen W A 1964. *Fundamentals of deformation processing* (Syracuse University Press) pp 259-292
[10] Warwick J O, Alexander J 1962/63. *J. Ins. Met.* **91** pp 1-10
[11] Fukui S, Yoshida K, Abe K 1960. *Sci. Papers Inst. Phys. Chem. Res.* **54** pp 199-205
[12] Pearce R 1991. *Sheet metal forming*, (Adam Hilger Series on new manufacturing processes and materials, IOP Publising Ltd)
[13] Keeler S P 1965. *Society of Automotive Engineers* No. 650535
[14] Keeler S P, Backofen W A 1963. *Trans. ASM* **56** pp 25-48
[15] Goodwin G M 1968. *Society of Automotive Engineers* No. 680093, pp 380-387
[16] Keeler S P 1968. *Circle Grid Analysis* (Livonia, MI: National Steel Corporation)
[17] Keeler S P 1968. *Society of Automotive Engineers* No. 680092
[18] Painter M J, Pearce R 1974. *J. Phys. D: Appl. Phys.* **7** p 992
[19] Kleemola H J, Kumpulainen J O 1980. *J. Mech. Work. Tech.* **3** p 289
[20] Hecker S S 1974 *Met. Eng. Q.* **14** p 30
[21] Ayres R A, Brewer E G 1979. *Society of Automotive Engineers*, No. 790741
[22] Tisza M, Kovács P Z, Lukács Z, Kiss A, Gál G 2015. *Materials Science Forum* **812** pp 271-275
[23] Kovács P Z, Tisza M 2008. *Materials Science Forum* **589** pp 233-238
[24] Tisza M, Kovács P Z and Lukács Z 2012. *Key Engineering Materials* **504-506** pp 863-868
[25] Nakajima K, Kikuma T 1971. *Yamata Technical Report* **284** pp 678-680
[26] Nakajima K, Kikuma T, Hasuka K 1972. *Yamata Technical Report* **264** p 141
[27] Marciniak Z, Kuczynski K 1967. *Int. J. Mech. Sci.* (Pergamon Press) **9** pp 609-620
[28] Swift H W 1952. *Journal of the Mechanics and Physics of Solids* **1** pp 1-18
[29] Hill R 1952. *J. Mech. Phys. Solids* **1** p 19
[30] Stören S, Rice J R 1975. *J. Mech. Phys. Solids* (Pergamon Press) **23** pp 421-441
[31] Chow C L, Jie M, Hu S J 2003. *Journal of Engineering Materials and Technology* **125** (Transactions of the ASME)
[32] Altan T, Tekkaya A E 2012. *Sheet Metal Forming Fundamentals*, (ASM International OH 44073 0002) pp 44-51
[33] Cayssials F 1999. *Proc. IDDRG Congress Brussel*, pp 443-454
[34] Cayssials F, Lemoine X 2005. *Proc. IDDRG Congress Besançon*, pp 1-8
[35] Abspoel M, Scholting M E, Droog J M M 2013. *Journal of Materials Processing Technology* **213** pp 759–769
[36] Levy B S, Van Tyne C J 2015. *Proc. IDDRG Shanghai*, pp 262-269
[37] Levy B S, Green D E 2002. *Analysis and Documentation of Laboratory Studies on the Enhanced FLC Concept* (Auto/Steel Partnership, Southfield, MI, USA)

IOP Conf. Series: Materials Science and Engineering **426** (2018) 012006 doi:10.1088/1757-899X/426/1/012006

Natural rubber/boehmite nanocomposites *via* latex compounding

P Berki[1], D Q Khang[2], N T Tung[2], L N Hai[3], T Czigány[1,4] and J Karger-Kocsis[1,4*]

[1] Department of Polymer Engineering, Faculty of Mechanical Engineering, Budapest University of Technology and Economics, Budapest, Hungary
[2] Department of Functional Polymers and Nanomaterials, Institute of Chemistry, Vietnam Academy of Science and Technology, Hanoi, Vietnam
[3] Department of Chemical Technology Development, Center for High Technology Development, Vietnam Academy of Science and Technology, Hanoi, Vietnam
[4] MTA-BME Research Group of Composite Science and Technology, Budapest, Hungary

E-mail: karger@pt.bme.hu

Abstract. Natural rubber (NR) latex was modified with unmodified boehmite alumina (BA) nanoparticles, added in NR up to 15 parts per hundred rubber (phr) amounts. Dispersion of the BA nanofillers was inspected in scanning electron microscopy (SEM). The tensile and fracture mechanical (J-integral approach) properties of the NR/BA nanocomposites were determined. Information on the rubber-BA and possible BA-BA interactions were deduced from dynamic mechanical analysis (DMA) and differential scanning calorimetric (DSC) tests. It was found that BA particles, through agglomerated, were well dispersed in the NR/BA nanocomposites up to a given threshold (7.5 phr). Improvements in the tensile (strength) and fracture mechanical properties (crack initiation- and propagation-related data) were found until this threshold BA content. DMA results confirmed enhanced stiffness and Payne-effect with increasing BA content both in the glassy and rubbery states and in the rubbery state, respectively. The amount of the bound NR fraction on the BA surface was rather small according to DSC results. It was thus concluded that unmodified BA acts as a semi-active nanofiller in NR.

1. Introduction

Nanocomposites can be easily produced using polymer latices and water swellable, water dispersible nanofillers. Note that polymer latex is an aqueous dispersion of nano-scaled polymer particles. Among the water swellable and dispersible nanofillers clays (of natural or synthetic origin), oxidized carbonaceous nanofillers (carbon nanotube, nanofiber, graphene), nanocellulose and boehmite should be mentioned. These nanofillers have been introduced in various thermoplastics *via* water-assisted compounding versions [1]. Many reports are available also on clay-containing thermoplastics making use of this technique [1]. Moreover, works were published also boehmite alumina (BA) filled nanocomposites using various thermoplastic matrices [2]. Introduction of aqueous dispersion of nanocellulose particles, whiskers is preferably combined with plasticization to produce thermoplastic starch [3]. The basic reason of the successful preparation of thermoplastic nanocomposites with clays and BAs through water-mediated melt compounding is that the concentration of these nanofillers in their aqueous dispersions may be as high as 10 wt%. This is highly beneficial for the fast evaporation of the water carrier during extrusion compounding based on the fact that the target nanofiller content in the related thermoplastic is at about 3 wt.% [1]. The water-mediated melt compounding technique is less suited for oxidized graphene, oxidized carbon nanotube, and the like, because their concentration in aqueous fluids is usually below 1 wt.% [4].

The introduction of the above mentioned nanofillers in rubber latices is straightforward because it is not concentration-dependent since the dispersing fluid for the nanofillers is the latex itself. Large body of works addressed the preparation of rubber nanocomposites *via* latex compounding, also referred to latex cocoagulation. As rubber latices mostly natural rubber (NR) and styrene-butadiene rubber (SBR), whereas as fillers clays [5, 6] and graphene oxide [7] were selected. Works published on NR/clay [5-6] and NR/graphene oxide [7, 8] nanocomposites, produced by the latex route, revealed a fine and homogenous dispersion of these nanofillers in the NR. The related dispersion was better than by traditional melt compounding [9]. This was associated with strong improvements in the mechanical properties compared to reference mixes produced by traditional melt compounding.

Boehmite aluminas (BAs) are oxide-hydroxides of aluminum, available in different shapes. BA may have very high specific surface area (>300 m^2/g) since the primary crystallite size may be as small as 4.5 nm which makes it ideal as a nanofiller for polymers [3]. The numerous hydroxyl groups on their surfaces guarantee the dispersion in aqueous fluids. The particle size of BA dispersed in aqueous media is, however, 4 to 10 times of its primary crystallite size. BAs with and without surface modifications were already incorporated in various thermoplastics and thermosets, as reviewed recently [3]. The modification of rubbers by BA is far less advanced compared to the companion thermoplastic- and thermoset-based nanocomposites. This is a surprising fact because many properties of BAs are in favour of application in rubbers. These features are: high specific surface (comparable to traditional fillers, such as carbon black and silica), different shapes (aspect ratios), surface basicity (unlike acidity, it does not influence the vulcanization substantially), versatile surface modification (to tailor the interfacial adhesion) and nanoscale dispersion in aqueous media (the possibility of using latex (pre)compounding). Researchers made use of various surface modification strategies of BA in works performed on rubber/BA nanocomposites [3]. By contrast, unmodified (pristine) BA was very rarely used as nanofiller for rubbers, and only one report is found for adopting the BA (though surface-modified) dispersion rubber latex [10].

Considering the above information our intention was to check whether the latex compounding method is applicable for the production of rubber/BA nanocomposites. In this work commercial BA was added in 0, 2.5, 5, 7.5, 10 and 15 parts per hundred rubber (phr) amounts to modify an NR latex that was coagulated and cured by sulfuric curatives. The microdispersion of the BA was studied by scanning electron microscopy. Properties of the NR/BA nanocomposites were assessed in dynamic mechanical thermal analysis, tensile and fracture mechanical (J-integral) tests. Differential scanning calorimetry served to estimate the bound (also termed to immobilized) NR fraction.

2. Experimental

2.1. Materials

Synthetic boehmite alumina (Disperal® 40) was kindly provided by Sasol Germany (Hamburg, Germany). Characteristics of this BA are: Al2O3 content 80 %, BET surface area: 100 m2/g, primary crystallite (120) size: 40 nm. NR latex with 60 wt. % dry content was purchased from Tay Ninh Rubber Company (Vietnam). The other chemicals such as sulfur (S) from Sae Kwang Chemicals Ind., Co. Ltd (South Korea), zinc oxide (ZnO) from Triveni Chemicals Company (India), stearic acid from PT Cisadane Raya Chemicals (Indonesia), dibenzothiazyl disulfide (DM), 2-benzothiazosulfenamide (CBS), polymerized 2,2,4-trimethyl-1,2-dihydroquinoline from TMQ, Kemai Chemicals Co. Ltd, (China), ethanol 96% v/v from Thinh Cuong Co. Ltd, (Vietnam) were used as received. NH3 solution (25%) was purchased from Guangdong Guanghua Sci.-Tech Co, Ltd. (China).

2.2. Preparation of the samples

Firstly, BA in different amounts was dispersed in distilled water setting a weight ratio of BA:H_2O = 1:15 by adding of NH_3 solution (25%) to adjust the pH of BA dispersion to pH = 10. The BA dispersion was stirred for 30 min by magnetic stirrer. Afterward, BA solution was further homogenised by sonication for 2 h to get a well-dispersed BA solution. Then, 60 ml NR latex was poured to the BA solution under magnetic stirring of 800 rpm and the mixing time was 1 h. Afterward, NR/BA nanocomposites were coagulated by adding ethanol and then washed with distilled water. Finally, NR/BA nanocomposites were dried in vacuum at room temperature for 1 day.

The latex precompounded NR/BA samples were mixed with sulfur, ZnO, SA, DM, CBS, TMQ according to the recipe in Table 1 on a laboratory two roll mill (Toyoseiki, Tokyo, Japan) at room temperature for 15 min. The samples were cured in a press (Toyoseiki, Tokyo, Japan) at 145 oC and 2 MPa pressure for 15 min producing sheets with 2 mm thickness.

Table 1. Designation and composition of the NR/BA mixes

Designation	Composition [phr]							
	NR latex (dried)	ZnO	Stearic acid	TMQ	CBS	Sulfur	DM	Boehmite (BA)
NR ref	100	5	1	1	1.5	2	0.5	-
NR-2.5BA	100	5	1	1	1.5	2	0.5	2.5
NR-5BA	100	5	1	1	1.5	2	0.5	5
NR-7.5BA	100	5	1	1	1.5	2	0.5	7.5
NR-10BA	100	5	1	1	1.5	2	0.5	10
NR-15BA	100	5	1	1	1.5	2	0.5	15

2.3. Characterization and testing

To get information on the BA dispersion in the NR the fracture surfaces of the nanocomposites were inspected in field-emission and scanning electron microscopes (FE-SEM: Hitachi S-4800, Tokyo, Japan and SEM: JEOL JSM 6380LA, Tokyo, Japan), respectively.

Tensile tests were performed on a Gotech AI 7000M (Taichung,Taiwan) machine at room temperature with crosshead speed 200 mm/min according to DIN 53504. The hardness of the specimens was measured by a Teclock JIS K7215A (TTS, Osaka, Japan) instrument.

Fracture mechanical tests were performed on single edge-notched tensile loaded (SEN-T) specimens. SEN-T specimens of 100 x 25 x 2mm dimension (length x width x thickness) with 10 mm initial notch length were loaded with 10 mm/min crosshead speed on the above mentioned Zwick testing machine. The crack tip opening displacement (CTOD) was followed by visual inspection using a digital microscope (Celestron 44302, USA). The camera was positioned in front of the crack in order to focus on the internal surface generated by blunting and/or growing of the crack. Prior to this test the crack surfaces were coated by talc for contrast. By analyzing the registered sequence of pictures the point where fracture started to propagate could be detected and the corresponding J-integral (J-critical) value determined by [11]:

$$J = \frac{\eta \cdot U}{t \cdot (W - a)} \tag{1}$$

where η is a geometry factor (0.9 in this case), U is the input energy (given by the area under the load-displacement curve up to the point considered), t is the thickness (≈ 2 mm), W is the width (25 mm) of the specimen and a is the initial crack length (10 mm). The J-integral was determined as a function of the crack tip opening displacement (CTOD) between 0.1 and 0.8 mm. J-critical was read at CTOD=0.1 mm, whereas the tearing modulus (T_J) was given by the slope of the linear regression fitted for the J vs CTOD data.

Dynamic-mechanical analysis (DMA) traces were registered on rectangular specimens in tensile mode at a static preload of 0.1 N with a superimposed sinusoidal 0.1% strain. The frequency was 10 Hz and the spectra were taken in a temperature range of -100 to 60 °C using a Q800 device of TA

Instruments (New Castle, DE, USA). The temperature ramp was 3°C/min. The DMA technique was used also to assess the Payne effect. It was determined in tensile mode at 30 °C using 10 Hz frequency and 0.01 N static preload with strain sweep from 0.01 to 10% strain (denoted as M0.01 and M10, respectively).

Differential scanning calorimetry (DSC) served to investigate the glass transition region of the neat and BA-containing NR systems. 3-5 mg samples were heated from -90 °C to room temperature at a rate of 5°C/min under nitrogen flushing in a Q2000 device of TA Instruments (New Castle, DE, USA). Change in the heat capacity was used to estimate the immobilized NR faction (see later).

3. Results and discussion

3.1. Particle characteristics

Figure 1 shows SEM pictures taken from the fracture surface of the NR nanocomposites containing 2.5 (NR-2.5BA) and 7.5 phr BA (NR-7.5BA), respectively. One can clearly see the average dispersed size of the BA agglomerates did not change significantly with the BA content, at least until 7.5 phr content. The size of the BA agglomerates is in the range of 200-300 nm. This range agrees very well with that of the Sasol's brochure (180-250 nm), measured in slightly acidic aqueous dispersion. A closer look on the BA agglomerates reveals even the presence of prismatic primary crystallites of low aspect ratio.

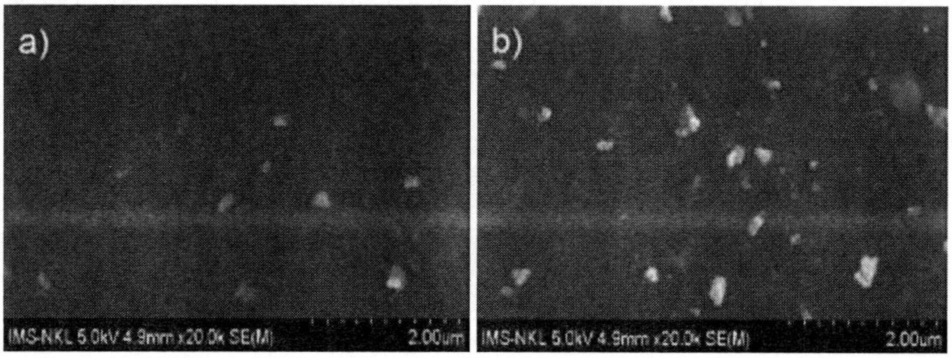

Figure 1. FE-SEM images showing the BA particles' dispersion in the NR-2.5BA (a) and NR-7.5BA (b) nanocomposites.

3.2. Mechanical and fracture mechanical properties

The mechanical (hardness, tensile strength, elongation at break) and fracture mechanical (J-critical, tearing modulus) of the unfilled (reference) NR and NR/BA nanocomposites are presented in Table 2.

The hardness increased monotonously with increasing amount of BA. The tensile strength went through a maximum (peaked at 7.5 phr) as a function of the nanofiller content suggesting agglomeration phenomena at high filler contents. The elongation at break decreased monotonously with the BA content. This is a usual feature when the amount of nanofillers is enhanced. The courses of J-critical and T_J as a function of the BA content were identical with that of the tensile strength, even the peak values at 7.5 phr BA agreed. This finding suggests that the dispersion of BA strongly affects both the J-critical and T_J. Inhomogeneous dispersion of BA particles owing to highly different agglomerates' sizes, associated with different matrix ligaments between them, facilitate the crack tip opening and growth.

IOP Conf. Series: Materials Science and Engineering **426** (2018) 012006 doi:10.1088/1757-899X/426/1/012006

Table 2. Mechanical and fracture mechanical properties of the NR/BA systems studied

Designation	Hardness [Sh °A]		Tensile strength [MPa]		Tensile strain [%]		J-critical [kJ/m²]		Tearing modulus, T_J [MJ/m³]	
NR ref	40	±1	19.01	±1.11	778	±20	3.85	±0.24	12.14	±0.78
NR-2.5BA	42	±1	25.82	±0.95	743	±15	5.11	±0.31	16.29	±0.81
NR-5BA	45	±1	27.80	±0.52	736	±19	6.21	±0.32	18.26	±0.66
NR-7.5BA	47	±1	29.53	±0.39	722	±21	6.82	±0.27	20.01	±0.55
NR-10BA	50	±1	24.70	±0.59	669	±34	6.58	±0.37	17.74	±0.76
NR-15BA	54	±1	20.90	±0.85	632	±32	5.19	±0.42	14.23	±0.82

3.3. Dynamic mechanical analysis

The storage modulus (E') vs. temperature and mechanical loss factor (tanδ) vs. temperature curves are depicted for the NR/BA nanocomposites in Figure 2.

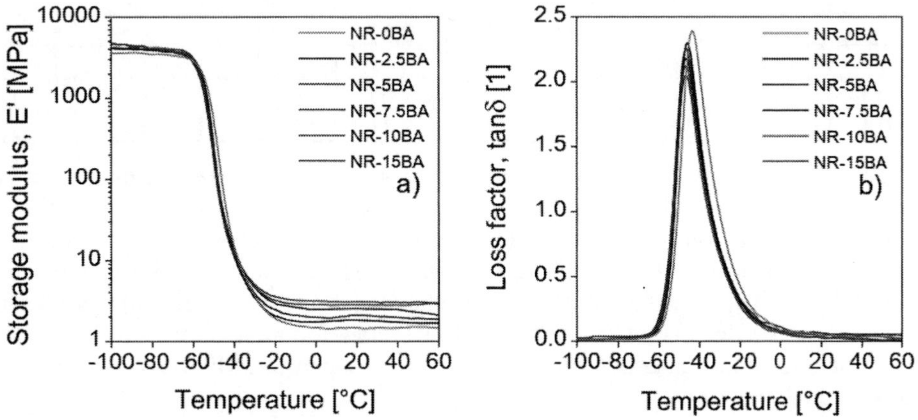

Figure 2. Storage modulus vs. temperature (a) and mechanical loss factor (tanδ) vs. temperature (b) curves for the NR/BA nanocomposites studied

One can notice in Figure 2a that both the glassy and rubbery moduli increased with increasing BA content. They were read at T_g-30 °C and T_g+30 °C, respectively (Table 3). This confirms the reinforcing effect of BA. The most surprising effect is that the tanδpeak was only slightly reduced with increasing amount of BA (cf. Figure 2b and Table 3). Modulus increase, reflecting the reinforcing effect, is usually associated with prominent decrease in the mechanical loss factor. Note that the glass transition temperature (T_g) values of the nanocomposites, determined as the peak temperature of tanδ, were slightly shifted toward lower temperatures (cf. Table 3). This was also unexpected and hints along with the high T_g peak values for a weak interphase. From the rubbery E'-moduli, read at the apparent mean molecular weight between crosslinks (M_c) was determined by the rubber elasticity theory:

$$M_c = \frac{3\rho \cdot R \cdot T}{E_{rubbery}} \tag{2}$$

where $E_{rubbery}$ is the modulus at $T=T_g + 30K$, ρ is the density (determined in a pycnometer using methanol), R is the universal gas constant (8.314 J/(K.mol)), and T is the absolute temperature. M_c is an apparent value because in the rubber nanocomposites it implies not only the crosslinking but also effects of the rubber–filler and filler–filler interactions. All the above mentioned data are tabulated in Table 3.

To characterize the filler-rubber interaction also the change in the mechanical loss factor (tanδ) can be considered. The reinforcing efficiency (R) in this case is given by [12]:

$$R = \frac{\tan\delta_{max,ref} - \tan\delta_{max}}{\tan\delta_{max,ref}} \tag{3}$$

where $tan\delta_{max,ref}$ and $tan\delta_{max}$ represent the peak values of the unfilled and filled rubber (nanocomposite), respectively. The larger is R, the higher is the rubber-filler interaction.

The above parameters are summarized in Table 3. Based on the data in Table 3 it can be concluded that BA acts as reinforcing nanofiller in NR. Its action, however, differs from that of traditional nanofillers (carbon black, silica) due to the fact that the T_g is slightly sifted toward lower temperature along with a small reduction of the height of the T_g peak. This is a peculiar feature that requires further investigations.

Table 3. DMA-related and Payne effect representing parameters for the NR/BA nanocomposites

Designation	T_g [°C]	tanδ max. [1]	R [1]	E'_{glassy} [MPa]	$E'_{rubbery}$ [MPa]	M_c (T_g+30°C) [g/mol]	E' at 0.01% strain (M0,01) [MPa]	Payne effect, M0,01-M10
NR ref	-43.6	2.39	0	3446	1.60	3876	1.73	0.72
NR-2.5BA	-45.9	2.30	0.038	3856	1.89	3327	2.02	0.96
NR-5BA	-46.6	2.26	0.054	3896	2.23	2855	2.54	1.04
NR-7.5BA	-46.6	2.18	0.088	3840	2.81	2311	2.95	1.39
NR-10BA	-46.8	2.12	0.113	4108	3.07	2160	3.52	1.89
NR-15BA	-46.8	2.05	0.142	4193	3.41	2015	4.75	2.43

Figure 3 displays the Payne-effect of the NR/BA nanocomposites. Note that this effect is related to the formation of a filler network [13]. The decrease in the tensile E'-modulus as a function of strain is due to a progressive destruction of the filler-filler interaction in the related nanocomposites. The Payne effect in this work was quantified by the modulus at 0.01% (M0.01), and modulus change between 0.01 and 10% (M0.01-M10) [14]. The related data are tabulated also in Table 3. It is noteworthy that M0.01 is a more sensitive parameter than M0.0.1-M10. M0.01 monotonously increased with increasing BA loadings. Practically the same information can be deduced when the M0.01-M10 data of the related nanocomposites are taken into account.

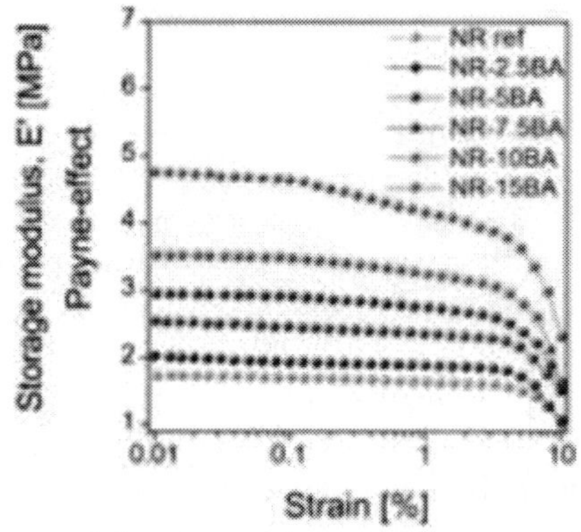

Figure 3. Storage modulus vs. strain curves measured in tension mode in DMA for the NR/BA nanocomposites

3.4. Differential scanning calorimetry

Based on the rubber-filler interaction reflecting DMA some change in the immobilized rubber layer's thickness is expected. Therefore, DSC tests were performed to estimate the amount of the immobilized rubber layer *via* [15]:

$$\Delta C_{pn} = \frac{\Delta C_p}{1-w} \quad (4)$$

$$\chi_{im} = \frac{\Delta C_{p0} - \Delta C_{pn}}{\Delta C_{p0}} \quad (5)$$

where ΔC_p is the heat capacity at the T_g jump, ΔC_{pn} is normalized to the polymer weight fraction, w is the weight fraction of filler, ΔC_{p0} is the heat capacity at T_g jump of the unfilled rubber. χ_{im} is the weight fraction of the immobilized polymer layer. Figure 4 shows characteristic DSC scans of selected NR/BA nanocomposites along with the related ΔC_p values. The DSC derived data are listed in Table 4.

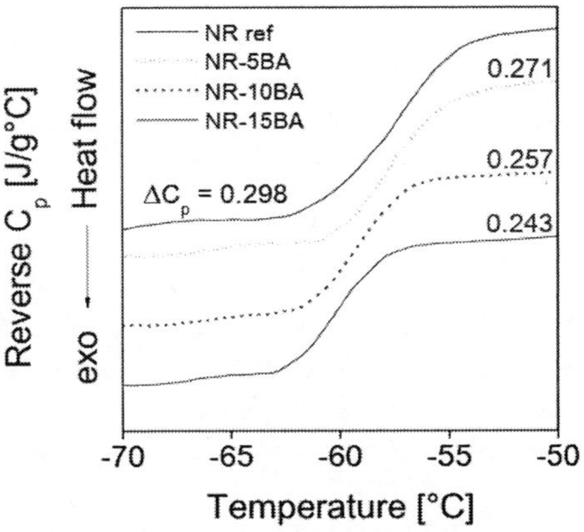

Figure 4. DSC scans in the glass transition (T_g) region of the NR and NR/BA nanocomposites containing 5, 10 and 15 phr BA amounts. Note that the Tg values deduced from DSC (cf. Figure 4) are lower than those from DMA (cf. Table 3) which is due to differences in these experiments (e.g frequency, sample amount).

Results in Table 4 are in line with the former results deduced from the DMA tests, i.e. small reduction in the height of the tanδ peak. The amount of the bound layer increased from ~5 to only ~7 wt% when the BA content was enhanced from 5 to 15 phr. This finding suggest that the BA nanoparticles worked as semi-active nanofillers.

Table 4. Heat capacity and bound layer amount for the NR/BA systems studied

Property [unit]	NRref	NR-5BA	NR-10BA	NR-15BA
Filler content [phr]	0	5	10	15
w, weight fraction [1]	0	0.043	0.083	0.120
ΔC_p, heat capacity [J/g°C]	0.298	0.271	0.2573	0.243
ΔC_{pn}, normalized heat capacity [J/g°C]	-	0.283	0.281	0.276
χ_{im}, weight fraction of the bound polymer layer [1]	-	0.049	0.058	0.074

4. Conclusions

This work devoted to study the properties of sulfur-cured natural rubber (NR)/boehmite alumina (BA) nanocomposites as a function of the BA nanofiller content. BA, being water dispersible, was mixed to the NR latex, followed by drying. These NR/BA premixes were compounded with sulfur curatives in two-roll mill before cured in a hot press.

Based on the results achieved the following conclusions can be drawn:

- BA particles were homogeneously and nanoscale dispersed in the NR/BA nanocomposites. The agglomerate size range (200-300 nm) was close to that of the in aqueous dispersion measured one (180-250 nm)
- hardness increased, whereas tensile strain decreased monotonously with increasing BA content indicating the reinforcing effect of BA. The tensile strength and the J-integral parameters related to crack imitation and propagation (J-critical and T_J, respectively) went through a maximum (peaked at about 7.5 phr) as a function of BA loading that was traced to the onset BA particles' agglomeration at higher BA contents
- DMA results indicated that with increasing BA amount both the glassy and rubbery moduli were increased. On the other hand, the peak height of the T_g-transition was only slightly reduced with

BA loading. The Payne-effect was enhanced with increasing BA loading, as expected. DSC tests indicated a small increase in the bound (immobilized) layer fraction with increased BA content. Based on these results it was concluded that unmodified BA acts as a semi-active reinforcement in NR.

Acknowledgements
This work was performed in the framework of a bilateral project between the Hungarian Academy of Science (Project number: NKM 55/2015) and Vietnam Academy of Science and Technology (VAST.HTQT.HUNGARY.01/16-17 - Institute of Chemistry), and partly also supported by the Hungarian Scientific Research Fund (OTKA 109409).

References

[1] Karger-Kocsis J, Kmetty A, Lendvai L, Drakopoulos S X and Barany T 2015 Water-Assisted Production of Thermoplastic Nanocomposites: A Review *Materials* **8** 72-95
[2] Hietala M, Mathew A P and Oksman K 2013 Bionanocomposites of thermoplastic starch and cellulose nanofibers manufactured using twin-screw extrusion *European Polymer Journal* **49** 950-6
[3] Karger-Kocsis J and Lendvai L Polymer/boehmite nanocomposites: A review *Journal of Applied Polymer Science* 45573-n/a
[4] Manek E, Berke B, Miklosi N, Sajban M, Doman A, Fukuda T, Czakkel O and Laszlo K 2016 Thermal sensitivity of carbon nanotube and graphene oxide containing responsive hydrogels *Express Polymer Letters* **10** 710-20
[5] Galimberti (Ed.) M 2011 *Rubber-Clay Nanocomposites: Science, Technology, and Applications* (Hoboken, New Jersey: Wiley)
[6] Gatos K G and Karger-Kocsis J 2010 *Rubber Nanocomposites,* ed S Thomas and R Stephen: John Wiley & Sons, Ltd) pp 169-95
[7] Varghese S and Karger-Kocsis J 2003 Natural rubber-based nanocomposites by latex compounding with layered silicates *Polymer* **44** 4921-7
[8] Wu X, Lin T F, Tang Z H, Guo B C and Huang G S 2015 Natural rubber/graphene oxide composites: Effect of sheet size on mechanical properties and strain-induced crystallization behavior *Express Polymer Letters* **9** 672-85
[9] Berki P, László K, Tung N T and Karger-Kocsis J 2017 Natural rubber/graphene oxide nanocomposites via melt and latex compounding: Comparison at very low graphene oxide content *Journal of Reinforced Plastics and Composites* **36** 808-17
[10] Florjanczyk Z, Debowski M, Wolak A, Malesa M and Plecha J 2007 Dispersions of organically modified boehmite particles and a carboxylated styrene-butadiene latix: A simple way to nanocomposites *Journal of Applied Polymer Science* **105** 80-8
[11] Ramorino G, Agnelli S, De Santis R 2010 Investigation of fracture resistance of natural rubber/clay nanocomposites by J-testing *Engineering Fracture Mechanics* **77** 1527-36
[12] Fu G J, Chang X Z, Mao J and Shi X Y 2016 Insights into the Reinforcement of Butyl Rubber by Carbon Black and Silica with the Aid of Their Dynamic Properties *Journal of Macromolecular Science Part B-Physics* **55** 925-36
[13] Roland C M 2013 *The Science and Technology of Rubber (Fourth Edition),* (Boston: Academic Press) pp 285-336
[14] Berki P, Göbl R and Karger-Kocsis J 2017 Structure and properties of styrene-butadiene rubber (SBR) with pyrolytic and industrial carbon black *Polymer Testing* **61** 404-15
[15] Zhong B C, Jia Z X, Luo Y F, Jia D M and Liu F 2017 Understanding the effect of filler shape induced immobilized rubber on the interfacial and mechanical strength of rubber composites *Polymer Testing* **58** 31-9

IOP Conf. Series: Materials Science and Engineering **426** (2018) 012007 doi:10.1088/1757-899X/426/1/012007

Simulation of crystallographic changes during recrystallization by one-dimensional cellular automaton

T Bubonyi[1], P Barkóczy[1]

[1] University of Miskolc, Miskolc-Egyetemváros, Hungary

E-mail: fembubo@uni-miskolc.hu

Abstract. Cellular automata are one of the most frequently used methods to simulate the recrystallization process. The cellular automata give the complex nature of this process with a simple implementation. One of the greatest effort in this field is to calculate the changes in crystallographic texture. One-dimensional cellular automata are introduced to calculate the kinetics of recrystallization. This study introduces a simple method which makes it possible to incorporate the texture components to the calculations.

1. Introduction

Cellular automaton is a widely used mathematical method [1], due to the simple implementation and complex behavior [2]. The method is originated to famous Hungarian mathematician John von Neumann [3]. The first application in physical metallurgy was made by Hesselbarth and Göbel [4]. The aim of this study was the simulation of recrystallization process. This solution was a two-dimensional deterministic automaton. Davies developed this method with the incorporation of the stochastic rules into the grain growth process [5, 6]. Schönfish studied further the stochastic nature of the stochastic, asynchronous and random-grid automata [7]. After Davies' work the main efforts were to couple the automata with crystal plasticity computations [8], simulate the dynamic recrystallization [9] and incorporate the solute drag to the simulations. In parallel during the research the effort to simulate crystallographic changes are clearly visible [10].

In this field there are two main directions: the study of local changes during nucleation and growth, and measuring and evaluating the texture of the metals in relation to the given technological parameters [11]. These two directions divided the simulations also into two groups. In one case the simulations handle few grains, and for example orientation imaging microscopy (OIM) gives the initial state of the calculations. The other methods take into consideration the local changes in higher abstraction level, and calculate the macroscopic texture, the variation of the mechanical parameters. The aim of the two group is the same: to get a deeper understanding about the crystallographic changes.

The cellular automaton is a discrete dynamic system. The examined space (universe) is divided into identical volume elements, called cells. The size, the properties and the behavior of the cells are totally the same. The properties of the cells are called states, which have a finite number of possibilities. The states are determined mostly by physical values, for example, the dislocation density. This is a continuous measure, but it clearly describes that the cell is either deformed or recrystallized. The physical interactions are described by local rules, these are discrete functions. The input parameters are the states of the given cells and the states of its neighbors. The output of these functions are the new states of the cells. The automaton tries to change the state of each cells, it builds a new universe. When every cell has been updated, one simulation step is finished. This covers the main problem of the cellular automata. The time and the characteristic distance is measured by discrete values, the number of steps and the number of cells.

It is necessary to determine the real time interval and distance during the simulation. This method is called scaling. There are several ways of scaling an automaton. The easiest way is scaling with global optimization methods [12].

The complexity of the automaton increases the computation time of scaling. One of the parameter which increase the complexity is the dimensionality. There are large differences between one- and two-dimensional automata [13].

The one-dimensional automata have been studied and classified by Wolfram [14]. Wolfram showed that even the simplest automaton has a complex behavior. According to this idea, one-dimensional cellular automata were developed to simulate recrystallization, allotropic phase transition and grain coarsening [15]. All automata are studied and scaled. It was found that the scaling of a one-dimensional automaton is a fast process, and it can be automatized.

The one-dimensional automata have several limitations, of course. But the calculations are fast, the automata are scaleable, and the calculations with a given resolution (number of cells) are robust. If the aim is the simulation of the effect of technological parameters on the crystallographic changes, the one-dimensional automata could be an ideal method. In this report a possible solution is introduced for the calculation of these changes.

2. Method

One-dimensional simulation of recrystallization [12] extends with the simulation of crystallographic changes. The main question is the description of the texture of the metals. Several ways can be used, like pole figures, orientation distribution functions (ODF) or the volume fraction of the main crystallographic components [16]. The one-dimensional cellular automaton can prove to be a fast calculation if it concerns a few 10s to 100s of grains. This amount of grains does not provide enough data to plot a valuable pole figure or ODF if it has an exact crystallographic orientation described by the Euler angles. The aim of this study is to calculate the changes of the crystallographic orientations during recrystallization. The scaling is performed based on the changes of the volume fractions of main texture components. Based on the texture measurement the volume fraction of the main components can be calculated. In case of EN-AW-3003 aluminum alloy in rolled (86% reduction) and annealed (280°C, 5h) state figure 1 shows the volume fractions. The exact results and methods are introduced in [19].

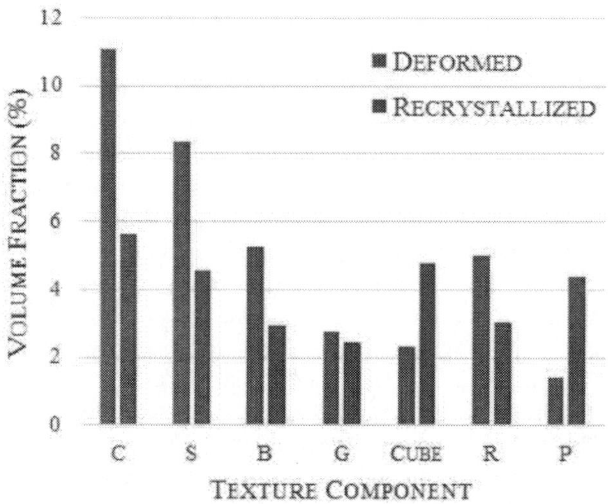

Figure 1. The volume fractions of the EN-AW-3003 aluminum alloy in deformed (86%) and annealed state (280°C, 5h) measured and published by Pethő [19].

First a deformed grain structure is needed with the measured volume fraction. The volume fraction determines the probability of the existence of a main crystallographic component in the given volume element. A recrystallization like automaton is developed to evolve the deformed grain structure. The crystallographic orientation of the different grains are determined at the nucleation process. A scale was built up by the volume fraction which was extended with the random component to 100%, and the texture component is determined with a simple Monte-Carlo algorithm [18].

In a one-dimensional automaton it is impossible to include all of the rules of nucleation process [17]. So, the similar scheme can be followed during simulation, but the scale of probability is calculated based on the recrystallized volume fractions. But in this case the simulation can be scaled by a measured kinetic function.

In the one-dimensional simulation the activation energy of the nucleation and growth are the two main kinetic parameters. In both cases (deformed structure generation and the simulation) the activation energies are constant values independent of the crystallographic orientations. Of course, these values can depend on the local orientation differences, but in this consideration this simple assumption gives a useful solution.

The disadvantage of this method is the number of grains. Because if only a few grains build up the universe the accuracy of the computation is small. Enough nuclei have to be formed to get the necessary result in case of all texture components. So, in this case a larger universe is needed as in a normal kinetic simulation. But this means a slower computation. However, the introduced method is faster than a two-dimensional computation, and easily take into account the partial annealing and the dynamic recrystallization too if enough measured data exist.

3. Results

The measured volume fraction of the main texture components is shown on Figure 1. Values of the deformed and the annealed states (4 an 5 hours) are used in the computation. The sequence of the volume fractions with the random component build a probability function. The sum of the volume fractions is extended by the random component to 100%. The probabilities of the deformed state are used in the computation of the deformed grain structure. The probabilities of the annealed state are used in the simulation of recrystallization. Based on the probability functions the deformed structure is calculated. The parameters of the simulation are: activation energy of nucleation: 71kJ/mol, activation energy of growth: 49kJ/mol, number of cells: 10000, temperature: 553K. The dataset comes from a kinetic analysis and scaling of the automaton of EN-AW-3003 aluminum alloy. The results are shown by figure 2.

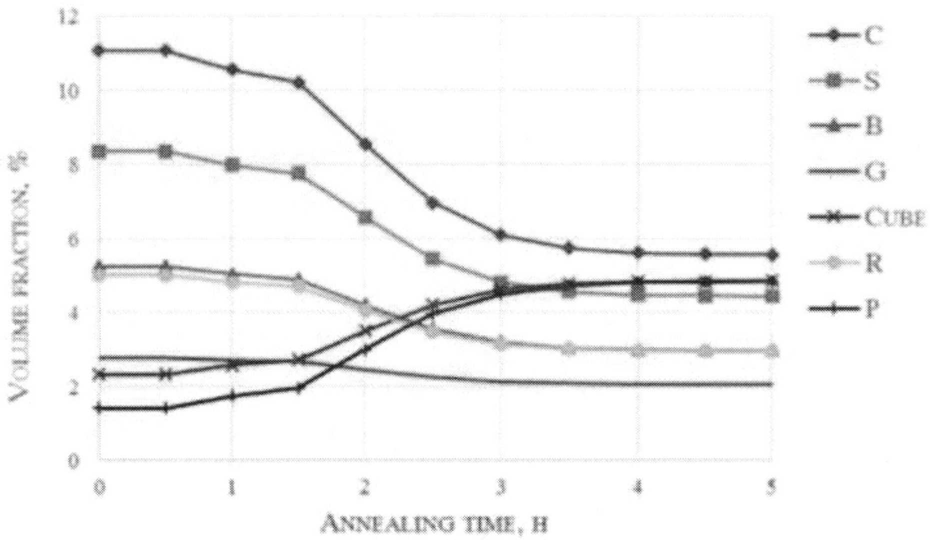

Figure 2. The calculated changes of volume fraction of the main texture component based on cellular automaton.

Figure 2 shows that it is possible to calculate the changes of the volume fractions based on one-dimensional cellular automaton. The approximation is not as sophisticated as in case of deep researches in this field, but it can be applied to calculate the effect of technological parameters on the crystallographic orientations.

4. Summary

One main task of the cellular automata simulations of recrystallization is to calculate the crystallographic changes during the phase transformation. One possible solution of the mentioned simulation is the one-dimensional cellular automaton. One-dimensional automata provide fast computation but have some restrictions. The calculation of the crystallographic changes is one apsect which seems to be lost in one-dimension. But in high abstraction level a possible solution can be found.

The volume fractions of main texture components can be calculated based on texture measurements. This is a discrete description of the crystallographic texture which fit with the nature of the cellular automata. The changes of the main texture components during recrystallization can be calculate based on the nucleation probabilities of different texture component. The probability of nucleation of a given texture component is equal to its volume fraction in the annealed state if a constant rate of growth is considered. These assumptions are proved by the one-dimensional cellular automata.

The calculation requires a larger number of nuclei than a simple kinetic calculation, which means a larger universe. A larger universe makes the calculation slower of course, but the one-dimensional automata remain faster than a sophisticated two-dimensional calculation. In the study of the effect of technological parameters the one-dimensional automata are a competitive solution.

5. Acknowledgments

The described article was carried out as part of the GINOP-2.3.2- 15-2016- 00027 "Sustainable operation of the workshop of excellence for the research and development of crystalline and amorphous nanostructured materials" project implemented in the framework of the Szechenyi 2020 program. The realization of this project is supported by the European Union. The described research was also supported through the National Research, Development and Innovation Office – NKFIH K119566 project.

Special thanks to Adrienn Hlavács, and Danel Pethő, for letting us use their measurement results.

References

[1] Bandini, S., Chopard, B., Tomassini, M. (2002). Cellular automata, 5th International Conference on cellular automata for research and industry. Geneva. Switzerland.
[2] Espericueta, R. (1997). Cellular Automata Dynamics. Bakersfield College.
[3] Israel, G., Gasca, A. M. (2009). The world as a mathematical game. John von Neumann and Twentieth centurí science. Science Networks. Historical Studies. Vol. 38.
[4] Hesselbarth, H. W., & Göbel, I. R. (1991). In Acta Metallurgica et Materialia Vol. 39 No. 9 pp. 2135-2143.
[5] Davies, C. (1997). Sripta Mater. Vol. 36. 35-40.
[6] Davies, C. (1999). Sripta Mater. Vol. 40. 1145-1150.
[7] Schönfisch, B. (1995). Propagation of front sin cellular automata, Physica D 80 p. 433-450
[8] Raabe,D.(2002). Advanced Engineering Materials, Vol. 4, No. 5, p. 255-267.
[9] Sarma,G. B.,. Radhakrishnan,B., Zacharia,T.(1998). Comp. Mater. Sci., Vol.12, p. 105-123.
[10] Crumbach, M., Goerdeler, M., & Gottstein, G. (2006). Modelling of recrystallisation textures in aluminium alloys: I. Model set-up and integration. Acta Materialia, 3275–3289.
[11] Rolett,A. D.(1997). Progress in Material Science, Vol. 42, p. 79-99.
[12] Szilvia Gyöngyösi, Péter Barkóczy: Scaling cellular automaton simulations of short-range diffusion processes MATERIALS SCIENCE FORUM 729: pp. 150-155.(2013)
[13] Gyöngyösi Sz., Barkóczy P.: Cellular Automata Simulation of the Allotropic Transformation of Uranium MATERIALS SCIENCE AND ENGINEERING: A PUBLICATION OF THE UNIVERSITY OF MISKOLC 41 pp. 39-50. (2016)
[14] Wolfram, S. (2002). A new kind of science.
[15] Szilvia Gyöngyösi, Péter Barkóczy: Simulation of grain coarsening using one-dimensional cellular automaton MATERIALS SCIENCE FORUM 752: pp. 217-222.(2013)
[16] Kocks, U. F., Tomé, C. N., & Weng, H. R. (1998). Texture and anisotropy. UK: Cambridge University Press.
[17] Humphreys, F. J. (2004). Nucleation in Recrystallization. Materials Science Forum Vols. 467-470, 107-116.
[18] Eckhardt, R. (1987). Stan Ulam, John von Neumann and the Monte Carlo Method. Los Alamos Science Special Issue.
[19] D. Pethő (2016). Thesis, University of Miskolc, http:\\midra.uni-miskolc.hu

IOP Conf. Series: Materials Science and Engineering **426** (2018) 012008 doi:10.1088/1757-899X/426/1/012008

Cobalt release of oxygen scavenger added multilayered PET bottles in various domestic solvents

Gy Czél[1] and F Hámoros[2]

[1] Professor: University of Miskolc, Institute of Ceramic and Polymer Engineering, Miskolc, Hungary
[2] Reserch fellow: Balint Analitics Ltd. Budapest, Hungary

E-mail: femczel@uni-miskolc.hu

Abstract. Cobalt catalyzed oxygen scavenger added multilayered PET bottles were inspected in our work. X-ray fluorescent spectrometry (XRF) technique was used to select the investigated bottles. Inductive coupled plasma mass spectrometry (ICP-MS) was applied to determine the cobalt content of different domestic solvents after extraction. Extraction experiments were separated into 48 hour extractions, Lakanen-Erviö extraction, and extraction kinetic experiments. The fruit brandy and the orange juice contained in the bottles were the two most successful agents dissolving cobalt without chemical reactions. 1 M NaOH was the agent dissolving the highest quantity of cobalt as a consequence of the hydrolysis of PET. Continuous cobalt dissolving over time could be experienced when NaOH agent penetrated into the grains of flakes. Reuse of active multilayered bottles for domestic solvents did not show harmful cobalt concentration, but the ground and oxygen scavenger added multilayered PET bottles showed cobalt concentration over the limit value.

1. Introduction

Nowadays, multilayered PET bottles are widely used for juice packaging [1]. These types of bottles, produced by three or five layered preform molding technology, give favorable packaging features. The layers made of different linear polyester compounds show passive barrier [2], and active blocking function. With this up-to-date technology, five layered PET bottles were made so that the outer and the inner layers were made of virgin PET raw materials, but the three layers situated in the middle were made of different blended polyesters. The middle layer, as an active blocking layer, serves as an oxygen scavenger layer. The function of the scavenger layer is that oxygen diffusion into the bottle is hindered by additional chemicals, e.g. cobalt salts. Due to this favourable oxygen scavenger feature, a wide range of oxygen sensitive products such as tomato-based products, UHT milk, fruit juice are often bottled into active oxygen scavenger layered PET bottles. Some difficulties arise during the recycling process operations of multilayered PET bottles. Currently, blended multilayered PET bottles and mono layered bottles undergo the same recycling procedure. Large amounts of bottles or regranulates are transported from Europe to countries in the Far East as raw material for the textile industry. A small number of ground or compressed bottles finish their life cycle in communal waste-piles, oceans, or living waters. Due to their practical form and strength, the PET bottle recycling process begins in the home when acidic, basic or alcoholic liquids are used to refill the bottles. In recycling sites the used PET bottles are selected by their colour, but they are ground, mechanically washed in common soda solvents, and the final steps are for them to be dried, re-granulated, and recrystallized. Tons of ground PET flakes, as ready to use recyclement, are on the market. Several technical publications explain the short lifetime and the appropriate process technology for multilayered PET bottles [1]. Due to the new and profitable feature of active and passive oxygen scavenger layer technology [2], only a few scientific publications are available to explain the chemical components and the way these special polyester blends operate [3,10]. The present publication aims to describe the cobalt release of the oxygen scavenger [4,5] added multilayered PET bottles exclusively in domestic solvents in order to verify the possibility of safe application following commercial use. Another aim is to define the quantity of extractable cobalt ions in order to assess the impacts of these extracted amounts on the environment. In the course of our series of experiments, cobalt catalyzed oxygen scavenger added PET bottles separated by X-ray fluorescent spectrometry (XRF) were tested by inductive coupled plasma mass spectrometry (ICP-MS). We determined the cobalt content of different domestic solvents. The extraction experiments were divided into 48-hour extractions, Lakanen-Erviö extraction, and extraction kinetic experiments.

2. The function and operational mechanism of active scavenger in PET bottles

The active scavenger layer applied in multilayer bottles consumes oxygen that permeates through bottles thereby hindering the adverse effects caused by oxygen. This system stays active until there is a sufficient amount of additive material in order to maintain an oxidation reaction. The oxygen scavenger system is a multicomponent system and the reaction itself is a radical reaction, i.e. oxidative degradation, and follows three steps analogous to free-radical polymerization reactions: initiation, propagation, termination. In the initiation step a free-radical is formed by a free-radical donor. Usually this is conducted by the addition of peroxides, hydroperoxides or α, α-azobisisobutyronitrile. However, it is not essential to use such initiators. The autoxidation reaction may be initiated by the presence of light or heat as well. In the propagation step the addition of oxygen to alkyl free-radicals and abstraction of hydrogen from the polymer-chain can be observed. The polymer should be unsaturated and oxidizable and should have a so-called oxygen scavenging moiety, e.g. olefin functional group. These functional groups form peroxyl (R-O-O·) and alkoxyl (R-O·) free-radicals in combination with oxygen. Then peroxides (R-O-O-R) and hydroperoxides (R-O-OH) are formed, and finally they decompose into carbonyl groups (>C=O). The saturation of such an active system takes place when these oxygen scavenging moieties are depleted. Oxidizable polymers are polyamides or 1,4-polybutadiene. Intramolecular and intermolecular oxidation can be assumed in the reaction. Transition metal ions have an important role in the redox-decomposition of alkyl hydroperoxides. Most often cobalt and magnesium are used, but silver, cerium, iron and copper are also suitable for such purposes. In the case of polymers, the presence of a transition metal complex leads to decomposition, and, therefore, the production rate of alkoxyl and alkylperoxyl radicals will be increased. Thus the reaction is catalyzed in the initiation and in the propagation steps. The presumable mechanism of the catalytic reaction is an oxidation-reduction catalytic cycle in which the valence state of metal ions changes back and forth. The most effective transition metal complexes are cobalt carboxylates. Such metal complexes are cobalt octoate, cobalt stearate and cobalt neodecanoate. There are several active oxygen scavenger systems available on the market: Amosorb™ DFC [6], Polyshield™, Aegis™ OX, Oxbar™ and Bindox™. Amosorb™ DFC is a PET masterbatch which can be added into any bottle grade PET during manufacture and the mixture thus obtained can be used to create an active scavenger layer.

3. Characteristics of cobalt and its salts, physiological, environmental impacts, and environmental limits

In nature, cobalt can be found in small quantities. It is present in the soil, natural bodies of water, plants, grain, and seeds, and it can be also found in animals. The concentration of cobalt in soil varies widely. It is usually about 1 to 40 mg/kg. Soils containing less than about 3 mg/kg of cobalt are considered cobalt-deficient. Plants growing in them do not have a sufficient amount of cobalt, and, therefore, these plants are not suitable to use as forage. In surface waters and ground waters cobalt can be found in the form of ions normally in the range of ppb. Due to certain industrial activities, the concentration of cobalt may increase in natural bodies of water, the air and the soil. Therefore, in such areas, authorities may require a sampling of these media and analysis of these samples at given times or time intervals. In the connection with human health, cobalt has beneficial and harmful effects. Cyanocobalamin, or vitamin B12, is an essential compound of cobalt needed to maintain human and also animal health. Vitamin B12, together with the other B7 vitamins, helps the body convert carbohydrates into glucose [9]. Vitamin B12 plays an important role in sustaining healthy nerve cells, controlling blood levels of homocysteine, aiding the production of RNA, DNA, and also red blood cells [7]. For humans, good dietary sources of vitamin B12 are meat and dairy products, eggs and fish. The Recommended Dietary Allowance of vitamin B12 for an adult of average body weight is 2,4 μg/day, which means 0,1 μg of cobalt per day. Cobalt has been found in most tissues of the human body, but the highest concentrations are located in the liver [8].

In recent years increasing attention has been paid to the quality and the extent of contamination, including monitoring the quantitative changes, of soil as a geological medium and also of natural waters. In Hungary, regulation No 6/2009. (IV. 14.) KvVM-EüM-FVM is currently in force with respect to the sampling and administration of analytical measurements as well as the limit values of different contaminants in the geological media and ground waters. According to this regulation, the relevant limit values for cobalt are as follows: in geological media: 30 mg/kg dry matter, hazard classification K2; in ground waters: 20 ppb, hazard classification K2 [11].

Cobalt is an off-white ferromagnetic metal. Its reaction with non-metallic elements leads to +2 and +3 oxidation states, but there are known examples where 0, +1 or +4 oxidation states occur. Its reaction with carbon monoxide at 200 bar and at 150°C results in dicobalt octacarbonyl $[Co_2(CO)_8]$. The color of aqueous solutions of Co(II) salts is typically pink due to the cobalt(II) hexahydrate $[Co(H_2O)_6]^{+2}$ ions, while the color of anhydrous cobalt salts are typically blue (Co^{+2}).

4. Experimental materials and methods

Knowing that the bottle production line middle layer had an active scavenger layer made by an AmosorbTM additive, bottles of high-fiber content orange juice having three-layers were purchased from consumer markets. The matrix of AmosorbTM was polyethylene terephthalate, the active component was 1.4- polybutadiene, and the free radical formation initiator was some kind of cobalt carboxylate. To increase the efficiency of the extraction and to copy the first step of the recycling process, we chose a grinding sample preparation. A Piovan's RSP 15/15 B2RX granulator (serial number: 110000002898) was used for grinding 40 pieces of one litre bottles of a well-known brand of 100% high-fiber orange juice. Before grinding we randomly selected 5 bottles to determine their average weight and cobalt concentration. The cobalt concentration was measured by a ED-XRF spectrometer. Based on the ED-XRF measurements, the Co concentration showed homogeneity. A sieve analysis was used to validate the grinding by using 4.00 mm, 3.15 mm, 2.00 mm, 1.00 mm, 0.5mm, and 0.315 mm sieves. From each sieve fraction a 0.5 gr sample was taken in order to determine the average cobalt concentration. The average cobalt concentration of ground bottles was 36.805 mg/kg. This value was higher than the cobalt limit value for geological media (30 mg/kg dry matter). The average cobalt content for a one litre bottle was calculated and is 1.014mg/bottle. The domestic solvents which were selected were ones that can be found in households and which may possibly come into contact with mono or multilayer active blocker bottles (detergents, fruit brandy, etc.). In some cases, we use more powerful solvents in our households than those which may be found in a waste landfill. Such solvents, 1 M NaOH, HCl, ammonia, etc., were used in our experiments. The parameters of the 48-hour extraction experiments and the used solvents are shown in Table 1. The density of the traditional fruit brandy and ethanol volume percentage was measured in the Food Laboratory Department of Bálint Analitics Ltd. During the extraction and sample preparation the weight was measured by a digital mass balance. Flakes were homogenized by a strong shake of the powder in the container bag. The various solvents were dosed by a graduated cylinder. A heated magnetic stirrer (with 12 blades) was applied for the extraction preparations. The experimental mixtures were stored at 40°C and 80°C during the night in an incubator without stirring. We found a milky precipitate during the extraction of the 1 M NaOH. Consequently, the concentration of the extract was measured by pre-filtration of the precipitate and by addition of the precipitate to the solution. This method was used only in the case of extraction at 40°C. After completing the extraction, the compounds were filtered by using a paper filter and the solutions obtained in such a way were measured by means of the ICP-MS method after the appropriate sample preparation methods. As can be seen from the results, solutions having a cobalt-concentration higher than that allowed by law were obtained after extraction in the case of 1 M NaOH, and, therefore, an experiment was performed on the filtered bottle flakes where a 48-hour extraction was performed again at 40°C by using a fresh dissolving agent. The *Lakanen-Erviö (LE)* extracting agent extracts ion-bound metals from the soil for which mobilization by the roots can well be modelled and the proportion that can potentially be taken by the plants through their roots can be subtracted.

Table 1. The solvents and extraction parameters

Solvents used	**PET flakes contents and the extraction parameters**
1 M NaOH	Temperature of solvents: T =40°C and T= 80°C
1 M HCl	
5% ammonia solution	
Distilled water	Extraction time: t_{extr} = 48 h
20% household vinegar	
Orange Juice (100%)	Mixing time: t_{mix} =16-17 h
Fruit brandy * Ethanol % (V/V): 49.986 Density: 0.9302 g/cm3	Incubation time: t_{incub} = 31-32 h
1% solution of washing powder (Aro Univerzális Mosópor)	Mixing speed: n= 500 rpm Amount of flakes used: m= 10 g
10X dilution of liquid detergent (Diversey Tip Professional -Kézkímélő mosogatószer)	Amount of solvents used: m= 100 ml
100X dilution of liquid detergent	*Density and ethanol content of fruit brandy were determined by Bálint Analitics Ltd.

The extract made by such an extracting agent shows the actual toxicity. This method can well be used to characterize the metal-contamination of contaminated soil- and waste-samples. The extracting agent we used is made as follows: The LE extracting agent contains 0.5mol/dm^3 ammonium-acetate, 0.5 mol/dm^3 acetic acid, and 0.02 mol/dm^3 EDTA (ethylene-diamine-tetra-acetic acid). Its pH is 4.65+/-0.05. 5.85 g of EDTA, 35 ml of 25% (m/m) ammonium-hydroxide and about 400 ml of water were measured in a flask with a volume of 1000ml. After the EDTA had completely dissolved in the ultrasonic tank, 57 ml of acetic acid was added to the compound. Then it was diluted to approximately 950 ml with water. The pH-value of the solution could be set to a value of 4.5+/-0.05 by using 3 M ammonium-hydroxide or acetic acid as needed. For making the extract, a 5 g sample was added to 50 ml of extracting agent and then it was mixed for an hour by a shaking machine. *Dissolution-kinetic tests* were performed in order to know the kinetics of dissolution. In the course of planning the experiments, we decided to use two types of dissolving agents: one, fruit brandy, through which the largest quantity of cobalt could be dissolved without causing a chemical reaction and another, 1 M NaOH, by which a chemical reaction takes place while dissolving the largest quantity of cobalt. The extraction parameters of dissolution kinetics are identical to the parameters of the 48-hour extraction experiments (10 g powder, 100 ml of dissolving agent, n=500rpm) though the experiment was performed only at 40°C. In the course of the test, dissolution was started simultaneously in 6 compounds. Then the concentration of cobalt was investigated in the individual solutions at given time periods (8, 24, 32, 48, 72 and 104 hours). The testing method was the same for both the fruit brandy and the 1 M NaOH dissolving agents.

5. ICP-MS sample preparation for determining the cobalt-content of PET flakes

Following homogenization, a sample containing approximately 0.5 g cobalt was measured in a glass tube by using a tare scale and then ~5 ml of 68% nitric acid and ~4 ml of 30% hydrogen peroxide were added to it. This mixture was then digested by a microwave process using a Milestone Ultrawave ECR type microwave destruction autoclave programmed for t=25 min; the starting pressure was p=40 bar and the value of T_{max}=220°C. After the whole quantity of flakes had been digested (dissolved, got into the solution), the compound was diluted to 25 ml in a volumetric flask by using distilled water. The solution prepared in such a way was injected into the instrument. Owing to the organic-matter content of orange juice-, washing- and dishwashing solutions, a microwave digestion process was also applied. The method of preparing the individual samples was identical to the process used for preparing the PET flakes, though it was enough to use a lower quantity of destruction agents including the quantity of nitric acid and hydrogen peroxide as well. 5 g of the each of the extracts remaining after the other dissolving agents was measured in separate beakers by means of a tare scale and they were evaporated until dry. If no precipitation was experienced, the beakers were rinsed with 3 ml of 68% nitric acid and they were washed into volumetric flasks of a volume of 25 ml. If there was precipitation, the compound was distilled by adding 5 ml of 68% nitric acid and 3 ml of 30% hydrogen peroxide until 1-2 ml liquid remained on the bottom of the beaker. Then the compound was washed into a volumetric flask having a volume of 25 ml without residue by means of distilled water. Such an extract was, for example, the vinegar extract – in this case a brown precipitation was experienced. A blind sample and a background sample were both prepared during the preparation of the extracts. The blind sample was the clean dissolving agent (e.g. 1 M NaOH, fruit brandy, orange juice) used during the extraction process. It was necessary for investigating the quantity of cobalt in the given clean dissolving agent (c_{blind}). The background sample was the compound of materials used during the preparation which can "incidentally" be contaminated ($c_{background}$). The background sample consisted of 5 ml of 68% (m/m) nitric acid and 4 ml of 30% hydrogen peroxide diluted to 25 ml. So the result measured in the given extract (c_{extr}) had to be corrected in each case as follows: $c_{korr} = c_{extr} - c_{blind} - c_{background}$. Both the earlier and the further results are the concentration values corrected in accordance with the above formula (c_{korr}). During the preparation of the individual samples, an internal standard solution had been added in the given quantity before the volumetric flasks were filled to the given marking in such a way that after filling them to the given marking, the scandium, rhodium, and lutetium concentrations of solution was 1 ppb. All cases Agilent 7500 ICP-Ms type mass-spectrometry instrument was used for measurement purposes.

6. Results of the extraction investigations

The results of cobalt dissolution tests performed by using different dissolving agents are shown in Table 2. The extraction experiments were performed at both 40°C and 80°C in order to reveal the effect of temperature increase regarding dissolution. In the course of the experiments performed at 40°C, a milk-white precipitation was experienced during digestion by nitric acid when preparing the extract obtained during the dissolution test performed by using 1 M NaOH. It was supposed that the precipitate could include some cobalt. So the cobalt-content of the extract was investigated in such a way that the precipitate was filtered and then the clean extract was prepared for an ICP-MS test. The other version of precipitate investigation was that the precipitate was digested by microwave destruction and later its solution was tested again.

6.1. Results of the 48-hour extraction experiments

Owing to the high degree of dissolution of 1 M NaOH, the dissolution test was performed again at 40°C in such a way that the extract was filtered after completing the 48-hour extraction (1/2) and a 48-hour extraction was performed again by adding fresh extracting agent to the filter cake that remained plastic flakes (2/2). The results of these experiments are also shown in Table 2. It was found that in the case of NaOH, in contrast to the other dissolving agents, not just surface dissolution takes place. Because PET is hydrolysed by NaOH, the dissolving agent can better diffuse into the structure of

IOP Conf. Series: Materials Science and Engineering **426** (2018) 012008 doi:10.1088/1757-899X/426/1/012008

material, so it can take in a higher quantity of cobalt into the solution than that experienced in the case of the other dissolving agents. The results of the dissolution tests are shown graphically in Fig. 1. The efficiency of extraction is determined by the percent of cobalt getting into the solution from the quantity of cobalt contained in the flakes used for extraction. It is calculated by the following relationship:

$$\eta_{ext} = ((0.1 \cdot c_{corr})/(c_{PET} \cdot m_{flakes}/1000)) \cdot 100 \quad (1)$$

Where, η_{ext}: efficiency of extraction as a percentage, c_{korr}: measured and corrected concentration of the given extract, c_{PET}: Cobalt-concentration of PET flakes, m_{flakes}: Mass of flakes used for extraction (~10 g).

Table 2. Results of the 48-hour extraction experiments

Solvent	Cobalt Concentr. 40°C (ppb)	Powder used for extraction 40°C (g)	Efficiency of extraction 40°C (%)	Cobalt Concentr. 80°C (ppb)	Powder used for extraction 80°C (g)	Efficiency of extraction 80°C (%)
1 M HCl	6.27	10.043	0.17	9.04	10.018	0.25
1 M NaOH	92.92	10.034	2.52	-	-	-
1 M NaOH precipitate	112.15	10.034	3.04	170.77	10.017	4.63
5% aqua ammonia	11.17	10.036	0.30	11.20	10.024	0.30
Distilled water	0.97	10.038	0.03	1.57	10.032	0.04
20% acetic acid	9.05	10.024	0.25	7.67	10.022	0.21
Orange juice	8.66	10.033	0.23	5.88	10.037	0.16
Fruit brandy	11.58	10.004	0.31	17.25	10.031	0.47
Washing powder 1%	4.35	10.011	0.12	7.31	10.001	0.20
Detergent 10X	3.25	10.007	0.09	9.93	10.027	0.27
Detergent 100X	2.68	10.021	0.07	7.27	10.018	0.20
1 M NaOH 1/2	118.37	10.030	3.21	-	-	-
1 M NaOH 2/2	183.97	-	-	-	-	-

It can be concluded from the results of the experiments seen in Fig. 1. that the dissolved quantity was not dangerous for consumers as far as the dissolving agents used in the household (water, fruit brandy, and orange juice) are concerned. The quantities of the agents dissolving under the influence of the higher temperature and shaking during delivery do not have any dangerous effect either. Certain unwholesome effects can only be observed in case of an intake of a quantity of 13.3-58.9mgCo/ kgbw, and—even in the worst case— the dissolution of merely 17.25 ppb of orange juice could be experienced. Therefore, it is considered that the quantity of intake/day cannot be more than 10-20 µg in case of consumption of an even a higher quantity; here it is necessary to take into consideration the fact that a surface dissolution takes place in the bottles while the flakes used for dissolution during the experiments have got a higher specific surface. The RDA for cobalt/day is 0.1 µg for an adult of average body weight (in the form of 2.4 µg B12), which is obtained for the human organism from vitamin B12 being the only source of cobalt.

6.2. Results of the Lakanen-Erviö dissolution test

The cobalt-concentration was 0 ppb in all the three Lakanen-Erviö extracts made simultaneously. As the LE extracting agent does not extract cobalt from PET, it can be stated that no serious cobalt contamination can be caused in the soil by the PET plastic flakes containing cobalt. Cobalt releases from the plastic matrix during the bio-degradation of PET, so it can continuously dissolve into the soil and into the ground water over the centuries. In addition, it can be supposed that the bottles being in landfills biodegrade faster owing to the more intensive circumstances: the pH of leachate of older landfills is characteristically alkaline, and, moreover, a higher temperature (40-70°C) can prevail in them during the bio-degradation of more waste-mass.

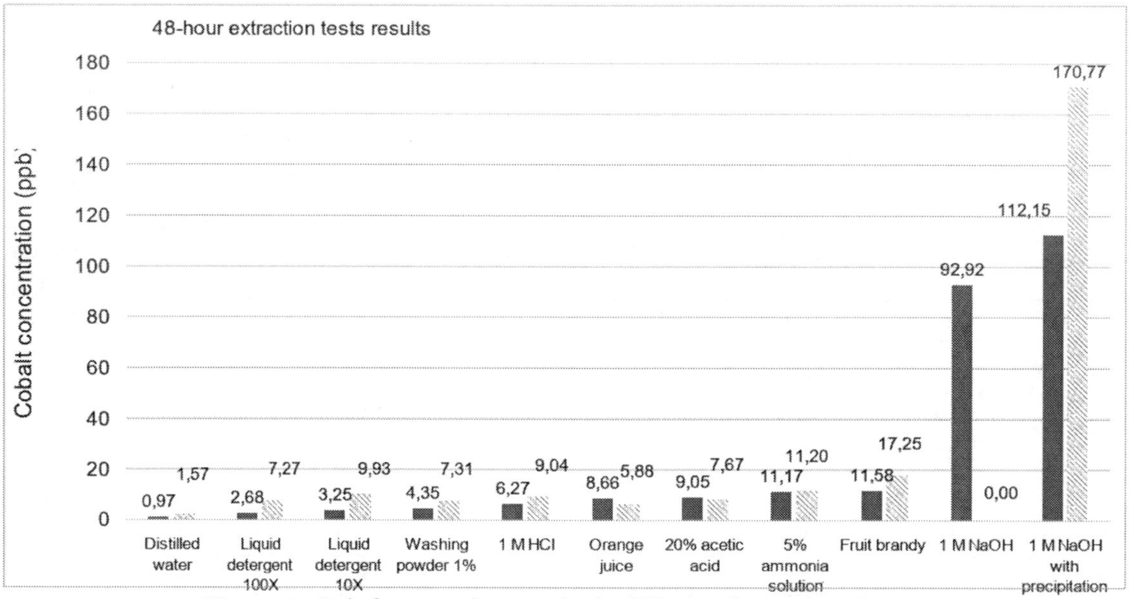

Figure 1. Cobalt extraction results in different domestic solvents (investigation performed at 40°C- solid, at 80°C – stripped)

6.3. Results of the dissolution-kinetic tests

As can be observed in Fig. 1, the dissolution of merely 0,97-11,58 ppb of dissolving agent could be experienced even in the case of fruit brandy considered as a dissolving agent that dissolves the highest quantity of cobalt without a chemical reaction in 40°C. The chemical reaction speed itself can well be observed in the case of the NaOH dissolving agent in Fig. 2.

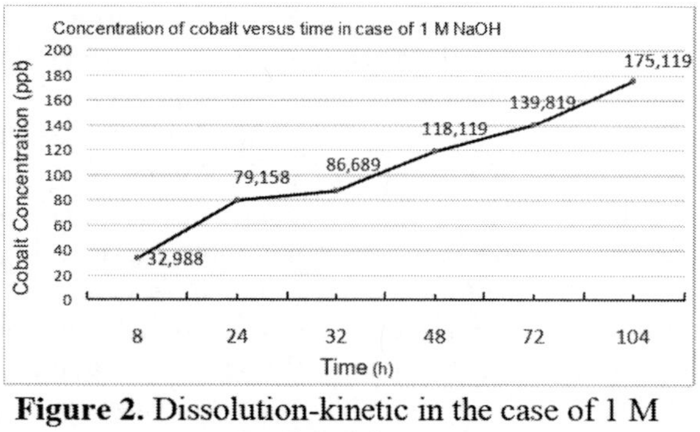

Figure 2. Dissolution-kinetic in the case of 1 M NaOH dissolving agent

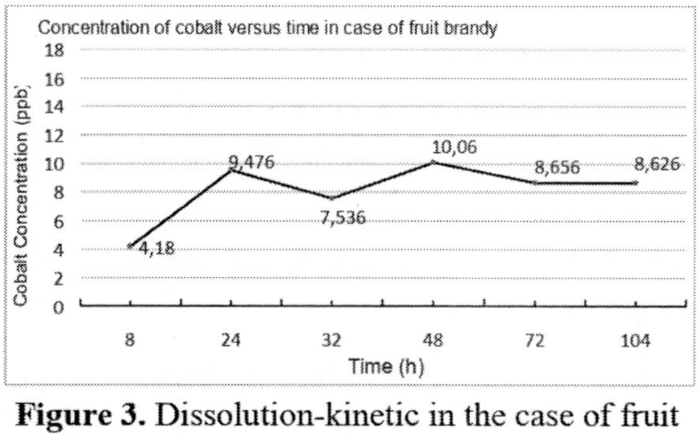

Figure 3. Dissolution-kinetic in the case of fruit brandy

PET is hydrolysed by soda-lye, so the surface structure of the individual grains is decomposed by it. In this case the dissolving agent is able to penetrate into the internal structure of the grains, and the dissolution speed keeps a nearly constant value over time as can well be observed in Fig. 2. The surface dissolution can also well be seen in Fig. 3 in the case of fruit brandy. Cobalt-segregation stops increasing in the solution after 24 hours. The dissolving agent is not able to penetrate into the internal structure of plastic grains after reaching a maximum concentration of 10,06 ppb, so further dissolution stops. Minor variations can be experienced between the individual measurement points in Fig. 3.

7. Summary

In our paper, cobalt-dissolution tests of PET bottles having an active oxygen scavenger layer were performed by dissolving agents used in the household. Similar to the first step of recycling of empty PET bottles, a sample was ground and swelled in order to estimate the extent of cobalt-dissolution from the additive material containing cobalt. In the course of the experiments using ICP-MS investigation technology, it was found that the average cobalt-content of three-layered PET-flakes containing Amosorb active oxygen scavenger was 36.805 mg/kg. The fruit brandy and the orange juice contained in the bottles were the two most successful agents dissolving cobalt without chemical reactions. The concentration of cobalt-extracts was 11.58 ppb and 8,66 ppb respectively with these agents. 1 M NaOH was the agent dissolving the highest quantity of cobalt. As a consequence of the hydrolysis of PET, a continuous dissolution over time could be experienced here when this soda agent penetrates into the grains of flakes. With the exception of NaOH, the dissolution of cobalt did not increase considerably by changing the temperature. It can be concluded from the aforementioned facts that the quantity of dissolved cobalt was not dangerous for consumers as far as the dissolving agents used in the household are concerned and applied by us in the form of flakes. Namely, some unwholesome effects could only be observed in the course of certain animal experiments when a quantity of 13.3-58.9 mgCo/kgbw (equal approximately 13.3-58.9 ppm of Cobalt) was taken. More dangerous effect could be arising in the case of natural water containing multi-layered PET flakes or bottles under long term deposition. We suggest that an individual collection or special recycling process technology need to be performed on these PET bottles. If not, the environmental impact increases under the influence of such active oxygen scavenger layered PET bottles. However, it can be stated also that the material of these kinds of PET bottles decomposes faster and the cobalt releases if they are placed in a communal landfill owing to the higher temperature and the presence of an alkaline solvents.

Acknowledgements
This research was supported by the European Union and the Hungarian State, co-financed by the European Regional Development Fund in the framework of the GINOP-2.3.4-15-2016-00004 project, aimed to promote the cooperation between higher education and industry.

References

[1] M Ramos, A Valdés, A C Mellinas, M C Garrigós. New Trends in Beverage Packaging Systems, A Review, 2015
[2] P Maul: Barrier Enhancement Using Additives, Pira International Conference, Brussels, Belgium, 2005
[3] Rotter et al. US Patent. Active Oxygen Scavenger Packaging, 2002 Patent No. US6406766
[4] Hua Li. Metal-Catalyzed Oxidation of Polybutadiene in Oxygen Scavenging Packaging Applications (Dissertation), 2010
[5] M Demicheva. Novel Oxygen Scavenger Systems for Functional Coatings, 2015 Degree Thesis. Plastics Technology. 2015
[6] Material Safety Data Sheet. Amosorb DFC, 2003 (MSDS 80-724-1; ColorMatrix Corporation, 3005 Chester Avenue, Cleveland, Ohio 44114;
[7] U.S. Department of Health and Human Services, Division of Toxicology: Toxicological Profile for Cobalt, 2004 https://www.atsdr.cdc.gov/toxprofiles/tp33.pdf
[8] http://www.umm.edu/health/medical/altmed/supplement/vitamin-b12-cobalamin
[9] K Gruber, C Kratky. Methylmalonyl CoA mutase, Chichester: John Wiley & Sons, Ltd; 2001.
[10] D Spaseska, M Civkaroska. Alkaline hydrolysis of poly-ethylene-terephthalate recycled from the postconsumer soft-drink bottles, 2010.

IOP Conf. Series: Materials Science and Engineering **426** (2018) 012009 doi:10.1088/1757-899X/426/1/012009

Distance measurements on compound workpiece with industrial computed tomography

Á Drégelyi-Kiss[1] and A Marczis[1]

[1] Óbuda University, Bánki Donát Faculty of Mechanical and Safety Engineering, Budapest, Hungary

E-mail: dregelyi.agota@bgk.uni-obuda.hu

Abstract. More recently, the industrial CT equipment is used not only for non-destructive analysis but to perform geometrical size evaluations. The three dimensional, optical measurements made by CT are popular because the measurement time is much more less than by using traditional 3D measurement machines, furthermore the inner geometries can be determined by non-destructive manner. In this article the measurement error and uncertainty are investigated in a particular case where the test piece consists of two materials, aluminium and polymer.

1. Introduction

Dimensional metrology has an important role in the quality control of manufactured (cut, milled or turned) parts. The tolerances of the characteristics (i.e. diameter or length) of these parts are usually within the range of 0.003 mm to 0.01 mm. There are several dimensional measurement equipments which are good for the control of these specifications such as simple devices (e.g. micrometer, dial calliper) or CMM machines with tactile or optical probes. Nowadays new technologies have been developed for dimensional measurements. One of them is industrial computed tomography (iCT) metrology using X-rays [1]. The advantages of CT technique are as follows: non-contact, capability of measuring both internal and external geometries, fast scanning time [2]. Due to the automation of the measurements and the requirements concerning faster measurement processes the dimensional measurements are preferred to apply CT equipments in industrial practice [2].

The use of iCT on dimensional metrology has a lot of advantages, but it has many disadvantages [3][4]. One of the main advantages of CT is to use non-destructive dimensional testing of parts and products. In the field of metrology this non-destructive measuring technology can be used to measure internal and external dimensions that can not be achieved with traditional coordinate measuring machines without cutting the test piece. CT measurements may in some cases be faster than touch probe measurements.

There are some problems with the calibration process of dimensional iCTs. The traceability of measurement results is not standardized [5]. There are some guidelines which refer to the use of iCTs for dimensional measurements [6][6]. The disadvantages of using iCT for dimensional measurements are as follows:

- there are no accepted, uniform test procedures;
- there are complex and countless influencing factors for measurements [11];
- due to measurement errors, reduced shape measurement capability [8];
- measurement uncertainty is unknown in most cases [9] [10][12];

- there are problems to test products with compound workpieces composed of different materials [4].

Industrial CT equipments are increasingly used to determine the dimensions of geometric features (e.g. diameter, position, cylinder) in the industry. The pieces typically consist of some metal and polymers, such as a hybrid car control unit consisting of an external pressure-resistant aluminium workpiece including electronic devices. There are problems with iCT dimensional measurements performed in case of internal dimensions where two materials meet. Within this research the performance, the measurement errors and uncertainty are investigated by using compound test workpieces constructed by us.

2. Materials and methods

2.1. Test workpieces

When selecting the materials of the test piece, the main criterion was to have two types of materials that can be scanned simultaneously and separately with CT. The two types of materials chosen were aluminium and POM (polyoxymethylene). Aluminium is advantageous for testing, because it has a small beam absorption factor during the use of X-ray and is used in a lot of applications in industrial practice. The AlMgSi1 alloy was chosen because of its easy machinability, availability. The other material was POM, which is a general technical plastic, widely used in industry and is easily obtainable. The POM can be used as gears, control cams, heavily loaded sliders, as various precision gauges in the machine industry, as an insulator in electrical engineering.

The design of the aluminium part is shown in Fig. 1. It contains three cylinders with 15 mm diameter and three cylinders with 20 mm diameter. These holes were prepared by H7 reamer process.

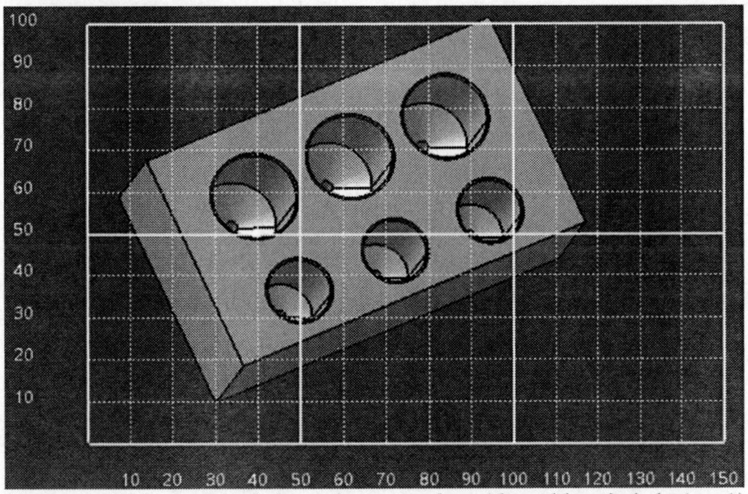

Figure 1. CAD model of aluminium test piece (the grid scale is in (mm))

During the design of the six plastic taps the main point was that the POM taps located in the aluminium body holes should be small, but not overlaped. The tolerance is thus 0 / -0.03 mm, which can be achieved even during turning manufacturing process.

2.2. Reference measurements

The reference measurements were performed at the Dimensional Metrology Lab , Óbuda University. The measurement device was Mitutoyo Crysta Plus 544, the diameter of the probe was 5 mm (Fig. 2). The recorded measurement points were 18, 36 and 72 for each holes and taps, respectively. The measurement points were equally distributed at 3 sections parallel to the z axis.

2.3. *iCT measurements*

Two types of industrial CTs were used during the experiments: GE phoenix v|tome|x m (at Continental Hungary Ltd.) and Werth HV-500 (Enerswiss Hungary Ltd.). The setting parameters are in Table 1. Measurements of the GE CT data were carried out using software VGStudio Max 2.2.2 from Volume Graphics, and were calculated on voxel data. The STL file was generated by GOM 2016 software from this scanned, reconstructed image. The output of the measurements performed by Werth CT was STL file. The philosophy of Werth does not allow us to examine the voxel data, only the generated STL files as in case of other 3D measurement techniques.

The purpose of this research is to compare the performance of the two iCT, and determine the measurement error. Therefore the same type of outputs is advantageous for comparison.

a)

b)

Figure 2. Reference measurements: a) for aluminum piece and b) for POM piece

Table 1. The iCT measurement settings

iCT	Materials	U (kV)	I (μA)	t_{exp} (ms)	No.$_{scan.}$
GE	Al	280	260	500	1800
	POM	250	250	500	1620
Werth	Al	190	450	1000	800
	POM	120	250	1000	800
	Al+POM	190	450	1000	800

2.4. *Design of experiments and evaluation method*

During experiments three different measurements were performed:

- measurement of one POM tap by GE and Werth CT three times, respectively
- measurement of the aluminium test piece by GE and Werth CT
- measurement of the aluminium and POM taps assembled by Werth CT (Fig. 3)

Figure 3. Measurement of the assembled test piece by Werth iCT

The outputs of measurements are given in STL files representing results about the geometry of the taps and holes. In order to determine the measurement error of the iCT dimensional measurements it is needed to make comparisons with the reference values. During the reference measurements the diameters of the holes and taps were determined in several ways.

- In case of holes (aluminium test piece) two evaluation methods were used to calculate the diameters of the cylinders: LSC (least square or Gaussian circle) and MIC (maximum inscribed circle) evaluations.
- In case of taps (POM test piece) similarly two evaluation methods were used for the diameters of the cylinders: LSC and MCC (minimum circumscribed circle) evaluation.

These evaluation methods were used in all cases.

The dimensional measurements and calculation of the diameters were evaluated by Kotem SmartProfile software. This software is able to handle the measurement points independently from the measuring device. One example for the output of the SmartProfile software is illustrated in Fig. 4.

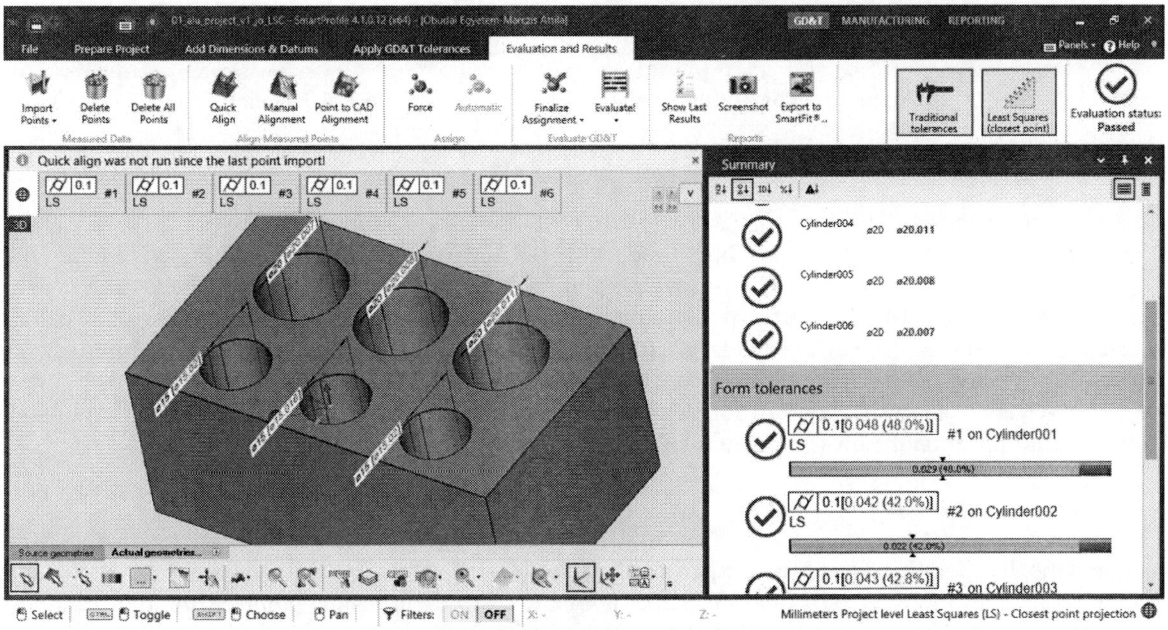

Figure 4. Results of the evaluation by SmartProfile (Werth CT measurement for the aluminium test piece, the evaluation method is LS, the diameters and the cylindricities are given)

The reference values for the diameters of the holes and taps were determined on the basis of 18, 36 or 72 touched measurements points. For each diameter calculation two types of evaluation methods (LS and MI; or LS and MC, respectively) were used. Therefore each diameter has 6 reference values.

Additionally, the purpose of this investigation was to determine the measurement error of the iCT measurements. For this, it was necessary to calculate the iCT measured diameter minus the reference value. With the use of Smartprofile the diameters can be calculated from the STL data originated from CT scanning process. The Smartprofile software do not use all the scanned points (which is 241934 points for a certain STL file, see Fig. 5), only some of them elaborated by CMM-like sampler. This means that during the calculation and evaluation process a limited amount of the measurement points are used. For the calculation of the diameters the CMM-like sampler points were the points from the reference measurements. From this data it is possible to determine the measurement errors for the diameters, i.e. iCT measured diameter minus reference diameter (for each measurement points: 18, 36 or 72 and evaluation: LSC, MIC or MCC).

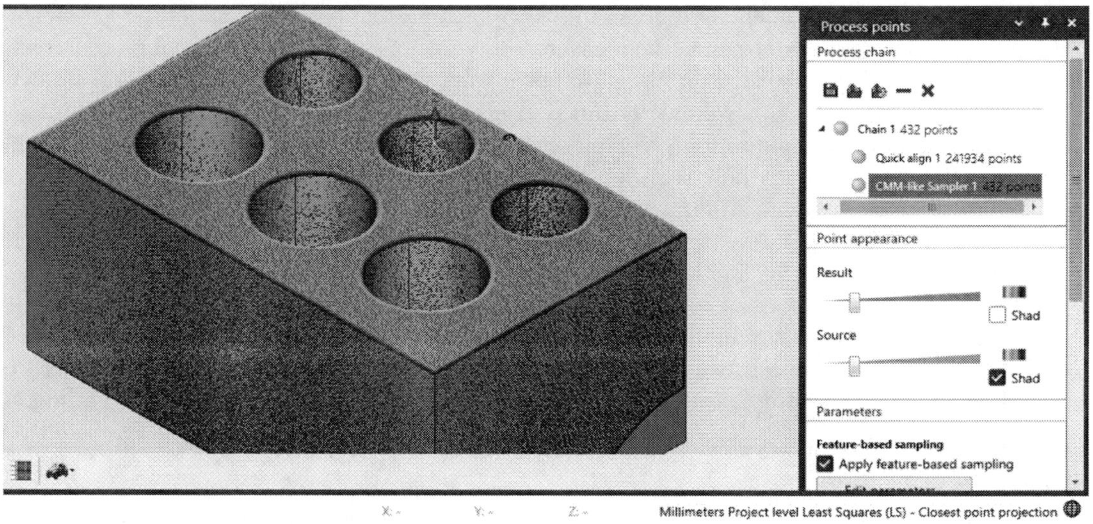

Figure 5. Fitting of the scanned iCT point cloud to the CAD model

3. Measurement results

3.1. Measurement of POM tap

The POM tap No. 6. was measured three times with GE and Werth CT. Three factors were taken into consideration, such as type of CT (two levels), number of measurement points (3 levels) and type of fitting (two levels). The main effects plot for the measurement error is shown in Fig. 6. It can be stated, that the measurement error is larger in case of GE CT comparing to Werth CT. Moreover, the type of the fitting method has a large impact on the measurement results.

The ANOVA analysis shows (Table 2) that the type of CT and the method of Fitting have large, significant impact on the measurement result, on the measurement error.

3.2. Measurement of aluminium holes

The aluminium test workpiece was measured by two CT equipments. The following factors were systematically changed during the experiments: diameter of the hole (at two levels: 15 or 20 mm nominal); type of iCT (GE and Werth); number of measurement points (for the reference measurement; 3 levels); fitting method (least squares of maximum inscribed circle). The results show (Fig. 7.) that there are negative measurement results in all cases, i.e. the measured value by the CT is less than the reference value. As a result of ANOVA analysis it is stated, that the type of CT and the fitting method have significant effect on the measurement error.

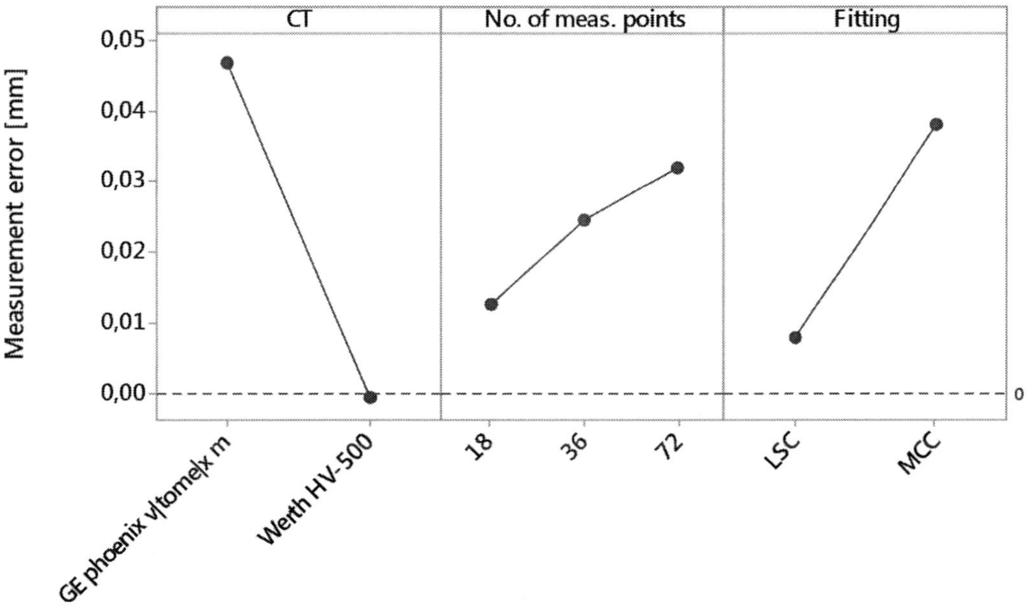

Figure 6. Main effect plot of POM tap No. 6.

Table 2. ANOVA results for the POM tap measurement

Source	DF	Adj SS	Adj MS	F-Value	P-Value
CT	1	0.020259	0.020259	40.52	0
No. of meas. points	2	0.002319	0.00116	2.32	0.118
Fitting	1	0.008281	0.008281	16.56	0
CT*No. of meas. points	2	0.001518	0.000759	1.52	0.238
CT*Fitting	1	0.008464	0.008464	16.93	0
No. of meas. points*Fitting	2	0.002278	0.001139	2.28	0.122
Error	26	0.012998	0.0005		
Lack-of-Fit	2	0.00109	0.000545	1.1	0.349
Pure Error	24	0.011908	0.000496		
Total	35	0.056118			

3.3. Measurement of the compound workpiece (aluminium and POM)

Finally, the diameters of the aluminium workpiece were measured in a case where the POM taps were within the holes (only in case of Werth CT). The comparison of the together (i.e. Al+POM) and the separately (only Al, previously mentioned case) measured results is given in Fig. 8. It is seen that there are large differences between the measurement error. The separate measurements have given much better results as the assembled version. It has to be emphasized the measurement errors are up to 160 μm.

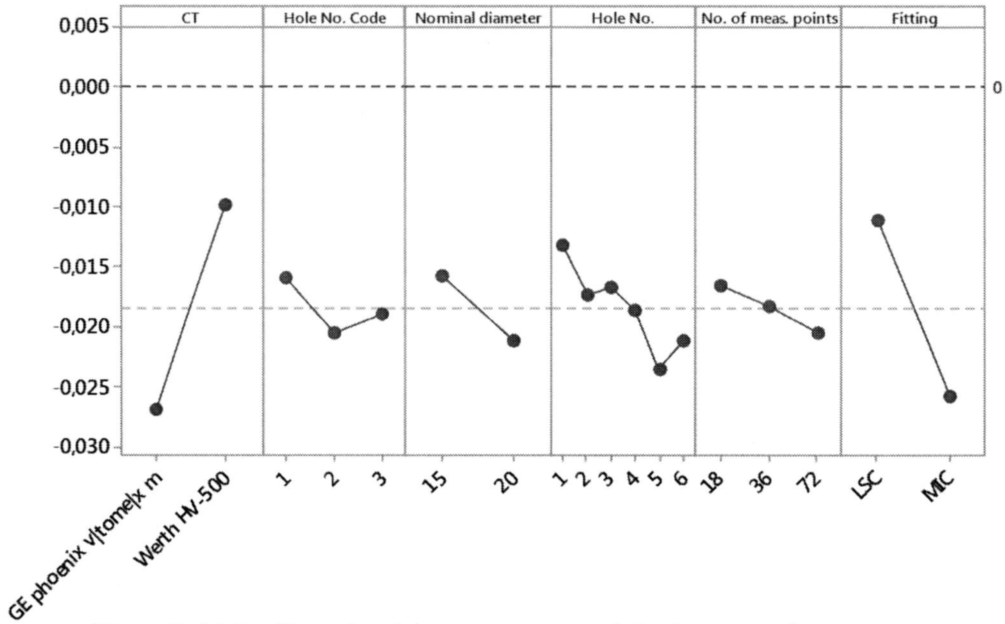

Figure 7. Main effects plot of the measurements of aluminum test piece

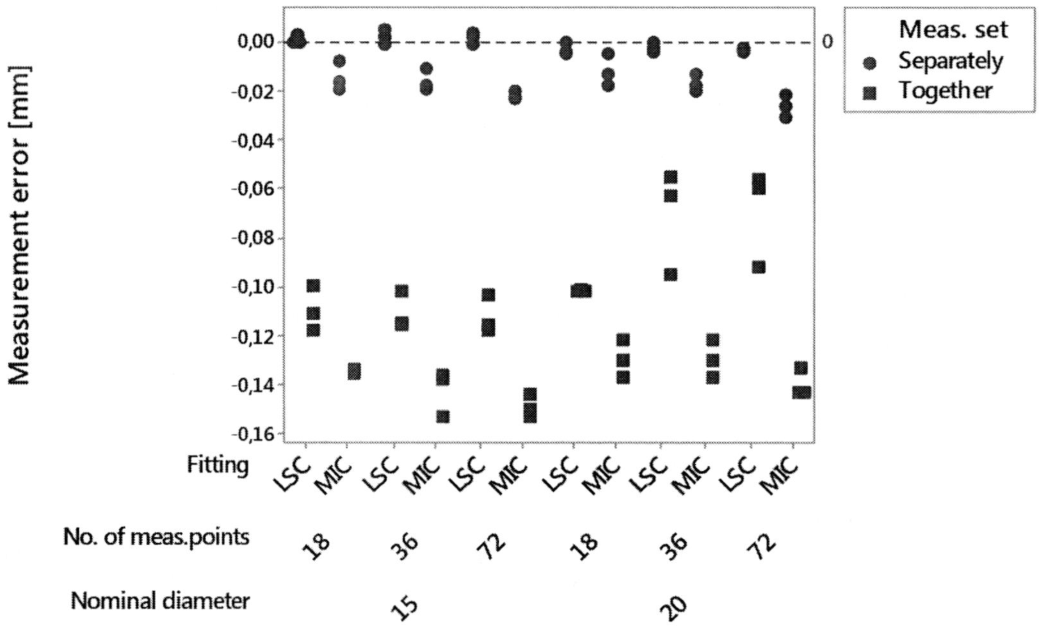

Figure 8. The diameters of the holes of Al test piece with Werth CT alone (separately) and together with the POM taps

4. Conclusion

During this study some preselected parameters were examined from the iCT dimensional measurement process in case of compound workpieces. From the investigations the following conclusion can be drawn:

- there are large differences between iCT equipments related to the accuracy;
- the fitting method has significant effect on the measurement error in case of tap and hole diameter measurements;
- the number of touched points for the reference measurements has not significant effect on the iCT dimensional measurement error;
- there are large difference in the measured diameter values depending on the type of test workpieces (one material or two materials).

Acknowledgements

The research was supported by the ÚNKP-17-IV-6 New National Excellence Program of the Ministry of Human Capacities. The authors would like to thank for the measurements for Continental Ltd. and Enerswiss; and for the use of SmartProfile software for KOTEM Ltd.

References

[1] Bartscher M, Illemann J, Neuschaefer-Rube U 2016 ISO test survey on material influence in dimensional computed tomography *Case Stud. Nondestr.Test. Eval.* **6** 79-92.
[2] Hiller J, Hornberger P 2016 Measurement accuracy in X-ray computed tomography metrology: Toward a systematic analysis of interference effects in tomographic imaging *Precis. Eng.* **45** 18-32.
[3] Kruth JP, Bartscher M, Carmignato S, Schmitt R, De Chiffre L, Weckenmann A 2011 Computed tomography for dimensional metrology *CIRP J. Manuf. Sci. Technol.* **60** 821-42.
[4] Cantatore A, Müller P 2011 *Introduction to computed tomography* (DTU Mechanical Engineering, Denmark)
[5] VDI/VDE 2617 Part 1.3 - 2011
[6] VDI/VDE 2630 Part 1.1 - 2014
[7] VDI/VDE 2630 Part 1.2 - 2010
[8] Xue L, Suzuki H, Ohtake Y, Fujimoto H, Abe M, Sato O, Takatsuji T 2015 A method for improving measurement accuracy of cylinders in dimensional CT metrology *Comput. Aided Des.* **69** 25-34.
[9] du Plessis A, Rossouw P 2015 X-ray computed tomography of a titanium aerospace investment casting *Case Stud. Nondestr.Test. Eval.* **3** 21-26.
[10] Ferrucci M, Ametova E, Carmignato S, Dewulf W. 2016 Evaluating the effects of detector angular misalignments on simulated computed tomography data *Precis. Eng.* **45** 230-241.
[11] Hiller J, Hornberger P. 2016 Measurement accuracy in X-ray computed tomography metrology: Toward a systematic analysis of interference effects in tomographic imaging *Precis. Eng.* **45** 18-32.
[12] Kiekens K, Welkenhuyzen F, Tan Y, Bleys P, Voet A, Kruth JP, Dewulf W 2011 A test object with parallel grooves for calibration and accuracy assessment of industrial computed tomography (CT) metrology *Meas. Sci. Technol.* **22**, 115502.

IOP Conf. Series: Materials Science and Engineering **426** (2018) 012010 doi:10.1088/1757-899X/426/1/012010

Determination of solid solution composition in Al alloys by electrical transport properties measurement and XRD method

Á Filep[1] and M Sepsi[1]

[1] University of Miskolc, Institute of Physical Metallurgy, Metalforming and Nanotechnology, Miskolc-Egyetemváros, Hungary

E-mail: femfilep@uni-miskolc.hu

Abstract. The non-heat treatable aluminium alloys usually need homogenization before hot rolling to decrease the inhomogenities. In the process the segregations are dissolving and small precipitations are forming. The progress of the processes strongly depend on the temperature and the duration of the heat treatment. Topic of the research was to describe the dissolution/precipitation processes during the heat treatment. It was carried out by two different methods. Using electrical transport properties (resistance and thermo-power) measurement the percentage of the dissolved elements were calculated, and by X-ray diffraction method the cumulative effect was determined through lattice parameter change. The results of the methods were compared regarding to the heat treatment parameters.

1. Introduction

After casting the non-heat treatable aluminium ingots, during the non-equilibrium crystallization inhomogenities can occur. [1] In this condition the material is not suitable for further forming processes, therefore homogenization is also needed. During this heat treatment process the segregations can dissolve into the solid solution. There are several methods to examine the result, e.g. microstructure analysis. Optical or scanning electron microscopy can be used to investigate the existence, size and shape of the segregations. By microprobe examination the solid solution concentration can be measured, but it can involve the secondary phases also. Using X-ray diffraction (XRD) method this can be avoided. If we know the correlation between the concentration and the lattice parameter changing, approximate calculations can be done. [2] Using the suitable equipment, the lattice parameter definition can be a fast measurement. Also a relatively fast and simple method to measure the resistance and thermo-power of the specimens, and it can provide information about each element in the solid solution. [3] It gives proper result in case of two solute elements, but approximate calculations can be done for multiple alloying elements too.

The goal of the study is to monitor the changing of the alloying elements in different aluminium alloys, after different heat treatment processes. XRD and electrical transport properties (resistance and thermo-power) measurements were carried out, and the results of the two methods were compared.

IOP Conf. Series: Materials Science and Engineering **426** (2018) 012010 doi:10.1088/1757-899X/426/1/012010

2. Electrical transport properties measurement

A known method was used to calculate the solute percentage of the elements in the solid solution. [3] It starts from the Matthiessen rule, which says for solid solutions:

$$\rho_{alloy} = \rho_0 + \Delta\rho \tag{1}$$

where

ρ_0 specific resistance of the pure metal (temperature dependent)

$\Delta\rho$ residual resistance of the alloying elements (independent from temperature)

For more alloying elements:

$$\Delta\rho = \sum_i a_i \cdot x_i \tag{2}$$

where

a_i the value characterizing the effect of the "i" alloying element

x_i the quantity of the "i" alloying element (wt%)

Normally the specific resistance depends on the sample geometry. But if it is measured on two different temperature, then it can be eliminated:

$$\frac{R_{T1} \cdot \frac{L}{A}}{R_{T2} \cdot \frac{L}{A}} = \frac{\rho_{T1}}{\rho_{T2}} = r \tag{3}$$

where

R_{Tx} the resistance measured on given temperature

L the length of specimen

A the cross-section area of the specimen

If we write Matthiessen rule for an etalon ($\rho_{E(T)}=\rho_{0(T)}+\Delta\rho_E$) and the unknown ($\rho_{x(T)}=\rho_{0(T)}+\Delta\rho_x$) samples, it can be written up:

$$r_x = \frac{\rho_{x(T1)}}{\rho_{x(T2)}} = \frac{\rho_{E(T1)}+\Delta\rho_x-\Delta\rho_E}{\rho_{E(T2)}+\Delta\rho_x-\Delta\rho_E} = \frac{\frac{\rho_{E(T1)}}{\rho_{E(T2)}}+\frac{\Delta\rho_x-\Delta\rho_E}{\rho_{E(T2)}}}{\frac{\rho_{E(T2)}}{\rho_{E(T2)}}+\frac{\Delta\rho_x-\Delta\rho_E}{\rho_{E(T2)}}} \tag{4}$$

Expressing δ like:

$$\delta = \frac{\Delta\rho_x-\Delta\rho_E}{\rho_{E(T2)}} \tag{5}$$

From Eq. (4) and (5):

$$r_x = \frac{r_E+\delta}{1+\delta} \tag{6}$$

$$\delta = \frac{r_x-r_E}{1-r_x} \tag{7}$$

By transforming the relationship:

$$\frac{\delta}{1+\delta} = \frac{\Delta\rho_x}{\rho_{0(T2)}+\Delta\rho_x} \tag{8}$$

where
$\rho_{0(T2)}$ specific resistance of the pure metal on T_2 temperature

Therefore a characteristic value can be determined, where no need to measure the exact geometrical sizes by measuring on two temperatures. The δ value has linear correlation with the solute alloying element percentage, and "δ/(1+δ)" is linear to the thermopower of the material, as known from the Nordheim-Gorter relationship (Figure 1). [4]

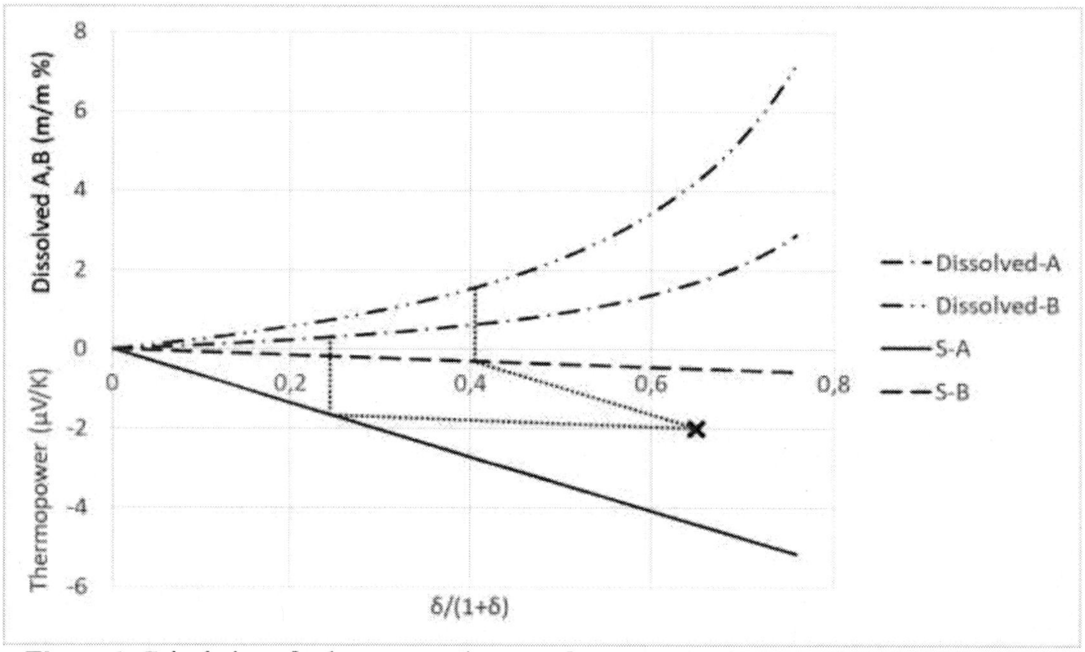

Figure 1. Calculation of solute content in case of two dissolved elements (S – thermopower)

3. Lattice parameter determination by XRD method

The dissolved elements change not only the electrical resistance, but also the lattice parameter. The Figure 2 shows how the alloying elements change the lattice parameter in binary systems of aluminium [2].

In multicomponent systems the influence of each element is separable with cross correlation calculations if enough measurement results are available. The lattice parameter determination is feasible with techniques based on diffraction (TEM, SEM, XRD, neutron diffraction). The precision lattice constant determination by XRD have been used for decades to examine the lattice parameter and perhaps this is still the most suitable method to investigate it, especially for powder specimens. The precision lattice parameter determination, the full pattern refinement, Rietveld analysis or any kind of method require a smooth diffractogram to determine the lattice parameter [5, 6]. It is possible to determine numerically the effect of each alloying element to the lattice parameter and electrical resistance if enough literary data are available. Conversely, the dissolved alloying elements concentration is predictable according to the change in the electrical resistance and lattice parameter.

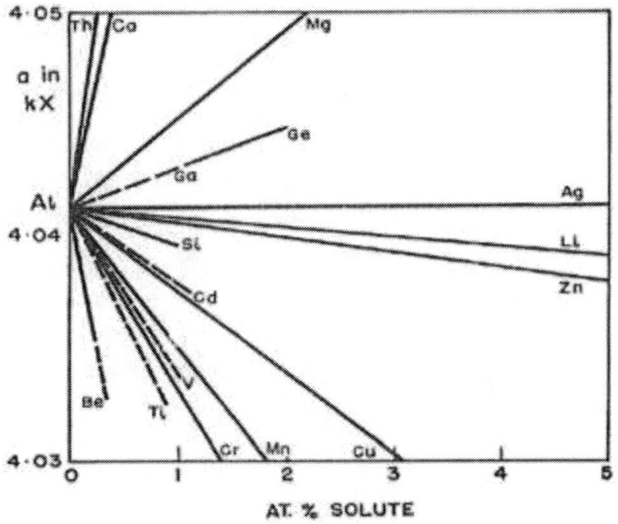

Figure 2. The influence of the alloying element to the lattice parameter in the binary systems of aluminium [2]

4. Experimental

Homogenizing heat treatments were carried out following six different processes. The first three occurred on lower temperature (LT) at 510°C for 1-5-10 hours. The rest were annealed first at higher temperature (HT), at 540°C for 4 hours, then the same 510°C, 1-5-10 hours long heat treatment were applied. The size of the specimens for electrical transport properties measurement were 1×4×120 mm, for XRD examinations it was 17×20×60 mm.

The composition of the investigated aluminium alloys (alloy A, B, C) are shown in Table 1. It is assumed, that Fe is segregated during solidification and the amount does not change because of the low solubility [7], therefore only the other two main elements were taken into account in every case (highlighted in the table).

Table 1. Composition of the examined alloys (m%)

	Al	Si	Mn	Fe	Mg	Zn	Pb	Cu	Cr	Ti	Cd	Ni	Zr
A	97,985	***0,23***	***1,1***	0,53	0,016	0,023	0,001	0,057	0,004	0,022	0,0005	-	-
B	96,89	***0,12***	0,08	0,26	***2,41***	0,01	0,002	0,044	0,16	0,012	-	-	-
C	97,97	0,06	***1,09***	0,2	0,006	0,02	0,01	***0,52***	0,001	0,125	0	0,003	0,003

4.1. Electrical transport properties measurement

The electrical resistance measurements were performed at -196°C (liquid nitrogen) and at 0°C (melting ice). The thermo-power measurements were carried out four times on each sample: twice (with different current direction) during "cooling" and twice during "heating". One side of the specimen was always at room temperature, while the other was at 0°C (melting ice) or ~45°C (hot water). Naturally during a given measurement the temperature difference was fixed.

During evaluation in some cases correction was necessary because the calculated values exceeded the maximum limit, which is impossible. It is caused by the segregations, which have effect on the resistivity but not on the thermo-power, therefore the "$\delta/(1+\delta)$" values were decreased by empirical amount. It is less precise, but it still gives a very good view on the progress of the process.

IOP Conf. Series: Materials Science and Engineering **426** (2018) 012010 doi:10.1088/1757-899X/426/1/012010

4.2. XRD method

Unfortunately, the conventional lattice parameter determination by XRD (Bruker D8 Advance with Euler cradle) was not applicable to our bulk casted specimens because of the diffraction pattern split due to the strong texture and very large grain size. Longer exposure time and rotation oscillation was also performed but there was not significant improvement in the quality of the results in an acceptable time range. The Figure 3 shows the diffractograms measured by the conventional XRD technique.

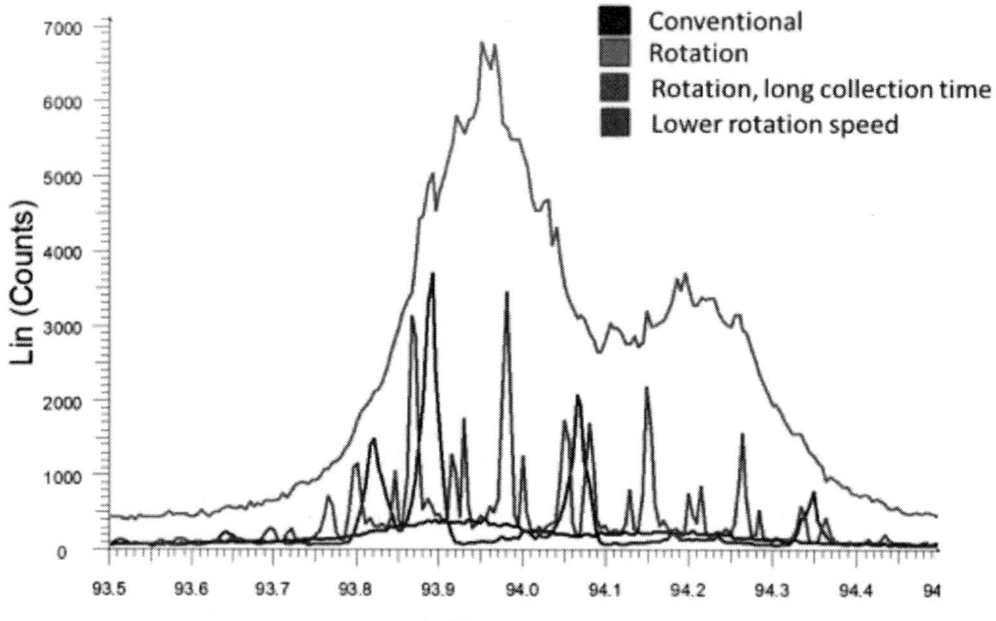

Figure 3. Split peaks measured by conventional technique

After several attempts it became clear the cause of the split is the texture and the grain size. To eliminate this peak splitting it was necessary to irradiate the largest area during one exposure, and collect the diffracted beam from as much direction as possible (from the same plane series). This problem was solved by nationally unique centreless X-ray diffractometer, the Stresstech G3R, in the Institute of Physical Metallurgy, Metalforming and Nanotechnology. There were several advantages to use this diffractometer. These devices produced for measuring residual stress, so they can measure the lattice distance changing very accurately and quickly. Two line detector collect the diffracted beams at the same time, which reduce the measuring time to 5 minutes (compared to the 16 hours measuring time with the conventional method). Since it is a centreless diffractometer, it is possible to use any kind of sample holder, in this case a modified high-speed rotation table was used as the Figure 4 shows. Applying this high-speed rotation table and linear oscillation the number of the grains fulfilled the Bragg diffraction requirements were increased, so it behaves just like a powder sample. The signals of the two detectors during one exposure were similar, so it confirms that the texture and the grain size effect was eliminated (Figure 5.). The details of the measurements were: Cr radiation source, 60 s exposure time, 28 kV accelerating voltage, 8 mA currents, ±4 mm linear oscillation, rotation speed of the table ~600 rpm. Details of the applied calculation: Peak fit with Pearson VII function, $K\alpha_2$ stripping for the signal of both detectors.

Figure 4. Centreless X-ray diffractometer with high speed rotation table

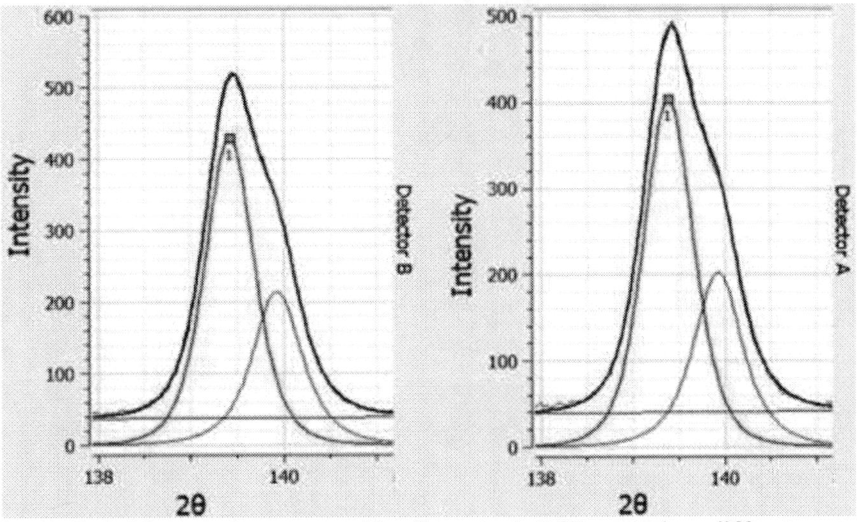

Figure 5. Spitless peaks measured by Stresstech G3R centreless diffractometer, {311} reflection on the two detectors

5. Results and discussion

The calculated values are shown on Figure 6. If we look first on diagrams on the left, it is clear, that the solute Mn amount decreases, because it forms segregations. This is visible on the microstructure, as Figure 7 shows. This segregation formation has no difference in the examined 1-10 hour range. The lattice parameter changing verifies this theory, but in some cases with higher fluctuation of the values.

The Mn-free alloy B has no big changes in the solute content except some Mg segregation and minor Si dissolution, which can cause the small lattice deformation measured by XRD.

The alloy C behaves like the alloy A in case of Mn segregations. But here some differences can be seen between the two heat treatment processes, if we look on the calculated Cu values. On lower temperature the amount decreases, some segregations can be formed. In contrast, on higher temperature the amount barely changes. There is no segregation process or they are dissolving during the heat treatment.

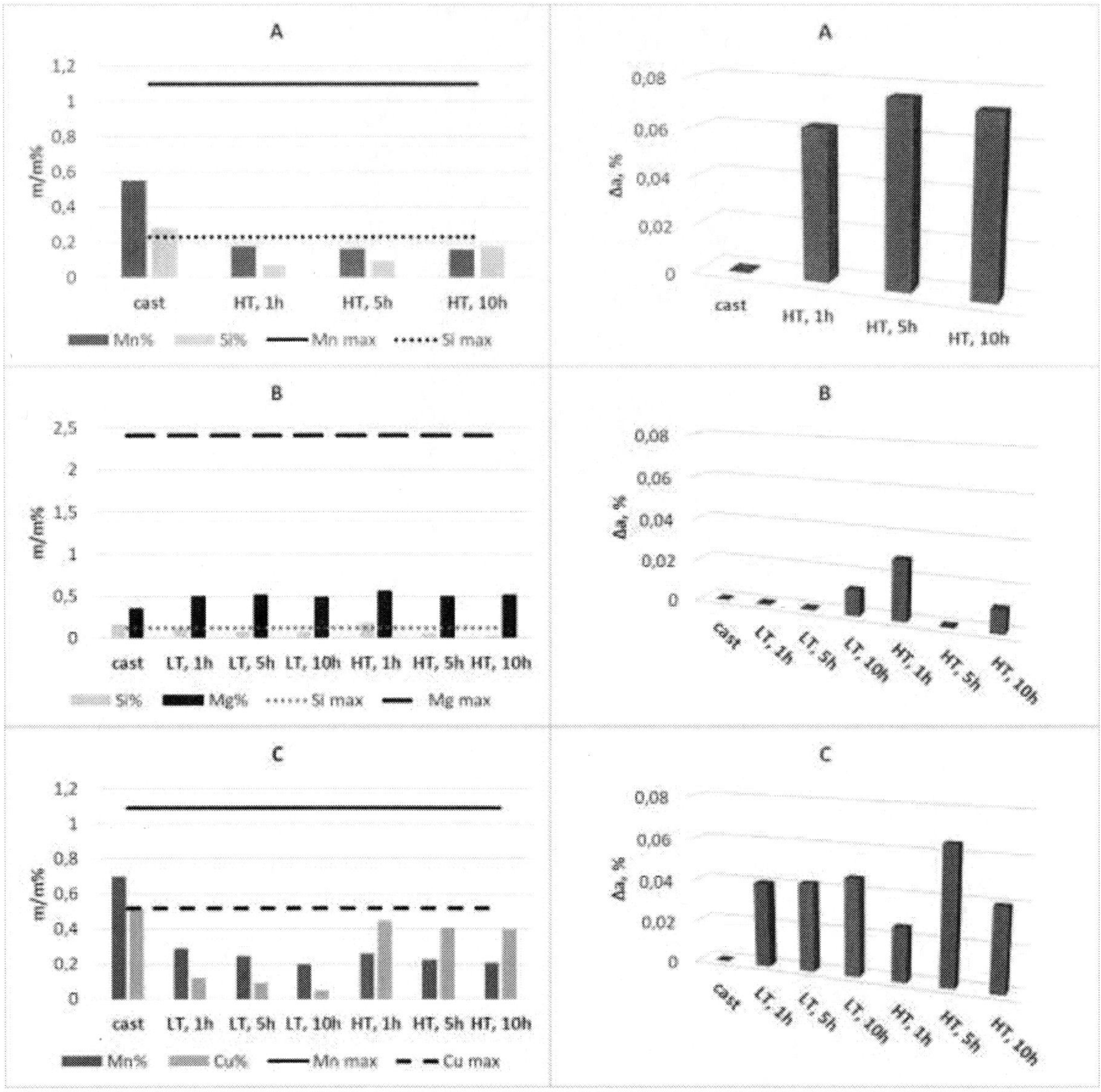

Figure 6. Calculated results in case of the different alloys
(left column: by electrical transport properties measurements, right column: by XRD method)

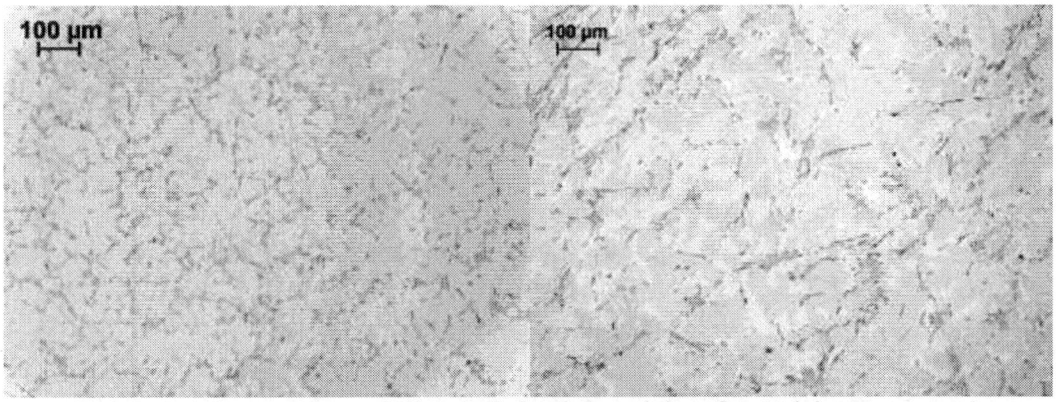

Figure 7. Microstructure of alloy A (left: as cast, right: after 10 h heat treatment)

6. Summary

Two different examination methods were carried out on Al alloys. The specimens were different materials, with various heat treated state. Electrical resistance measurements were carried out at -196°C (liquid nitrogen) and at 0°C (melting ice), then thermo-power examination on the same samples. Using the Matthiessen's rule and the Nordheim-Gorter relationship the alloying element content in the solid solution can be determined, therefore the homogenization process can be evaluated. The samples after the same processes were measured by X-ray diffraction method. From the peak positions the lattice parameters were calculated. During a 540°C annealing followed by a 510°C process the percentage of the manganese segregation can be detected. It was observed, that the homogenizing time in 1-10 h range has no effect on the dissolved manganese content, in Al-Mn + Si/Cu alloys. In case of Cu the segregation can be seen only after the process without the 540°C annealing. The used heat treatment parameters had only minor effect on the dissolved Mg-Si content in the alloy B. The previous observations were confirmed by the XRD examinations. Therefore the lattice parameter change calculation (beside the electrical transport properties measurement) can be a suitable method to follow the homogenizing heat treatment process, by determining this way the cumulative effect of the alloying elements.

Acknowledgements

The authors would like to thank Ferenc Tranta for the help in measurements and evaluation of thermopower and resistance results. The article was carried out as part of the GINOP-2.2.1-15-2016-00018 project implemented in the framework of the Szechenyi 2020 program. The realization of this project is supported by the European Union. M. Sepsi's research was supported through the New National Excellence Program of the Ministry of Human Capacities (Hungary).

References

[1] Mondolfo F E 1976 Aluminium alloys: Structure and properties *Butterworths*
[2] Pearson W B and G. V. Raynor 1958 A handbook of lattice spacings and structures of metals and alloys *Elsevier* pp 346.
[3] Tranta F 2014 A new method to examine the annealing of two alloying elements of 3004-grade cast aluminum by resistivity and thermoelectric power *Material Science Forum* **752** pp 147-156
[4] Lok Z J, Flemming A J E, Hamerton R and Zwaag S 2002 Changes in Fe and Si solute levels during deformation of AA1200 *Materials Science Forum* **396** pp 457-462.
[5] Klug H P and Alexander L E 1974 X-Ray diffraction procedures, 2nd edition *New York - London: A Wiley-Interscience Publication* pp 566-612.
[6] Razik N A 1985 Precise lattice constants determination of cubic crystals from X-ray powder diffractometric measurements *Applied Physics A* **37** pp 187–189
[7] Altenpohl D 1965 Aluminium und Aluminiumlegierungen *Springer-Verlag*

IOP Conf. Series: Materials Science and Engineering **426** (2018) 012011 doi:10.1088/1757-899X/426/1/012011

Self-organising formation of fullerene molecules from graphene patterns

D Fülep[1], I Zsoldos[1] and I László[2]

[1] Széchenyi István University, Győr, Hungary
[2]Budapest University of Technology and Economics, Budapest, Hungary

E-mail: zsoldos@sze.hu, fulep@sze.hu

Abstract. The self-organised formation of C_{60} and C_{70} from well designed graphene patterns was proved by molecular dynamic simulations. In this paper graphene nanopattern models are shown which can be used for self-organised formations of larger fullerene molecules. The rules of building of the nanopatterns are defined. Examples are shown for the self-organised processes in case of larger fullerene molecules.

1. Introduction

Lately, more and more researchers are paying attention to the self-organizing behaviour of the graphene patterns. As there is missing only an order of magnitude for obtaining atomic precise nanopatterns [1], theoretical calculations are motivated to simulate the formation of 3D nanostructures from graphene patterns. In general, they are looking for the answers given, regarding the different interventions and interactions, with molecular dynamics simulations, mainly in the form of upwinding (curling) on pre-designed formations.

It was shown that owing to the van der Waals interaction, graphene nano-ribbons wind up, in a self-organizing manner, on nanoscale metallic wires [2]. In case of graphene ribbons, an upwinding, taking place in a self-organizing manner, was also examined where, instead of a straight metal wire, a carbon nanotube was the van der Waals interaction provider factor, triggering the upwinding [3]. The self-organising upwinding was triggered by fullerene molecules [4], as well. The deliberately created defects on the graphene ribbons were examined, with molecular dynamics simulation as well, as factors causing the folding of the graphene ribbons [5]. Cone-like self-organizing folding of slashed graphene platelets was studied [6]. It was proved by molecular dynamics calculations that C_{60} and C_{70} fullerene molecules can be formed from one graphene pattern [7-9]. It was also proved that zigzag and armchair nanotubes can be obtained from graphene patterns. As the driving force was generated with the help of a half fullerene the final tube was closed with a half fullerene at one end [9]. Creating nanotubes from two nanoribbons is important because in this way open-ended nanotubes can be produced [9]. The production of nanotubes from two parallel nanoribbons is not a trivial task. First the possibility was published [10-11], and after that the topological and energetic conditions for obtaining perfect tubes were published [12-13].

In this paper we show that, similarly to the case of the C_{60} and the C_{70}, large fullerene molecules can be formed from well-designed graphene patterns by self-organising way.

2. Method

In our molecular dynamics simulations the carbon-carbon interaction were calculated with the help of Density Functional Theory adjusted Tight Binding method (DFT-TB) [14]. The time step was $\Delta t=0.7$ fs and Verlet algorithm [15] was used in calculating velocities of the carbon atoms. The environmental temperature was controlled with the help of Nosé-Hoover thermostat [16-17]. Applying the Nosé-Hoover thermostat made it possible that we could start our simulation with randomly determined initial velocities of the carbon atoms. The corresponding T_{init} atomic temperature was different of the environmental temperature of the simulation. This initial temperature was taken from the range [1000, 1100] Kelvin.

3. Examples for self-organising formation of fullerene molecules

In this chapter results are shown which were run by molecular dynamics simulations for the self-organising formation of larger fullerene molecules.

In Figure 1 snapshots from the simulation for the case of C_{88} molecule can be seen. The input model (Fig. 1a) consists of 88 carbon atoms, so the graphene pattern has 88 vertices. The 3D structure starts to develop so that several chemical bonds is coming into existence between atom pairs being close to each other at the perimeter, the new bonds compose pentagons with the neighbouring bonds, therefore the pattern starts to curve (Fig. 1b). Subsequently the number of the new bonds and so the new pentagons is rising, therefore the structure is curving more and more largely (Fig. 1c-d), in the end the structure is closing up and the C_{88} molecule is being born (Fig. 1e).

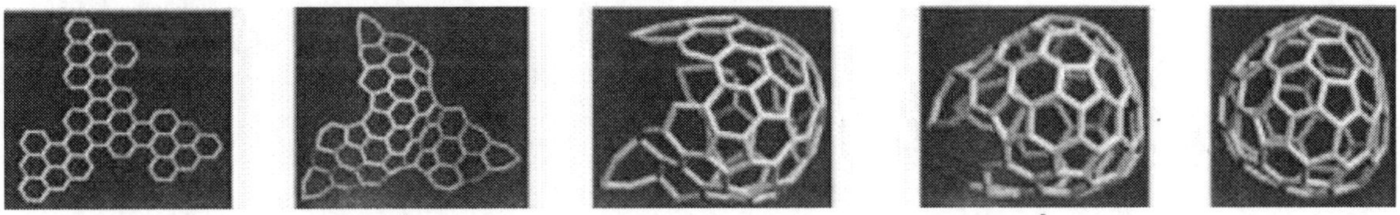

a b c d e

Figure 1. Snapshots from the simulation of self-organising process of C_{88}

The process in the case of the C_{100} molecule is very similar.

In Figure 2 snapshots from the simulation for the case of C_{100} molecule can be seen. The input model (Fig. 2a) consists of 100 carbon atoms, accordingly the graphene pattern has 100 vertices. The 3D structure starts to develop so that several chemical bonds is coming into existence between atom pairs being close to each other at the perimeter, the new bonds compose pentagons with the neighbouring bonds, therefore the pattern starts to curve (Fig. 2b). Subsequently the number of the new bonds and so the new pentagons is rising, therefore the structure is curving more and more largely (Fig. 2c-d), in the end the structure is closing up and the C_{100} molecule is evolving (Fig. 2e).

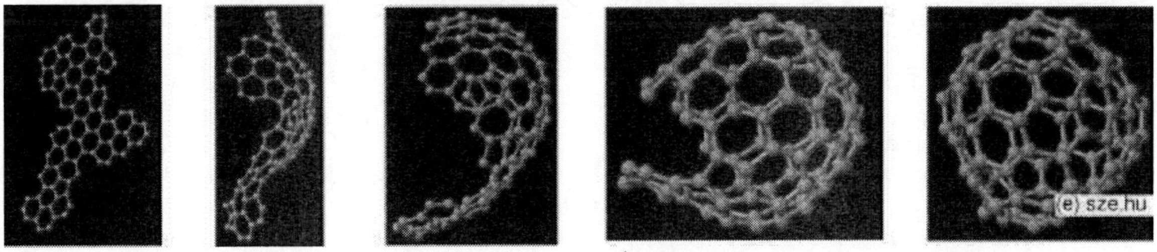

a b c d e

Figure 2. Snapshots from the simulation of self-organising process of C_{100}

IOP Conf. Series: Materials Science and Engineering **426** (2018) 012011 doi:10.1088/1757-899X/426/1/012011

4. Rules of planning graphene patterns for self-organising formations of fullerene molecules

The planning of the graphene patterns (cutting from the graphene sheet) has the following rules:

1. The number of the vertices in the graphene pattern has to be equal to the number of the atoms in the planned fullerene molecule.
2. The pattern has to consist of only hexagons.
3. The pattern has to be contiguous.
4. Curving up the pattern the planned fullerene structure has to recover.

These rules seem to be simple however they can be fulfilled not so simple way. In Figure 3 patterns planned for the formation of C_{140}, C_{240} and C_{540} fullerene molecules are shown. It is clear that there are a number of different patterns having e. g. 140 vertices. Only few of them fulfil the condition of 4, that is to say, the structure of C_{140} recovers only in case of few patterns after curving up the pattern. The structure recovering with the curving up of the pattern of Figure 3a is shown in Figure 3d. The patterns shown in Figure 3b-c are planned for C_{240} and C_{540}, that is, after curving up them the structure of C_{240} and C_{540} recover.

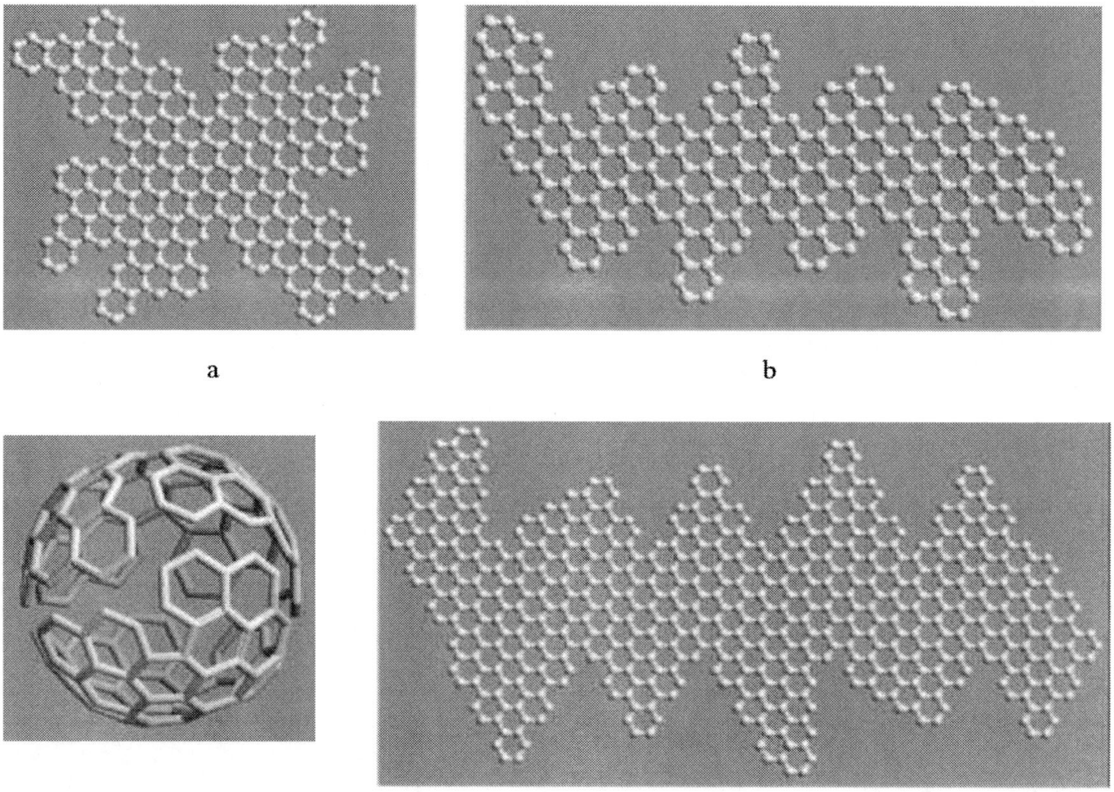

Figure 3. Graphene patterns for self-organising formation of C_{140}, C_{240}, C_{540} (a-c) and the curved up pattern of the C_{140} (d)

5. Conclusions

According to the results of molecular dynamics simulations, similarly to the self-organising formation of C_{60} and C_{70} (which was already shown earlier [7-9]) larger fullerene molecules can develop from pre-designed graphene patterns, as well.

IOP Conf. Series: Materials Science and Engineering **426** (2018) 012011 doi:10.1088/1757-899X/426/1/012011

Acknowledgement
The support of the grant with identification number EFOP-3.6.1- 16-2016- 00017 and GINOP-2.3.4-15-2016-00003 is gratefully acknowledged.

References
[1] P. Nemes-Incze, G. Magda, K. Kamarás, L.P. Bíró: Crystallographically selective nanopatterning of graphene on SiO2, Nano Research 3:110-116, 2010.
[2] Y. Li, Self-assembly of graphene nanoribbon ring on metallic nanowire, Physica E 56:277-282, 2014.
[3] Y. Wang, H.F. Zhan, C. Yang, Y. Xiang, Y.Y. Zhang, Formation of carbon nanoscrolls from graphene nanoribbons: A molecular dynamics study, Computational Materials Science 96:300-305, 2015.
[4] J.W. Feng, H. M. Ding, Y. Q. Ma, Self-assembly of fullerenes and graphene flake: A molecular dynamics study, Carbon 90:34-43, 2015.
[5] J. Wallace, L. Shao, Defect-induced carbon nanoscroll formation, Carbon 91: 96-102, 2015.
[6] W. Li, X. Zheng, B. Liu, X. Sun, T. Wang, J. Zhang, Y. Yan, A novel method for designing carbon nanostructures: Tailoring-induced self-scrolling of graphene flakes, Carbon 89:272-278, 2015.
[7] I. László, and I. Zsoldos, Molecular dynamics simulation of carbon nanostructures: The C_{60} buckminsterfullerene. Physica Status Solidi B 249:2616-2619, 2012.
[8] I. László, and I. Zsoldos, Molecular dynamics simulation of carbon nanostructures: The D5h C_{70} fullerene. Physica E 56:422-426, 2014.
[9] I. László, I. Zsoldos, Graphene-based molecular dynamics nanolithography of fullerenes, nanotubes and other carbon structures, Europhysics Letters 99:63001, 2012.
[10] S. S. Han, K. S. Lee, H.M. Lee, Nucleation mechanism of carbon nanotube, Chemical Physics Letters 383:321-325, 2004.
[11] L. He, J.Q. Lu, H. Jiang, Controlled Carbon Nanotube Junctions Self-Assembled from Graphene Nanoribbons, Small 5:2802- 2806, 2009.
[12] D. Fulep, I. Zsoldos, I. Laszlo: Topological and energetic conditions for lithographic production of carbon nanotubes from graphene, Journal of Nanomaterials, Volume 2015, Article ID 379563, 2015.
[13] D. Fulep, I. Zsoldos, I. Laszlo: Self-organised formation of nanotubes from graphene ribbons: A molecular dynamics study, Materials Research Express 3, Article ID:105044, 2016.
[14] D. Porezag, T. Frauenheim, T. Köhler, G. Seifert and R. Kaschner, Phys. Rev. B, 51:12947-12957, 1995.
[15] L. Verlet , Phys. Rev. 159:98-103, 1967.
[16] S. Nosé, Mol. Phys. 52: 255-268, 1984.
[17] W.G. Hoover, Phys. Rev. A, 31:1695- 1697, 1985.

IOP Conf. Series: Materials Science and Engineering **426** (2018) 012012 doi:10.1088/1757-899X/426/1/012012

Improving the HAZ toughness of Q+T high strength steels by post weld heat treatment

M Gáspár[1] and R. Sisodia[2]

[1]Assistant professor, University of Miskolc, Institute of Materials Science and Technology, Miskolc-Egyetemváros, Hungary
[2] PhD student, University of Miskolc, Institute of Materials Science and Technology,Miskolc-Egyetemváros, Hungary

E-mail: gasparm@uni-miskolc.hu

Abstract. Steel producers are continuously developing the mechanical properties and improving the weldability of high strength steels. Quenched and tempered steels belong to the one of the highest strength grades of structural steels with outstanding toughness characteristics due to the high temperature tempering. During fusion welding the thermal cycle irreversibly changes the microstructure and the mechanical properties of the base material, therefore an inhomogeneous heat-affected zone (HAZ) forms. In the HAZ hardened and softened zones occur. Due to the thermal cycles experiences during welding, these HAZs can exhibit significant losses in toughness when compared to the base metal. In real welded joints the HAZ properties can be analysed by conventional material tests to a limited degree; therefore physical simulators have been developed for the detailed examination of different HAZ areas. In the present research work the HAZ properties of a 960 MPa yield strength quenched and tempered steel (S960QL according to EN 10025-6) are investigated. Since the toughness reduction can be only partially handled by the adjustment of welding parameters, the possibility of local post-weld heat treatment was examined. Based on preliminary physical simulations and welding experiments a medium heat input gas metal arc welding technology ($t_{8/5}$ = 15 s) was selected for the HAZ simulations. The welding and the post-weld thermal cycles were determined according to the Rykalin 3D model. The effect of post weld heat treatment on the properties of the selected coarse-grained (CGHAZ), intercritical (ICHAZ) and intercritically reheated coarse-grained (ICCGHAZ) zones were investigated by electron microscopic, hardness tests and instrumented Charpy V-notch impact tests. The materials tests showed significant improvement of the toughness properties especially in ICHAZ due to the post-weld tempering, whilst the softening was acceptable.

1. Introduction

Nowadays the application ratio of high strength steels is continuously increasing. Due to their outstanding mechanical properties, especially their strength properties, significant weight reduction can be achieved. Besides decreasing operational costs due to the energy saving in mobile structures, thinner plates and smaller cross sections result in savings in the amount of base and filler materials applied. Because of the above-mentioned advantages the development of strength characteristics of steels is in the research focus of steel and welding consumable producers [1]. Besides the outstanding yield strength the new generations of high strength steels have increased toughness and acceptable ductility properties [2]. When the possible applications of high strength steels are considered their weldability and the behaviour of welded components during the total lifetime of the welded joint cannot be neglected [3], [4]. Therefore, in parallel to their development as base materials there is extensive research related to their welding. For the demanded intensive spread of high strength steels the possible advantages related to their fatigue resistance should be also exploited; however, the effect of mismatch ratio on the fatigue behaviour, further complicates this goal as the available filler metals are limited in the higher categories [5].

Quenched and tempered (Q+T) high strength steels belong to the highest steel grades of structural steels. They have generally a tempered martensitic microstructure due to the water cooling used in the quenching cycle and to the high temperature tempering applied after quenching. In order to realize the quenching condition in the whole cross section, alloying components (Cr, Mo) are added to the steel, which moves the continuous cooling temperature (CCT) curves to the right. Microalloying elements (Nb, V and Ti) are also used in order to ensure and preserve a fine grain microstructure; however, their amount and ratio can strongly and not always positively affect the toughness properties in the different HAZ areas [6]. The tempered, fine-grained microstructure results in high toughness at negative temperatures (even at -40 °C); however the welding thermal cycle can be detrimental to this ideal microstructure. The outstanding strength and toughness properties of quenched and tempered high strength steels cannot be adequately preserved during welding due to the irreversible microstructural changes in the HAZ. Cold cracking, softening and the reduction of toughness properties can also happen due to the effect of the welding heat input [7]. Controlled heat input is indispensable during welding; whilst adjustment of welding parameters has little effect on the toughness reduction, however the HAZ width can be minimized. Using advanced welding technologies (e.g. electron beam welding), that minimally damages the special microstructure of the base material in the HAZ and ensure a productive solution for their joining may have wider application in the future [8].

In industrial applications gas metal arc welding (GMAW) is the most commonly used welding technology; however, the heat input cannot be significantly minimized, since sufficient penetration and productivity are important requirements in the industry [9]. The present research aims to investigate a possible local post-weld heat treatment solution for the improvement of HAZ properties. In real welded joints the HAZ toughness can be analysed by conventional material tests only to a limited extent due to the complex microstructure, and thus physical simulators (i.e. Gleeble) were developed for the examination of different HAZ areas [10], [11]. The present paper examines the effect of post-weld heat treatment on the critical HAZ areas of the quenched and tempered steels.

2. Structure of heat-affected zone during fusion welding

The special structure of HAZ in single and multipass welded joints, including the formation of subzones is presented in [12]. In quenched and tempered high strength steels the toughness can significantly decrease in the coarse-grained (CGHAZ) and the intercritical zones (ICHAZ) compared to the base material. In multipass welded joints the intercritically reheated coarse grained zone (ICCGHAZ) may have even a lower toughness than the abovementioned zones. These three areas are considered critical in terms of HAZ toughness.

Next to the fusion line the material will be heated far above A_{c3} temperature and therefore homogeneous austenite forms. During grain coarsening the peak temperature is above 1100 °C where the grains start to exponentially grow in the function of presence of different microalloying elements [13]. There are two reasons for the decreased toughness of this zone in quenched and tempered high strength steels. First, the grain size can be more than 10 times higher than of the base material (>100 μm). The reason originates from the alloying elements resulting in a hard, lath martensitic microstructure. Therefore in many cases this area has the lowest toughness within the welded joint. Besides the weld metal, CGHAZ has the highest risk of cold cracking since the hydrogen can diffuse from the fusion line to the brittle, coarse-grained microstructure.

Further from the fusion line in ICHAZ, where the peak temperature during the thermal cycle is between A_{c1} and A_{c3}, the austenitic transformation just partially take place and thus an exceptionally heterogeneous microstructure forms. Transformed parts at the boundaries of original grains generally have a higher carbon content, since austenite has higher carbon-dissolving ability in this temperature range. In Q+T high strength steels the austenitic parts transform to a more brittle microstructure than the base metal, which is mostly martensite. Retained austenite can be often observed near the brittle martensitic islands, so these areas are called together M-A constituents. The transformed parts between the relatively softened microstructure mean local brittle zones in the welded joint [1] [13].

In the case of multipass welded joints a combination of CGHAZ and ICHAZ can evolve, when the second heat cycle reheats the primer coarse grains between A_{c1} and A_{c3}. In Q+T steels this local zone can have the lowest toughness in the welded joint, since the disadvantageous properties of CGHAZ and ICHAZ meet here. The toughness of ICCGHAZ is determined by the tempered coarse grained martensite and the amount, distribution, type and hardness of austenitized parts. In real welded joints the unfavourable properties of ICCGHAZ are less harmful, since this zone just forms locally whilst ICHAZ can be found in the whole plate thickness [12].

3. Investigated material and the experimental program

In the present study the highest steel grade of EN 10025-6 standard, S960QL is investigated in the aspect of the microstructural changes in the HAZ of GMAW joints. It is important to note that higher steel grades are already available on the market, although they have not been included in the governing standard yet [2], [8]. The chemical composition of the investigated base material is shown in Table 1 and the mechanical properties are summarized in Table 2.

Table 1. Chemical composition of the investigated S960QL (m%)

S960QL	**C**	**Si**	**Mn**	**P**	**S**	**Cr**	**Ni**
	0.16	0.23	1.25	0.009	0.001	0.20	0.05
Mo	**V**	**Ti**	**Cu**	**Al**	**Nb**	**B**	**N**
0.592	0.042	0.003	0.01	0.056	0.015	0.001	0.0036

Table 2. Mechanical properties of the investigated S960QL base material

$R_{p0.2}$ (MPa)	**R_m (MPa)**	**A_5 (%)**	**$CVE_{-40\,°C}$ (J)**
1014	1053	14	75

Applying the HAZ test on Gleeble the desired heat-affected zone can be precisely and homogeneously created in a volume sufficient for further material tests, e. g. Charpy V-notch impact or CTOD tests [10]. Although several welding thermal cycle models are available in the QuickSim software developed for the simulator, the GSL programs were manually written in our case, using the time and temperature values determined by Rykalin-3D model [14]. This model describes the temperature field generated by a moving spot-like heat source on the surface of a semi-infinite body. In this case 3D thermal conductivity is dominant whilst surface heat transfer (convection) is negligible.

In our earlier physical simulation tests [15] we found that the toughness in the HAZ decreased sharply (almost independently from the applied welding parameters) in the $t_{8/5}$ cooling time (cooling time between 850 and 500 °C) interval (5-30 s) of conventional arc welding processes. Therefore $t_{8/5}$ = 15 s (the middle value of the previously tested interval) cooling time was selected for simulating the effect of post-weld heat treatment on CGHAZ, ICHAZ and ICCGHAZ. Regarding the selection of the peak temperatures 1350 °C (safely under the nil-strength temperature (NST) of the investigated steel where the material can not bear any mechanical loading [16]) was set for the CGHAZ simulation in order to allow the largest possible grains to form in the HAZ. Related to ICHAZ, 775 °C was selected based data in the literature [10] and previous experiments [15]. In all cases a T_{max} = 650 °C tempering thermal cycle, similarly to the high temperature tempering during the base material production, was applied for investigating the effect of post-weld heat treatment (PWHT). In industrial practice the arc of GTAW (gas tungsten arc welding) equipment can be used for the improvement of fatigue properties of welded joints by remelting (or just reheating) the weld-base material transition [17]. The idea was to analyse whether local and relatively short PWHT by arc can have a positive effect on the toughness of the heat-affected zone. In Figs. 1-3. the determined thermal cycles applied during the HAZ and PWHT physical simulations are presented.

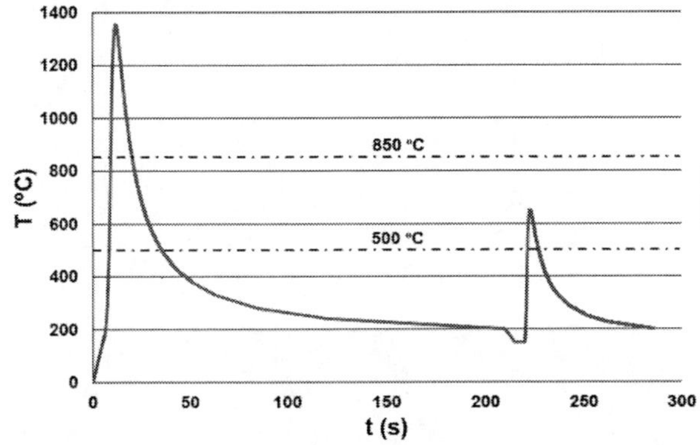

Figure 1. Thermal cycle for PWHT of CGHAZ

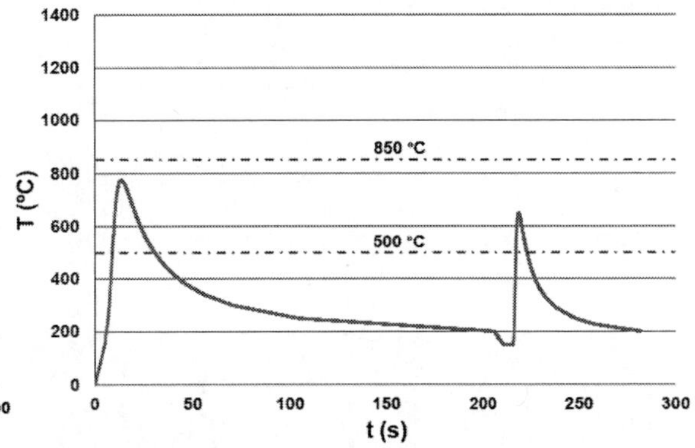

Figure 2. Thermal cycle for PWHT of ICHAZ

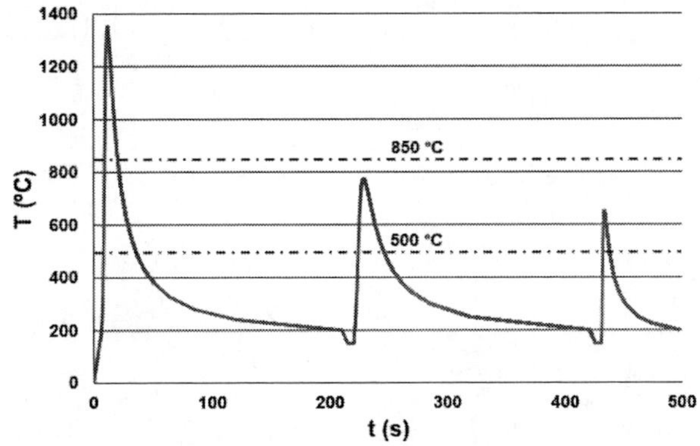

Figure 3. Thermal cycle for PWHT of ICCGHAZ

The thermal cycles were realized in the centre of 10×10×70 mm specimens, manufactured from the 15 mm thick base material. Due to the control process of the simulation, K-type thermocouples were welded to the surface of the samples and joined to the simulator. Each thermal cycle was simulated on six specimens each. From each series five samples were used for the instrumented Charpy V-notch impact test and the rest for hardness testing.

4. Results and discussion

4.1. Scanning electron microscopy (SEM) tests

A ZEISS EVO MA10 scanning electron microscope was used for the examination; results are illustrated in Figs. 4-9. Samples were coated with a thin gold layer in order to increase picture quality due to the resin surrounding the specimen. All microscopic tests verified that the desired heat-affected zones were simulated. Regarding the CGHAZ, a martensitic microstructure with large (>100 μm) prior austenite grain size were measured. In ICHAZ fine M-A islands formed as the result of partial austenitic transformation, while the middle of the original grains was tempered. In ICCGHAZ similar microstructural changes can be identified with essentially ten-time-larger grains. The tempering of lath martensite is shown in Fig 6, while tempered M-A islands can be identified in Figs 8. and 10.

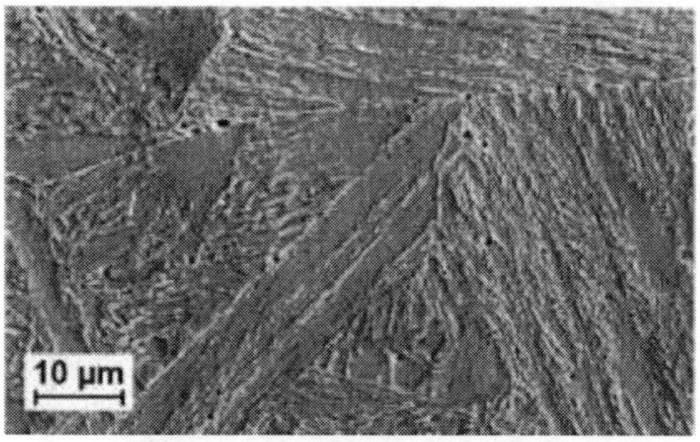

Figure 4. CGHAZ (2% Nital)

Figure 5. CGHAZ with PWHT (2% Nital)

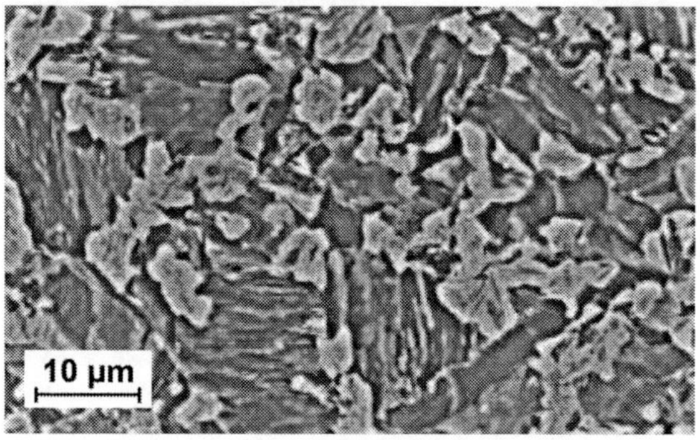

Figure 6. ICHAZ (2% Nital)

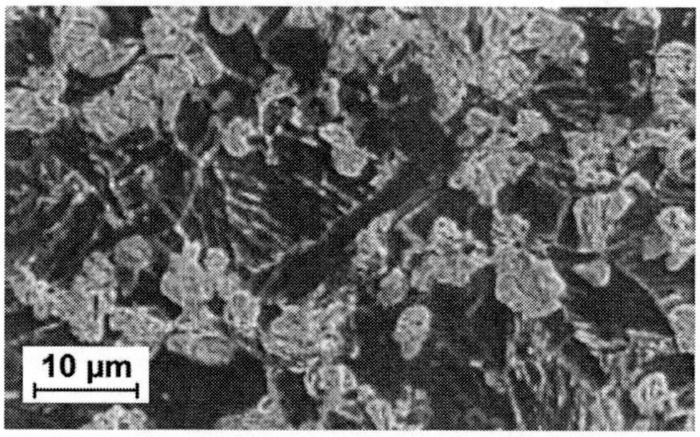

Figure 7. ICHAZ with PWHT (2% Nital)

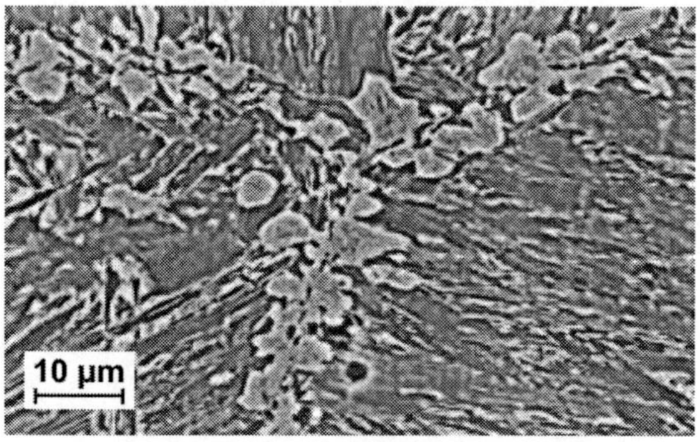

Figure 8. ICCGHAZ (2% Nital)

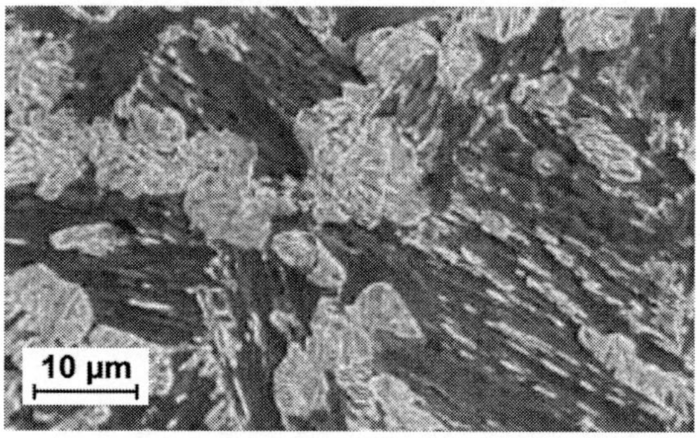

Figure 9. ICCGHAZ with PWHT (2% Nital)

4.2. Hardness tests

Five hardness measurements were made on the surface of the medium cross section of the samples by a Reicherter UH 250 universal hardness tester. The average values and the standard deviations values of hardness tests are summarized in Table 3. It may be important to note that the hardness of the base material was 335±5 HV10, whilst the governing EN 15614-1 standard allows maximum 450 HV10 for the welded joints of Q+T high strength steels.

Table 3. Effect of PWHT on the average hardness

	Hardness (HV10)		
	CGHAZ	**ICHAZ**	**ICCGHAZ**
without PWHT	409±6.8	323±5.4	344±7.6
T_{PWHT} = 650 °C	346±6.4	298±4.3	328±9.2

From Table 3 it can be seen, that the tempering heat cycle reduced the hardness of all zones. The most significant reduction was noticed in CGHAZ, where originally higher than 400 HV10 hardness was measured. Due to the PWHT the hardness decreased to the level of base material which can be favourable in terms of cold cracking sensitivity. In the case of ICHAZ and ICCGHAZ the hardness decreased under the level of the base material, although this softening cannot be considered critical, since in multipass welded joints of the same material generally similar hardness values are measured at the root side.

4.3. Instrumented Charpy V-notch impact tests

Instrumented Charpy V-notch impact tests (according to EN ISO 14556) were performed for analysing the supposed positive effect on toughness of PWHT. Standardized 10×10×55 mm specimens with a V-notch were manufactured from the Gleeble samples. Measurements were done by a PSD 300 instrumented impact testing equipment. The absorbed energy values (CVE) with the standard deviations (D) are presented in Fig. 10.

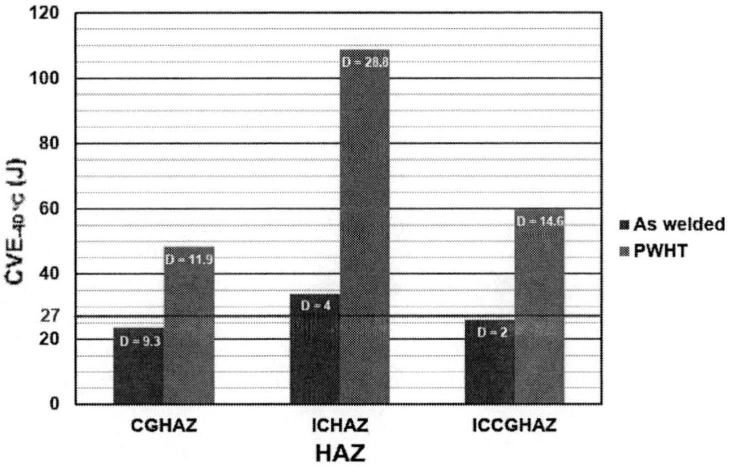

Figure 10. Results of Charpy V-notch impact test in the different HAZ's

According to EN 10025-6 and EN 15614-1 standards the toughness of the heat-affected zone of S960QL should be higher than 27 J at -40 °C. Due to the 650 °C tempering cycle the impact energy was doubled in CGHAZ despite the large grain size and tripled in ICHAZ. In the case of ICCGHAZ the improvement was also significant. By the application of instrumented impact testing the force – displacement (F-s) diagrams (Figs. 12 and 14) can be determined, and based on the measured maximum force value the impact energy can be divided according to the absorbed energy needed for crack initiation (W_i) and crack propagation (W_t) [18]. The reduction of the ratio of W_i compared to the Charpy-V energy (CVE) means that more energy is absorbed for crack propagation, therefore the toughness improves. In the base material crack initiation (W_i) was 25.9%, indicating the high toughness of the quenched and tempered microstructure; however, in the investigated HAZ areas this value was systematically higher. The average percentage of the absorbed energy for W_i compared to CVE in the investigated HAZ areas is summarized in Table 4.

Table 4. Average percentage of the absorbed energy for crack initiation (W_i) compared to CVE

	without PWHT	**T_{PWHT} = 650 °C**
CGHAZ	90%	71%
ICHAZ	78%	38%
ICCGHAZ	88%	68%

These values verify the positive effect of PWHT on the toughness of critical HAZ areas, especially in ICHAZ During the instrumented impact tests the registered force-displacement diagrams at tempered ICHAZ included few instable crack propagation stages (from the five specimens three did not include any at all). As an example, the determined force-displacement diagrams during the instrumented Charpy V-notch impact test of normal and tempered ICHAZ are shown in Figs. 11 and 12.

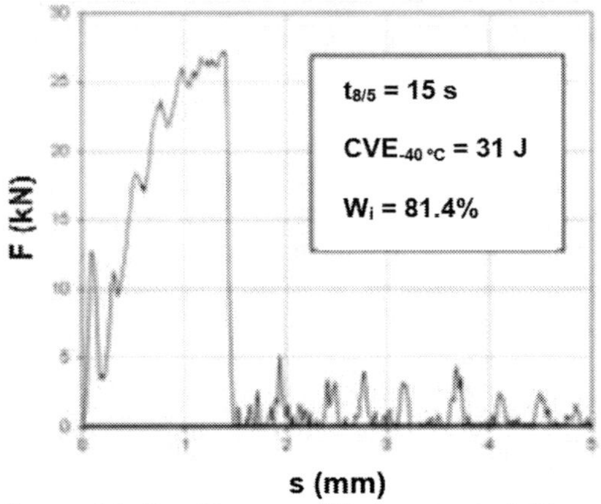

Figure 11. F-s diagram of instrumented Charpy V-notch impact test on ICHAZ

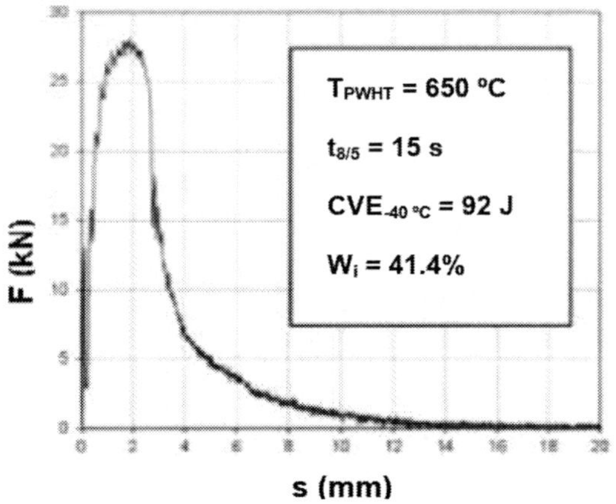

Figure 12. F-s diagram of instrumented Charpy V-notch impact test on ICHAZ with PWHT

Although the presented PWHT only influences the part of the heat-affected zone that is close to the surface, the improvement can be still relevant for the total lifetime of the welded joint. At the root side of the multipass welded joints HAZ tempering always automatically happens due to the heat input related to the further layers and therefore the tempering effect of filler passes. Because of this, HAZ toughness can be mostly critical at the face side, where the highest hardness values are generally measured. As could be seen above, a local tempering heat cycle can significantly increase the toughness at this crucial part of the HAZ, which can be also combined with the improvement of fatigue properties of the welded joints.

5. Conclusion

By applying post-weld heat treatment the toughness in the heat-affected zone of quenched and tempered high strength steels can be significantly improved. After welding at the HAZ a local and short heat input, which should be strictly kept under A_1 temperature, can effectively increase the toughness of CGHAZ, ICHAZ and ICCGHAZ. The highest improvement can be expected in ICHAZ, where the toughness characteristics approached the properties of the base material during the physical simulation experiments.

Acknowledgement

The presented research work was supported by the European Union and the Hungarian State, co-financed by the European Regional Development Fund in the framework of the GINOP-2.3.4-15-2016-00004 project, aimed to promote the cooperation between the higher education and the industry. The authors are grateful to the Institute of Physical Metallurgy, Metalforming and Nanotechnology at the University of Miskolc for ensuring the scanning electron microscope during the evaluation of HAZ tests.

IOP Conf. Series: Materials Science and Engineering **426** (2018) 012012 doi:10.1088/1757-899X/426/1/012012

References

[1] Porter D A 2015 Weldable high-strength steels: challenges and engineering applications Portevin lecture IIW International Conference on High Strength Steels – Challenges and Applications, Helsinki Finland 0102

[2] Weglowski M 2012 February Modern toughened steels – their properties and advantages, *Biuletyn Instytutu Spawalnictwa* pp 25-36

[3] Lukács, J 2013 Dimensions of Lifetime Management *Mat. Sci. For.* Vols. **473-474** pp 361-368

[4] Alus H Katz Y Maros M B Tóth L: Some insights into the remote strain vs. fatigue life relationship *Journal of Mat. Pro. Tec.* **157-158** pp 16-22

[5] Dobosy Á and Lukács J 2017 The effect of filler material choice on the high cycle fatigue resistance of high strength steel welded joints, *Mat. Sci. For.* Vol. **885** pp 111-116

[6] Spachinger J S Ernst W and Enzinger N 2017 Influence of Ti on the toughness of the FGHAZ and the CGHAZ of high-strength microalloyed S700MC steels *Wel. in the Wor.*, Vol. **61**., Issue 1. pp 1117-1131

[7] Dobosy Á Lukács J 2015 Welding properties and fatigue resistance of S690QL high strength steels, *Mat. Sci. For.* Vol. **812** pp 29-34

[8] Blacha S Weglowski M S Dymek S and Kopyscianski M 2017 Microstructual and Mechanical Characterization of Electron Beam Welded Joints of High Strength S960QL and Weldox 1300 Steel Grades *Arc. of Met. and Mat.* Vol. **62** Issue 2 pp 627-634

[9] Májlinger K Kalácska E Spena P R 2016 Gas metal arc welding of dissimilar AHSS sheets *Mat. and Des.* Vol. **109** pp 615-621

[10] Laitinen R Porter D A Karjalainen L P Leiviskä P and Kömi J 2013 Physical Simulation for Evaluating Heat-Affected Zone Toughness of High and Ultra-High Strength Steels *Mat. Sci. For.* Vol. **762** 711-716

[11] Adonyi Y 2006 Heat-affected zone characterization by physical simulations *Wel. Jou.* Vol. **85** Issue 10 pp 42-27

[12] Davis C and King J 1994 Cleavage initiation in intercritically reheated coarse-grained heat affected zone: Part I. Fractographic evidence *Met. and Mat. Tra. A* **25** 563–573

[13] Bhadesia H K D H and Honeycombe R W K 2006 Steels Microstructure and Properties Third Edition Elsevier Linacre House Hordan Hill Oxford OX2 8DP UK

[14] Rykalin N N 1953 Teplovie processzi pri szvarke Vüpuszk 2 *Izdatelsztvo Akademii Nauk SzSzSzR*, Moscow 56

[15] Gáspár M Balogh A Lukács J 2017 Toughness Examination of Physically Simulated S960QL HAZ by a Special Drilled Specimen *Lecture notes in mechanical engineering* **F12** pp 469-481

[16] Kuzsella L. Lukács J. Szűcs K. 2013 Nil-strength temperature and hot tensile tests on S960QL high-strength low-alloy steel *Pro. Proc. and Sys.* Vol. **6** No. 1. pp 67-78

[17] Sjors H J Van Es H Kolestein R J M Pijpers F S K Bijlaard 2013 TIG-dressing of High Strength Steel Butt Welded Connections – Part 1: Weld Toe Geometry and Local Hardness *Pro. Eng.* **66** pp 216-225

[18] Lenkeyné Bíró Gy Winkler S Tóth L and Blauel J G 1997 Investigations on the brittle to ductile fracture behaviour of base metal, weld metal and HAZ material by instrumented impact testing, 1st Int. Conf. on Welding Technology, Materials and Material Testing, Fracture Mechanics and Quality Mgmt., Wien Austria pp 423-432

IOP Conf. Series: Materials Science and Engineering **426** (2018) 012013 doi:10.1088/1757-899X/426/1/012013

Energy-based evaluation of hardness testing with discrete element method

Z Gyurkó[1] and R Nemes[1]

[1] Department of Construction Materials and Technologies, Budapest University of Technology and Economics, Budapest, Hungary

E-mail: gyurko.zoltan@epito.bme.hu

Abstract. Present paper focuses on the static hardness testing of porous construction materials. A series of experiments were conducted by the authors, during which the main strength properties of twelve different construction materials (normal and high strength concrete, concrete with supplementary cementing materials (SCM), limestones, ceramics, polymer concrete, etc.) were tested. Compressive strength, Young's modulus and Brinell hardness of every material was measured and based on an energy-based approach a relationship between the variables was defined. Brinell hardness test is a commonly used method for estimation of strength of solid materials (e.g. metals), however in case of porous materials contradictory results can be found in the literature. Present research is aiming to find a general formula for porous materials with as less restrictions as possible. The Brinell hardness of the materials was determined with DSI (Depth Sensing Indentation) method, which is an updated form of the classical Brinell hardness test. The output of DSI test are the loading-unloading curves belonging to a predefined maximal loading force. Based on these figures the elastic, dissipated and total strain energy can be determined and a formula can be given to estimate strength parameters based on them. The results were validated by using DEM (Discrete Element Method), mainly focusing on concrete and the effect of SCMs. The model was set up based on the experimental data and DSI test was performed with the model as well. The results of the model were compared to the laboratory measurement results.

1. Research objectives

Present research focuses on the static hardness testing of porous construction materials. A series of experiments were conducted by the authors, during which the main strength properties of fifteen different construction materials were tested (Fig. 1). Static hardness testing methods were developed for testing of metals, especially steel. The expressions, which were derived to define hardness, were defined based on the measurements performed on steel samples. The intention of the authors was to investigate the hardness of other construction materials besides steel. First, three metals (aluminum, mild steel, reinforcing bar steel (S235)) were subjected to the tests. Based on literature data, it is well known that hardness testing is applicable for metals. Then the tests were performed on a wide range of concretes (normal strength concrete, high-strength concrete, concrete containing supplementary cementitious materials (SCMs), polymer concrete). Concrete is the material used in the largest quantity by the construction industry, that's why it was the first choice for testing. It was seen that hardness testing is applicable for materials with silicate frame structure (concretes), thus other calcium silicate hydrate (CSH) based products were applied for testing (cellular concrete block (Ytong), sand-lime brick (Silka)).

Finally, natural stones and ceramics, which can be used as masonry material were subjected for testing. Summarized, the tested materials were the following:

- **Metals**: aluminum, mild steel, reinforcing bar steel (S235).
- **Concretes**: normal strength concrete, high-strength concrete, concrete containing SCMs, polymer concrete.
- **Other CSH based materials**: cellular concrete block (Ytong), sand-lime brick (Silka).
- **Ceramics**: clinker brick, facebrick, clinker tile.
- **Stones**: porous limestone and compact limestone, rhyolite tuff.

Figure 1. The investigated materials (bottom row: three metals, Ytong block, three ceramics, sand-lime brick, polymer concrete (150×150×150 mm); top row: three concrete and three stone specimens)

The tested properties were the following for every material:

- density,
- particle size distribution,
- compressive strength,
- Young's modulus,
- and Brinell hardness.

Based on an energy-based approach a relationship between the variables was defined. It was aimed to find a formula, which can be applied for a wide range of construction materials. To get a deeper understanding on the processes inside the material Discrete Element Method (DEM) was applied, which is a powerful tool for testing the macro-behavior of porous materials and with its application the processes inside the material can be visualized, which can lead us to the better understanding of the phenomena.

2. Literature review

Hardness testing was the first material testing in the engineering practice, starting with the scratching hardness testing methods in the 17th century [1-3]. The research of the theoretical background of hardness was initialized by the pioneering work of Heinrich Hertz in the 19th century [4]. Hertz's proposal formed also the basis of the indentation hardness testing methods by Brinell (1900), Rockwell, (1920), Vickers (1924) and Knoop (1934) [5]. All these methods are measuring the size of a residual plastic deformation in the tested specimen caused by the indenting load. Different indenter geometries (pyramid, spherical etc.) can be applied, however from these the spherical indenters can be used for testing both ductile materials (e.g. metals) and brittle materials (e.g. ceramics). The response of materials to the indentation test includes elastic (reversible) and plastic (irreversible) deformations as well as forming of cone cracks in brittle materials; therefore, the definition of the term 'hardness' is not evident [6]. Lately a computer aided version of hardness testing was developed and successfully applied on porous materials as well [7]. The method is called Depth Sensing Indentation (DSI) test and it is mostly applied to evaluate mechanical properties, like compressive strength or Young's modulus [8-11]. DSI method allows researchers to study the penetration process on an energy basis.

3. Experimental and numerical methods

A wide range of construction materials were applied for material testing. The research is mainly focused on the Depth Sensing Indentation (DSI) testing, which is an updated form of the classical Brinell hardness test (Fig. 2). Brinell hardness test is a commonly used method for estimation of strength of solid materials (e.g. metals), however in case of porous materials contradictory results can be found in the literature [12]. Brinell hardness test (and thus DSI) is a static (very low speed) hardness test method [13]. The main advantage of DSI that it is a non-destructive testing (NDT) method, meaning that during the test does not occur global failure in the material and it can be used for further testing.

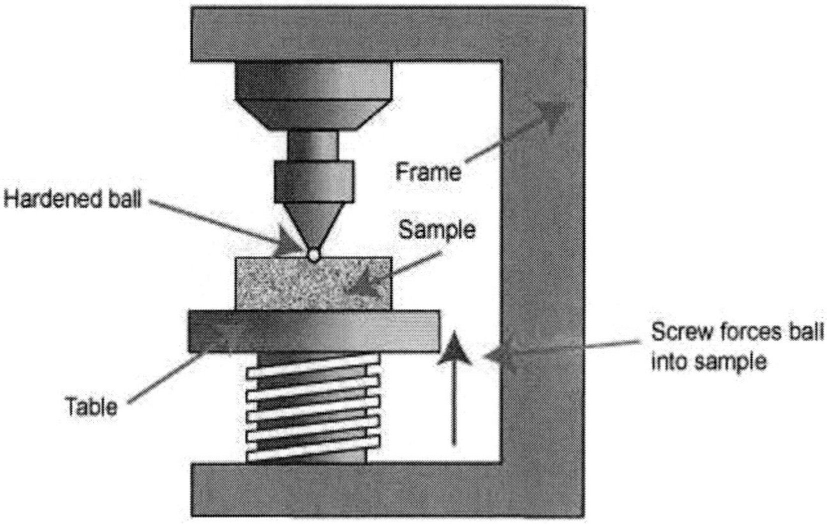

Figure 2. Schematic figure of a Brinell-hardness tester [13]

During the DSI test a polished steel sphere penetrates into the flat surface of the tested material, under a specified maximum load. A theoretical output of the DSI test can be seen in *Fig. 3*, which shows the relationship of the loading force and indentation depth (or indentation diameter). Based on that the hardness can be defined as follows:

$$HB = \frac{F}{D\pi h} \tag{1}$$

where F is the loading force in Newton, D is the diameter of the indenter in millimeters and h is the indentation depth in millimeters.

The DSI test has two phases, the loading phase and the unloading phase. Between the two phases, 30 sec holding time was applied. During the loading period, the indenter body penetrates into the surface of the sample at a constant rate until reaching the maximum value of the load. During that phase elastic and plastic deformations as well occurs in the material. During the unloading period, the indenter body moves away from the sample with the same rate. Elastic deformation of the material occurs during unloading and displacements are formed in the opposite direction, than formed during loading. During the unloading phase, the indentation depth is decreasing and in case of a perfectly elastic material it would reach 0. It is theoretically possible to find the elastic properties, including Young's modulus, of materials from the unloading curve of the indentation characteristics. In the literature only a few suggestions can be found for the calculation of Young's modulus based on indentation hardness and most of them applies the Boussinesq problem expressed by Sneddon. Sneddon defined the load vs. indentation depth functions for a linear elastic half-space for various types of indenter bodies [14].

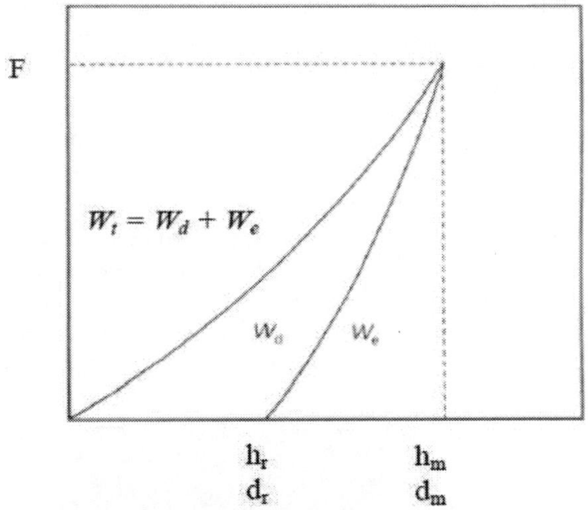

Figure 3. Theoretical relationship of loading force, indentation diameter (or depth) and indentation work

The area under the loading curve in Fig. 3 is equal to the work that is invested until reaching the maximum penetration depth that deforming the material elastically and plastically, agglomerate and set up cracks in it. This work is called the total indentation work:

$$W_t = \int_0^{h_m} F_1 \cdot dh \tag{2}$$

where W_t is the total work, h_m is the maximum indentation depth, F_1 is the value of the loading force in the loading phase. The area under the unloading curve is called the elastic indentation work. This is the amount of work that is recovered during the unloading:

$$W_e = \int_{h_m}^{h_r} -F_2 \cdot dh \tag{3}$$

where W_e is the elastic work, h_m is the maximum indentation depth, h_r is the residual indentation depth, F_2 is the value of the loading force in the unloading phase. The difference between the two works gives the dissipated energy during the indentation:

$$W_d = W_t - W_e \tag{4}$$

After rearrangement the following expression can be written:

$$\frac{W_d}{W_t} + \frac{W_e}{W_t} = 1 \tag{5}$$

So the depth sensing indentation (DSI) test may be suitable for addition to the introduction of hardness metrics, the energy based examination of the entire load-unload process. The DSI tests were performed on a Zwick Z050 computer-controlled universal tensile testing machine.

Besides the hardness testing compressive strength test was performed on all materials with an Alpha 3-3000 S hydraulic press. Before the tests the specimens were dried to lose their water content which are not in a chemical bonding. For the investigations specimens were prepared, which have the same size and shape (150×150×150 mm cube). After that these specimens were loaded up to failure with the same loading rate (5 kN/sec) and the maximum force was recorded. The compressive strength results were used to compare them with the hardness test results and to verify the numerical models.

The test methods were modelled with numerical methods. The aim of the numerical modelling was to generate more results without laboratory experiments. As it was mentioned in *Section 1*, DEM was applied to model the materials. The Discrete Element Method (DEM) is a family of numerical methods for computing the motion and effect of a large number of small particles. Any type of discrete element model is made up of two basic components: the elements and the contacts between them. In the present case a model was set up, which applies rigid elements that represented the aggregates of the material. As it was shown in [15], Brinell hardness testing can be modelled with DEM. Similar models, as presented in [15] were set up and DSI test performed on them. The particle distribution of the models followed the particle size distribution of the original material (e.g. in case of normal concrete: 0-4 mm: 40%; 4-8 mm: 22%; 8-16 mm: 38%). Several different test specimens were created with the same particle size distribution but with different arrangement of the particles. The parameters of the models were set up based on the laboratory compressive strength results. Finally similar curve (Fig. 3) could be plotted from the software as in case of the laboratory tests. A sample model can be seen in Fig. 4.

Figure 4. DEM model of a concrete cube

4. Results and discussion

In Fig. 5 some typical results of the DSI tests can be seen. Based on these figures the materials were arranged into three groups:

- elastic materials (clinker tile, clinker brick),
- elasto-plastic materials (compact limestone, high-strength concrete, polymer concrete, normal-strength concrete, rhyolite tuff, facebrick, concrete with SCM, metals)
- and plastic materials (sand-lime brick, porous limestone, cellular concrete).

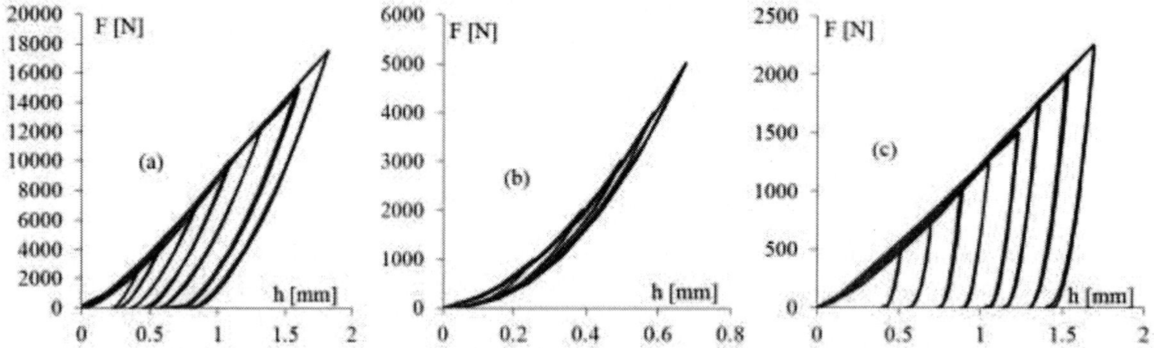

Figure 5. DSI test results: (a) normal strength concrete – elasto-plastic, (b) clinker tile – elastic and (c) porous limestone - plastic

The arrangement was done based on the shape of the loading-unloading curves and the slope of the loading and unloading paths. If, the slope of the unloading path was close to a vertical line, the material was considered as plastic material, while the material was considered to be elastic if the loading and unloading paths were close to each other. Between them the materials were considered to be elasto-plastic. Metals behavior is elasto-plastic, as it is well known based on a wide range of literature data and the method of static hardness testing was developed for such materials. The elastic and dissipated energy both contributes significantly in the total indentation energy in case of metals (e.g. in case of S235 steel W_t=25124 Nmm, W_e=17762 Nmm, W_d=7361 Nmm). Based on the results it can be observed that other materials, mostly concretes show similar behavior, as it can be seen in *Fig. 5*. Even polymer concrete, which is an unconventional type of concrete, fits into the group of elasto-plastic materials.

The investigations covered a wide range of construction materials. The lowest body density was 516 kg/m^3 (cellular concrete block), while the largest was 2502 kg/m^3 (compact limestone).

The curves can be described best with the indentation work, which is equal to the area under the curves. Based on the results, it is shown that the theoretical expression (Eq. 5) is true for these measurements. This supports the validity of the measurements.

The relationship of the elastic and dissipated energy with other material properties (Young's modulus, compressive strength, Brinell-hardness) was investigated. Based on the results, it could be observed that the elastic energy and the compressive strength (presented in Table 1.) has a relationship with high correlation, as it can be seen in *Fig. 6*. This could be written in a form of a power function.

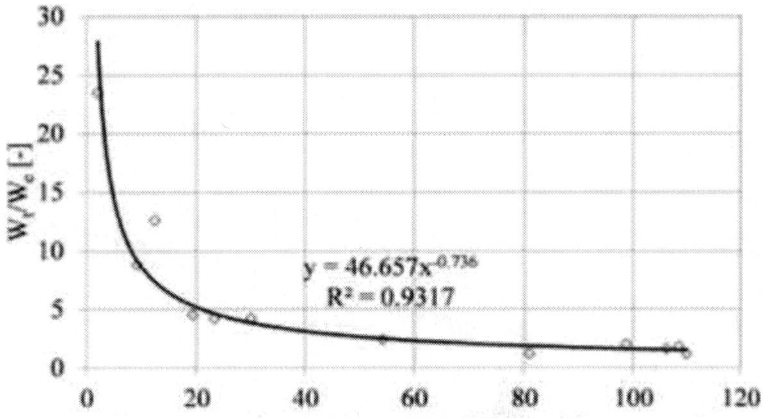

Figure 6. Relationship of the compressive strength and the ratio of the total and elastic energy

Table 1. Compressive strength and Young's modulus of the investigated materials

Material	*Compressive strength [N/mm²]*	*Young's modulus [N/mm²]*
Aluminum	*350*	*70000*
Mild steel	*420*	*200000*
Reinforcing steel bar	*600*	*210000*
Normal strength concrete	*54.2*	*34578*
High strength concrete	*106.2*	*43802*
SCM concrete	*72.5*	*36914*
Polymer concrete	*98.7*	*40000*
Cellular concrete block	*2.02*	*500*
Sand-lime brick	*19.3*	*2000*
Clinker brick	*80.9*	*35000*
Facebrick	*23.3*	*3000*
Clinker tile	*233*	*65000*
Porous limestone	*12.4*	*5127*

Compact limestone	*108.5*	*56402*
Rhyolite tuff	*30*	*10206*

It was also observed that there is a relationship between the elastic energy and the Brinell-hardness of the materials, as it can be seen in Fig. 7. The Brinell hardness (HB) is written in function of the ratio of the total (W_t) and the elastic (W_e) indentation energy (Eq. 6). This relation was valid for the experimental and numerical results coming from the DEM software as well. The results of the DEM analysis correlated well with the experimental data. The results indicate that there is a strong correlation between the Brinell hardness and the elastic indentation energy of the porous materials. Eq. 6 was applicable for all investigated materials, except the clinker tile, which behaved fully elastically within the applied load range (*Fig. 5*). Thus it could be stated that an expression was derived between the Brinell hardness and elastic indentation energy of plastic and elasto-plastic materials. Based on our experiment the value of parameter '*k*' in the expression was found to be 3894.2 as it can be seen on Fig. 7.

$$HB = k \cdot \frac{1}{\left(\frac{W_t}{W_e}\right)^2} \quad (6)$$

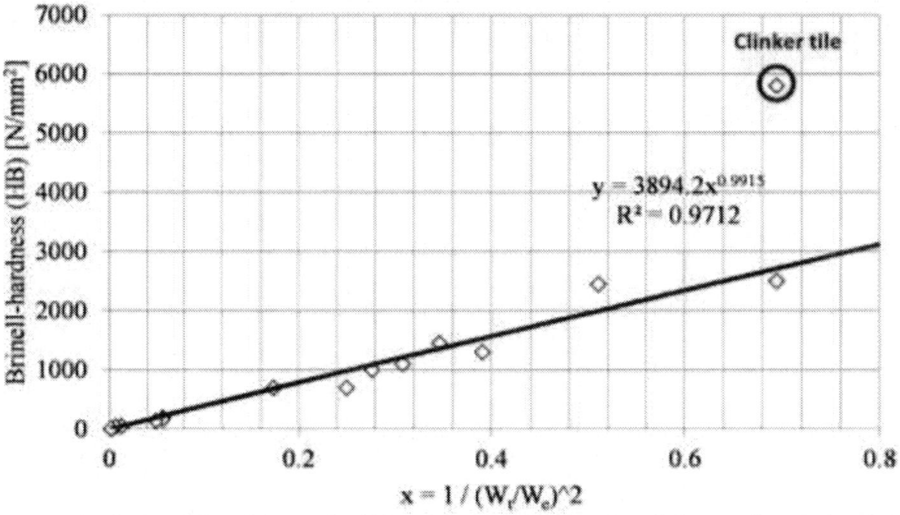

Figure 7. Relationship of the Brinell-hardness and the ratio of the total and elastic energy

5. Conclusions

Wide range of construction materials (concretes, stones, metals, etc.) were subjected to different material tests (compressive strength, Young's modulus, Brinell hardness) and the results were analyzed with an indentation energy based method. The aim was to find relationship between hardness of the material and other mechanical parameters. Besides the laboratory experiment numerical tests were performed as well, using Discrete Element Method (DEM). Depth Sensing Indentation tests were performed to determine the hardness of the materials. Based on the results the materials were arranged into three groups: elastic, elasto-plastic and plastic (Fig. 5).

The energy-based analysis of the DSI results indicated that there is a linear relationship between the Brinell hardness and the elastic indentation energy of elasto-plastic and plastic construction materials. The measurement results indicate that based on the found expression a materials property can be defined, which is valid for all elasto-plastic and plastic materials.

Acknowledgements
Authors are grateful to the Hungarian Scientific Research Fund (OTKA) for the financial support of the OTKA K 109233 research project. Special thanks to the Itasca Consulting Group for providing the software used during the project.

References

[1] Barba A A 1817 Arte de los metales *Reprint* (Lima)
[2] Réaumur R A F 1722 L'art de converter le fer forgé en acier *French Academy of Sciences (Paris)*
[3] Mohs F 1812 Versuch einer Elementar-Methode zur Naturhistorischen Bestimmung und Erkennung von Fossilien (University of Graz)
[4] Hertz H 1881 Über die Berührung fester elastischer Körper *Journal für die reine und angewandte Mathematik* **5** pp 12-23
[5] Fischer-Cripps A 2000 Introduction to Contact Mechanics *Mechanical Engineering Series (Springer)* p 221
[6] Szilágyi K, Borosnyói A and Dobó K 2011 Static indentation hardness testing of concrete: A long established method revived *Építőanyag - Journal of Silicate Based and Composite Materials* **63(1-2)** pp 2-8
[7] Glinicki M C A and Zielinski M 2004 Depth-sensing indentation method for evaluation of efficiency of secondary cementitious materials *Cement and Concrete Research* **34** pp 721–724
[8] Tezcan J and Hsiao K J 2008 Nondestructive evaluation of material strength using depth-sensing indentation *Engineering Structures* **30** pp 2206–2210
[9] Ye D, Xu H, Feng X, Xu Y and Xiao L 2016 Depth-sensing indentation-based studies of surface mechanical behavior and fatigue damage evolution of an austenitic stainless steel subjected to cyclic straining *Materials Science & Engineering A* **650** pp 38–51
[10] Chen P, Han Q, Ma T and Lin D 2015 The mechanical properties of shale based on micro-indentation test *Petroleum Exploration and Development* **42 (5)**
[11] Gubicza J, Juhasz A, Arato P, Szommer P, Tasnadi P, Voros G 1996 Elastic modulus determination from depth sensing indentation test *Journal of Materials Science Letters* **15** pp. 2141-2144
[12] Brinell J-A 1901 Mémoire sur les épreuves á bille en acier *Communications presentés devant le congrés international des méthodes d'essai des matériaux de construction* **2** pp 83-94
[13] Chandler H 1999 Hardness testing *ASM International* p 192
[14] Sneddon I N 1965 The Relation between Load and Penetration in the Axisymmetric Boussinesq Problem for a Punch of Arbitrary Profile *International Journal of Engineering Science* **3(1)** pp 47-57
[15] Gyurkó Z and Borosnyói A 2015 Brinell-hardness testing and discrete element modelling of hardened concrete *Építőanyag - Journal of Silicate Based and Composite Materials* **67** pp 8-11

IOP Conf. Series: Materials Science and Engineering **426** (2018) 012014 doi:10.1088/1757-899X/426/1/012014

Effect of locally increased melted layer thickness on the mechanical properties of laser sintered tool steel parts

I Hatos[1], I Fekete[1], T Ibriksz[1], B Kocsis[1], A L Nagy[2] and H Hargitai[1]

[1] Széchenyi István University, Department of Materials Science and Technology, Győr, Hungary

[2] Széchenyi István University, Department of Internal Combustion Engines, Győr, Hungary

E-mail: hargitai@sze.hu

Abstract. Additive technologies have several advantages over conventional manufacturing, such as the freedom of geometry of the products and internal structures. There are also some limitations and problems, deriving from stopping the process during the production. By restarting the process, the building often continues with a thicker starting layer due to the deposition of two or more layers. The effect of skipped melting of layers is investigated in this paper. Maraging steel powder (MS1) was used in direct metal laser sintering (DMLS) process to produce samples with increased thickness of melted layers. The layer thickness was increased in 20 μm steps up to 160 μm with 0.5 mm offset between the increased thickness layers. Porosity caused by the uneven melting was measured by optical microscope, mechanical tests were carried out to quantify the effect of skipped layers and fractured surfaces were observed under SEM. We have found that the yield strength and tensile strength are not affected if the layer thickness is slightly increased locally in the laser sintered part, while even a small increase in porosity greatly reduces the total elongation of the specimen. The decrease of impact energy due to the porosities shows similar correlation with the decrease of percentage elongation at break. However, the Charpy impact test is much more sensitive to layer skipping, the lack of melted layers lowers the impact strength significantly.

1. Introduction

Additive technologies make it possible to produce almost any geometry of products and internal structures. This manufacturing freedom is also used in the design of injection molding dies. Curved cooling lines that can follow the geometry of the part (conformal cooling) produces better quality product with shorter cycle times than traditional tools with straight drilling [1][2].

The final part quality is influenced by various factors: the powder material, exposure parameters such as laser power, scan speed, layer thickness and exposure-strategies, as well as the inert gas flow and temperatures in the process chamber [3]. Problems occurring during the process can result in job failure, create porosities or other defects. Due to the layer-wise nature of the process, these defects will not always be visible once the part production is completed. By introducing an on-line process monitoring system, the part quality can be monitored during the build [4].

Especially when printing large pieces, the production process sometimes needs to be interrupted due to lens cleaning, build powder refilling, collision or other technical problems. When restarting the process, changing the temperature of the system or moving the table may result in a thicker starting layer thickness. Inadequate powder deposition settings, uneven heights of hybrid parts, or table movements due to collisions may cause the laser to melt more than the optimal amount of metal powder. The effect of printing parameters, especially the layer thickness on the quality of printed parts are discussed in several literature.

The influence of layer thickness with 304 steel in the range of 20-40 μm has shown that the ductility and tensile strength are independent of slice thickness [5]. Samples were fabricated with layer thickness of 30, 50 and 70 μm and laser scanning speed set at 70 and 90 mm/s, at layer thickness of 70 μm porosity increases and cracks started to form which decreased strength and ductility [6].

During the manufacturing of molds by laser sintering, production parameters are not changed, the parts are produced with uniform layer thickness. It has been found that the available literature does not deal with the problem often encountered in practice when parts of a workpiece have an increased build thickness. During printing the part is under periodic reheating, the top layer and a couple of layers below that are remelted [7].

In our experiment, we made specimens from 1.2709 metal powders with DMLS technology. In the specified cross section of the specimens, the construction layer thickness was changed in the range of 20-160 μm and the effect of the increased layer thickness on the porosity and mechanical properties of the test specimens were investigated.

2. Experimental

2.1. Sample manufacturing

For the experiments „MaragingSteel MS1" steel powder – a regular material of the injection moulds having conformal cooling channels – was used to prepare samples by direct metal laser sintering. EOSINT M270 (200W) equipment was used by applying EOS MS1_Surface 1.0 parameter setting and 20 μm layer thickness.

We printed rectangular prism test pieces to quantify the porosity by skipping layers. Increased thickness of melted layers was created; the thickness was increased in 20 μm steps with a 0.5 mm offset between the increased thickness layers. Cylindrical Ø8 mm tensile test bars and 10x10 mm Charpy impact specimens with V notch were made with different layer thickness in the middle layer of the specimens. The samples were numbered 1 to 8, the test layer thickness corresponded to the multiplication of the number of the sample and the original layer thickness. (For the specimen No. 7 the average thickness of the specimen to be melted in the middle is 7×20 μm = 140 μm.)

2.2. Test methods

Tests for determining mechanical properties were carried out, tensile tests according to ISO 6892-1 B method on an INSTRON 5582 testing machine and Charpy V-notched impact tests according to ISO 148-1 standard at room temperature.

Porosity of the increased layer thicknesses were measured on micrographs. The sample was ground and polished. Panorama image was taken with a Zeiss Axioimager M1 microscope. We determined the amount of pores on the microscopic image with an in-house developed software for quantitative analysis. The surface morphology studies were conducted using a HITACHI 3400 scanning electron microscope (SEM).

3. Results

3.1. Porosity

A cross-section perpendicular to the construction direction of the test specimen was analysed after polishing. The microscopic investigation for porosity analysis was carried out by scanning the entire surface of the part and panorama image was created. The porosity was determined by an in-house-developed software. To quantify the degree of porosity, we marked areas with the same size around the skipped layers. The resulted porosity is the proportion of porosity measured on the reference surface to the size of the reference surface (Figure 1).

The region with the higher layer thickness, No. 6-8 has clearly visible lines of pores. While looking at the layers with lower layer thickness, No. 5 and below, the pores are difficult to see on the microscope image.

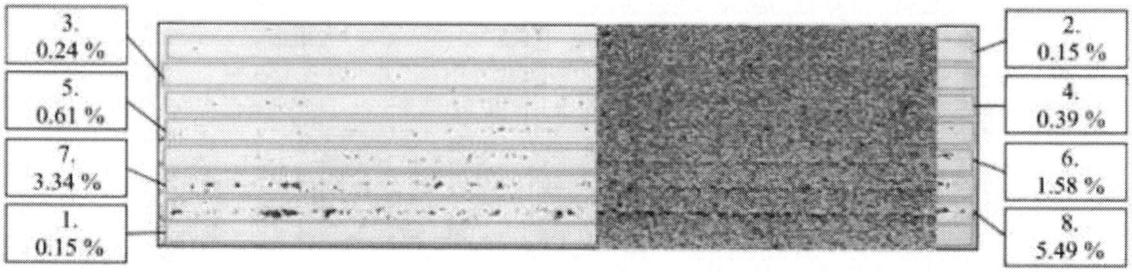

Figure 1. Panorama image of the polished cross section of the laser sintered part containing skipped layers with different thicknesses (20-160 μm) with the porosity measurements of the layers. The image of the rough outer surface was overlaid on the micrograph.

Analysing the porosity values, it can be concluded that skipping 1-2 layers does not cause measurable increase in porosity on the tool surface. However, in the case of Nr. 6, by skipping more layers and melting 120 μm thickness of powder, the porosity has multiplied. Moreover, by increasing the number of skipped layers, porosity increases exponentially. In the area of layer with higher thickness, it can be clearly seen that large-scale structures are created instead of prior uniform melting. The material structure is non-continuous; the input heat energy cannot melt the metal powder and in the area of scanning boundaries, significant porosity remains (Figure 2).

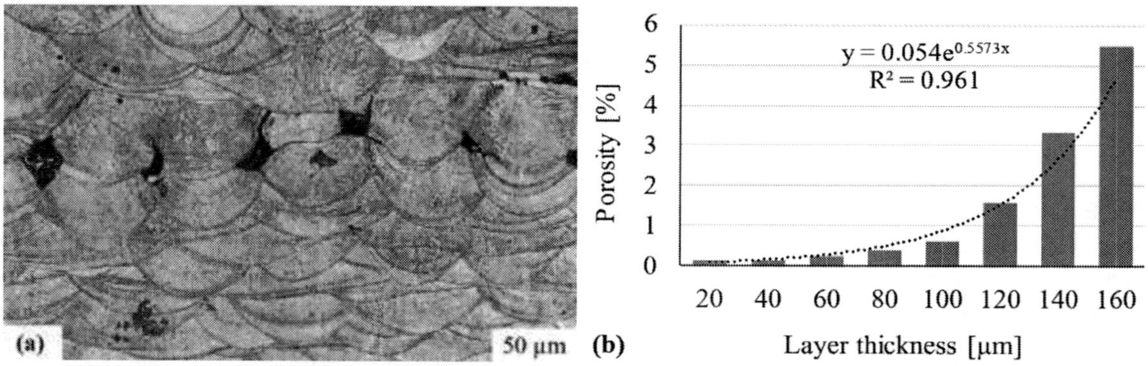

Figure 2. (a) Microstructure and location of the pores in specimen Nr. 7 (140 μm), (b) calculated porosity values on the cross section (b)

3.2. Mechanical properties

Tensile test measurements were carried out at room temperature using a non-contact video extensometer to measure strain. In our study, three specimens for each modified layer thickness were tested and from the load-extension curve the Yield strength, Ultimate tensile strength and percentage elongation after fracture (A_{40}) and percentage non-proportional elongation at maximum force (A_g) were determined. The resulted strength values and the test curves are given in Figure 3.

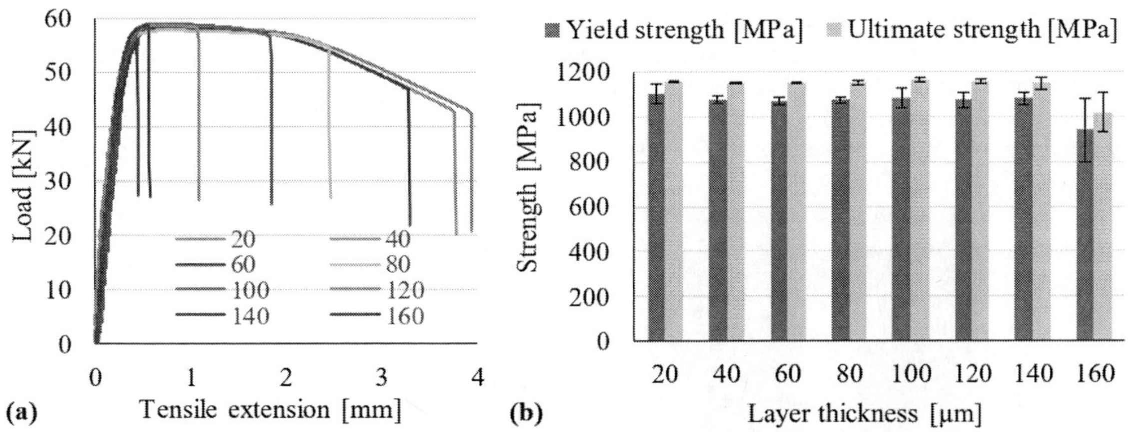

Figure 3. (a) load-extension curves of DMLS specimens with modified layer thicknesses (20-160 μm) and (b) calculated yield strength and ultimate tensile strength values.

Based on the results it can be established that the yield strength and tensile strength are not affected if the layer thickness is slightly increased locally in the laser sintered part. A significant decrease in the yield strength and tensile strength values was found in case of No. 8 specimen, in which a 160 μm layer was melted in one step. By comparing the test curves, it can be established that all specimens having different skipped layer thickness show the same load-extension values up to the failure as those made by constant layer thickness (20 μm). It can be stated that at static load, the maximal 5.5% local porosity created in the specimen does not affect the elastic deformation ability of the specimen.

The elongation values determined during the test as a function of the porosity can be seen in Figure 4. There is a well-defined correlation between the porosity and elongation values, the elongation decreases by increasing the porosity. Even a small increase in porosity greatly reduces the total elongation of the specimen. This can be explained by the smaller cross-section of the porous part, and higher local stresses generated around the porosity acts as starting points of failure. Linear relationship was found between the percentage non-proportional elongation at maximum force and the porosity. The more porosity in the material the shorter the uniform deformation is (Figure 4.b).

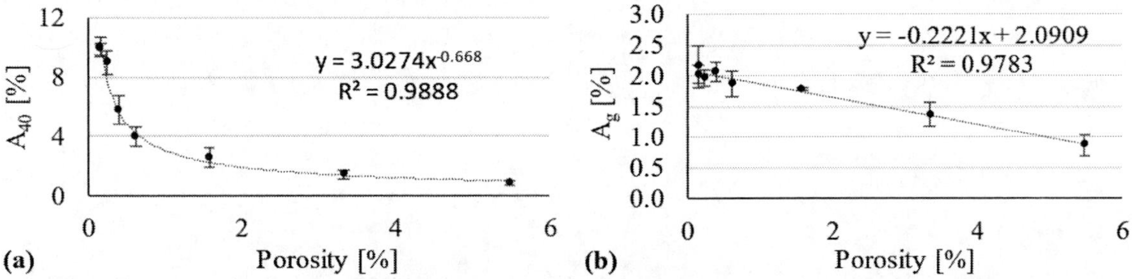

Figure 4. (a) percentage elongation at break and (b) percentage non-proportional elongation at maximum force as a function of porosity

Based on the strength and elongation values it can be stated that the specimens with 40 μm and 60 μm local layer thickness and the specimen built with 20 μm constant layer thickness showed no significant difference under static loading.

The V-notched Charpy test specimens of 10×10 mm cross section were tested at room temperature. The increased layer thickness was created in the centerline of the V-notch. Significant differences were found not only in the values of the impact energies but also in the quality of the fractured surface. Increasing porosity values showed less ductile behavior and this nature can be clearly seen on the fractured surface, failure appears dull and rough to the naked eye, with clear amounts of gross plastic deformation (Figure 5). As it can be seen in Figure 5.b. the energy absorbed by the specimens during fracture and the porosity shows similar correlation (exponential) as the percentage elongation at break (tensile test) and porosity. However, the Charpy test is much more sensitive to layer skipping and a minimal amount of layer thickness increase significantly lowers the impact energy of the material.

(a)

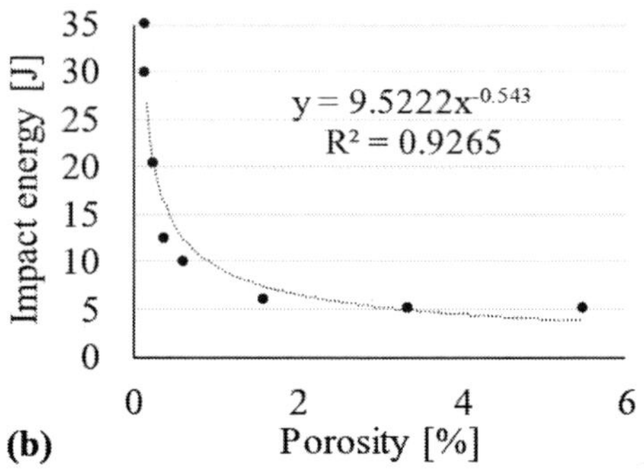

(b)

Figure 5. (a) Fractured surface of Charpy specimens No.1 and No.8; (b) Effect of the porosity on the impact energy absorbed during Charpy test

The fractured surfaces of the broken Charpy impact test samples No.1 having uniform 20 μm layer thickness and No. 8 containing a 160 μm thick layer in the middle were analysed in details by scanning electron microscope to characterize the mode of fracture (Figure 6.).

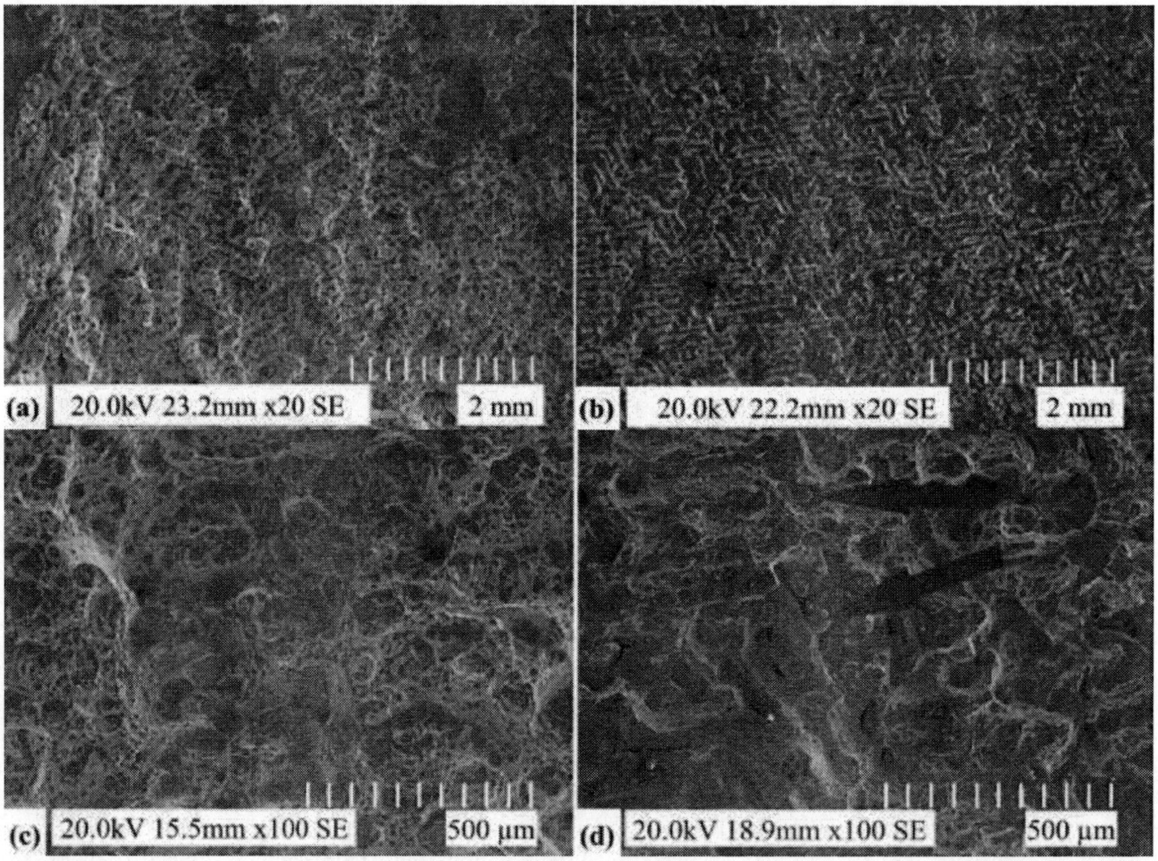

Figure 6. SEM image of fractured surface of Charpy specimens No.1 (a) and (c) and No.8 (b) and (d)

At lower magnification quite different topography can be observed in case of the lowest and highest melted layer thickness (Figure 6. a-b). In case of the specimen built with uniform layer thickness ductile fracture occurred by very fine dimples and micro-void nucleation and coalescence on the whole surface (Figure 6. c). Although, the fractured surface of the broken Charpy test specimen showed a brittle-like behaviour with no visible plastic deformation compared with the uniform layered sample, the SEM image shows that there are mostly ductile regions on the broken surface. However, the lack of fusion and visible cracks on the fracture surface function as stress concentrators. This is the main reason for the lower impact energies and gives ground to describe this kind of fracture as ductile (in those regions where the metal powder was totally melted) with local elements of brittle fracture.

Due to the high layer thickness and thus the inadequate melting several big holes can be discovered on the surface. Some powder particles in these areas were not melted and did not coalesce to the wall of the holes (signed with the arrows in Figure 6.d.).

4. Conclusions

Direct metal laser sintering has growing interest in tool manufacturing due to its several advantages over conventional technologies. However nearly the same mechanical properties can be achieved by proper heat treatment methods and the required surface quality can be achieved by post manufacturing with conventional technologies, there are also some limitations and problems in practical use. The sintering method often stop during the production and after restarting it often continues with a thicker starting layer due to the deposition two or more layers.

In our experiments, the effect of skip melting of layers during DMLS process was investigated by producing samples from maraging steel powder (MS1) with increased thickness of melted layers. The standard layer thickness was 20 μm, which was increased in steps up to 160 μm with 0.5 mm offset between the increased thickness layers. Porosity caused by the uneven melting has been inspected by optical microscope, and quantified by panorama microscopic images. Tensile and Charpy impact specimens containing a layer having different thickness in the middle were produced and tested to analyse the effect of skipped layers and the fractured surfaces were also observed under SEM.

It has been concluded that skipping one or two layers does not cause measurable increase in porosity, while skipping more layers and melting a 120 μm thick layer, the input heat energy cannot melt the metal powder and significant porosity remains. The porosity increased exponentially by increasing the melted layer thickness. The morphology was locally different in the area of the thicker melted layer; large-scale structures were created instead of prior uniform melting.

The yield strength and tensile strength were found not to be affected if the layer thickness is slightly increased locally in the laser-sintered part, while even a small increase in porosity greatly reduces the total elongation of the specimen. The decrease of impact energy due to the porosities shows similar correlation with the decrease of percentage elongation at break. However, the Charpy impact test is much more sensitive to layer skipping; the lack of melted layers lowers the impact strength significantly. Although the surface of the broken Charpy test specimen showed a brittle-like behaviour with no visible plastic deformation compared with the uniform layered sample, the SEM inspections indicate that there are mostly ductile regions on the broken surface between the porous parts of the surface.

Acknowledgements
The research presented in this paper was supported by EFOP 3.6.1-16-2016-00017, and it was carried out as part of the GINOP-2.3.4-15-2016-00003 project.

References
[1] Agazzi A, Sobotka V, LeGoff R and Jarny Y 2013 *Applied Thermal Enginnering* **52** 170
[2] Zink B and Kovács J G 2017 *Heat and Mass Transfer* **86** 101
[3] Grünberger T and Domröse R 2014 *Laser Technik Journal* **11/2** 40
[4] Clijsters S, Craeghs T, Buls S, Kempen K and Kruth J-P 2014 *Int J Adv Manuf Technol* **75** 1089
[5] Guan K 2013 *Materials & Design* **50** 581
[6] Abd-Elghany K and Bourell D L 2012 *Rapid Prototyping Journal* **18/5** 420
[7] Knowles C R, Becker T H and Tait R B 2012 *South African Journal of Industrial Engineering* **23/3** 119

IOP Conf. Series: Materials Science and Engineering **426** (2018) 012015 doi:10.1088/1757-899X/426/1/012015

Developing a glass fibre sensor for polymer technology applications

G Hegedűs[1] and T Czigány[1, 2]

[1] Budapest University of Technology and Economics, Faculty of Mechanical Engineering, Department of Polymer Engineering, Budapest, Hungary
[2] MTA–BME Research Group for Composite Science and Technology, Budapest, Hungary

E-mail: czigany@eik.bme.hu

Abstract. The goal of our research is to prove that the single-mode optical glass fibre used in telecommunications and in the optical loss test set used for the characterization of telecommunication networks is suitable for the structural health monitoring of polymer composites made from unidirectional e-glass fabric (UD). We built optical fibre into UD specimens and analysed the relationship between the power of the emitted light from the optical fibre, and the deformation of the specimen. Based on these results, a cost-effective measurement method can be achieved.

1. Introduction

Polymers are gaining in popularity and are developing fast in the 21st century. There are different kinds of polymers with different properties, but the fastest growing segment is the composites, which is used to reduce cost and mass [1]. Composites require new non-destructive test methods for structural health monitoring [2] due to their different structure from that of metals and because the matrix and reinforcement behave differently under loads [3, 4]. Several methods of structural health monitoring of composites are known, some of which use optical fibres [5]. Optical fibres are widely used to transmit digital information. Due to their advantageous properties (low mass, small size, high flexibility, heat and corrosion resistance etc.), these fibres can be easily embedded into composites of a polymer matrix [6]. As a result of external deformation, some properties of the light transmitted with the optical fibre changes [7], which can be measured with an optical instrument connected to the fibre, and therefore, the load on the structure can be identified [8, 9]. These measuring instruments are far more complicated and costly than the optical fibre itself, and optical loss test sets are widely used in the telecommunications to characterize the quality of optical networks. Attenuation (a) is the ratio of the optical power of incident light (P_i) and emitted light (P_e), expressed in decibels [10] (1):

$$a = 10 \log P_i/P_e \ \text{[dB]} \quad (1)$$

An optical loss test set continuously shows the attenuation of the optical fibre with reference to a reference value. The ratio of the optical power of incident and emitted light (P_i/P_e) can be easily calculated from these measured attenuation values, which shows the intensity of the emitted light in percentage, with reference to a reference value.

IOP Conf. Series: Materials Science and Engineering **426** (2018) 012015 doi:10.1088/1757-899X/426/1/012015

Our goal is to identify the deformation of a composite of a polymer matrix with the help of a general purpose optical fibre and an optical loss test set, which are widely used in the telecommunications. The intensity of light emitted by the optical fibre changes as a result of the deformation of the structure and it can be measured with an optical loss test set.

Our partial test result carries the promise of the development of a novel cost-effective polymer materials test method, and even a procedure usable in the industry.

2. Materials and methods

The optical fibre used is a G.652.D single-mode glass fibre of a diameter of 125 microns (Corning, USA). It has a core of a 9 micron diameter, it is sensitive to deformation and the fibre end is perpendicular. The optical fibre was tested embedded into a composite specimen, which was reinforced with unidirectional glass fabric. The coating of the fibre section which was embedded into the specimen was removed with an optical fibre stripper. The matrix around the fibre provided mechanical protection for the fibre and a stripping was performed to reduce the diameter of the fibre as much as possible in order to preserve the continuity of the composite structure. The section of the fibre outside the composite specimen was not stripped of its coating to providing some mechanical protection for the fibre. Making the specimen, we put the fibre between two layers of UD fabric, parallel to the fibre bundles (figure 1). The matrix material was an unsaturated polyester resin (AROPOL M105 TB, Ashland S.p.A., Italy). Initiator was added to the resin (PROMOX P200TX, PROMOX SRL, Italy) which was 1.5% of the mass of the resin.

The full length of the specimen was 250 mm and the embedded length of the optical fibre was 140 mm. The thickness of the specimens was 1.6±0.2 mm.

We used a cleaver (Fujikura, CT-30, USA) and a fibre welder (Fujikura, FSM 12 S, USA) in connecting the optical fibres. One end of the optical fibre was illuminated with a light source (AFL Telecommunications, OLS7 FTTH UCI, USA) of a wavelength of 1550 nm, and at the other end the intensity of the emitted light was measured with an optical power meter (AFL Telecommunications, OPM5-4D, USA). The optical fibre and the composite specimens were deformed in a tensile tester (Zwick, BZ050/TH3A, Germany) at a tensile speed of 0.4 mm/minute and with a clamping length of 185 mm.

Figure 1. The specimen (a) and the measurement layout (b) (1–composite specimen, 2–direction of reinforcing fibre bundles, 3-optical fibre, 4–light source, 5-optical power meter, 6-clamps of the tensile tester, 7-values measured by the tensile tester, 8-video camera)

The optical power meter cannot collect the data but displays the actual attenuation value. In order to assign the emitted light intensity of the measured displacement and force values, we recorded the values displayed by the tensile tester and the optical power meter. These recordings made it possible to connect the intensity values calculated from the attenuation values to the values had been recorded by the tensile tester.

3. Results and discussion

The goal of the tests was to prove the viability of the procedure. The optical fibre went out of the specimen in front of the clamps of the tensile tester so that the clamping did not influence the measurements. As a results of loading a change in attenuation was recorded with the above-mentioned method.

The power of emitted light showed greatly different values and tendencies during the tensile tests of various composite specimens. A characteristic measurement is shown in figure 2.

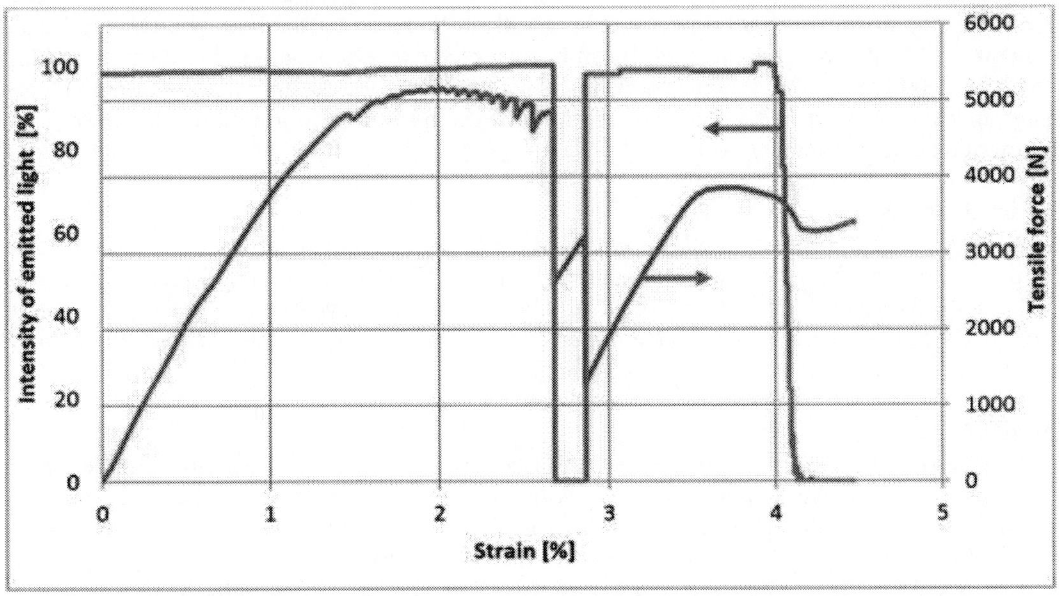

Figure 2. The relationship between the intensity of emitted light from the optical fibre and the strain of the specimen

The differences of the values appearing during the tests can be explained with the selected materials. During the tensile test, some fibres of the UD fabric were separated from each other; some bundles were nearly broken, and others were almost unloaded. The layout of the specimen was the reason for that the individual fibre bundles had not been prestressed to the same extent and transversal bundles did not hold them together, either. The optical fibre connected to the nearby fibre bundles through the matrix. Coating of the optical fibre was removed but optical fibre-matrix adhesion was still far less than the adhesion between the fibre bundles and the matrix. When the fibres of fibre bundles had been separated from each other or had been broken, the connections between the optical fibre and the nearby fibre bundles disappeared in several tests, and even in extreme cases the optical fibre was "torn out" of the specimen. The diagram shows that the optical fibre was connected to two fibre bundles more strongly. When the first of these fibre bundles had been broken at the strain of 2.7%, it caused the breaking of the optical fibre as well. The broken ends of the fiber moved away from each other and the optical fiber was not able to emit light. Due to the breaking of the other connected fiber bundle, the environment of the optical fiber was restored to its unloaded state from the loaded state and the broken optical fiber ends have been reconnected again within the specimen. As a result, the power of emitted light has restored to over 98% of incident light power. At a strain of 4%, the optical fibre was in a loaded state again. The breaking of the specimen caused the separation of the optical fibre ends permanently. The diagram shows that a load on the fibre bundle that is in connection with the optical fibre has a great influence on the power of light emitted by the optical fibre.

4. Conclusions

The tests showed that in embedding the optical fibre, the type of reinforcement the proper connection of the optical fibre and the fibre bundle must be taken into account and in this way the composite structure must be ensured. In the specimens of the tests, the optical fibre is connected to certain fibre bundles, and therefore it was able to provide information about the health of that fibre bundle (which was not the same as the health of the whole specimen). For that reason the UD reinforcement is not a good choice – the optical fibre must be connected well to the whole specimen and has to be loaded together with the structure. In spite of that fact, the test series provided some very useful results – it showed that the transmitted light power can be greatly influenced by the load, which is in connection with the health of the structural element. It can be observed as well that after loading stops, the amount of the emitted light can go back to nearly its initial value (if the optical fibre did not break as a result of loading), therefore measuring the intensity of emitted light the *in-situ* information about the current state of deformation can be obtained.

Acknowledgments

This research was supported by the New National Excellence Program ÚNKP-17-3-I. of the Ministry of Human Resources. The article was supported by the OTKA (K 116070 and K120592) and NVKP (NVKP_16-1-2016-0046) projects of the National Research, Development and Innovation Office (NKFIH). The authors wish to thank Péter Dobos (ATL Kft), Balázs Bartalos, Tamás Rácz and Bence Péter for their help in performing the measurements.

References

[1] Szakács J and Mészáros L 2017 Effect of fiber contents on fatigue behavior of injection molded polyamide 6 matrix composites *Per. Poly. Mech. Eng.* **61** 6–10
[2] Hegedűs G, Sarkadi T and Czigány T 2017 Light transmission characteristic of reinforcing glass fibres used in polymer composites *Materials* **10** 637 p9
[3] Saeed M U, Li B B, Chen Z F and Cui S 2016 Self-healing of low-velocity impact and mode-I delamination damage in polymer composites via microchannels *Express Polym. Lett.* **10** 337-348
[4] Muñoz-Vélez M F, Valadez-González A and Herrera-Franco P J 2017 Effect of fiber surface treatment on the incorporation of carbon nanotubes and on the micromechanical properties of a single-carbon fiber-epoxy matrix composite *Express Polym. Lett.* **11** 704-718
[5] Gholamzadeh B and Nabovati H 2008 Fiber optic sensors. *Int. J. of Electric., Comp., Energ., Electron. and Commun. Eng.* **2** 1107-1117
[6] Di Sante R 2015 Fibre optic sensors for structural health monitoring of aircraft composite structures: Recent advances and applications *Sensors.* **15** 18666-18713
[7] Glisic B 2013 Distributed fiber optic sensing technologies and applications – an overview. *ACI Spec. Publ.* **292** 1-18
[8] Glisic B and Inaudi D 2007 Fibre optic methods for structural healt monitoring *John Wiley & Sons Ltd.* New York
[9] Guo H, Xiao G, Mrad N and Yao J 2011 Fiber optic sensors for structural health monitoring of air platforms *Sensors* **11** 3687-3705
[10] Grattan K T V and Meggitt B T 1995 Optical fiber sensor technology *Springer-Science+Business Media* B.V, Dordrecht

IOP Conf. Series: Materials Science and Engineering **426** (2018) 012016 doi:10.1088/1757-899X/426/1/012016

The effect of alloying elements for the hot workability of AlMn1 aluminium alloy

T Hegyes[1] and P Barkóczy[2]

[1] Product engineer, Arconic Köfém Kft., Székesfehérvár, Hungary
[2] R&D manager, FUX Zrt, Miskolc, Hungary

E-mail: tibor.hegyes@arconic.com, peter.barkoczy@fux.hu

Abstract. There is a wide range of applications of aluminium alloys that contain manganese alloying element. This is mainly due to its good durability, corrosion resistance and cost effectiveness. The biggest proportion of manganese alloying element is in solid solution during the casting process and solidification. Moreover the remaining manganese content can be found also in the casted aluminium in precipitation form. Because of high production volume it is essential to optimise the production process in every respect to assure more economical production. Our experiment focused on one of the most important production steps of rolled products that is hot rolling. During hot forming the plastic deformation and the softening processes takes place in parallel in a complex system. In our experiment we carried out physical simulation with Gleeble thermomechanical simulator. Three aluminium-manganese alloys were compared with different chemical compositions. We applied hot compression tests and then we studied the measurement results.

1. Introduction

There is wide range of applications of aluminium alloys that contain manganese alloying element because it has good formability and corrosion resistance. These are sold and used in high volume around the word in rolled forms like strip, sheet and shate. They have many industrial applications for example cooking and automotive industry where these alloy is used for heat changer products.

One method to produce rolled products is to use semi-continuous cast ingots for it [1, 3]. These ingots are fabricated after casting to guarantee a suitable geometry, and a clean surface composition. To reach the required thickness the ingot is rolled. The reduction of thickness in an economic way can be fulfilled by hot rolling where the work hardening [2, 5] and the annealing of the aluminium [4, 6] occur in the same time or closely. High thickness reductions can be reached by applying smaller forming forces during hot rolling with less energy consumption. For hot rolling it is necessary to heat up the material to hot rolling temperature. It is essential to apply such a temperature during preheating cycle where the homogenization of the ingot can be achieved. At solidification micro-segregations exist in the material that causes concentration differences. At the appropriate homogenization temperature these concentration differences start to equalize and this heat treatment refer to sure hot deformation at hot rolling [5, 6].

Work hardening effect starts during metal forming. Dislocation movements support the plastic deformation in metals. Dislocation is a one dimension lattice defect. During plastic deformation the number of dislocations are multiplied. When the number of them such a volume that they hinder of each other in their movement the material getting harder and for further forming of material it is essential to apply higher forming force that generates higher stress in the material.

When the temperature is sufficient, in relation to amount of strain the annealing starts in the material. These two phenomena change the dislocation structure and we talk about recovery and recrystallization. Dislocations are organized at recovery. Aluminium tends to recovery such that dislocations are easily moved from their position. These organisations of dislocations can be small as they compose lines and set up low angle grain boundaries and in the original grains a new sub-grain structure develops. Because of the organization of dislocation structure these dislocations do not hinder each other with high effort, hence the necessary stress is decreasing for the deformation. If we measure stress during deformation in terms of the applied strain in that situation the recovery compensates the work hardening effect when the recovery is achieved then ready state stress develops.

The other process that causes annealing is the recrystallization. In the deformed material volume fraction small nucleuses are developed then they are growing. If we study the stress for deformation as a function of the strain, in that case the curve has a local maximum. This is the location where the recrystallised grains are growing continuously and in parallel the deformed volume fraction is decreasing and the stress achieves steady state values. Nucleuses develop mostly on the grain borders so therefore the result at the local maximum of the curve a typical necklace structured grains development where the recrystallized new grains are located on the formed grain borders. When these above mentioned phenomena occur during hot forming we are talking about dynamic recovery (Fig. 1.b) and dynamic recrystallization (Fig. 1.c).

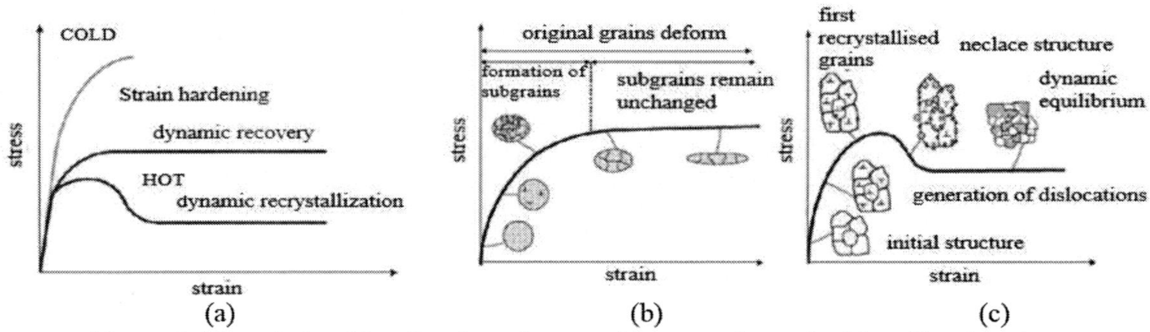

Figure 1. The relationship of strain and stress: (a) comparison of cold and hot forming, (b) dynamic recovery and (c) dynamic recrystallization [7]

Aluminium-manganese alloys belongs to workable and non-precipitation hardening alloys. During preheating of the material up to hot rolling temperature the manganese starts to precipitate from the solid solution. The kinetics of precipitation process has influence for the size of new phases and also for those processes that happen during hot rolling. Manganese element has a special feature as it has low solid solubility in the aluminium. During the solidification the manganese precipitates with the iron atoms in the form Al6(Mn,Fe) that is an eutectic intermetallic phase. The volume, size and the distribution of intermetallic phases and the manganese volume in solid solution are changing during solidification as a function of the cooling rate. The ratio of Fe-Mn in the intermetallic phase changes during homogenization and also the manganese in solid solution starts to precipitate mostly in Al_6Mn form. The developed microstructure, concentration of alloying elements, the size and distribution of precipitations and grain structure after homogenization is the initial state at the beginning of hot rolling.

During hot rolling the work hardening and the start of the regeneration process of crystals determine the properties of the microstructure that is very important to take into consideration during the technology preparing. In case of aluminium alloys at dynamic annealing the dynamic recovery and the dynamic recrystallization are realised in the same order of magnitude rate.

These two phenomena determine together the hot rolled grain structure. Further processes like cold rolling and heat treatment change the hot rolled microstructure but even so the hot rolled microstructure still influences the final property of the end-product. The hot compression test of these alloys is the method to study this complex system with different parameters. The main important parameters are the forming temperature, strain and strain rate.

2. Material and methods

In our experiment three types of AlMn1 aluminium alloy were compared. The chemical composition of them were different as shown in Table 1. EN AW 3003 aluminium alloy was our base material and we compared the other ones against it.

Table 1. Chemical composition of the measured material (m%)

	Alloy code	Si%	Fe%	Cu%	Mn%	Mg%	Ti%
EN AW 3003 (base)	A	0-0,6	0-0,7	0,05-0,2	1-1,5	-	-
Difference from EN AW 3003	A1	+0,5	-	+0,5	+0,1	+0,2	+0,1
Difference from EN AW 3003	A2	-0,2	-0,3	+0,5	-	-	+0,1

Physical simulation was carried out in this experiment. Uniaxial hot compression measurements were taken by Gleeble thermomechanical simulator. We studied the microstructure changes during hot deformation by taking into consideration the effect of main alloying elements which are manganese, cooper, magnesium and silicon. For this experiment samples were taken from cast aluminium and were subjected to homogenization heat treatment in laboratory furnace. We applied two different homogenization characteristics for comparison. Samples were in the furnace at 510 [°C] for 1 hour and the other was at 540 [°C] for 6 hours holding time, then at the end of the cycle they were cooled down in cold water to room temperature. From the homogenized material smaller 16x27 mm cylindrical test specimens were prepared for the hot compression test that were formed at 380 [°C] and at 500 [°C] by using 0,2 and 20 [1/s] strain rates. Specimens were heated up at a 2 [°C/sec] rate to the hot forming temperature and were held at this temperature for 90 seconds before the forming. Every specimen was compressed by 1,5 strain and at the end these were quenched in cold water to cool down to room temperature in order to freezing the microstructure. The temperature was controlled during the tests and the strain and the stress values were registered. Specimens were prepared for metallographic examinations. Microsection were taken perpendicular to the forming direction then polished and etched. Struers Lectro-Pol 5 type of electropolishing device was used for etching with applying special Barker type of liquid. Pictures were taken by Zeiss AxioVert40 type of optical microscope in polarized light with Zeiss ICc camera.

3. Results and discussion

Based on the experiment results we studied the effect of each of the applied measurement parameters for the hot workability and for the developing new microstructures by taking into consideration the specimens preheating circumstances known as homogenization heat treatment. During the experiment the registered compressing stress was plotted in accordance with the strain. At the first step we analysed the effect of the two different homogenization treatment processes for the hot workability. The strength of homogenized specimens at 540 [°C] was decreased for every alloy and at every hot compression test with compared with the homogenized specimens at 510 [°C]. It is shown in Fig. 2.

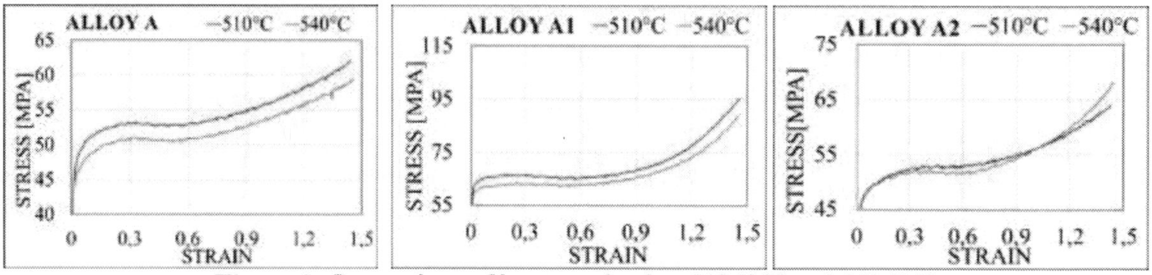

Figure 2. Comparison of homogenization 510 [°C] and 540 [°C]:
The stress curves of tested alloys with 380 [°C] and 0,2 [1/s] strain rate.

The reason for this is that the higher temperature and it's holding time during homogenization treatment has influence for the microstructure of the cast specimen. It is changed. The manganese volume that is in solid solution in the aluminium matrix starts to precipitate in the measured alloys as shown in Fig. 3.

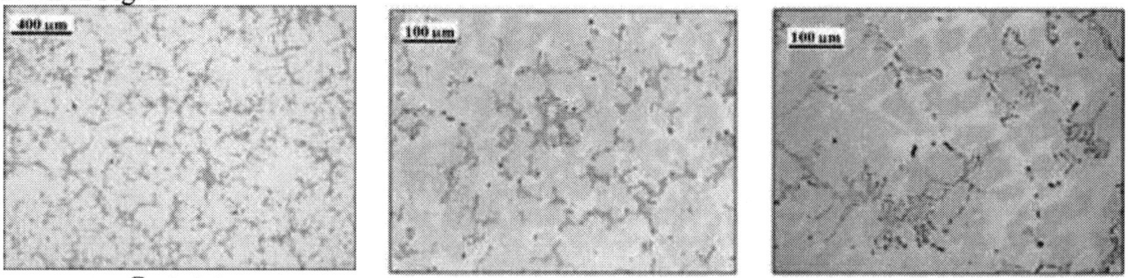

Cast state Homogenized on 510 [°C] Homogenized on 540[°C]

Figure 3. Optical microscopic microstructure of A1 alloy.

The higher temperature and it's holding time supports the diffusion of atomic manganese element. By reason of the higher energy state the size and the structure of precipitation is different now along with the intermetallic phases which are developed during solidification. These together cause the changing of the strength of material. The volume of manganese in solid solution starts to decrease with increasing the temperature so thus the solid-solution hardening effect of manganese are getting deceased when precipitations are appeared. These phenomena cause the decreasing stress as shown in Fig. 2.

In further diagrams in this article the stress curves will be shown two forming temperatures and strain rates applied to the base alloy "A" is marked by dashed curves for easiest segregation (Fig. 4-7.)

As a first step the stain rate effect was studied. On the evidence of the curves it can be determined for the same alloy and hot compression temperature and with the smallest strain rate the stress of the measurement is the lowest. But conversely the higher strain rate causes higher stress in the formed material. By increasing the strain rate the tool of the equipment facing with higher resistance. If we are increasing the temperature then the stress is continuously decreasing for the same alloy with the same strain rate hence the same deformation can be reached by lower forming force. It is seen that the highest temperature and small strain rate gives the lowest stress results, whereas the lowest forming temperature and highest strain rate cause the highest stress in the material that resulting in two or three times higher strength. This phenomena is representative for the whole experiment.

From the workability point of view it is possible to make obvious distinction between alloys. Increasing alloying elements generated higher stress in our experiment. The basic "A" alloy can be formed easiest if it is followed by "A2" and finally "A1" alloy. The last one has the highest Si, Mn, Mg and Ti alloying element. From the stress curves the difference can be seen between alloys. The continuous curves of "A2" and "A1" specimens are above of the base "A" alloy with dash curves almost in every case. There were only two exemptions are which will be discussed later (Fig. 6. and 7. black curves).

IOP Conf. Series: Materials Science and Engineering **426** (2018) 012016 doi:10.1088/1757-899X/426/1/012016

Independently from the alloys with 380 [°C] and 20 [1/s] parameters the "A1" and "A2" alloy curves divide wider from "A" alloy dashes curves. These hot compression parameters caused the biggest difference in stress result comparing these alloys. With increasing forming temperature up to 500 [°C] and with decreasing the strain rate the strength of "A1" and "A2" alloys approach the results of the base "A" specimen. The curves of stress values are close to each other and the order of magnitude is similar at the same location.

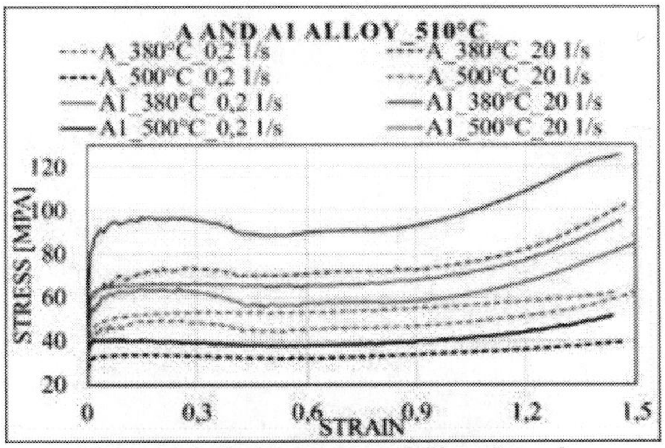

Figure 4. Comparison of alloy A and A1_homogenisation: 510 [°C]

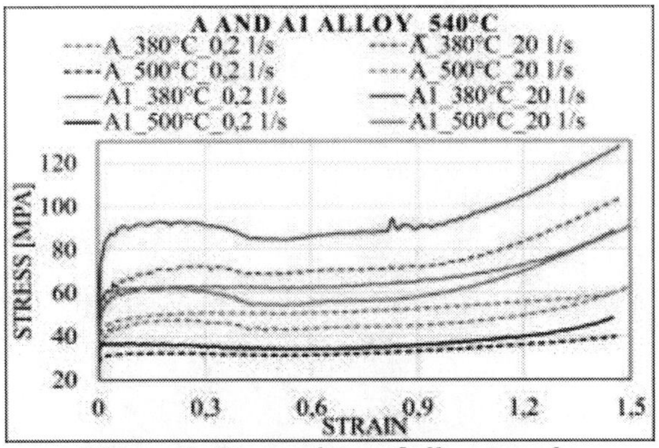

Figure 5. Comparison of alloy A and A1_homogenisation: 540 [°C]

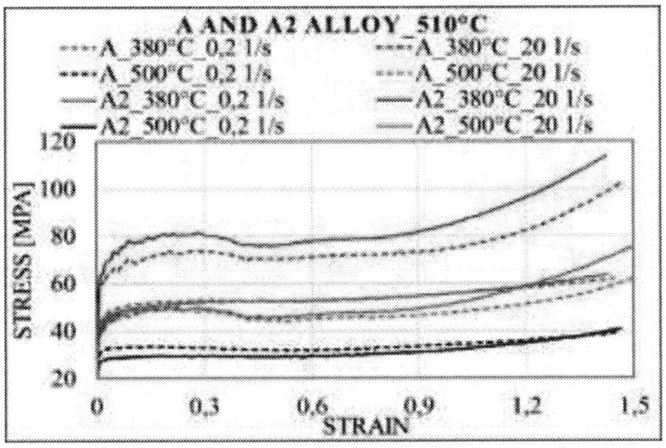

Figure 6. Comparison of alloy A and A2_homogenisation: 510 [°C]

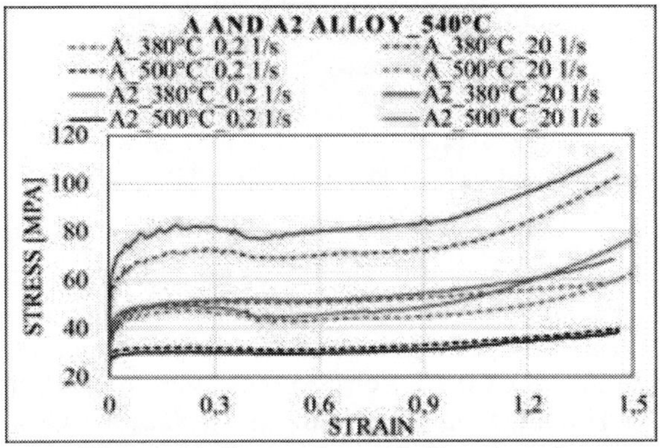

Figure 7. Comparison of alloy A and A2_homogenisation: 540 [°C]

Alloying elements also hinder the movement of dislocations so where dissimilar atoms are located in the basic aluminium matrix these area of the material are intensely getting harder during plastic deformation. At these areas where the dislocation density and the lattice distortion is high during hot rolling the dynamic recovery starts sooner and moreover it is getting softer. The curves at 500 [°C] and 0,2 [1/s] parameters on Fig. 6 and 7. don't have the same trend like the others. The curves of "A2" alloy is working with lover stress values. It is assumed that the dynamic recrystallization started with these forming parameters because the deformation occurs with lower forming when compared with specimen "A2". The difference here is low. On the optical microscopic picture of the specimen it can be seen inconspicuously as new nucleus and recrystallized small grains are developed under hot forming Fig. 8 and 9.

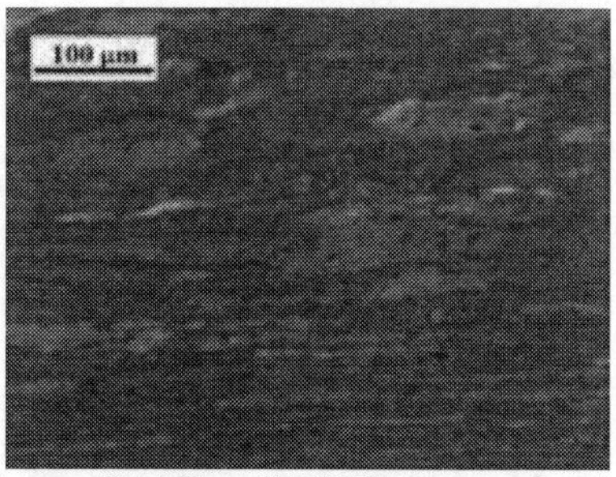

Figure 8. Alloy "A2" homogenization at 510 [°C] hot forming at 500 [°C] with 0,2 [1/s] strain rate

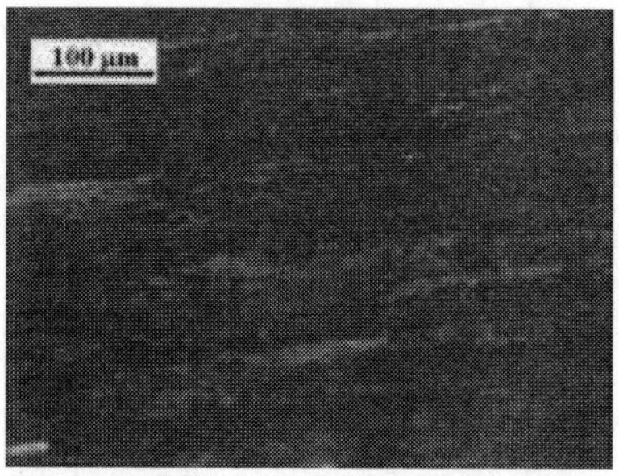

Figure 9. Alloy "A2" homogenization at 540 [°C] hot forming at 500 [°C] with 0,2 [1/s] strain rate

For the same temperature but with higher stain rate the registered stress values were with about 20 [MPa] higher. The microstructure of these specimens show that the grain structure is more formed and a shape that represents a typically cold formed structure (Fig. 10 and 11).

Figure 10. Alloy "A2" homogenization at 510 [°C] hot forming at 500 [°C] with 20 [1/s] strain rate

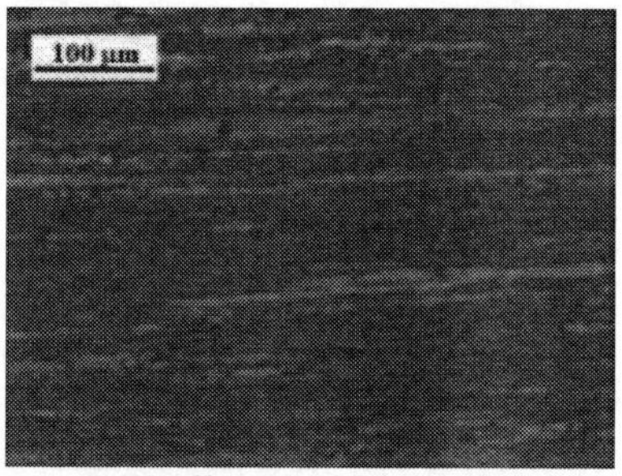

Figure 11. Alloy "A2" homogenization at 540 [°C] hot forming at 500 [°C] with 20 [1/s] strain rate

Curves become flattened with increasing strain. The measured stress didn't grow intensively. With these applied parameters dynamic recovery occurs rather in the material than the recrystallization. Dynamic recovery can't be seen on the microstructure on these pictures with this magnification.

We took into consideration the microstructure and noted that the nucleation and dynamic recrystallization started only from 500 [°C] with 0,2 [1/s] parameters in case of every measured alloy (Fig. 8; 9; 12; 13; 14 and 15). Besides the formed grains the new modified grain structure can be also seen.

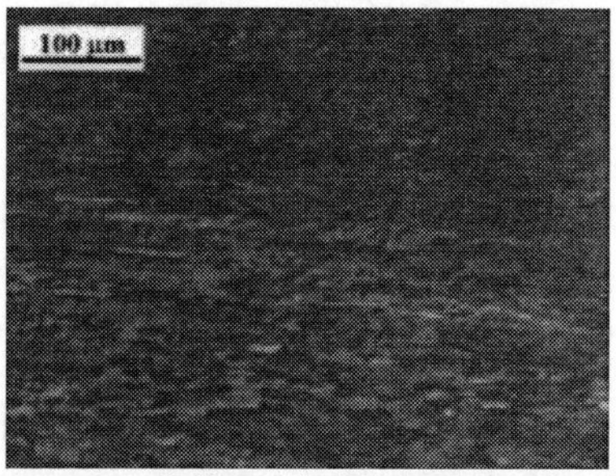

Figure 12. Alloy "A1" homogenization at 510 [°C] hot forming at 500 [°C] with 0,2 [1/s] strain rate

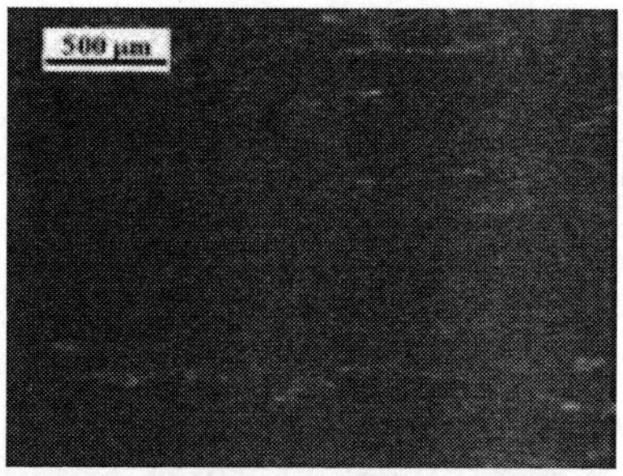

Figure 13. Alloy "A1" homogenization at 540 [°C] hot forming at 500 [°C] with 0,2 [1/s] strain rate

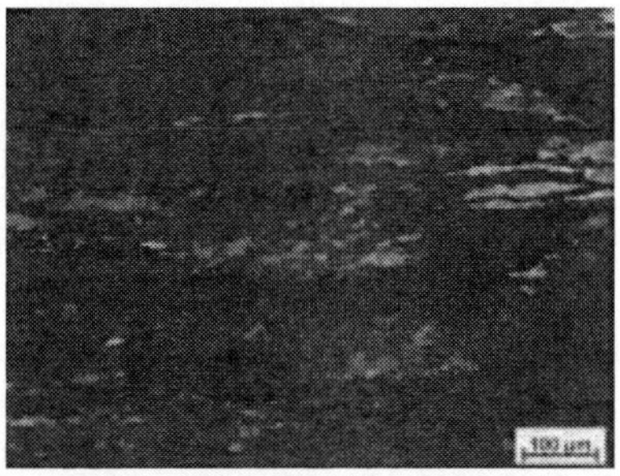

Figure 14. Alloy "A" homogenization at 510 [°C] hot forming at 500 [°C] with 0,2 [1/s] strain rate

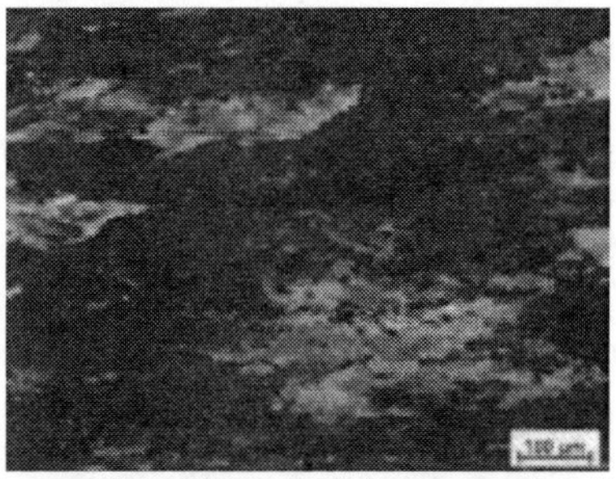

Figure 15. Alloy "A" homogenization at 540 [°C] hot forming at 500 [°C] with 0,2 [1/s] strain rate

The microstructure refers to cold deformation in the case of other hot forming parameters (Fig. 10; 11; 16 and 17).

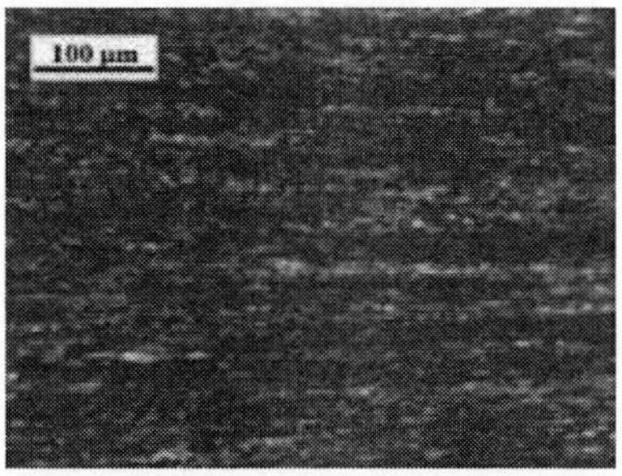

Figure 16. Alloy "A" homogenization at 510 [°C] hot forming at 380 [°C] with 0,2 [1/s] strain rate

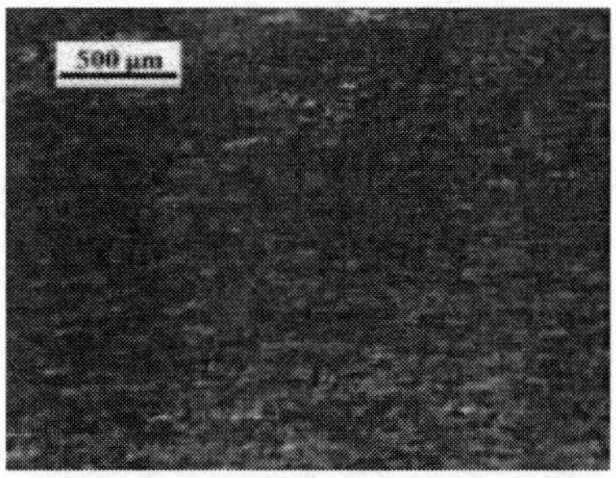

Figure 17. Alloy "A1" homogenization at 510 [°C] hot forming at 380 [°C] with 0,2 [1/s] strain rate

If we compare the trend of the evaluated curves with the theoretical curves on Fig. 1.b it can be determined as in our experiment that instead of dynamical recrystallization dynamic recovery is established. The dynamic recovery under hot compression compensates the hardening effect in the specimen when it starts. This can be seen in the curves when at about 0,1 strain the stress values jump up suddenly where the dynamic recovery starts up and the stress is closely steady-state. On Fig. 1.c the stress results have a significantly strong decreasing in the curve that represent the dynamic recrystallization. This is the state when dynamic recrystallized grains develop in the formed material and with these phenomena the material is getting softer. In some case our curves are similar but these are coming from the unsteadiness of measurement. Our curves have a local maximum but the stress decreasing is not so significant as show in Fig. 1.c. Spectacularly it happened only in our experiment at 20 [1/s] strain rate. We suppose that it is coming from the applied high strain rate which has higher effect with the amount of friction between the sample and the working tool of the device. In our case there is no dynamic recrystallization that can be seen in the microstructure in Fig. 10 and 11. Because of friction the trend of our curves in some cases jumps up from strain 0,9-1,2 values. At this stage at higher reduction of the specimen it is getting wider because of the material flow and there is bigger surface area of the specimens in contact with the tool.

4. Summary

There is wide range of applications of aluminium alloys that contain manganese alloying element, thanks to it's good workability and corrosion resistance. One possibility is to produce rolled products using semi-continuous cast ingots for it. The manganese has a feature of small solid solubility in the aluminium thus during solidification the intermetallic phase in the aluminium is configured with the iron. Some volume of manganese stays in solid solution. It is essential to heat up the cast ingot for hot rolling. In accordance with the pre-heat treatment before hot rolling the remaining manganese starts to precipitate from solid solution that has later high effect for the further material property. In our experiment we studied the effect of preheating (homogenizing heat treatment) and we carried out hot compression experiments to simulate the hot rolling in laboratory conditions. Different forming temperatures and strain rates were applied. Based on the recorded data and with experimental results three aluminium alloy were compared. During hot forming of aluminium-manganese alloys the dynamic recovery is the typical phenomena and in our experiment it was dominant. The nucleation of new grains and the recrystallization started only from 500 [°C] temperature with 0,2 [1/s] strain rate in case of every alloy. The stress result was smaller during hot compression test in case the of higher temperature homogenization. With increasing alloying elements of the specimens harder material is generated, however the homogenizing heat treatment has a stronger effect for the strength. Some alloying elements assist with the precipitation of AlMn intermetallic phases and in this case the difference can be higher.

Acknowledgement

This work has been carried out as part of the GINOP-2.2.1-15-2016-00018 project. The realization of this project is supported by the European Union and co-financed by the Hungarian Government.

References

[1] Köves E 1984 *Aluminium Kézikönyv* (Budapest: Műszaki könyvkiadó)
[2] George E. Totten and D.Scott MacKenzie 2003 *Handbook of Aluminium* (Physical Metallurgy and Processes vol 1) (New York: Marcel Dekker)
[3] George E. Totten and D.Scott MacKenzie 2003 *Handbook of Aluminium* (Alloy Production and Materials Manufacturing vol 2) (New York: Marcel Dekker)
[4] Cotterill P. and Mould P. R. 1982 *Recrystallization and Grain Growth in Metals*, (London: Surrey University Press)
[5] Verhoeven J. D. 1975 *Fundamentals for Phisical Metallurgy* (New York: Willey & Sons)
[6] Porter D. A. and Easterling K. E. 1996 *Phase Transformation in Metals and Alloys* (London: Chapman & Hall)
[7] Wilson P. 2013 *Recent developments in the study of recrystallization* (InTech)

IOP Conf. Series: Materials Science and Engineering **426** (2018) 012017 doi:10.1088/1757-899X/426/1/012017

The effect of Mn content on the texture of aluminium alloys in cold rolled and annealed state

A Hlavacs[1], D Petho[1], DA Angel[1], E Nagy[2], V Mertinger[1], M Benke[1]

[1] University of Miskolc, Institue of Physical Metallurgy, Metalforming and Nanotechnology, Miskolc-Egyetemváros, Hungary
[2] MTA ME Materiel Science Research Group, Miskolc-Egyetemváros, Hungary

E-mail: femhadri@uni-miskolc.hu

Abstract. The crystallographic texture greatly influences the properties of the cold rolled and hot rolled semi- products. As semi- products undergo further shaping (and heat treatment) before reaching their final condition, it is important to know their crystallographic texture. During our research, three different alloy aluminium alloys were tested (3003, 3103, unalloyed Al sheet) with different Mn content. In the first step the initial hot-rolled sheets were cold rolled to ~1 mm thickness without intermediate annealing. In the second step the sheets were cold rolled to 3 mm and annealed. In parallel, an intermediate annealing was applied at ~3 mm thickness for the 3103 alloy and subsequently cold rolled to 1 mm and annealed. The annealings were performed at 300°C for 3 hours in an air furnace. The resulting textures are characterized by the volume fractions of the main texture components, which were calculated from the orientation distribution function (ODF). The dominant texture components were determined for the different alloys in different states.

1. Introduction

When some fraction of the grains within a polycrystalline metallic structure is oriented in one or more preferred directions, the crystal structure is called textured. The isotropy of the macroscopic properties depend on the texture (for example Young's modulus, Poisson's ratio, formability, magnetic, electric properties etc.) [1, 2]. Thus, texture greatly influences the formability of semi-products therefore knowing the crystallographic texture is essential to. In most cases, the rolling process is followed by an annealing heat treatment. After improper annealing, the deformation texture may be preserved, furthermore, it can even be strengthened. For other cases, recrystallization texture can be developed which, in some cases can be stronger than the deformation texture [3-7]. In case of deep-drawing, the most spectacular failure caused by texture is earing. The semi-product used for deep drawing is prepared by the combination of rolling and annealing steps. The stronger is the crystallographic texture, the stronger the anisotropy will be and because of this, different magnitude of earing will occur at deep drawing [8, 9]. Aluminium alloys are the most commonly used materials for rolling. The 3xxx aluminium alloys are typically used as raw materials for deep-drawing technologies. The aim of this research is to describe the effect of Mn content on the variation of crystallographic texture during cold rolling and annealing treatments of aluminum sheets.

2. Experiments

A research plan was established to investigate the texture variation through the applied thermomechanical treatments of the examined Al alloys. Figure. 1 shows the established research plan. The anisotropy of the cold rolled sheets and the softened sheets were examined by X-ray diffraction.

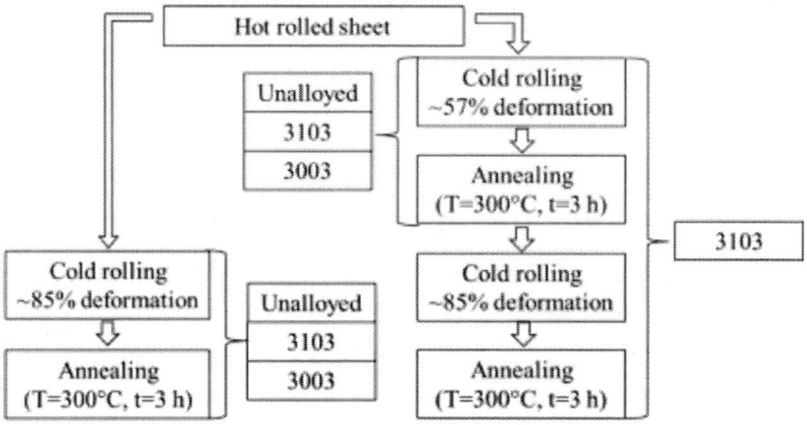

Figure 1. Research plan

2.1. Rolling

The examined aluminium sheets were obtained in hot rolled state with ~8 mm (unalloyed), 6 mm (3003) and 7 mm (3103) thickness from Arconic-Köfém Ltd. The sheets were cold rolled in multiple rolling steps to different thicknesses using a Von Roll experimental duo roll stand. The same deformations were used for all three raw materials. The deformations were calculated with eq (1).

$$D = \frac{h_0 - h}{h_0} * 100 \quad (1)$$

Table 1. shows the number of rolling passes (marking) and the thickness of the examined samples. The 0 passes correspond to the initial, hot rolled sample.

Table 1. The rolling steps and the thickness of the samples during cold rolling

	Number of passes	**0**	**1**	**2**	**3**	**4**	**5**	**6**	**7**	**8**
	Deformation [%]		**17,86**	**11,30**	**18,63**	**25,78**	**16,88**	**19,53**	**24,76**	**30,32**
Unalloyed	Thickness [mm]	8	6,57	5,83	4,74	3,52	2,93	2,35	1,77	1,23
3003	Thickness [mm]	6	4,93	4,37	3,56	2,64	2,19	1,77	1,33	0,93
3103	Thickness [mm]	7	5,75	5,10	4,15	3,08	2,56	2,06	1,55	1,08

2.2. Annealing

After the cold rolling, the sheets were annealed at 300°C for 3 hours in an air furnace. In the case of sheets without intermediate annealing, the annealing was the final step. For samples treated with intermediate annealing, the annealing was also performed after the fourth rolling step.

2.3. Texture examination

The texture examinations were carried out with a Bruker D8 Advance X-ray diffractometer equipped with an Eulerian cradle. The examination parameters were: CoKα radiation, 40 kV tube voltage 40 mA tube current. The samples were tilted from 0° to 75°. The defocusing (absorption) error was corrected with aluminium powder. The measured pole figures were evaluated and the ODF was synthetized using the equipment's own, TexEval software. Volume fractions of the main rolling and recrystallization texture components were calculated with 15° deviation (Δ). Table 2. summarizes the main rolling and recrystallization texture components with their Miller indices and Euler angles according to the Bunge system [9-11].

Table 2. The designations, Miller indices and Euler angles of the main texture components after rolling and recrystallization (approximated) [9-11]

Designation	Miller indices {*hkl*}<*uvw*>	Euler angles φ_1	Φ	φ_2	
C	{112}<111>	90°	30°	45°	*Rolling Texture Components*
S	{123}<634>	59°	34°	65°	
B	{011}<211>	35°	45°	0°/90°	
G	{011}<100>	0°	45°	0°/90°	*Recrystallization Texture Components*
Cube	{001}<100>	0°	0°	0°/90°	
R	{124}<211>	53°	36°	60°	
P	{011}<122>	65°	45°	0°/90°	

3. Results

3.1. The effect of Mn content

Figure 2. shows the texture component volume fractions of the unalloyed aluminium sheet in the cold-rolled and annealed state (thickness is 3 and 1 mm). It can be observed that the rolling texture-components are dominated after cold rolling. After higher deformation, a stronger rolling texture was obtained (1 mm thickness versus 3 mm thickness). In the case of 3 mm thickness, the volume fraction of the strongest recrystallization (Cube) and rolling (C) texture components are roughly the same after annealing.

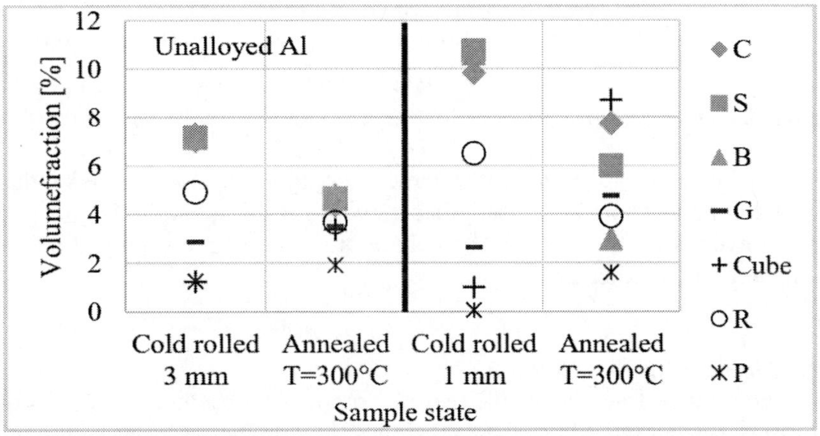

Figure 2. The volume fractions of texture components in the case of unalloyed aluminium

Figure 3 shows the results of the 3003 aluminium sheet in the cold-rolled and annealed state (thickness is 3 and 1 mm). The volume fractions of rolling texture-components are getting stronger as the deformation increases. After annealing, the values of the volume fractions of rolling and recrystallized components approach each other.

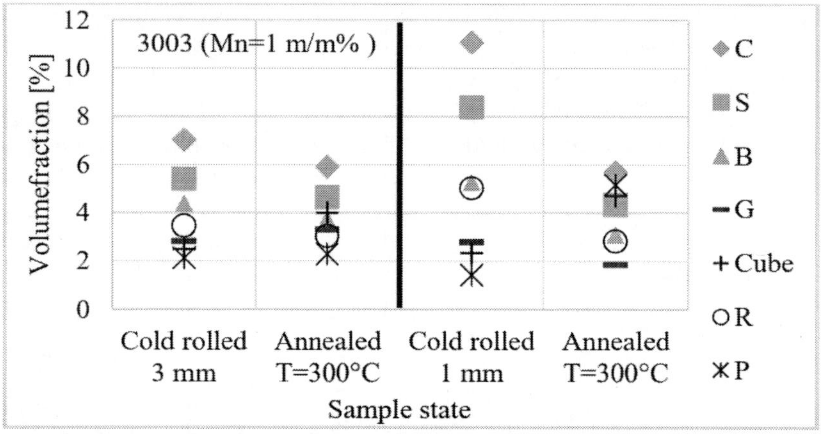

Figure 3. The volume fractions of texture-components in the case of 3003 aluminium

Figure 4 shows the texture components of the 3103 aluminium sheet in the cold-rolled and annealed state (thickness is 3 and 1 mm). At 3 mm thickness, a weak texture is formed after cold rolling. In the case of 1 mm thickness, a high rolling texture is obtained. After annealing, the values of the volume fractions of rolling and recrystallized components approach each other.

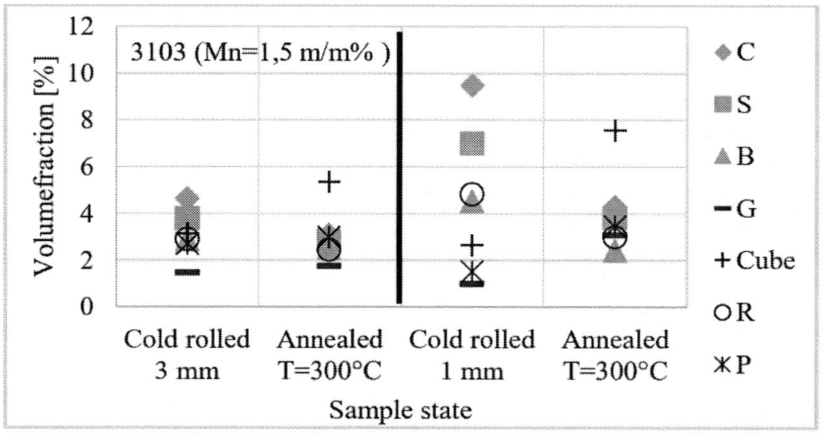

Figure 4. The volume fractions of texture-components in the case of 3103 aluminium

Comparing Figs. 3-5 it can be seen that as the Mn content increases to 1.5 m/m%, the volume fractions of rolling texture components are decreased, thus, a weaker rolling texture is obtained compared to the other alloys with lower Mn content. Also, the volume fractions of recrystallization texture components are smaller compared to the other two alloys.

3.2. The effect of intermediate annealing

Figure 5 shows the volume fractions of the texture components on the 1 mm sheet which was treated without intermediate annealing, while Figure 6 shows the volume fractions of the texture components on the 1 mm sheet which was treated by intermediate annealing. After rolling steps, a weaker rolling texture formed compared to the previous case. After annealing the Cube texture component is dominated, but it is weaker than the previous case.

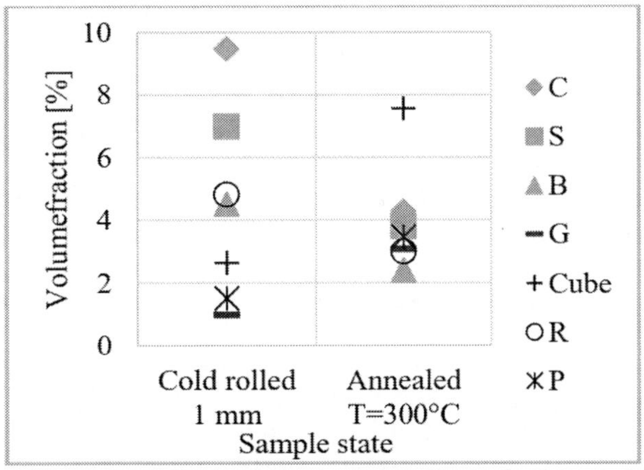

Figure 5. The volume fractions of the texture components on the 1 mm sheet which was treated without intermediate annealing

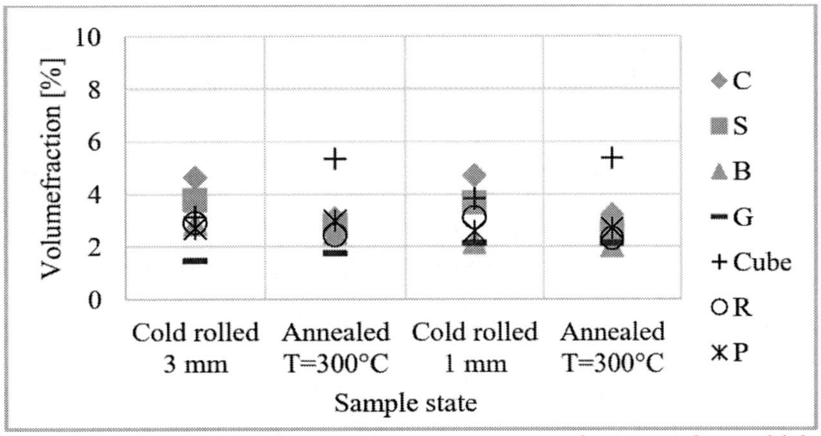

Figure 6. The volume fractions of the texture components on the 1 mm sheet which was treated with intermediate annealing

4. Conclusions

Based on the results it can be stated that the increase of the Mn content decreases the volume fraction of rolling texture components after cold rolling. Also, it decreases the volume fractions of recrystallization texture components after each thermomechanical treatment scenario. The Cube texture component is the most characteristic of recrystallization texture components for all three alloys.
With intermediate annealing, a smaller texture, that is, favourable rolling and recrystallization textures can be achieved for subsequent applications compared to treatment without intermediate annealing.

Acknowledgement

The described article was carried out as part of the GINOP-2.2.1-15-2016-00018 project implemented in the framework of the Szechenyi 2020 program. The realization of this project is supported by the European Union.
The described research work was supported through the National Research, Development and Innovation Office – NKFIH K119566 project.

References

[1] Dr. Bárczy P and Dr. Fusch E 1981 Metallográfia I. *Tankönyvkiadó*
[2] Kocks U F, Tomé C N and Wenk H-R 1998 Texture and Anisotropy *Cambridge University Press*
[3] Wagner P and Lücke K 1994 Quantitative correlation of texture and earing in Al.alloys *Materials Science Forum* Vols 157-163 pp 2043-2050
[4] Albou A, Raveendra S, Karajagikar P, Samajdra I, Maurice C and Driver J H 2010 Direct correlation of deformation microstructures and cube recrystallization nucleation in aluminium *Scripta Materialia* 62 pp 469-472
[5] Liu W C and Morris J G 2007 Recrystallization textures of the M{113}<110> and P{011}<455> orientations in a supersatureted Al-Mn alloy *Scipta Materialia* 56 pp 217-220
[6] Liu W C, Li Z and Man C-S 2008 Effect of heating rate in the microstructure and texture of continuou cast AA 3105 aluminium alloy *Materials Science and Engineering A* 478 pp 173-180
[7] Oscarsson A and Bate P 1996 Texture variations through the thickness of AA3004 and its effect on earing behaviour *Material Science Forum* Vols. 217-222 pp 559-564
[8] Engler O, Löchte L and Hirsch J 2007 Through-process simulation of texture and preporties during the thermomechanical processing of aluminium sheets *Acta Materialia* 55 pp 5449-5463
[9] Engler O and Hirsch J 2002 Texure control by thermomechanical processing of AA6xxx Al-Mg-Si sheet alloys for automotive applications- a review *Materials Science and Engineering* A336 pp 249-262
[10] Kocks U F, Tomé C N and Wenk H-R 1998 Texture and Anisotropy Preferred Orientations in Polycrystals and their Effect on Materials Properties *Cambridge University Press*
[11] Engler O and Randle V 2010 Introduction to Texture Analysis Macrotexture *Microtexture and Orientation Mapping* Second edition CRC Press

IOP Conf. Series: Materials Science and Engineering **426** (2018) 012018 doi:10.1088/1757-899X/426/1/012018

The mechanical properties of polyethylene-terephthalate (PET) and polylactic-acid (PDLLA and PLLA), the influence of material structure on forming

T Horvath[1], M Kalman[2], T Szabo[3], K Roman[4], G Zsoldos[5] and M Szabone Kollar[6]

[1] Research and Development Engineer at ÉMK LTD, Sajóbábony, Hungary
[2] Professor of Institute of Ceramic and Polymer Engineering, University Of Miskolc, Miskolc-Egyetemváros, Hungary
[3] Associate Professor and Director of Institute of Ceramic and Polymer Engineering, University Of Miskolc, Miskolc-Egyetemváros, Hungary
[4] PhD student, University Of Miskolc, Miskolc-Egyetemváros, Hungary
[5] Senior Lecturer of Institute of Ceramic and Polymer Engineering, University Of Miskolc, Miskolc-Egyetemváros, Hungary
[6] Senior Lecturer of Institute of Ceramic and Polymer Engineering, University Of Miskolc, Miskolc-Egyetemváros, Hungary

E-mail: horvath.tibor@emkkft.hu

Abstract. Those plastics that have been used by food industry as packaging (especially in sweet industry) have strict requirements. Most of these plastic constructions are primary or secondary packaging with direct or indirect contact with foodstuff. Over the appropriate mechanical properties, these packagings need to be safe enough to be used as direct contact material without any dangerous material migration or transfer. Recently the polyethylene-terephthalate (PET) is that material which is widely used for this kind of purpose and application. To substitute the PET, only that kind of alternative material which are fulfilling the requirements of applications and compliant to the relevant regulations can be applied.
The cost-efficient application of plastics strongly depends on the mechanical properties of the material. In light of these facts, both types of polylactic-acid (PDLLA, PLLA) and the PET were tested. Comparing their mechanical properties PLA could be used as a real alternative of PET.

1. Introduction

In the new generation of packaging materials the environmental friendly approach is more highlighted than in the last few decades. Over that, the growing importance of food safety initiated new requirements and regulations to the packaging materials. Although the traditional packaging still can be used but new alternatives need to be found which is more suitable to the getting stricter regulations. The scoped field is the sweet industry which uses a huge amount of plastic as primary (those packagings which is part of individual product) or secondary (use for logistic purposes) packaging construction with non-food contact/indirect food contact/direct food contact types.

Regarding the plastic packaging constructions due to the new challenges the best alternatives could be the bio-sourced biodegradable plastic types.

"In general the biobased polymers can be divided into three main categories like a.) Polymers directly extracted/removed from biomass; b.) Polymers produced by classical chemical synthesis using renewable biobased monomers; c.) Polymers produced by microorganisms or genetically modified bacteria. With classical chemical synthesis (category b.) a wide spectrum of bio-polyesters can be produced" [1]. The polylactic-acid (PLA) is one of these numerous alternatives.

From environmental performance point of view the PLA (compared to PET) has lower values in almost all impact categories (CO_2 emission, human toxicity, fossil energy demand) except of eutrophication and acidification potential [2].

As the scoped company mostly uses virgin polyethylene-terephthalate (PET) a special alternative was requested to be found with appropriate mechanical properties. The lactic acid based PLA is a linear aliphatic thermoplastic polyester. The lactic acid can be obtained from fermentation of plant sourced starches. The extracted starches chemically converted into dextrose that fermented to lactic acid followed by polycondensation or Ring Opening Polymerization (ROP) process to produce polylactic-acid. Two stereoisomers exists for lactic as L and D isomers, therefore three PLA alternatives distinguished as PDLA made from D lactic, PDLLA which contains both L and D isomers, and pure L isomer built PLLA. The PLA is biodegradable, renewable, environmental friendly material and has good thermal processability so it seemed to be a potential substitute of PET but to verify the usability of PLA, the mechanical properties of both materials had to be compared [3, 4].

Table 1. Physical and chemical parameters of the PLA and PET foils [5]

Properties	PDLA	PLLA	PDLLA	PET
Crystalline structure	Crystalline	Semi crystalline	Amorphous	Amorphous and semi-crystalline
Melting temperature (T_m)/ °C	180	180	variable	260
Glass transition temperature (T_g)/ °C	50-60	50-60	variable	67-81
Half-life in 37°C normal saline	4-6 month	4-6 month	2-3 month	700 year
Elongation at break/ (%)	20-30	20-30	variable	230
Breaking strength/ (g/d)	4-5	5-6	variable	5,3

2. Experimental

2.1. Material and methods

PET (virgin and recycled) and PDLLA samples were provided by Pro-Form Kft. and the PLLA sample was purchased from Goodfellow Cambridge Ltd. The thickness of the preformed PDLLA and preformed PET film were 300 μm, in case of PLLA foil the thickness was 50 μm. The glass transition temperatures (T_g) of materials have been preliminary determined by DMA analysis with results as 57°C for PDLLA and 62°C for PLLA as well as 82°C for polyethylene terephthalate. For visual examination and follow up of deformation range the PET film surface screen was printed with a 1 mm splitting grid. In case of PDLLA film a 10 mm splitting grid was used. In the pre-conditioning phase of test all samples have been pre-heated. The pre-treatment temperatures (to release the remained forming stress in material) were slightly above the specific transition temperatures, 100°C for PET, 70°C for PDLLA and PLLA.

IOP Conf. Series: Materials Science and Engineering **426** (2018) 012018 doi:10.1088/1757-899X/426/1/012018

The preheating time was 2 minutes for all of three samples. After the pre-treatment the mechanical measurements were carried out at room temperature.

The mechanical properties were determined with tensile test. The test performed on INSTRON 5566 testing machine. The measurements procedure followed the ASTM D389 standard. The test speed was 100 mm/min at room temperature (23±1°C). During the tests the 150%, 200% and 250% stretching levels were determined and adjusted. For purpose of further characterization of the materials DSC131 evo machine was used with 10°C as heating/cooling rate.

2.2. Comparison of mechanical properties of polyethylene terephthalate and polylactic acid

During the mechanical tests, the PET, PDLLA and PLLA samples were initially formed in a relatively high force range. In the first stage of the shaping the weak secondary bonds that cross connected the molecular chains are progressively disrupted and the collapsed molecular chains were oriented in the direction of forming force. This orientation process could be also observed in the further stages of tests but to ensure the adjusted stretch value smaller force was sufficient to realize it. At higher forming ratios (deformation from 200% up to 250%) the molecular chains reached the maximum of possible orientation. The final stage of mechanical test was the plastic material's creeping which was realised by the movement of mass centre of molecular chains. At lower forming ratio only the shorter molecular chains' mass centre could be moved (PET) while the longer ones (PLLA, PDLLA) stayed in positions. It was concluded that the PET material permanent creeping at lower forming force was caused mainly by the structure that built by shorter molecular chains.

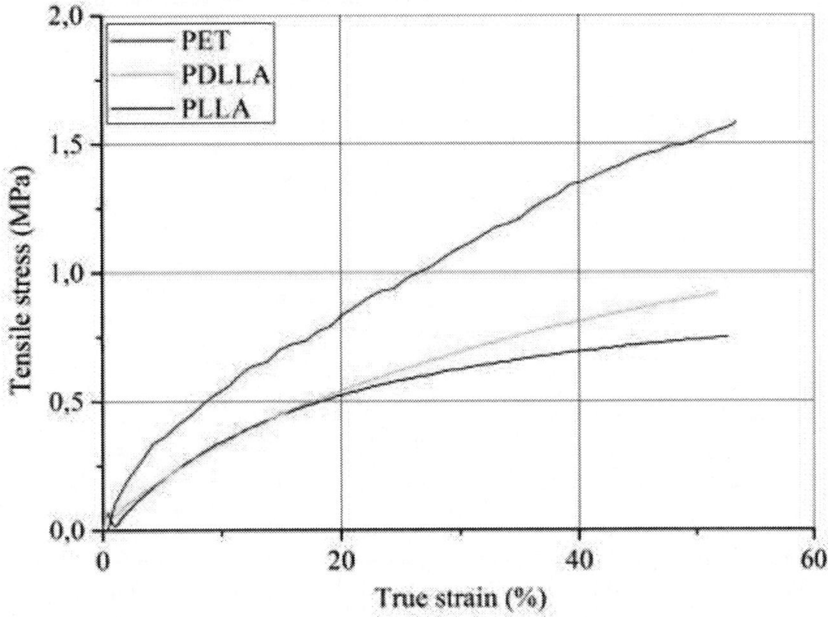

Figure 1. Results of tensile test of PET, PDLLA and PLLA foils at deformation of 150%

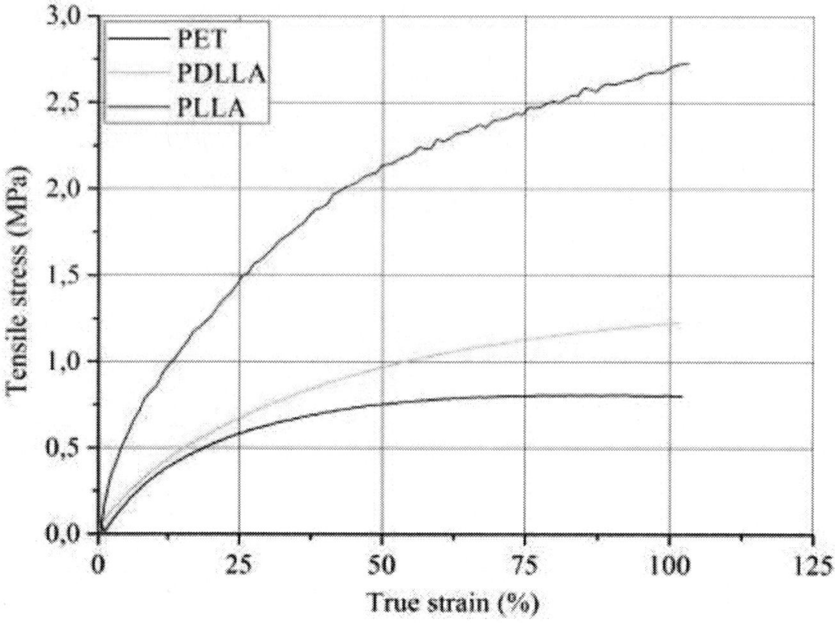

Figure 2. Results of tensile test of PET, PDLLA and PLLA foils at deformation of 200%

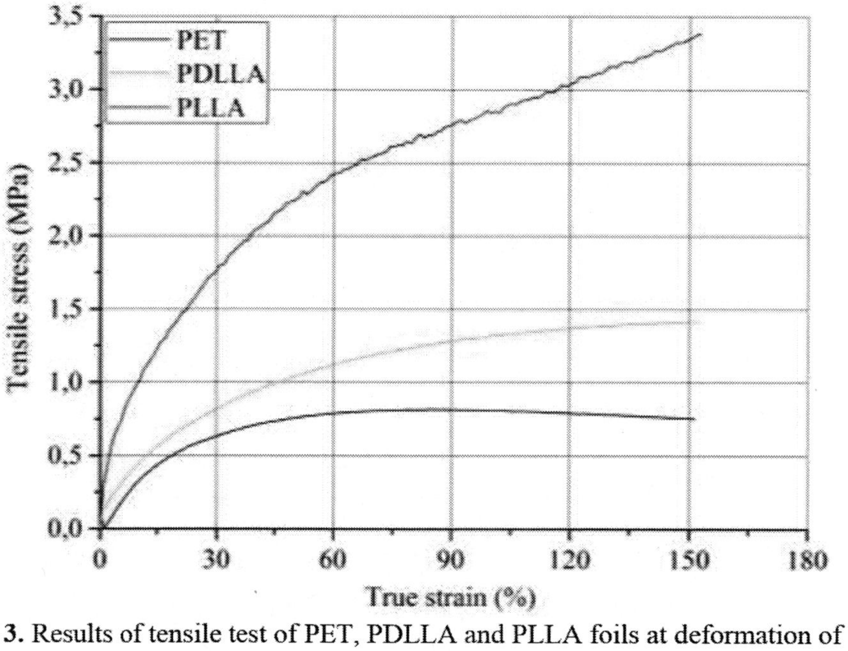

Figure 3. Results of tensile test of PET, PDLLA and PLLA foils at deformation of 250%

The result of tests defined that the PLLA required the highest forming force. The PLLA needs 224% (average) of force, the PDLLA needs 52% (average) higher forces to reach the same deformation value compared to PET.

Table 2. Average molar mass of PET [6], PDLLA [7] and PLLA [8] foils

Properties	PET	PDLLA	PLLA
Average molar mass (M_w)/ g/mol	30 000 – 80 000	100 000 – 180 000	100 000 - 300 000

This referred to longer molecular chains as well as better filling of amorphous parts.

It can also be caused by the partially crystalline structure. Larger filler amorphous molecules and highly folded crystalline parts need greater forces to form orientation.

Table 3. Tensile stress maximums of PET, PDLLA and PLLA foils

Strain %	PET (MPa)	PDLLA (MPa)	PLLA (MPa)
150	0.74	0.90	1.51
200	0.80	1.22	2.71
250	0.78	1.42	3.36

2.3. DSC analysis of polyethylene terephthalate and polylactic acid

During the function analysis the PET and PDLLA and the pure L isomer built PLLA samples were tested on DSC131 evo machine. Regarding the PDLLA this material contains both D and L lactic acid isomers and constructed dual crystal structure under appropriate conditions. However, this dual crystal structure is also disadvantage of the material, as the different isomers cannot build a homogeneous crystal structure. It makes crystallization difficult and the whole process slow.

Table 4. DSC measurement data of analysis of PET, PDLLA and PLLA foils

Material Type	**PET**	**PDLLA**	**PLLA**
Heating/Cooling ratio (°C/min)	10	10	10
Heat Range (°C)	25 - 275	25 - 275	-125 - 245
Heat Ramp	Linear	Linear	Linear

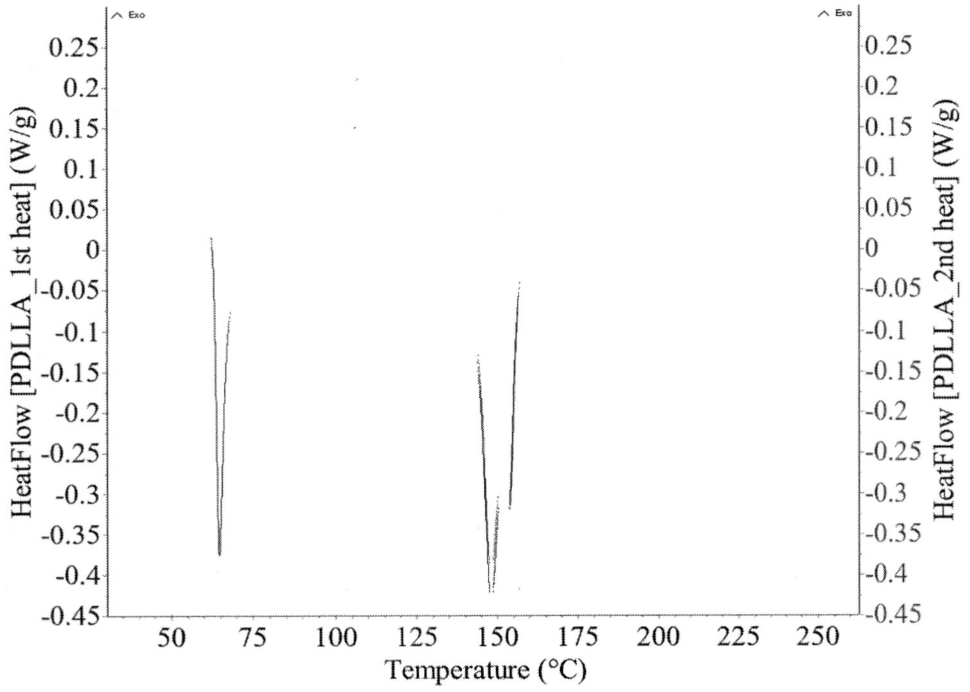

Figure 4. DSC curves of PDLLA foil (section I. first heating run section II. second heating run)

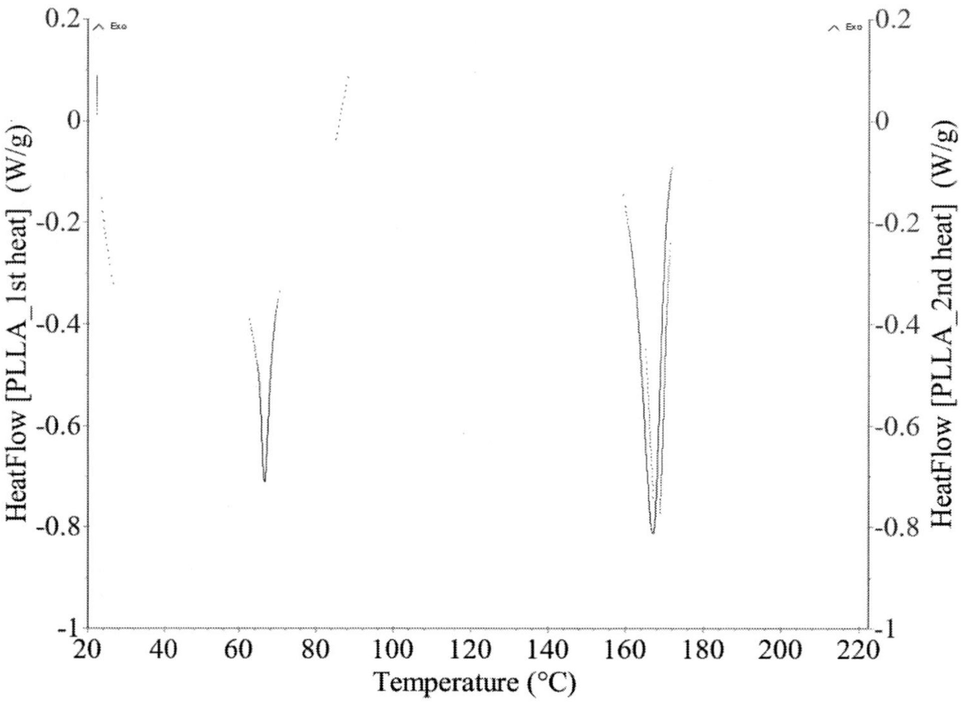

Figure 5. DSC curves of PLLA foil (section I. first heating run section II. second heating run)

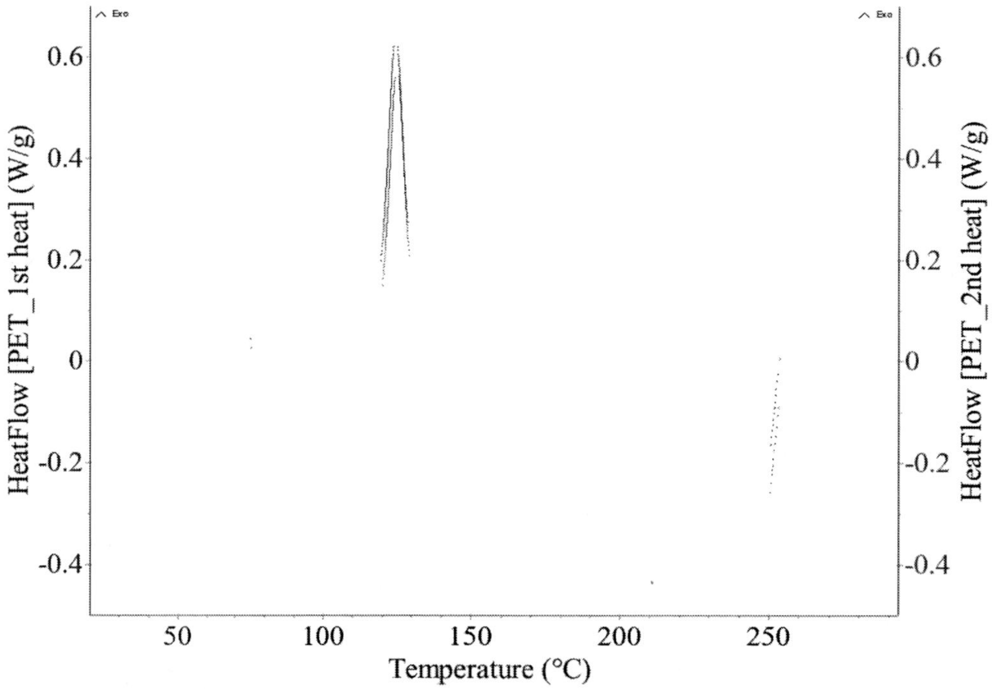

Figure 6. DSC curves of PET foil (section I. first heating run section II. second heating run)

The measurements showed that the material contained two distinct crystalline fractions of heterogeneous crystal structure. This is confirmed by the two sub-functions obtained as a result of the separation and partitioning of the decomposition curve. During the DSC analysis, it was also found that the crystalline fraction of the material is constantly changing in proportion due to the degree of deformation. Theoretically the purity of the material (proportion of virgin / recycled material contents) also significantly influences the crystal structure of the material but this aspect is going to be investigated only at further stage of recent research [9].

Table 5. Results of DSC analysis of PET, PDLLA and PLLA foils

Material Type	PET	PDLLA	PLLA
Heat of crystallisation (J/g)	-27.03	-23.55	-24.07
Onset temperature of crystallisation (°C)	130.54	101.47	80.44
Heat of decrystallisation (J/g)	38.19	26.80	37.03
Onset temperature of decrystallisation (°C)	230.56	143.06	159.77
Percentage of crystallinity (%)	7.97	3.5	13.94

The pre-forming of the material has effect on the material's structure as due to the stretch, the collapsed molecules are oriented towards the moulding force. However, the mentioned orientation supports the formation of a crystalline structure at appropriate heat level, but above a boundary the crystalline structure also breaks down as the folded molecular chains are partially oriented in the axis of the forming force.

The rate of cooling also highly influences the degree of crystallization because the high rate cooling prevents the crystallization. With lower cooling speed, higher the ratio of the crystalline fraction. Thus, by stretching the material, orientation can be evolved in the material structure, which supports the formation of the crystalline structure during the cooling [10-14].

3. Results

Mechanical- and structural properties of the PET and PLA foils (PDLLA, PLLA) were tested and determined by mechanical and DSC analysis. Based on the experimental analysis, the PLA's (polylactic-acid) better material structure is confirmed to ensure better mechanical properties to forming it in higher load range compared to the PET. Regarding the material structure it was found that the PLLA and PET structures were more stable than the PDLLA. As results of DSC analysis the PDLLA contains both lactic-acid (L and D) isomers, which causes irregularity of molecular chains that highly influences to the final material structure and physical properties. In case of PLLA the crystallization process is more easily controllable as this material contains only L type of lactic-acid isomers. Due to this the physical cross-linking between the molecular chains ensuring more stable material structure of PLLA. That is why the mechanical properties (tensile strength, stiffness) of PLLA are higher compared to PDLLA.

Recently the PET is the most popular plastic in the packaging area of sweet industry. The mechanical analysis confirmed that the PDLA and PLLA can be used as a packaging material for foodstuff because this PLA material is fulfilling the complex requirements of primary and secondary packagings of sweet industry. As the PLA plastic is synthesized from bio-sources, its degradation time is relatively shorter, it is more environmental friendly solution compared to the PET.

4. Conclusion

The PLA as bio-sourced and biodegradable material was found as applicable alternative of recently and widely used PET material in food industry area. Due to the different material structures which have been confirmed during the analysis, the PLA has better mechanical parameters than PET material does as the PLLA needs 224% (average) of force, the PDLLA needs 52% (average) higher forces to reach the same deformation value compared to PET. In addition the PLA's environmental performance is better compared to the PET as the PLA has lower overall environmental impacts.

In case of PLA, the PLLA which contains only L type of lactic-acid isomers was found as more advantageous material compared to the L and D type isomers built PDLLA. The reason of difference is the basic structures diversity of materials as the percentage of crystallinity was 13.94% for PLLA and 7.97% for PDLLA which ensuring better mechanical parameters to PLLA. The sweet industry mainly uses the polymers as primary (direct/indirect food contact) and secondary (indirect food contact) packagings of the products therefore both types of PLA can be used for the same purpose and application.

References

[1] Weber C J, Haugaard V, Festern R, Bertelsen G 2002 Production and applications of biobased packaging materials for the food industry *Food Additives and Contaminants,* 172-177

[2] Papong S, Malakul P, Trungkavashirakum R, Wenunun P, Chom-in T, Nithitanakul M, Sarobol E 2014 Comparative assesment of the environmental profile of PLA and PET drinking water bottles from a life cycle perspective *Journal of Cleaner Production,* **65.,** 539-550

[3] Buzási L 2016 A műanyag csomagolószer gyártás helyzete Magyarországon *Polimerek,* **2. évf,** ***9* *szám.***

[4] Muller J, González-Martínez C, Chiralt A 2017 Combination of Poly(lactic) Acid and Starch for Biodegradable Food Packaging *Materials (Basel)*, **10 (8),** 952-974

[5] Xiao L, Wang B, Yang G, Gauthier M 2012 Poly(Lactic Acid)-Based Biomaterials: Synthesis, Modification and Applications, *Biomedical Science, Engineering and Technology,* 247-282

[6] Dr. Bánhegyi Gy 2005 Poli(etilén-tereftalát) (PET) újrafeldolgozása a tulajdonságok javításával – *Műanyagipari szemle* **5.**

[7] http://www.bmg-inc.com/en/prod_and_res/bio_degmer/pdlla.html

[8] Ikada Y, Tsuji H 2000 Biodegradable polyesters for medical and ecological applications, *Macromol. Rapid Commun.* **21,** 117–132

[9] Ehrenstein G W, Riedel G, Trawiel P 2004 Thermal Analysis Of Plastics *Carl Hanser Verlag GmbH & Co.KG, Munich.* 236-275

[10] Dr. Czél Gy, Kollár M 2008 Anyagvizsgálati Praktikum *Sunplast,*

[11] Bodnár I 2002 Potenciálisan biodegradábilis, politejsav bázisú polimerek szintézise és vizsgálata - Doktori (Ph. D.) értekezés *Debreceni Egyetem, Alkalmazott Kémiai Tanszék, Debrecen*

[12] Pukánszky B 1995 Műanyagok *Budapesti Műszaki Egyetem Vegyészmérnöki Kar, Műegyetemi Kiadó*

[13] D. Battegazzore, S. Bocchini, A. Frache 2011 Crystallization kinetics of poly(lactic acid)-talc composites *eXPRESS Polymer Letters* **5 (10),** 849-858

[14] Zsoldos G 2012 UHMWPE - Biopolimer Felületének módosítása polimerizációs technológiákkal – PhD értekezés *Miskolci Egyetem*

IOP Conf. Series: Materials Science and Engineering **426** (2018) 012019 doi:10.1088/1757-899X/426/1/012019

The investigation of drilling an aluminium matrix composite reinforced with iron hollow spheres

R Horváth[1], E Nagyné Halász[2], F Oláh[3]

[1]Associate professor, [2]Master teacher, [3]Demonstrator, Óbuda University, Donát Bánki Faculty of Mechanical and Safety Engineering, Institute of Materials and Manufacturing Sciences, Budapest, Hungary

E-mail: horvath.richard@bgk.uni-obuda.hu

Abstract. Since composites have numerous advantages, they have a prominent role in manufacturing. That is the reason why they might substitute conventionally used materials. Nowadays cutting processes are among the most commonly used finishing technologies. In the case of composites the cutting tools have to meet particular requirements. This paper represents the machinability examinations of a special type of composites – iron hollow sphere-filled aluminium alloy matrix (AlSi12) – during drilling. The input cutting parameters, the feed (*f*, mm) and the cutting speed (v_c, m/min), were varied at a wide range according to a design of experiment (a full factorial design was used). Drilling studies were carried out under dry conditions and with internal cooling lubrication as well. The axial (feed) force (F_f, N) was measured. Based on the experimental results, the different drilling conditions were compared. In addition, a proposal is made for cutting this kind of material. The joint properties of the aluminium matrix and the iron hollow spheres during cutting were examined by scanning electron microscopy.

1. Introduction

Syntactic metal foams possess both the characteristics of known metal foams and the advantageous properties of composites. These materials have been researched more and more recently. They have excellent mechanical properties. Their energy absorption, compressive strength and damage localization capability are outstanding, while their density is low, thanks to the light metal matrix and the hollow spheres.

In an aluminium matrix the reinforcing material can be basically divided into two groups: metal and ceramic hollow spheres [6]. For example, Hollomet Gmbh. [2] sells ceramic hollow spheres (Globocer, GC), and iron hollow spheres (Globomet, GM).

The material properties of such materials are much researched [7]. The upsettability of these materials is very good. Orbulov and Ginsztler [10] examined the upsettability of three different metal matrix syntactic foams (the matrix was Al99.5 and reinforcement was three different types of ceramic hollow spheres). The three kinds of spheres mostly differed from each other in diameter (100 μm; 150 μm; 1450 μm). The results showed that the size of the spheres and the temperature of testing had a considerable effect on compressive strength. Smaller thin-walled hollow spheres produced higher compressive strength. The tests were carried out at room temperature and at 220 °C. The higher temperature reduced compressive strength by about 30%. Kiser et al. [4] investigated the upsettability of composites of aluminium matrix and ceramic hollow sphere reinforcement. They used a composite of A201 aluminium alloy and 55% Al_2O_3 ceramic hollow spheres, and a composite of A360 aluminium matrix and 60% ceramic hollow spheres. Their results showed that the energy absorption capability of the tested composites largely depend on the parameters and conditions of loading. Such materials can be very effectively used against local penetration, such as bullets. Szlancsik et al. [12] examined the upsettability of composites of different matrix materials (Al99.5, AlSi12, AlMgSi1 and AlCu5) and Globomet reinforcement. The obtained upsetting curves (gentle slopes) indicate that the composite has high energy absorption capability. They also found that the matrix material and heat treatment has a considerable effect on the mechanical properties of the composite, therefore with the proper choice of matrix and heat treatment, mechanical properties can be varied in a wide range according to requirements. Orbulov and Ginsztler [11] examined the upsettability of composites of four matrix materials (Al99.5, AlSi12, AlMgSi1 and AlCu5) and ceramic hollow sphere reinforcements of two diameters (100 μm, 150 μm). They also found that smaller spheres lend higher compression strength and stiffness to the composite than larger spheres.

Karabulut and Karakoç [3] examined the machinability (milling) of open-cell aluminium alloy-based SiC foam composite with an uncoated carbide tool. The tests were conducted with the use of a Taguchi L_{27} full-factor orthogonal array and milling parameters were optimized for surface roughness. The results showed that the feed rate was the milling parameter that affected surface roughness most significantly. Krajewski & Nowacki [5] examined the geometric structure of an AlSi-SiC composite after EDM cutting, waterjet cutting (with and without abrasive blasting), laser and air plasma cutting, and cutting with a band saw and circular saw. They characterized the surfaces with numerous 2D and 3D roughness parameters and based on the results, they determined optimal cutting conditions. Fakhri et al. [1] examined the relationship between porosity and cutting force. They found that the resultant cutting force has high correlation with porosity (ratio of solid material). They also examined the behaviour of aluminium foam with unconventional cutting technology. B.S. Yilbas et al. examined the size of cut surfaces in the case of laser cutting with triangles [13] and circular holes [14].

This paper presents the drilling tests of a eutectic aluminium matrix metal hollow sphere-reinforced composite (in dry conditions, and with internal cooling and lubrication.

2. Materials and tools used

We used a eutectic Aluminium-Silicon alloy (AlSi12) as matrix and Globomet of 100% Fe as reinforcement. The aluminium alloy consisted of Al = 87.1 vol%, Si = 12.8 vol% and Fe = 0.1 vol%. The iron hollow spheres were Globomets [2] sold by Hollomet GmbH, and had a characteristic diameter of 1.4–1.8 mm and a wall thickness of 0.2 mm. Manufacturing was done with infiltration [8]. The finished specimens (Figure 1) had outer sizes of 35 × 45 × 85 mm.

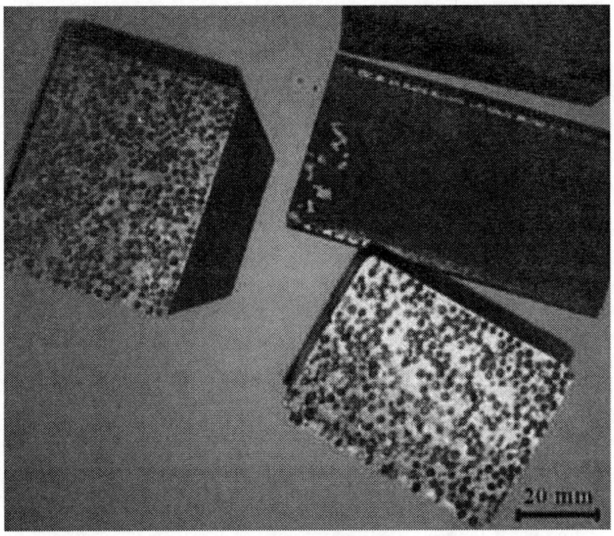

Figure 1. Specimens prepared for the drilling test

The cutting tests were performed with a MAZAK VCN 410A-II vertical machining centre (n_{max} = 12000 1/min; P_{max} = 11 kW).

The drilling test were executed with Dream Drills D5433100 5XD uncoated carbide tools [9] (Ø10 mm), which have cooling holes.

Forces in the feed direction were measured with a Kistler 9257b multicomponent dynamometer and a KISTLER 5019 multichannel charge amplifier. The measured forces were processed with the DynoWare program.

The cutting parameters (both the cutting parameters of drilling and cutting speed – v_c, m/min, and feed – f, mm were varied at four levels: v_c = 80, 120, 160, 200 m/min; f = 0.1, 0.2, 0.3, 0.4 mm. The 16 measurement points obtained this way can be seen in Table 1. The cutting tests were performed both dry and with internal cooling and lubrication.

Table 1. The measurement points of drilling experiments and their cutting parameters

Experimental runs	v_c (m/min)	f (mm)
1.	80	0.1
2.	80	0.2
3.	80	0.3
4.	80	0.4
5.	120	0.1
6.	120	0.2
7.	120	0.3
8.	120	0.4
9.	160	0.1
10	160	0.2
11.	160	0.3
12.	160	0.4
13.	200	0.1
14.	200	0.2
15.	200	0.3
16.	200	0.4

A 4 vol% oil emulsion (AGIP AQUAMET 4 HS-BAF) was used as coolant and lubricant.

3. Results And Discussion

In the drilling tests we recorded F_f forces in direction *z*. We performed a significance test to see which cutting parameters have a significant effect on the measured F_f force component. The test showed that only feed had a significant effect on the F_f force component. For this reason, this paper presents the analysis of the effect of feed.

3.1 The effect of feed on forces in the feed direction

Figure 2 shows the average of the measured forces when lubricant-coolant was used and in dry conditions. It can be clearly seen that the forces are much higher in dry cutting than when lubricant-coolant was used.

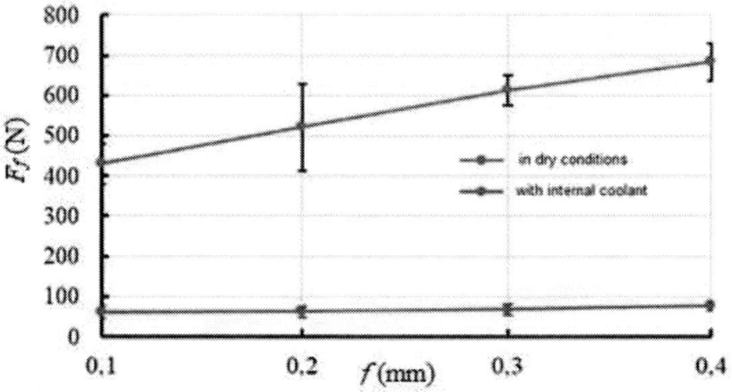

Figure 2. Average F_f force values as a function of f feed

A paired t-test was performed on the data obtained, which showed that on average the force requirement of feed is 484.8 N higher in the case of dry cutting (Figure 3).

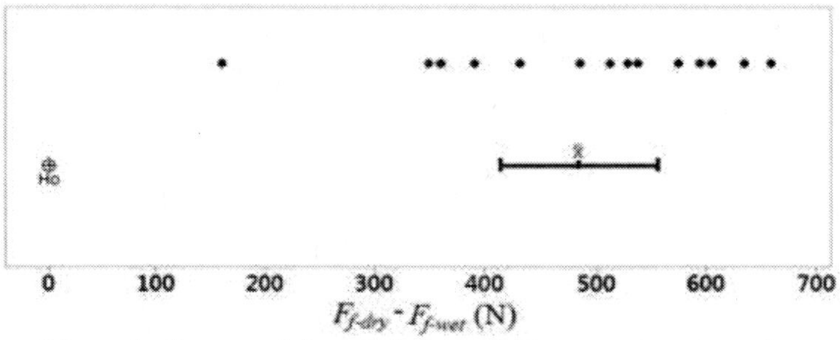

Figure 3. Results of the paired t-test of the axial force differences between the wet and dry drilling

3.2 Built Up Edge (BUE) and burr formation analysis

After finishing the holes, the tool was checked and it was found that during dry cutting a BUE was always found, while when the tool was lubricated, BUE wasn't formed. Figure 4 shows the tool with the BUE and the lubricated tool without a BUE.

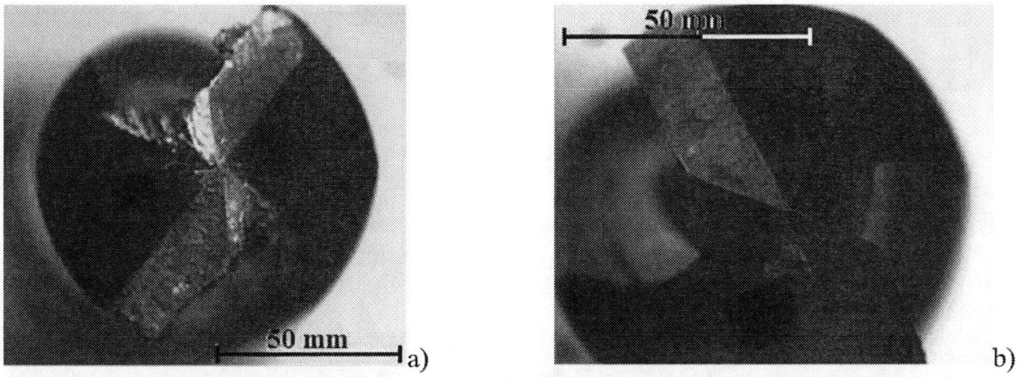

Figure 4. Built Up-Edge: a) on the tool with BUE after the dry drilling process and b) the tool without it after lubricated drilling

After this the holes and the burrs were inspected on the side the tool exited the sample. During dry cutting, burr was formed at this side in every case, but with a coolant-lubricant this did not happen (Figure 5).

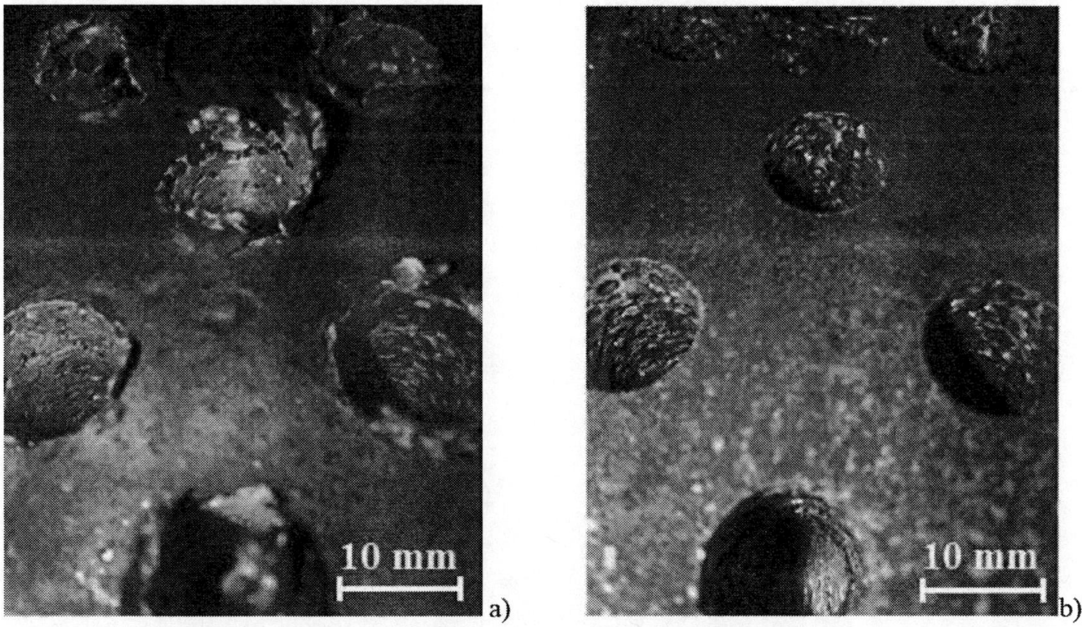

Figure 5. Burr on the exit sides of the holes: a) during dry conditions and b) during lubricated cutting

3.3 Examining the drilled surfaces

Examining the holes, we found that in the case of dry cutting the matrix material was always smeared and the Globomets were torn out. On the other hand, with internal cooling and lubrication, cutting conditions were better, the matrix was cut cleanly and the Globomets were cut through. In the case of internal cooling and lubrication, far fewer Globomets were displaced. Figure 6 shows the microscope images of the walls of the holes.

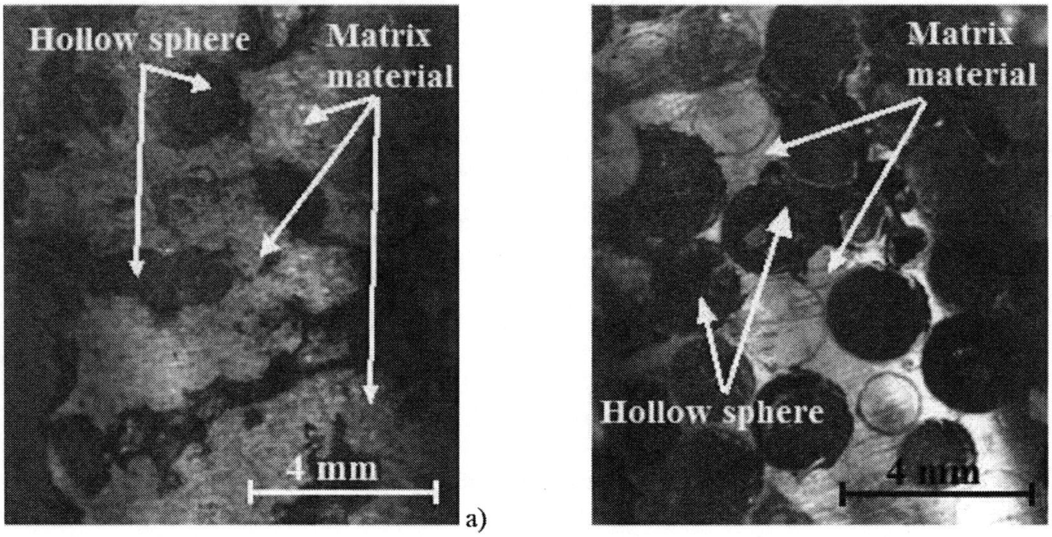

Figure 6. Inner surfaces of the holes: a) made without lubricant and b) made with internal cooling-lubrication

The drilled surface of the holes was tested at several points by scanning electron microscopy. The volume percentage of the Fe, Al and Si atoms was analyzed on the drilled surface (Figure 7). The chemical composition of the measurement points in Figure 7 can be seen in Table 2.

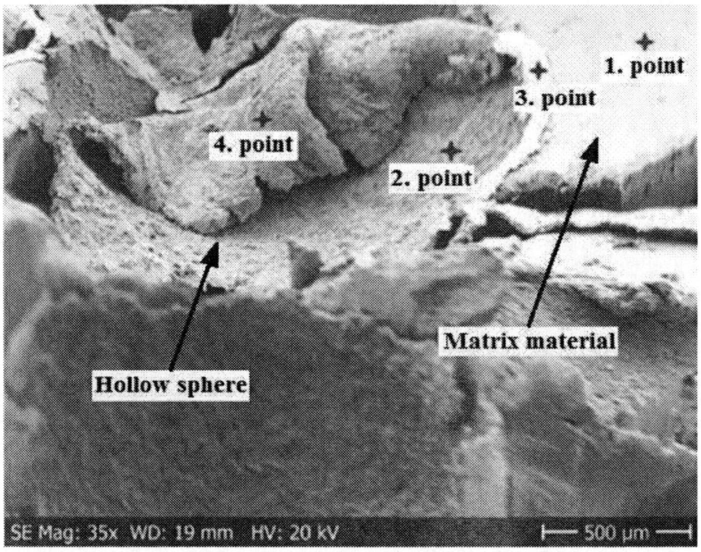

Figure 7. SEM image of the inner surface of the drilled hole (internal cooling-lubrication)

Table 2. The measured chemical composition (by EDS method) of the material on the drilled surface according to Figure 7.

Measurement points	Al (vol%)	Si (vol%)	Fe (vol%)
1.	83.55	14.65	1.80
2.	2.84	0.51	96.65
3.	14.00	1.46	84.54
4.	26.17	0.64	73.2

At the first measurement point only matrix material was found, as expected.

At the second measurement point the inner surface of a cut sphere was analyzed, where almost only iron was found.

The third measurement point shows the outer surface of the cut sphere. In this case, mostly iron was found, as in the previous point, but the spectroscopy showed a small amount of aluminium, as well. The reason for this, that the iron hollow sphere bended on the matrix material where the measurement was performed.

Point 4 analyzes the outer surface of a reinforcing sphere, which was bent during drilling. A significant amount of aluminium was found here. Fig. 7 shows that the two sides of the sphere were affected differently by the tool; the upper side of the sphere was bent because of the feed but the exit side is cut through. This phenomenon can be explained with the fact that when the sphere was bent, the tool was in contact with the outer surface of the sphere, but when the tool exited the sphere, it was in contact with the inner surface of the sphere.

4. Conclusion

We performed drilling experiments on a metal matrix syntactic foam under dry drilling conditions and with internal cooling in the tool. The parameters of feed and cutting speed were varied in a wide range. The following conclusions can be drawn from the tests:

Of the cutting parameters, only feed has a significant effect on the feed direction forces (F_f) both under the conditions of dry cutting and with internal cooling-lubrication. The effect of cutting speed on feed direction forces is negligible.

In the case of dry cutting, feed direction (axial) forces could be 8-10 times more (the average difference is 484 N between the dry and wet cutting process).

In dry cutting a built-up edge was formed on the tool edges in every test run, and burr was found on the exit sides of the holes, while with the use of coolant-lubricant, these effects were not experienced.

Dry cutting of these kinds of materials is not recommended because of the increased forces, the burr on the exit side of the hole, and the built-up edge. The uncut spheres reduce surface quality, the quality of the hole and tool life.

Under the conditions of internal cooling both the matrix and the reinforcing spheres were cut but some of the hollow spheres were cut through and some were deformed depending on the orientation. This was proved with electron microscopy and spectroscopy analysis.

Acknowledgements

This research was supported by the New National Excellence Program (code: ÚNKP-17-4-I-OE-779/60) of the Ministry of Human Resources.

The authors wish to thank GranTOOL Kft. for providing the tools necessary for the tests.

References

[1] Fakhri M A, Bordatchev E V, Tutunea-Fatan O. R. 2012 An image-based methodology to establish correlations between porosity and cutting force in micromilling of porous titanium foams *The International Journal of Advanced Manufacturing Technology* **60**(9-12) 841-851

[2] http://www.hollomet.com/produkte.html

[3] Karabulut Ş, Karakoç H 2017 Investigation of surface roughness in the milling of Al7075 and open-cell SiC foam composite and optimization of machining parameters *Neural Computing and Applications* **28**(2) 313-327

[4] Kiser M, He M Y, Zok F W 1999 The mechanical response of ceramic microballoon reinforced aluminum matrix composites under compressive loading *Acta Materialia* **47**(9) 2685-2694

[5] Krajewski S, Nowacki J 2015 Structure of AlSi–SiC composite foams surface formed by mechanical and thermal cutting *Applied Surface Science* **327** 523-531

[6] Májlinger K, Kalácska G, Orbulov I N, Zsidai L, Bozóki B, Keresztes R 2017 Global Approach of Tribomechanical Development of Hybrid Aluminium Matrix Syntactic Foams *Tribology Letters* **65**(1) 16
[7] Májlinger K 2015 Wear properties of hybrid AlSi12 matrix syntactic foams *International Journal of Materials Research* **106**(11) 1165-1173
[8] Májlinger K, Bozóki B, Kalácska G, Keresztes R, Zsidai L 2016 Tribological properties of hybrid aluminum matrix syntactic foams *Tribology International* **99** 211-223
[9] Ø10 DreamDrills (D5433100) GranTool datasheet http://www.toolontool.hu/DNNGranTool/PDF/YG-1_Drill_D5433.pdf
[10] Orbulov I N, Ginsztler J 2012 Compressive behaviour of metal matrix syntactic foams *Acta Polytechnica Hungarica* **9**(2) 43-56
[11] Orbulov I N, Ginsztler J 2012 Compressive characteristics of metal matrix syntactic foams *Composites Part A: Applied Science and Manufacturing* **43**(4) 553-561
[12] Szlancsik A, Katona B, Bobor K, Májlinger K, Orbulov I N 2015 Compressive behaviour of aluminium matrix syntactic foams reinforced by iron hollow spheres *Materials & Design* **83** 230-237
[13] Yilbas B S, Akhtar S S, Keles O 2013 Laser cutting of triangular geometries in aluminum foam: Effect of cut size on thermal stress levels *Optics & Laser Technology* **48** 523-529
[14] Yilbas B S, Akhtar S S, Keles O 2013 Laser hole cutting in aluminum foam: Influence of hole diameter on thermal stress Optics and Lasers in Engineering **51**(1) 23-29

IOP Conf. Series: Materials Science and Engineering **426** (2018) 012020 doi:10.1088/1757-899X/426/1/012020

The influence of BN additives on the phase composition, microstructure and mechanical properties of 316L steel consolidated by spark plasma sintering

P Jenei[1], Cs Balázsi[2], Á Horváth[2], K Balázsi[2] and J Gubicza[1]

[1] Eötvös Loránd University, Department of Materials Physics, Budapest, Hungary
[2] Hungarian Academy of Sciences, Centre for Energy Research, Budapest, Hungary

E-mail: jenei@metal.elte.hu

Abstract. Blends of commercial 316L steel and boron nitride (BN) were consolidated using Spark Plasma Sintering (SPS). The components were mixed by high energy milling (using attritor) before sintering. The aim was to study the effect of BN on the microstructure and the mechanical properties. Two composites containing 0.5 and 2 wt.% BN were produced beside a pure 316L specimen. It was found that bimodal grain size distribution formed during the production. A bcc α'-phase formed during milling beside the main fcc γ-phase. The addition of BN resulted in a slightly higher dislocation density and a smaller grain size in the fine grained parts of the samples. At the same time, the elastic modulus and the strength determined by bending test were reduced, most probably due to the weak bonding between the 316L and the BN grains, and the lower fraction of α'-phase.

1. Introduction

Ceramic dispersion strengthened steels are promising structural materials in new generation nuclear reactors and future fusion power plants [1–4]. Since the nanoscale (~10 nm) ceramic particles are thermally stable, therefore it can obstruct the movement of dislocations and prevent the grain growth even at high temperatures, thereby hindering both recovery and recrystallization. Hence this steels have excellent strength at high temperatures (even under neutron radiation [5]), however the low ductility and fracture toughness limit their industrial applications [6]. Beside the oxide particles, other filler materials can also be used for the improvement of the functional properties of steels [4,6,7]. For instance, boron nitride (BN) additive can increase the absorption of neutrons in the steel composites [8].

The steel composites strengthened by particles are usually produced by powder metallurgy methods [1,2,9,10]. The components of the matrix and the disperse phase are blended by mechanical milling, which is usually followed by a high-temperature consolidation process. During milling the initially small grain sized particles are fragmented and a part of them can dissolve into the matrix [9]. The microstructure of the matrix and the disperse phase depends strongly on the circumstances of the production of the material, therefore they determine the mechanical properties. The components are often consolidated by Spark Plasma Sintering (SPS) [1]. During SPS pulses of high current density are used to facilitate consolidation, thus the sintering time and temperature decrease, therefore much smaller grain size can be achieved compared to conventional sintering methods [11].

In the present study, the effect of BN additive on the microstructure and the mechanical behavior of 316L steel was investigated. The samples were produced by powder metallurgy using high energy milling and SPS. The microstructure was investigated by electron backscattered diffraction (EBSD) and X-ray line profile analysis (XLPA). The mechanical behavior was characterized by microhardness measurement and three point bending test. The effect of the phase composition, the porosity, the grain size and the dislocation density on the mechanical properties will be discussed.

2. Materials and experimental procedures

2.1. Sample preparation

The samples were processed from a commercial 316L austenitic stainless steel powder (manufacturer: Höganäs) with the composition of Fe – 16.8 % Cr – 12 % Ni – 2.5 % Mo – 1.5 % Mn – 0.6 % Si – 0.0025 % C (wt.%). The disperse phase in the composites was an ultrafine-grained (UFG) boron nitride (BN) powder obtained from a hexagonal boron nitride (hBN) (manufacturer: H.C. Starck). First, the starting 316L powder was intensively milled in a DMQ-07 attritor (manufacturer: Union Process) at 2800 rpm for 10 hours in propanol. In this machine, a stainless steel setup with delta disk agitators were used. The diameters of the stainless steel grinding balls were 1 mm. After sieving in a 100 micron mesh, the powder was milled in a dry environment for 1 hour at 600 rpm in 01-HD/HDDM type attritor (manufacturer: Union Process).

The UFG BN powder was prepared by mechanical milling. First, the commercial hBN powder was milled intensively in a high efficient attritor mill (01-HD/HDDM type attritor) equipped with zirconia agitator delta discs and zirconia grinding media (1 mm in diameter) in the presence of ethanol at 4000 rpm for 10 h. After wet milling, a dry milling process was applied at 4000 rpm for 5 h. Finally, a UFG BN powder with an average grain size of ~200 nm was achieved, as can be seen in figure 1a.

After this milling process, 0.5 and 2 wt.% BN were added to the 316L powder and this powder blend was further milled in the 01-HD/HDDM type attritor equipped with stainless steel tank, delta disk agitators and grinding media with 1 mm in diameter. The high energy milling was performed at 4000 rpm for 3 hours in ethanol in order to achieve an efficient dispersion of the boron nitride in the commercial 316L powder. Finally, the milled powders were sieved in a 100 micron mesh. A reference 316L powder was also processed without the addition of BN powder. However, this powder was subjected to the same milling process as the blends of 316L and BN powders. The milling process resulted in flat particles with the thickness of about 200 nm (see figure 1b).

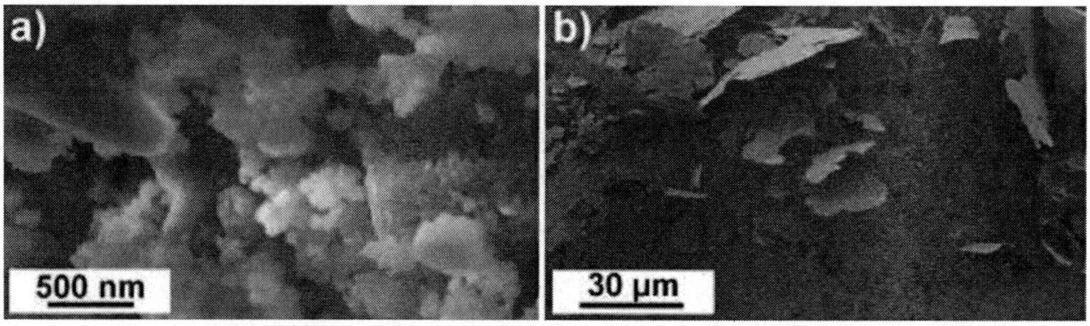

Figure 1. SEM images taken on the milled a) BN and b) 316L+2BN powder.

The powders were sintered by SPS in vacuum at 900 ± 10°C for 5 minutes under a pressure of 50 MPa. SPS-processing was carried out by an SPS-7.40MK-VII machine (manufacturer: SPS Syntex Inc.) installed in Istanbul Technical University using a current of 20,000 A. Disks with 50 mm in diameter and 5 mm in thickness were obtained after the consolidation process. In the following, the samples containing 0, 0.5 and 2 wt.% BN are referred to as 316L, 316L+0.5BN and 316L+2BN, respectively.

2.2. Microstructure investigations

The microstructure of the samples was studied by X-ray diffraction, Scanning Electron Microscopy (SEM) and EBSD. The mass density of the sintered samples was measured in water using Archimedes' principle. The phase composition of the samples was investigated by a Philips Xpert ($\lambda = 0.15418$ nm) X-ray powder diffractometer. The microstructure of the main γ-phase was studied by XLPA. The X-ray line profiles were measured by a high-resolution diffractometer using $CoK\alpha_1$ ($\lambda = 0.1789$ nm) radiation. For this measurement, the samples were electropolished in an A8 electrolytic solution at room temperature at a fixed voltage of 10 V for 30 s. Two-dimensional imaging plates detected the Debye-Scherrer diffraction rings. The line profiles were evaluated by the Convolutional Multiple Whole Profile (CMWP) method. The details of the CMWP procedure can be found in Refs. [12,13]. In this method, the experimental pattern is fitted by the convolution of the instrumental pattern and the theoretical size and strain line profiles. The instrumental pattern was measured on a LaB_6 line profile standard material. The theoretical profile functions used in this fitting procedure are calculated on the basis of a model of the microstructure, where the crystallites have spherical shape and log-normal size distribution, and the lattice strains are assumed to be caused by dislocations. As an example, the CMWP fitting for the sample 316L+2BN in a logarithmic intensity scale is shown in figure 2. The open circles and the solid line represent the measured data and the fitted curve, respectively.

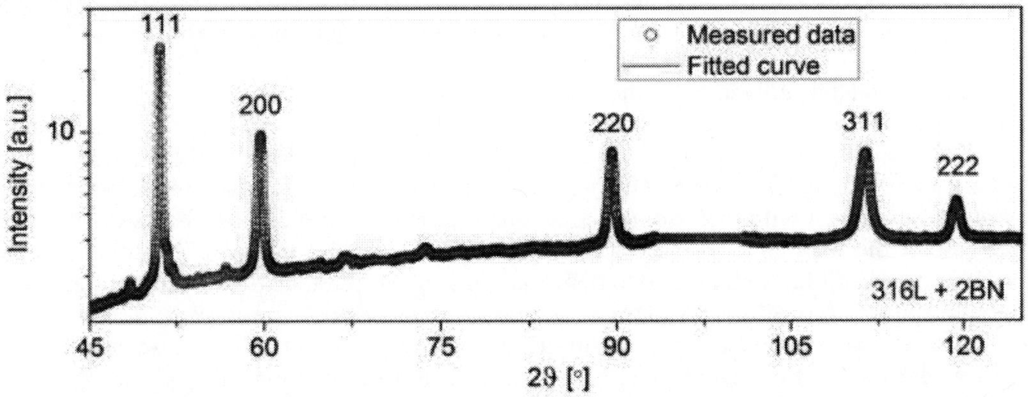

Figure 2. CMWP fitting of the X-ray diffraction pattern taken on the sample 316L+2BN. The open circles and the solid line represent the measured data and the fitted curve, respectively.

The EBSD analysis was performed using an FEI Quanta 3D dual beam SEM equipped with an EDAX type EBSD system. The surface of the sample was mechanically polished (steps: P1000, P2500, P4000, 1μm ALOX, Mastermet2 suspension (about 30 nm SiC, manufactured by Stuers)). The grain size was determined from the EBSD scans using OIM software from TexSEM Laboratories. The morphology of the powder particles was studied by SEM (Zeiss-SMT LEO 1540 XB and Jeol JSM-25-SIII).

2.3. Mechanical tests

The Vickers hardness of the samples was measured on electropolished surfaces using a Zwick Roell ZHμ Vickers indenter with an applied load of 300 g and a dwell time of 10 s. For the bending test, rectangular bars with the dimensions of $4 \times 4 \times 50$ mm^3 were cut from the sintered specimens using a water-jet cutting process. The three-point bending strength was determined by an Instron 1112 tensile/loading machine equipped with a data acquisition system.

3. Results and discussion

3.1. Phase composition and porosity

In all specimens, the main phase was an fcc Fe-based alloy (denoted as γ-phase) but in addition bcc Fe-based alloy (denoted as α'-phase), Fe_3O_4 and $Cr_{23}C_6$ phases were also identified. As an example, figure

3a shows a part of the X-ray diffraction pattern obtained for the 316L+0.5BN sample. It should be noted that BN was not detected, most probably due to the fragmentation of the BN particles during high-energy milling as well as the very small amount of this phase.

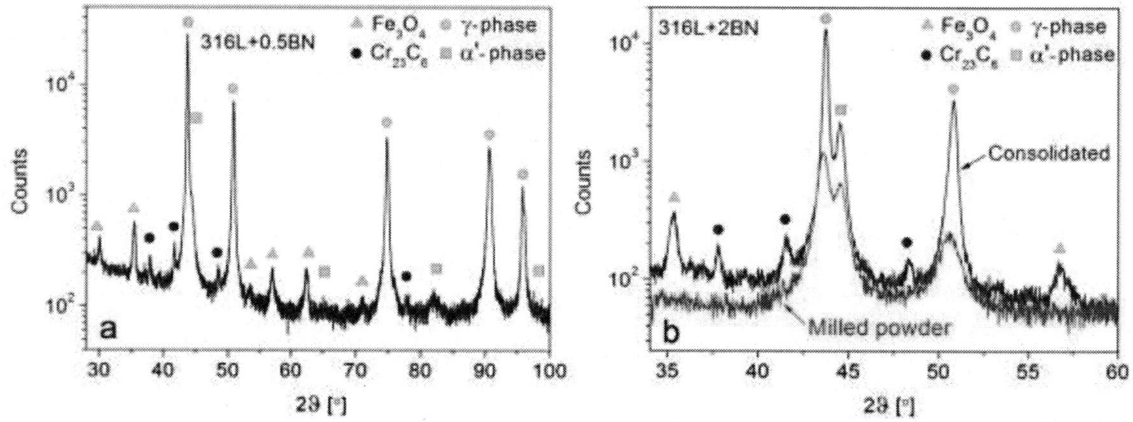

Figure 3. (a) X-ray diffractogram for the sample 316L+0.5BN. (b) A part of the X-ray diffractogram obtained for the sample 316L+2BN before and after consolidation.

In order to reveal the influence of the additives on the phase composition, the fractions of the integrated intensities for the different phases in the X-ray diffractograms were determined and listed Table 1 for all samples. In all milled powders (irrespectively of the additives), the fractions of γ-phase and α'-phase were 0.73 and 0.27, respectively (see Table 1). It is noted that the intensity fractions of the phases are not equivalent with their volume fractions. The addition of BN to 316L steel resulted in a slight increase of the fraction of γ-phase while the amount of the secondary α'-phase decreased. The α'-phase was formed during milling due to deformation induced martensitic transformation. However, at the high temperature of SPS the γ-phase is more stable, therefore there is an α' to γ reverse phase transformation as it can be seen in figure 3b. In the milled powder, beside the γ-phase only a secondary α'-phase was detected. At the same time, after consolidation the fraction of α'-phase decreased and Fe_3O_4 and $Cr_{23}C_6$ peaks appeared on the diffractogram. It seems that the addition of BN promotes the α' to γ transformation. A reasonable explanation of this effect may be that the BN grains serve as nuclei in the phase transformation during sintering.

Table 1. The fractions of the integrated intensities for the different phases and the porosity of the consolidated samples.

Sample	γ-phase	α'-phase	Fe_3O_4	$Cr_{23}C_6$	Porosity [%]
Milled powders	0.73 ± 0.02	0.27 ± 0.02	-	-	-
316L	0.81 ± 0.05	0.14 ± 0.02	0.04 ± 0.01	0.01 ± 0.01	5.4 ± 0.3
316L+0.5BN	0.93 ± 0.03	0.03 ± 0.01	0.03 ± 0.01	0.01 ± 0.01	4.4 ± 0.3
316L+2BN	0.88 ± 0.05	0.07 ± 0.02	0.03 ± 0.01	0.01 ± 0.01	5.3 ± 0.3

The porosity was calculated from the difference between the measured and calculated densities. In the case of the composite specimens, the calculated density was obtained as an average of the theoretical densities of 316L iron (7.99 g/cm^3) and BN (3.45 g/cm^3) weighted by their mass fractions. Table 1 shows that BN have no significant effect on the porosity of the consolidated samples.

3.2. Microstructure

3.2.1 EBSD investigation. The SEM investigations on the polished surfaces revealed that all samples exhibit bimodal grain structures as can be seen in Fig 4b, e and h. In the matrix, the light and dark regions correspond to coarse-grained (CG) and fine-grained (FG) regions, respectively. Coarse particles were also observed in the initial 316L powder. The EBSD images in figure 4 show that both regions are fragmented into grains. The grain sizes were evaluated from these images and listed in Table 2. It can be seen that the grain size is much higher in the CG regions for all samples. The BN additive resulted in a lower grain size as the BN grains hinder the recovery and recrystallization of the microstructure during sintering.

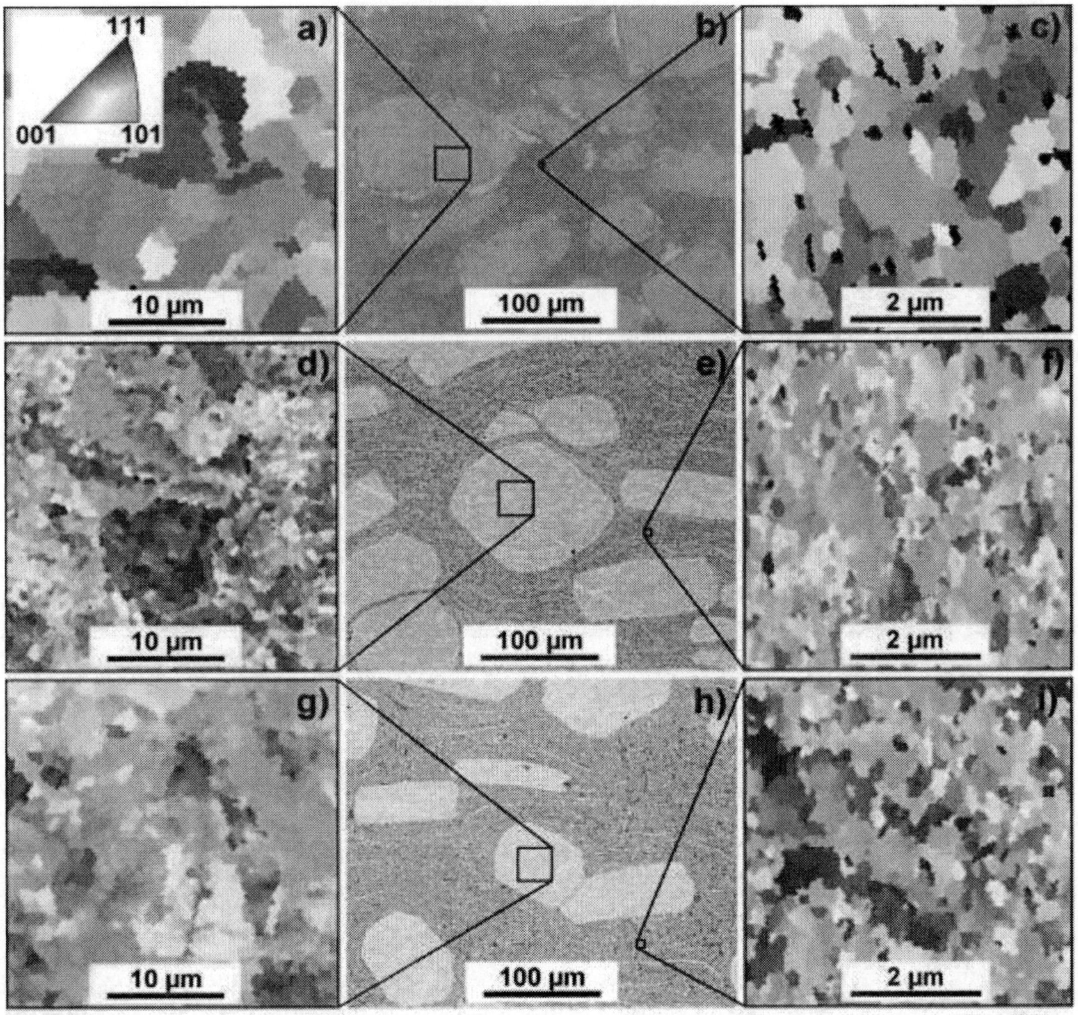

Figure 4. EBSD and SEM images showing the morphology of the grains in the CG and FG regions for the sample (a-c) 316L, (d-f) 316L+0.5BN and (g-i) 316L+2BN. The color code of the standard stereographic triangle inset in (a) illustrates the grain orientations in the gamma-phase. The black regions in Fig. 4c are occupied by α'-phase.

Table 2. The grain size values in the fine-grained (d_{FG}) and the coarse-grained (d_{CG}) regions determined by EBSD. The dislocation density (ρ_{UFG}) in the ultrafine grained (UFG) regions of γ-phase as obtained by XLPA, and the intensity fraction of the UFG parts of the microstructures (f_{UFG}). The total dislocation density (ρ_{total}) in the γ-phase was determined as the product of ρ_{FG} and f_{FG}.

Sample	d_{FG} [μm]	d_{CG} [μm]	ρ_{UFG} [10^{14} m^{-2}]	f_{UFG}	ρ_{total} [10^{14} m^{-2}]
316L	0.44	3.66	6.3 ± 0.7	0.68 ± 0.05	4.3 ± 0.8
316L+0.5BN	0.23	0.93	5.0 ± 0.6	0.65 ± 0.05	3.3 ± 0.7
316L+2BN	0.28	1.71	7.5 ± 0.9	1	7.5 ± 0.9

3.2.2. X-ray line profile analysis. The dislocation density was determined for the γ-phase and listed in Table 2. It should be noted that for the pure 316L and 316L+0.5BN specimens, the intensity distribution along the Debye–Scherrer (DS) rings was inhomogeneous and numerous sharp intensity spots were visible as it can be seen in figure 5b. These sharp peaks were as narrow as the instrumental broadening ($\Delta(2\theta)$=0.03°); therefore they were not evaluated for the microstructure. These sharp intensity spots are related to regions in which the crystallite size is larger and the dislocation density is lower than the detection limits of XLPA for the present experimental setup (800 nm and 10^{13} m^{-2}, respectively).

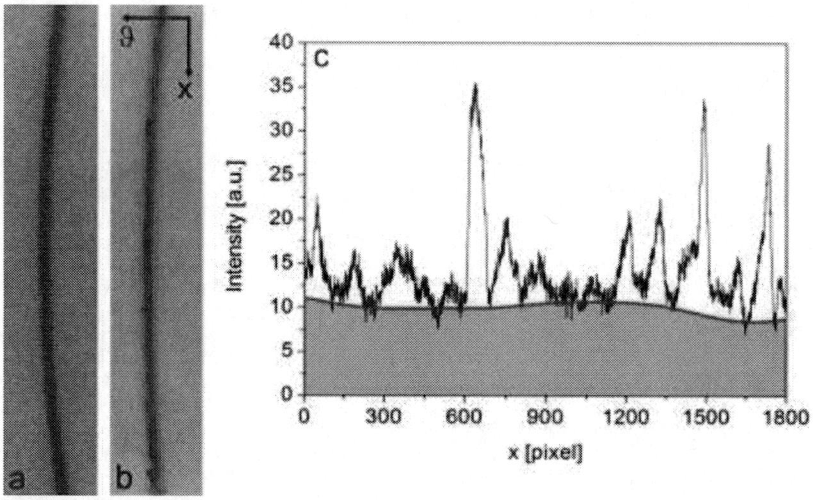

Figure 5. Debye–Scherrer diffraction rings of reflection 200 for the samples (a) 316L+2BN and (b) 316L+0.5BN. (c) Intensity distribution in direction *x* in figure (b) after subtracting the background. The red zone corresponds to the homogeneous and broad ring segments in (b) scattered from the UFG region.

Due to the spotty diffraction rings for the pure 316L and the 316L+0.5%BN samples, only the broader and more homogeneous parts between the large intensity spots were evaluated by XLPA. As a consequence, the results obtained for these two samples characterize only the ultrafine grained (UFG) part of the samples. It should be noted that the UFG region may be located in either the CG or the FG area. The total dislocation densities in the whole materials can be calculated by considering the UFG fractions in the samples. This quantity can be estimated as the fraction of the intensities of the homogeneous parts in the entire DS ring. The intensity was summed in direction 2ϑ (see figure 5b), and these intensity values were plotted as a function of the coordinate *x*, as shown in figure 5c for reflection 200 of the sample 316L+0.5BN. Thereafter, a spline was fitted to the parts of the intensity distribution which are free of sharp intensity peaks (see figure 5c). The area under this spline corresponds to the intensity scattered from the UFG parts of the sample. The fractions of the UFG parts for the pure 316L and the 316L+0.5BN samples are listed in Table 2. For the 316L+2BN sample, the DS rings were homogeneous (see figure 5a) therefore for this specimen the microstructural parameters determined by XLPA characterize the whole specimens (i.e., the fraction of the UFG parts is 1). Then, the total dislocation density was calculated as the product of the dislocation density in the UFG regions and their intensity fraction. Table 2 shows that the addition of 0.5 wt.% BN did not yield a considerable change in the total dislocation density. At the same time, 2 wt.% BN resulted in a higher dislocation density than in the addition-free 316L steel due to the pinning effect of BN particles.

It should be noted that there is an apparent discrepancy. It can be seen in Table 2, that the grain size measured by EBSD is higher for the sample 316L+2BN than in the specimen 316+0.5BN, while the DS rings are homogeneous in the sample 316L+2BN and spotty in the specimen 316L+0.5BN. This can be explained by the fact that the grains in EBSD investigation were defined as the regions bounded by high angle grain boundaries with misorientation angles higher than 15°. In the case of XLPA, the crystallites are the domains which scatter X-rays coherently. As the coherency of X-rays breaks even if they are scattered from volumes having quite small misorientations (1–2°), the crystallites correspond rather to the subgrains in the microstructures [14]. Therefore, most probably the subgrain structure is finer in the 316L+2BN sample, which can not be evaluated by EBSD.

3.3. Mechanical properties
Table 3 shows the average hardness measured for the FG and CG regions for all samples. In addition, the Young's modulus and the three-point bending strength are also listed in Table 3. Generally, the hardness is higher in the FG parts of the samples due to the lower grain size. It can be seen that the addition of 0.5% BN has no significant effect on the mechanical properties of 316L steel. At the same time, 2% BN resulted in a reduction of the Young's modulus and the bending strength. This can be explained by the lower fraction of α'-phase and the weak bonding between the BN and the 316L grains, which can promote crack formation and propagation during the bending test.

Table 3. The mechanical properties of 316L samples sintered with and without BN additive.

Sample	Young's modulus [GPa]	Three-point bending strength [MPa]	HV [GPa] Coarse-grained	HV [GPa] Fine-grained
316L	183 ± 5	1324 ± 28	2.55 ± 0.13	3.80 ± 0.10
316L+0.5BN	174 ± 2	1265 ± 58	2.65 ± 0.14	3.41 ± 0.15
316L+2BN	166 ± 2	1066 ± 45	2.36 ± 0.12	3.30 ± 0.15

It is worth to note that despite the very different microstructures in the studied samples, the hardness values only slightly vary with increasing the BN content. There are four main microstructure features which can influence considerably the mechanical properties: (1) the phase composition (the α'-phase content in our case), (2) the dislocation density in the main γ-phase, (3) the grain size and (4) the porosity. Table 4 shows the relative contributions of the different microstructural features to the hardness.

Table 4. The relative hardening effects of the microstructural features in the samples. The more '+' icons indicate a stronger hardening effect as compared to the other samples.

Hardening effects	316L	316L+0.5BN	316L+2BN
α'-phase fraction	+++	+	++
Relative density	+	++	+
Dislocation density	++	+	+++
Grain size	+	+++	++

It can be seen that for the sample 316L+0.5BN, the porosity, the total dislocation density in the γ-phase and the grain size were only slightly lower than in the pure 316L sample. Therefore, the smaller hardness in the FG region can be attributed to the higher γ to α'-phase ratio. It is well known that the γ-phase is softer than the α'-phase in 316L steel. For the sample 316L+2BN, the porosity is the same as for the pure 316L steel but the dislocation density is higher and the grain size is lower due to the addition of BN. Despite the much higher dislocation density, the hardness of the specimen 316L+2BN is slightly lower than for the sample 316L which indicates that the softening effect of the larger γ-phase fraction overwhelms the hardening effect of the higher dislocation density and the lower grain size. Comparing the two composites, for the sample 316L+2BN the hardening effects of the dislocation density and the α'-phase content were higher, while the contributions of the grain size and the porosity to the hardness were lower than for the specimen 316L+0.5BN, which can explain their similar hardness values.

4. Summary
The effect of BN addition on the microstructure and the mechanical properties of commercial 316L stainless steel was studied. Samples containing 0, 0.5 and 2 wt.% BN were produced using high energy milling and SPS method. The following results were obtained:

1) A high fraction (27-28 %) of α'-phase formed beside the major γ-phase phase during the milling process. During sintering at high temperature, the fraction of the α'-phase decreased due to a reverse martensitic phase transformation. It seems that the BN additive promotes this transformation since the fraction of α'-phase is considerably lower in the composites. Fe_3O_4 and $Cr_{23}C_6$ phases were also formed during the consolidation process.

2) All samples show inhomogeneous grain structure. Large particles can be seen in the fine grained matrix. The larger particles are fragmented into smaller grains. The addition of 0.5% BN decreased the grain size in both regions, but further BN addition did not result in a finer grain structure. The microhardness of the fine grained part was higher (irrespectively of the BN content) duo to the much lower grain size.

3) The addition of 0.5% BN did not result in a significant change in the dislocation density. The increase of the BN content from 0.5 to 2% yielded a higher dislocation density due to the hindering effect of BN particles on the annihilation of dislocations during sintering. The microhardness, the bending strength and the elastic modulus slightly decreased with increasing the BN content. This effect can be attributed mainly to the lower α'-phase fraction and the weak bonding between the 316L and the BN grains.

Acknowledgment

The authors are indebted to the National Research, Development, and Innovation Office Grant nos. K-109021, PD-121049 and MTA EK project „Nanostructural ODS steel development" for financial support. The authors thank to preparation of samples to V. Varga, SEM investigation to L. Illés, EBSD measurements to Á. Szabó and sintering to Prof. F. C. Sahin. JP acknowledges the financial support of the Hungarian Ministry of Human Capacities under Grant no. NTP-NFTÖ-16-0957.

References

[1] Balázsi C, Gillemot F, Horváth M, Wéber F, Balázsi K, Sahin FC, Onüralp Y and Horváth Á. 2011 *J. Mater. Sci.* **46**:4598
[2] Mukhopadhyay DK, Froes FH and Gelles DS 1998 *J. Nucl. Mater.* **258-263**:1209
[3] Renzetti R, Sandim HRZ, Bolmaro RE, Suzuki P and Möslang A 2012 *Mater. Sci. Eng. A* **534** 142
[4] Zhong Z, Hinoki H, Park YH and Kohyama A 2010 *Fusion Eng. Des.* **85** 992
[5] Monnet I, Dubuisson P, Serruys Y, Ruault MO, Kaïtasov O and Jouffrey B 2004 *J. Nucl. Mater.* **335** 311
[6] Mahathanabodee S, Palathai T, Raadnui S, Tongsri R and Sombatsompop N 2014 *Wear* **316** 37
[7] Patankar SN, Chandrasekaran M and Tan MJ. 2000 *J. Mater. Sci. Lett.* **19** 613
[8] Narayana Rao M. 2011 *Energy Procedia* **7** 199
[9] Alinger MJ, Odette GR and Hoelzer DT 2009 *Acta Mater.* **57** 392
[10] Rahmanifard R, Farhangi H and Javad A 2015 *J. Alloys Compd.* **622** 948
[11] Orrù R, Licheri R, Locci AM, Cincotti A and Cao G. 2009 *Mater. Sci. Eng. R Reports* **63** 127
[12] Ribárik G, Gubicza J and Ungár T 2004 *Mater. Sci. Eng. A* **387-389** 343
[13] Ungar T, Gubicza J, Ribárik G and Borbély A 2001 *J. Appl. Crystallogr.* **34** 298
[14] Ungár T, Tichy G, Gubicza J and Hellmig RJ 2012 *Powder. Diffr.* **20** 366

IOP Conf. Series: Materials Science and Engineering **426** (2018) 012021 doi:10.1088/1757-899X/426/1/012021

TIG and MIG welding of high strength Cr-Mn and Cr-Ni alloyed austenitic stainless steel combinations

E Kalácska[1], B Varbai[1], J Ginsztler[1] and K Májlinger[1]

[1] Budapest University of Technology and Economics, Department of Materials Science and Engineering, Budapest, Hungary

E-mail: welding@att.bme.hu

Abstract. To substitute the relative expensive Ni alloying in stainless steels manganese alloying can be used. The welding of Cr-Mn alloyed steels however is not very well documented. In our current research two Cr-Mn alloyed austenitic stainless steel grades (1.4371 and 1.4376) were welded together via tungsten inert gas welding (TIG) and metal inert gas welding (MIG). Also for comparison different joint combinations with other Cr-Ni alloyed steel grades (1.4318 and 1.4301) were welded. The different steel grade showed quite the different behaviour in the joint combinations. The Mn-content in the base materials increased the weld metal hardness and tensile strength of the joint and decreased the fracture elongation values. Also the Mn-alloyed grades had significant higher grain coarsening in the heat affected zones than the Cr-Ni alloyed.

1. Introduction

As the influence of automotive, oil and gas, and chemical industries is constantly growing, high strength steels, aluminum alloys and corrosion resistant steels gain more attention [1–9]. In these industries austenitic stainless steels (ASSs) are the most widely used corrosion resistant grades [10, 11].

The conventional ASSs, such as 1.4301 and 1.4404 have a relatively high nickel alloying content (Ni = 8–10 wt.%) next to chromium (Cr ~ 18 wt.%). However, recently in order to substitute the relatively expensive Ni in ASS grades manganese (Mn) and nitrogen (N) alloying is used [11, 12]. The austenite forming effect of high Mn (up to 9 wt.%) and N content (up to 0.3 wt.%) also comes together with higher mechanical strength [13–15]. Comparing to the conventional austenitic stainless steels, which has yield strength ($R_{p0.2}$) is around 200 MPa, the Mn and N alloyed grades reach above 600 MPa $R_{p0.2}$. Despite of the lower base material price and higher strength of Cr-Mn alloyed ASSs, the corrosion resistance is usually lower [16–19] and their weldability is not well documented in the literature.

The mechanical properties of low Ni ASSs after welding is found to be sensitive to the grain size and structure [20–22]. Lower heat input resulted smaller dendrite sizes and higher mechanical strength. The mechanical strength of welded low Ni ASS grades also depends on the used filler material. Researchers [23] found that using conventional filler wires (e.g. ER 308 LSi) resulted no decrease of mechanical strength of the welded joint. With the growing application of low Ni austenitic grades, their dissimilar joints to conventional high strength or corrosion resistant steels also gaining interest [24–26]. Generally, the dissimilar welding of conventional austenitic and low Ni, high N grades can be successfully made [27], this area still has a lot of research interest.

2. Materials and methods

In our current research the two used Cr-Mn alloyed ASS's were 1.4371 and 1.4376, both are additionally alloyed with nitrogen to increase yield and tensile strength and corrosion resistance. For comparison the Cr-Ni alloyed 1.4318 grade (with additional N alloying) and the most common 1.4301 normal strength austenitic steel grade were used. For both TIG and MIG welding tests the same grade (1.4316) austenitic filler was used. The microstructure of the sheets can be seen in Figure 1. All grades were fully austenitic, the high strength grade with relatively small grain size (Table 1). The chemical compositions, Cr- and Ni-equivalent and mechanical properties of the used steel grades are also listed in Table 1

Figure 1. Microstructure of the used ASS sheets

Table 1. Main properties of the used ASS sheets and filler metal

Steel grade	Chemical composition (wt%)											Mechanical properties				Grain size
	C	Cr	Ni	Mn	Si	Mo	Cu	N	Fe	Cr_{eq}*	Ni_{eq}*	$R_{p0.2}$** (MPa)	R_m** (MPa)	A** (%)	Hardness (HV0.5)	(µm)
1.4371	0.05	15.8	4.4	7.7	0.5	0.1	0.4	0.2	bal.	16,7	15,8	620	760	25	277±10	12
1.4376	0.11	17.1	3.8	7.2	0.6	-	0.5	0.2	bal.	18,0	16,8	405	740	40	251±10	11
1.4318	0.05	16.5	7.3	1.6	0.5	0.2	0.3	0.2	bal.	17,5	15,6	350	660	35	205±2	10
1.4301	0.05	18.3	8.6	1.8	0.4	0.2	0.4	-	bal.	19,1	11,0	190	620	35	187±5	22
1.4316^W	0.02	19.9	9.1	1.8	0.9	0,1	0,3	-	bal.	21.3	10,6	390	590	35	195±4	-
1.4316^R	0,02	19.8	9.8	1.9	0.6	0,1	0,1	-	bal.	20.9	11.3	390	590	35	195±5	-

^{W} wire, ^{R} rod, *according to DeLong, ** nominal values

For the welding tests the sheets were butt welded without chamfering with ~ 1 mm gap. The sheet thicknesses were approx. 2 mm and were mechanically clamped during welding.

For the MIG welding tests ESAB LUD 450 machine was used with direct current electrode positive polarity (DC+). For shielding gas 4.6 Ar (99.996 % Ar) was used on the face and the root side with the flow rates 10 and 8 $l \cdot min^{-1}$ respectively. As filler 1.4316 grade Ø 1.2 mm wire was used.

For the TIG welding tests ESAB Aristotig 250 machine was used with direct current electrode negative polarity. For shielding gas 4.6 Ar was used on the face and the root side with the flow rates 10 and 8 $l \cdot min^{-1}$, respectively. As filler 1.4316 grade Ø 2.4 mm filler rod was used.

The measured current (I), voltage (U), welding speed (v) and the calculated heat input values (Q) for the different joint combinations for both welding processes are listed in Table 2. For the calculation of the heat input values thermal efficiency factors according to EN 1011-1:2009 standard k = 0.8 and k = 0.6 were used for MIG and TIG welding, respectively.

From the welded sheets cross sectional metallography and tensile specimens were cutout, perpendicularly to the weld seam.

The quasistatic tensile tests (according to EN ISO 6892-1 standard) were done by MTS 810 universal materials testing machine on proportional ($l_0 = 10 \times d_0$) flat specimens (excess weld bead was grinded down), where the weld centerline was located in the middle of the tensile specimen. From each joint combinations three tensile tests were done. Tensile strength of the joints (R_m), and necking at fracture (Z) were determined. The fracture elongation of the whole specimen ($A_{11.3}$), the fracture elongation of the two sheet sides of the joints ($A_{5.65}$) and the uniform elongations (ε) of the sheet materials were also determined.

The standard metallography specimens were etched with Kalling reagent (2 g $CuCl_2$ + 40 ml HCl + 40 ml C_2H_6O) for 20 s. The images were taken by Olympus PM3 microscope. The grain sizes of the different heat affected zones; HAZ 1 at the weld pool and HAZ 2 further away near to the base material were also determined.

On the metallography specimen micro-Vickers hardness measurements (HV0.5) with Buehler 1011 type microhardness tester were done and the average hardness of the base materials and different joint parts; HAZ 1, HAZ 2 and the weld metal (WM) were measured.

Table 2. Parameters and heat input values for the MIG and TIG welding of the different joint combinations

Joints		MIG welding				TIG welding			
1. Steel grade	2. Steel grade	I (A)	U (V)	v ($cm \cdot min^{-1}$)	Q ($kJ \cdot mm^{-1}$)	I (A)	U (V)	v ($cm \cdot min^{-1}$)	Q ($kJ \cdot mm^{-1}$)
1.4371	**1.4371**	78	19.3	26.7	0.27	75	11.7	10.7	0.30
1.4371	**1.4376**	82	19.3	23.5	0.32	75	11.5	11.6	0.27
1.4371	**1.4318**	75	19.5	23.7	0.30	75	11.5	8.0	0.39
1.4371	**1.4301**	74	19.4	23.8	0.29	75	11.7	8.4	0.38
1.4376	**1.4376**	73	19.3	22.3	0.30	75	11.3	10.5	0.29
1.4376	**1.4318**	76	19.4	23.4	0.30	75	11.5	10.0	0.31
1.4376	**1.4301**	77	19.3	24.3	0.29	75	11.5	9.3	0.33

3. Results and discussion

The visual inspection of the weld beads and the cross sectional metallography showed macroscopically defect free joints for all steel combinations of MIG and TIG welding as well. Therefore these joints were further investigated.

3.1. Microstructural properties

The microstructural examinations (e.g. Figure 2, 3.) showed also no weld defects of any kind. The WM had relative large dendritic grains. To investigate the grain size evolution after welding a normalized (to the initial grain size of the base metals shown in Table 1) grain size of the HAZs is shown in Figure 4.

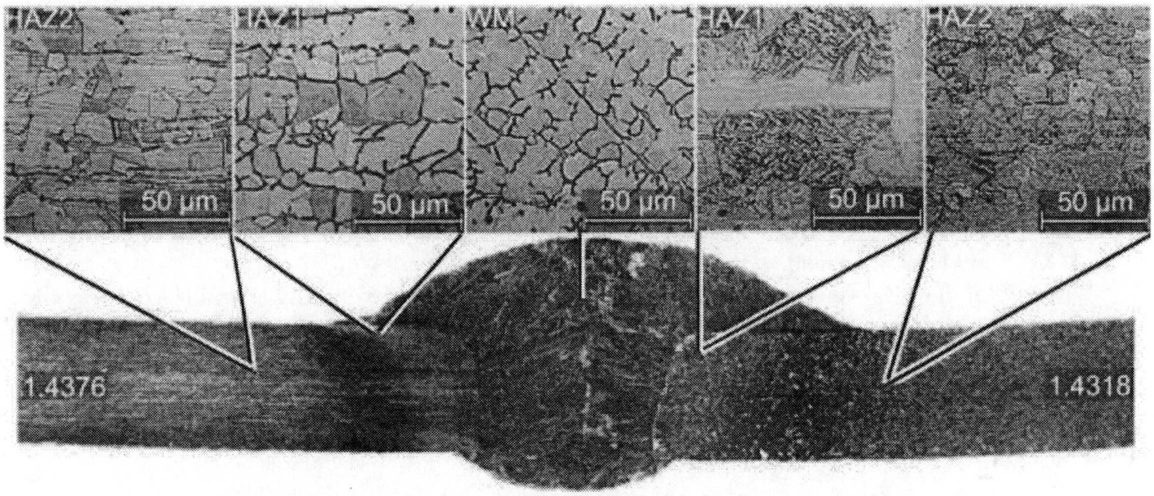

Figure 2. Macroimage of a MIG welded joint in cross section and microstructure images of the different joint zones

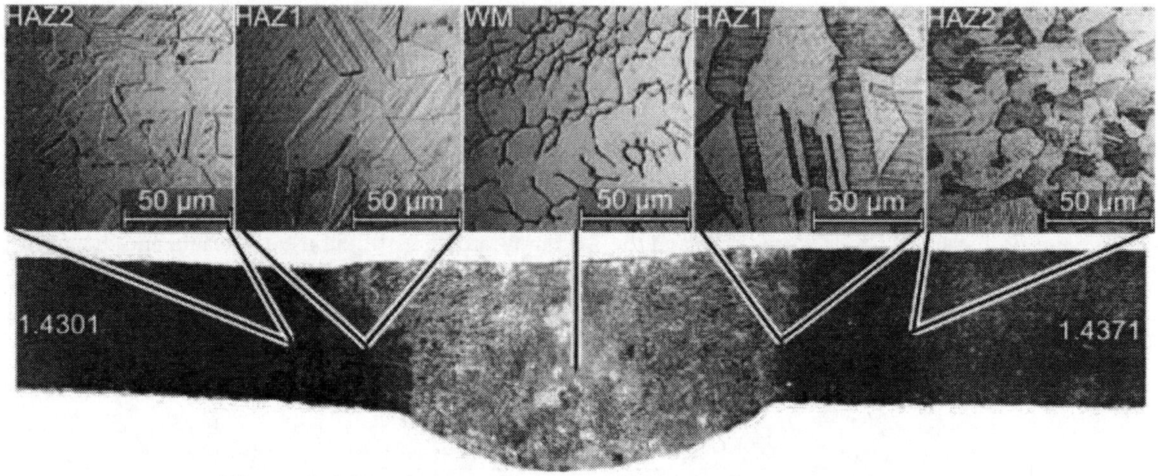

Figure 3. Macroimage of a TIG welded joint in cross section and microstructure images of the different joint zones

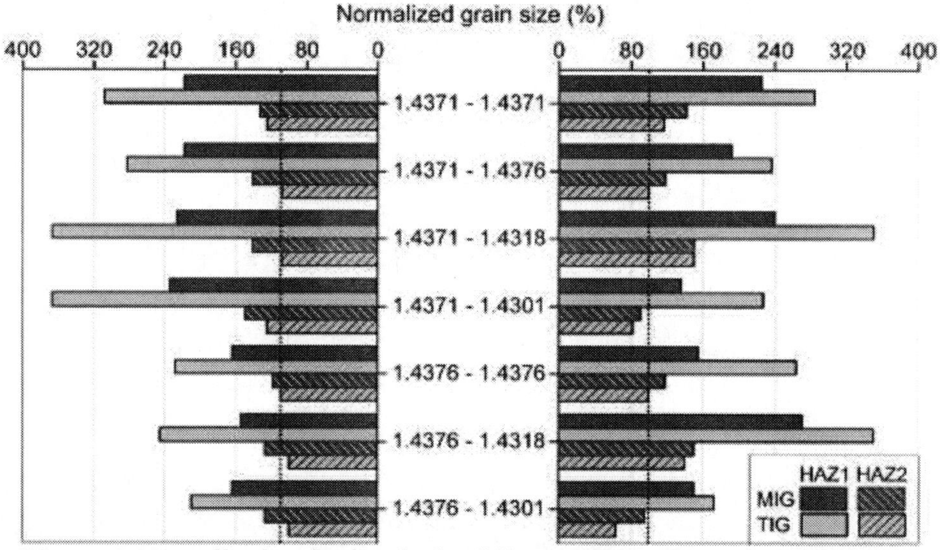

Figure 4. Normalized grain sizes in the different HAZs of the MIG and TIG welded joint combinations (vertically dotted lines corresponds to 100%)

In case of MIG welding the following trends can be observed. The highest grain coarsening occurred in the 1.4318 steel grade followed by the two Cr-Mn alloyed ones (1.4371 and 1.4376) and the normal strength 1.4301 had even smaller grains in the HAZs as the initial grain size. This trend was observed in HAZ 1 and HAZ 2 as well.

In case of the TIG welded specimens the grain coarsening order in HAZ 1 and HAZ 2 was the same with the same trends as in case of MIG welded ones (1.4318 > 1.4371 > 1.4376 > 1.4301). The absolute values were also greater than in case of MIG welding due to a "slightly" higher heat input which is also suggesting that the standardized thermal efficiency factor $k = 0.6$ for the TIG welding seems to be too low. Also the increment of grain sizes due to higher heat input seem to be the highest in case of the two Cr-Mn alloyed steel grades.

3.2. Hardness of the different joint parts

To investigate the effect of dilution on the weld metal hardness the normalized WM hardness (to the nominal hardness of the WM shown in Table 1) was measured (Figure 5).

In case of MIG welding all specimen showed slightly decreased hardness of the weld metal, only the 1.4371-1.4371 joint had higher hardness. However the variation between the normalized WM hardnesses were between 93–102% clear trends can be observed between the base materials Mn-content and the WM hardness. The higher the Mn-content of the two base materials the higher values of the of the WM hardness were measured. The joint with the normal strength (1.4301) steels had the lowest hardness values.

The same trends can be observed in case of the TIG welded joints, in some cases the hardness values were even higher than the MIG welded specimens.

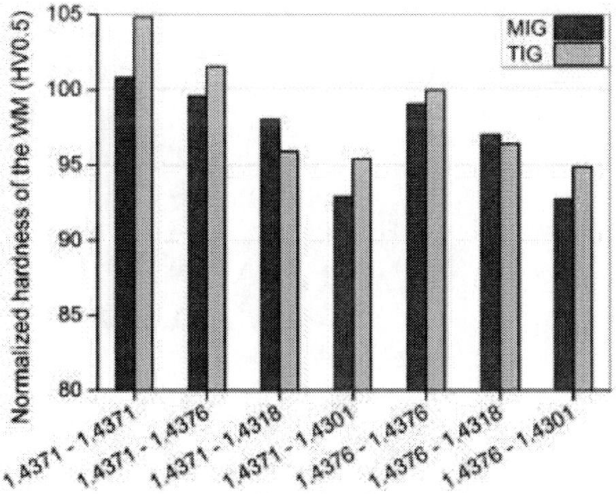

Figure 5. Normalized hardness values in the different HAZs of the MIG and TIG welded joint combinations

The increase in the WM hardness is likely the effect of Mn dilution in the WM thus Mn increases the austenite phase hardness through substitutional solid solution.

To investigate the effect of heat input in the HAZs hardness the normalized hardness values of the different joint combinations are shown in Figure 6.

In HAZ 1 (adjacent to the solidified WM) also clear trends can be observed with the Mn-content of the different steel grades for both welding methods. The greatest softening was in the 1.4371 steel grade with the highest Mn-content (1.4376) followed by the other Cr-Mn alloyed grade, than the normal strength Cr-Ni alloyed 1.4301 steel and the 1.4318 one.

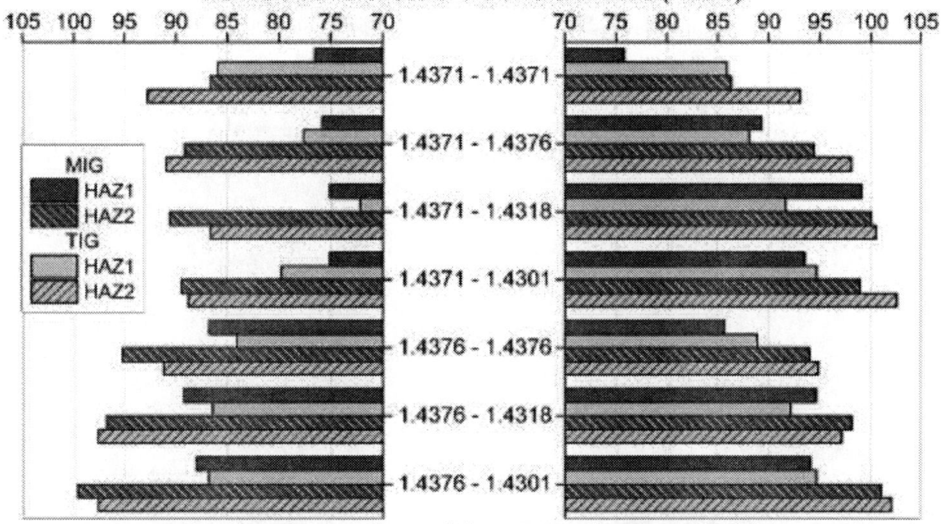

Figure 6. Normalized hardness values of the MIG and TIG welded joint combinations

In HAZ 2 also the 1.4376 steel had the greatest softening followed by the 1.4371 and the Cr-Ni alloyed had about the same hardness near to the initial base material hardness.

The large softening in the HAZ 1 cannot be attributed alone to the grain coarsening because the 1.4318 steel had the greatest grain coarsening and yet the smallest hardness decrease.

3.3. Tensile properties

In all tensile specimens the necking occurred in the WM, where the fracture of the joints also occurred. Therefore the differences in the tensile strength of the specimen can be attributed to the dilution of the base metal sheets and the filler material. The mean values of the tensile strength of the different joint combinations are illustrated in Figure 7.

In case of MIG welding the R_m values varied relatively small (between 645–675 MPa), nevertheless analogue to the WM hardness (Figure 5.) the same correlation was found. The Mn-content of the base materials increased the R_m values.

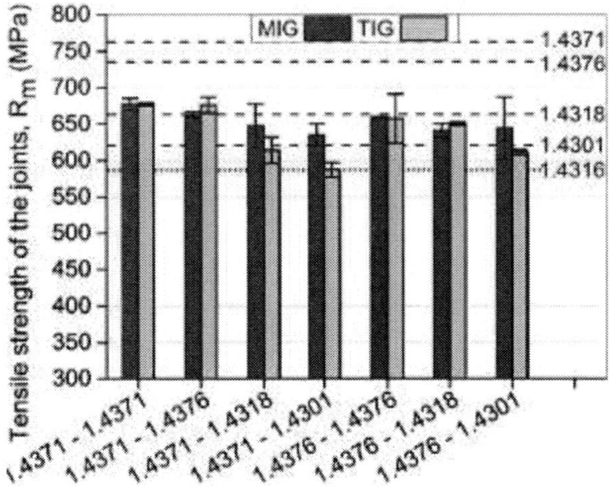

Figure 7. Tensile strength (R_m) of the MIG and TIG welded joint combinations

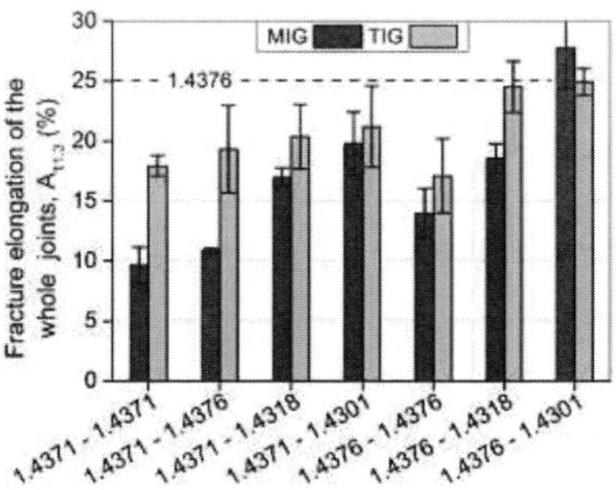

Figure 8. Fracture elongation ($A_{11.3}$) MIG and TIG welded joint combinations

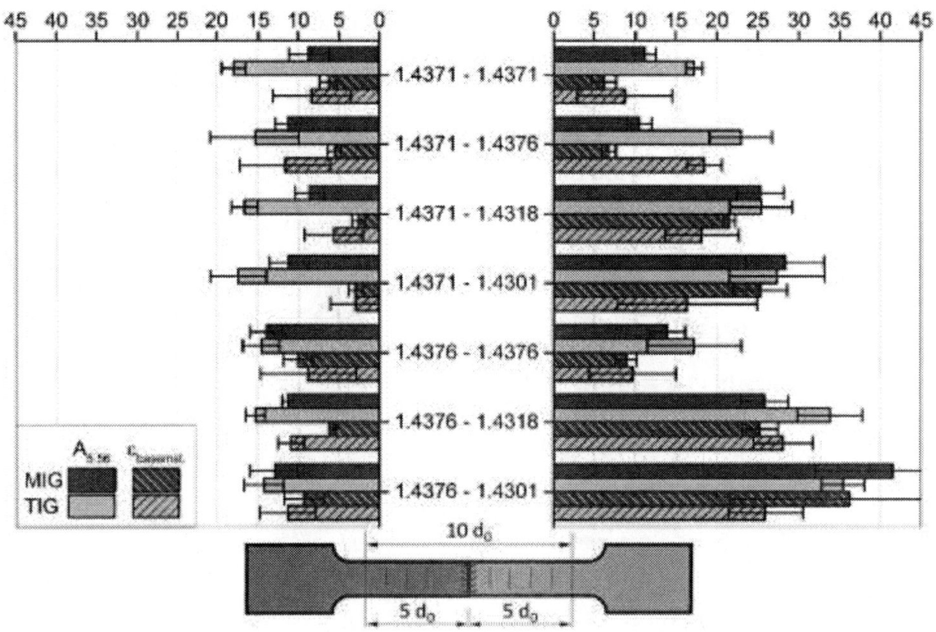

Figure 9. Fracture elongations ($A_{5.65}$) of the two sides of the MIG and TIG welded joint combinations and the uniform elongations (ε) of the base materials in these joints

In case of the TIG welded joints, R_m values varied more (575–675 MPa), and the same trends can be observed as in MIG welding.

The increase of the R_m values is also a result of the effect of Mn dilution in the WM, because Mn increases the austenite phases strength through substitutional solid solution.

To investigate the tensile response of the joined steel combinations, the fracture elongation of the whole specimen was determined (Figure 8). The $A_{11.3}$ values showed significant decrease with the increase of the Mn-content in the base materials in the joint combinations for both MIG and TIG welding.

Because all joints have at least two different base material parts (the dissimilar joint combinations three) it is interesting to see the deformations in these parts separately (Figure 9).

Du Toit and Steyn found [28] that the 1.4301 grade ASS has greater strain hardening exponent than the Cr-Mn alloyed 1.4373 grade. In case of MIG welding, we found the least deformation ($A_{5.65}$ and ε) occurred were the two Cr-Mn alloyed grades 1.4376 ≤ 1.4371 followed by the high strength Cr-Ni alloyed grade (1.4318) and the normal strength 1.4301 ASS grade. It is interesting, because the greatest hardness decrease was found in the HAZs of the Cr-Mn alloyed grades, and the grain coarsening was also high at these ASSs. These facts indicate that the strain hardening exponent of the Cr-Mn-N alloyed grades must be higher than the other used steel grades.

The necking values at fracture, and sides of the fracture within the WM are listed in Table 3. The necking values showed relative large scatter and no trends could observed, but the fracture side of the tensile specimen clearly showed the tendency that the favored fracture side was on the base material side with higher Mn-content.

Table 3. Necking at fracture, and sides of the fracture of the tensile specimens of the MIG and TIG welded joints

1. Steel grade	Fracture side						2. Steel grade	Necking at fracture Z (%)	
	MIG			TIG				MIG	TIG
1.4371	◀	◀	◀	◀	◀	●	**1.4371**	45±4	43±1
1.4371	◀	◀	◀	◀	◀	▶	**1.4376**	45±1	48±9
1.4371	◀	◀	◀	◀	◀	▶	**1.4318**	44±3	51±5
1.4371	◀	◀	◀	◀	▶	▶	**1.4301**	48±1	49±1
1.4376	◀	◀	●	◀	▶	▶	**1.4376**	39±4	56±6
1.4376	◀	◀	◀	▶	▶	▶	**1.4318**	45±2	47±2
1.4376	◀	▶	▶	▶	▶	▶	**1.4301**	46±5	45±4

Place of the fracture within the WM: ◀ 1st steel grade, ● WM center line, ▶ 2nd steel grade side

4. Conclusions

According to our investigations the following conclusions can be drawn:

- The Cr-Mn-N alloyed 1.4371 and 1.4376 high strength austenitic stainless steel grades can be welded together with good joint quality also in combination with Cr-Ni alloyed high strength 1.4318 and normal strength 1.4301 austenitic steel austenitic stainless steel grades.
- Using 1.4316 filler material with MIG process (between 0.27–0.32 $kJ \cdot mm^{-1}$ heat input range) and with TIG welding (between 0.27–0.38 $kJ \cdot mm^{-1}$ heat input range) macroscopically and microscopically defect free joints can be welded.
- In the different steel grades the order of grain coarsening for both MIG and TIG welding in both HAZs were: 1.4318 > 1.4371 > 1.4376 > 1.4301.
- The weld metal hardness increased with the Mn-content of the base materials through dilution.
- The tensile fracture occurred in every case within the weld metal, the yield strength of the joints increased with the Mn-content of the base materials through dilution.

- In HAZ 1 the hardness of the base materials decreased, the normalized hardness values and decreased with higher Mn-content of the base material and the smallest decrease was in the 1.4318 steel grade observed.
- The fracture elongations of the welded specimens were the smallest with the joints of the highest Mn-content in the base material.
- The fracture elongations measured on the two sides of the joint combinations were also the smallest by the two Cr-Mn alloyed grades 1.4376 ≤ 1.4371 followed by the high strength Cr-Ni alloyed grade (1.4318) and the normal strength 1.4301 ASS grade.
- The uniform elongation of the base materials measured on the two sides of the joint combinations showed also the same order: 1.4376 ≤ 1.4371 < 1.4318 < 1.4301.

With our research the mechanical properties of dissimilar joints, made of these austenitic stainless steel grades can be predicted.

Acknowledgements
This research was supported by Aperam and Outokumpu Distribution Hungary Kft. with the steel materials. This paper has been supported by the János Bolyai Research Scholarship of the Hungarian Academy of Sciences grant number: BO/00196/16/6 and by the National Research, Development and Innovation Office - NKFIH, OTKA PD 120865 (K. Májlinger).

References
[1] Russo Spena P et al 2016 *Steel Res. Int.* **87** 1592–600
[2] Russo Spena P et al 2015 *Steel Res. Int.* **86** 1480–9
[3] Gáspár M and Balogh A 2013 *Design, Fabrication and Economy of Metal Structures* (Berlin, Heidelberg: Springer Berlin Heidelberg) pp 463–8
[4] Lukács J and Gáspár M 2014 *Adv. Mater. Res.* **891–892** 563–8
[5] Lukács J 2010 *Procedia Eng.* **2** 1201–10
[6] Balogh A et al 2012 *Production Processes and Systems* **5** 79–90
[7] Boillot P and Peultier J 2014 *Procedia Eng.* **83** 309–21
[8] Haraszti F and Kovács T 2017 *IOP Conf. Series: Materials Science and Engineering* **175** 1–4
[9] Oyedemi K et al 2017 *Period. Polytech. Mech. Eng.* **61** 296
[10] Pauze N et al 2010 *Duplex World 2010 Conference* (Beaune, France)
[11] Charles J et al 2009 *Rev. Métallurgie* **106** 90–8
[12] Charles J 2007 *Rev. Métallurgie* **104** 308–17
[13] Kalácska E et al 2017 *YPIC 2017: 3rd Young Welding Professionals International Conference. 120 p.* pp 20–6
[14] Speidel M O 2006 *Materwiss. Werksttech.* **37** 875–80
[15] Pistorius P C and Toit M 2010 *The Twelth International Ferroalloys Congress; Sustainable Future* pp 911–8
[16] Khobragade N N et al 2014 *Trans. Indian Inst. Met.* **67** 263–73
[17] Shi F et al 2016 *Corros. Sci.* **107** 49–59
[18] Ha H Y et al 2016 *Mater. Charact.* **119** 200–8
[19] Sudhakaran R et al 2014 *Procedia Eng.* **97** 790–9
[20] Urade V P and Ambade S P 2016 *J. Mater. Sci. Eng.* **5** 1–6
[21] Vashishtha H et al 2014 *ISIJ Int.* **54** 1361–7
[22] Chuaiphan W and Srijaroenpramong L 2014 *J. Mater. Process. Technol.* **214** 402–8
[23] Vashishtha H et al 2016 *Sci. Technol. Weld. Join.* **21** 331–7
[24] Kalácska E et al 2017 *Mater. Sci. Forum* **885** 80–5
[25] Broggiato G B et al 2015 *Procedia Eng.* **109** 356–63
[26] Chandra-Ambhorn S et al 2012 *Adv. Mater. Res.* **538–541** 1464–8
[27] Vashishtha H et al 2017 *J. Manuf. Process.* **25** 49–59
[28] Du Toit M and Steyn HG 2012 *J. of Mat. Eng & Perf.* **21(7)** 1491-5

IOP Conf. Series: Materials Science and Engineering **426** (2018) 012022 doi:10.1088/1757-899X/426/1/012022

Systematic study of structural changes in the vicinity of indentation marks with HR-EBSD

Sz Kalácska[1] , Z Dankházi[1] and I Groma[1]

[1] Eötvös Loránd University, Department of Materials Physics, Budapest, Hungary

E-mail: kalacska@metal.elte.hu

Abstract. In crystalline materials a detailed understanding of the effects of external deformation is essential to develop new materials. The mechanical properties are highly influenced by the collective motion of lattice defects (mainly dislocations). The primary task of the research is to gain more experimental knowledge on the stress and strain distribution evolving in the crystal due to external deformation. In the reported investigation an indenter was pressed into the sample to a depth of a few micrometers. After the removal of the head high resolution electron backscatter diffraction (HR-EBSD) measurements were carried out in the vicinity of the mark, giving the 3D depth distribution of stress values.

1. Introduction

Plastic deformation of crystalline materials typically occurs by the collective motion of dislocation ensembles. In a macroscopic sample the large number of grains and moving dislocations generally produce smooth and reproducible stress–strain response allowing highly accurate predictions of the material's properties. In contrast, on the scale of a single grain, the inhomogeneities in the dislocation structure can be in the order of the grain size, leading to a stochastic activation of dislocation avalanches producing discontinuous response. The effect of dislocations is localized to an extremely small volume, so most of the time one can only investigate the cumulative influence of numerous defects.

The effect of crystal orientation on the indentation response is widely studied on single crystal titanium [1], nickel [2], copper [3] and even on perovskite oxide ceramic materials [4]. Synchrotron X-ray microdiffraction measurements show that indentation hardness increases with decreasing indentation depth on copper single crystal [5]. In these measurements Berkovich or conical indenters were used. For the better understanding of the indentation process it is important to get a detailed knowledge about how the dislocation cell structure forms, how it evolves due to external deformation, and inspect the stress state developing around the mark. As a model material we used a previously heat treated copper (Cu) single crystal, that has been studied extensively because of its unique properties like cubic symmetry, relatively low Young's modulus and ease of preparation.

Traditional 3D electron backscatter diffraction (EBSD) measurements, using conical indenter, were published in Ref. [3], showing size effect at small indents due to the geometrically necessary dislocations generated during the indentation. In the present work we aim to investigate the 3D distribution of strain developing under a Vickers indenter. In order to achieve this, around the indentation mark, a series of thin layers perpendicular to the sample surface were removed by focused ion beam (FIB). High resolution electron backscatter diffraction (HR-EBSD) measurements were carried out in each layer giving the maps of the ε_{ij} strain and σ_{ij} stress tensors.

High resolution EBSD is a cross-correlation based evaluation technique. It is a widely accepted methodology for strain calculation from the relative deformation of the EBSD patterns obtained at different points. The development of the HR-EBSD technique is mainly related to Wilkinson and his co-workers [6]. The evaluation, which was executed by a program developed by one of the presenting authors [7], enables us to calculate the relative distortion of the lattice and investigate the local stress variations. The distortion and stress values can be plotted as a function of the position on the sample surface. Combining these 2D maps leads to a 3D map of the structural changes occurred due to plastic deformation. This so called "slice and view" method is a novel approach in HR-EBSD measurements.

2. Methods

Using FIB technique thin layers can be removed from the surface of the material in a well-controlled and precise way. With the application of preliminary determined ion current it is possible to produce smooth and EBSD-ready surfaces. By repeating the slicing and EBSD mapping a relatively high volume (in the order of tens of micrometers) can be explored.

The measurements were performed on a previously heat treated and undeformed copper single crystal. The sample was cut from an oriented ingot by electrical discharge machining. The orientations and the dimensions of the specimen are indicated in Fig 1a). After cutting, the samples were electropolished, and heat treated in a vacuum furnace to remove any excess distortions that may have accumulated during the cutting. After the heat treatment the sample was electropolished again with Struers D2 formula.

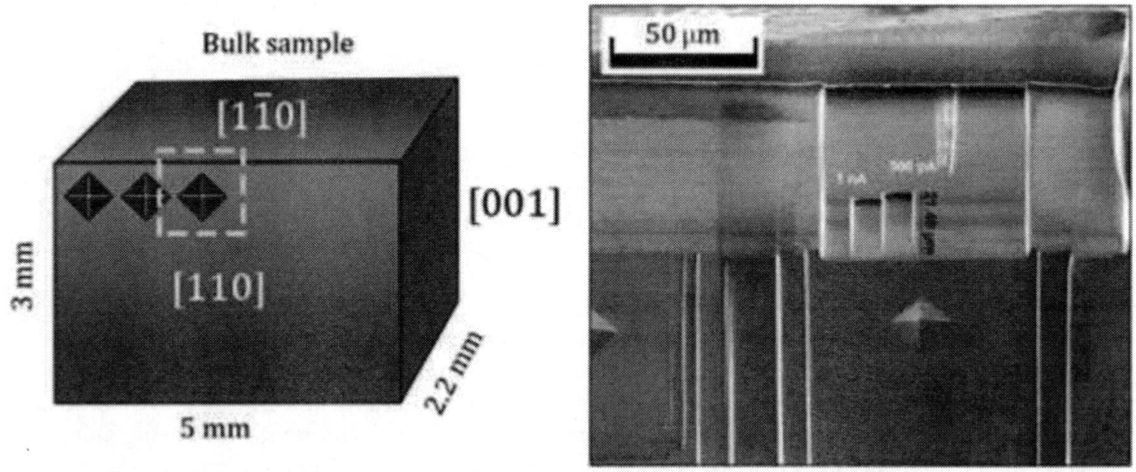

a) b)

Figure 1. a) The schematics of the specimen shown with the crystal orientations. On the sides the dimensions of the bulk sample is noted, and the coloring of the surfaces indicates the orientation. The three blue squares represent the position of the indentation marks. The yellow dashed area indicates the photographed area in b). An indentation mark with an estimated diagonal of 20 μm shows the initial arrangement before the slicing. This mark was only used for testing the FIB currents to check the quality of the diffraction patterns. The middle mark was used for the actual measurement.

IOP Conf. Series: Materials Science and Engineering **426** (2018) 012022 doi:10.1088/1757-899X/426/1/012022

2.1. Preparation of the indentation mark

In order to create a perfectly leveled and flat surface for the indentation marks, focused ion beam was used to form a plateau (with a normal vector parallel to the [110] crystal direction) near the sample's edge. The side of the sample was also cut by FIB to create a surface perpendicular to the plateau. As the quality of the surface is crucial for the accuracy of the evaluation, this newly formed surface was the one measured later by EBSD.

Since focused ion beam milling creates an amorphous layer on the surface that disturbs or even totally hides the diffraction patterns, to minimize this effect, during the FIB processing the ion current was decreased as much as it was necessary to be able to collect good quality diffraction patterns from the ion-treated surface. The biggest challenge of the preparation was to create perfect conditions for the EBSD measurements.

Three indentation marks were placed on the sample (100 μm from each other) by a *Zwick/Roell ZHμ Indentec* Vickers hardness testing device. The smallest available load (HV 0.01) was chosen while the Vickers head was pressed into the sample for 10 seconds. The marks were approximately 20 μm wide (see in Fig 1 b)).

2.2. EBSD measurements

To be able to make both the FIB slicing and the EBSD measurements we mounted the sample on a 45° sample holder. This allowed us to rotate the specimen between the ion slicing and the EBSD mapping positions. For the manipulation and measurements an FEI Quanta 3D FEG dual-beam SEM equipped with an EDAX Hikari EBSD camera was used. Parallel slices of the material were cut with a 30 kV / 15 nA ion beam. Due to the relatively big size of the indentation mark an area of 30 μm × 30 μm was chosen to be scanned by EBSD at 20 kV. Fig 2. shows the surface view of the indentation mark, indicating how the slices perpendicular to the surface were made. Each slice was 1 μm thick, and 30 slices were measured (therefore the reconstructed volume was 30 μm × 30 μm × 30 μm in size) by EBSD with a step size of 200 nm on a square grid.

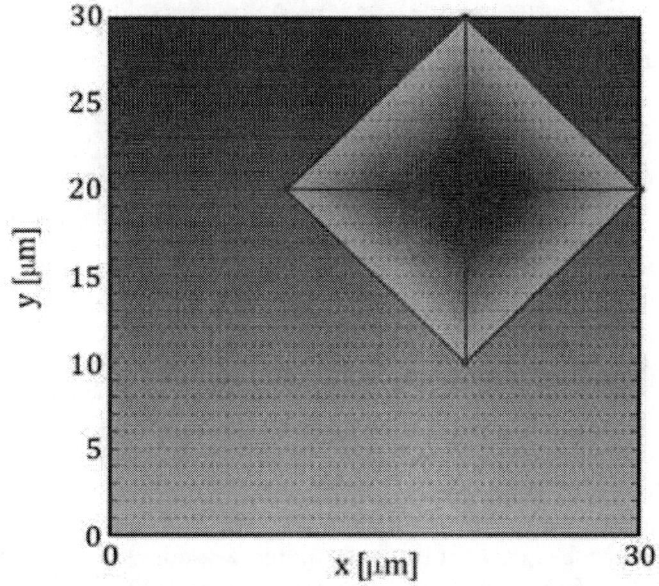

Figure 2. The schematics of the FIB slicing. A volume of 30 μm × 30 μm × 30 μm was mapped, including the indentation mark on the surface. The slicing was performed along the *z* axis, and the cuts are indicated by the dotted lines. The EBSD mapping was performed on the *x* - *z* surface.

IOP Conf. Series: Materials Science and Engineering **426** (2018) 012022 doi:10.1088/1757-899X/426/1/012022

The total number of registered EBSD patterns is close to 7×10^5 in a volume of about 3×10^4 μm^3 . Both the cutting of a slice and the creating an EBSD map took half an hour to perform. The whole measurement required 4 days to finish.

The EBSD maps were then evaluated by the *Strain_calc* program developed at the Eötvös Loránd University by one of the authors of this paper. The program is capable of calculating the tensor elements of the distortion, strain and stress matrices. During the cross-correlation-based evaluation the diffraction patterns from the scanned area are compared to a reference patterns (ideally measured at a stress-free point). As the effect of the indentation mark is only confined to a small volume, therefore the reference was recorded on the same surface but further away so it represents the initial (pre-deformed) stress state. All of the 30 slices were evaluated the same way, using the same reference.

3. Results

To investigate the residual stress distribution that formed after the indenter was removed from the surface, the $\sigma_{ij}(x,z)$ stress maps were generated. In Fig 3. only the σ_{12} stress component is plotted as a function of the position (x,z). The maps generated from the other components show similar behavior, therefore they are not shown here.

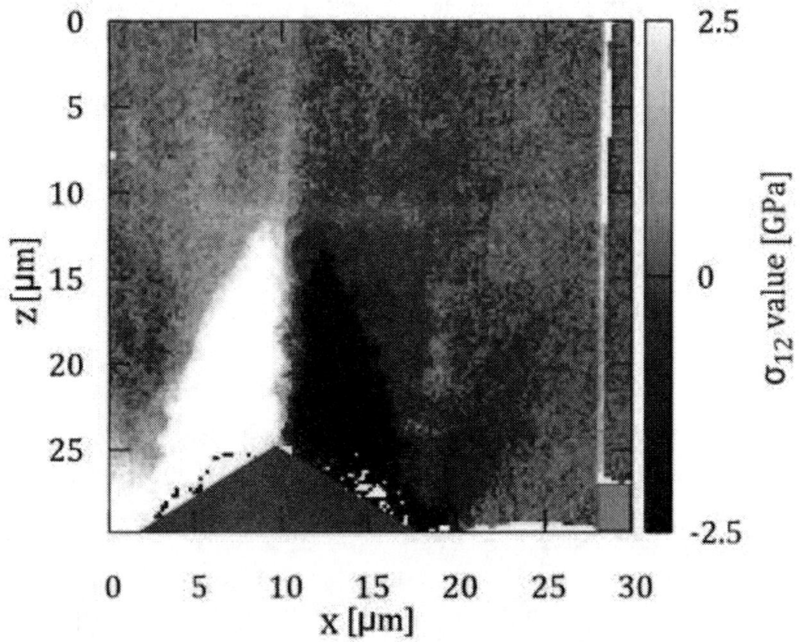

Figure 3. The σ_{12} stress component map measured on the 21st slice.
The red triangle indicates the cross-section of the indentation mark. The green rectangle indicates the FIB-milled ditch used for 3D alignment.

After comparing the results given by the σ_{ij}components some global statements can be made. The volume that was mostly affected by the deformation lies beneath the indentation mark. The σ_{11}, σ_{12} and σ_{13} stress maps show opposite signs of stress levels in these regions, while the σ_{22} and σ_{23} component maps indicate stresses of the same sign. Similar behavior was observed in the earlier GND measurements of *Reuber et al.* [8] along the cross-section of an indentation mark, and also in the strain distribution measured on the surface by *Britton et al.* [9,10].

For the 3D reconstruction of the indentation mark the slices were loaded into the *Amira 3D* software provided by FEI. The 3D model was created by the assembling of the aligned stress maps. The alignment was carried out with the help of a previously FIB-milled ditch on the surface (which can be seen in Fig 3. bottom right corner, green marker). The shapes of the surface corresponding to constant stress levels are seen in Fig 4.

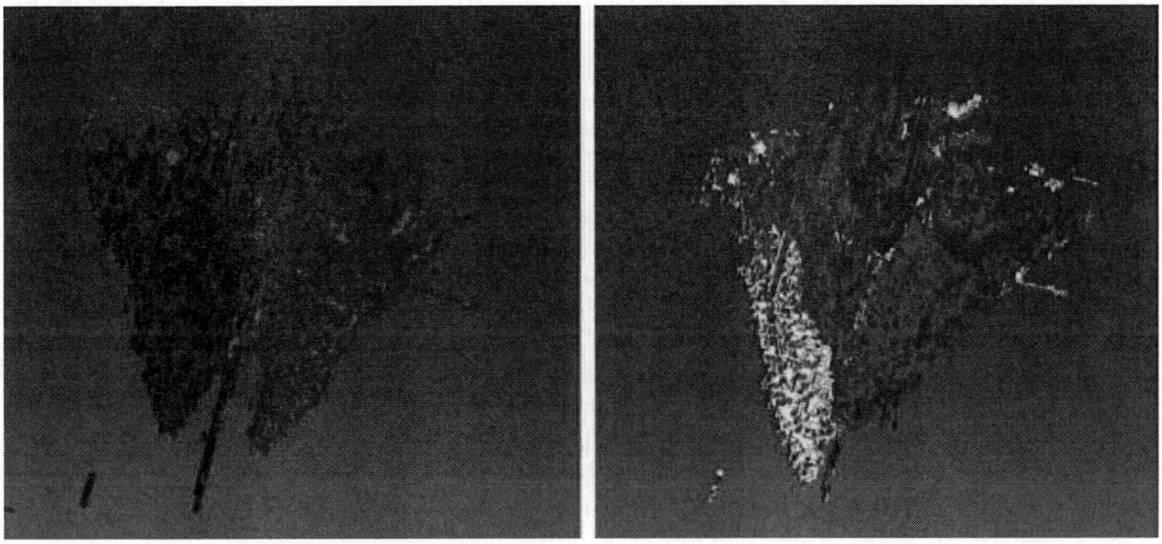

a) **b)**

Figure 4. Constant stress levels at 90% of their maximum values:
(a) σ_{11} component – purple; σ_{12} component – blue; (b) σ_{13} component – yellow; σ_{23} component– red.

4. Summary

The "slice and view" method combines the ability of the FIB slicing technique to manipulate the material on a micron scale, and the HR-EBSD method that can calculate the local stress distribution. With this combined method the structural changes in the vicinity of an indentation mark were investigated. A 3D map of the σ_{ij} stress tensor was determined. We found good agreement with previously obtained results, namely that σ_{11}, σ_{12} and σ_{13} stress maps show antisymmetric stress levels in these regions, while σ_{22} and σ_{23} component maps indicate stresses of the same sign. The combined method proposed is a new approach to investigate the deformation and stress distribution in small volumes. It opens the possibility to investigate the microdeformation properties of different problems, like microbending, micropillar compression, etc.

Acknowledgements
The present work was supported by the National Research, Development and Innovation Fund of Hungary (contract number: NVKP_16-1-2016-0014). IG and SzK have been supported by the National Research, Development and Innovation Found of Hungary (project no. NKFIH-K-119561). SzK was supported by the Hungarian Scholarship ÚNKP-16-3 New National Excellence Program of the Ministry of Human Capacities, No. ELTE/8495/53(2016).

References

[1] T.B. Britton, H. Liang, F.P.E. Dunne, A.J. Wilkinson. The effect of crystal orientation on the indentation response of commercially pure titanium: experiments and simulations. Proc. R. Soc. A (2010) 466:695.

[2] J.W. Kysara, Y. Saitoa, M.S. Oztopa, D. Leeb and W.T. Huhc. Experimental lower bounds on geometrically necessary dislocation density. International Journal of Plasticity (2010) 26(8):1097.

IOP Conf. Series: Materials Science and Engineering **426** (2018) 012022 doi:10.1088/1757-899X/426/1/012022

[3] E. Demir, D. Raabe, N. Zaafarani and S. Zaefferer. Investigation of the indentation size effect through the measurement of the geometrically necessary dislocations beneath small indents of different depths using EBSD tomography. Acta Materialia (2009) 57(2):559.
[4] F. Javaid, E. Bruder and K. Durst. Indentation size effect and dislocation structure evolution in (001) oriented SrTiO3 Berkovich indentations: HR-EBSD and etch-pit analysis. Acta Materialia (2017) 139:1.
[5] G. Feng, A.S. Budiman, W.D. Nix, N. Tamura, and J.R. Patel. Indentation size effects in single crystal copper as revealed by synchrotron X-ray microdiffraction. Journal of Applied Physics (2008) 104:043501.
[6] T.B. Britton and A.J. Wilkinson. High resolution electron backscatter diffraction measurements of elastic strain variations in the presence of larger lattice rotations. Ultramicroscopy (2012), 114:82.
[7] Sz. Kalácska, I. Groma, A. Borbély and P.D. Ispánovity. Comparison of dislocation density obtained by HR-EBSD and X-ray profile analysis. Applied Physics Letters (2017) 110:091912.
[8] C. Reuber, P. Eisenlehr, F. Roters and D. Raabe. Dislocation density distribution around an indent in single-crystalline nickel: Comparing nonlocal crystal plasticity finite-element predictions with experiments. Acta Materialia (2014), 71:333.
[9] A.J. Wilkinson and T.B. Britton. Strains, planes, and EBSD in materials science. Materials Today (2012), 15(9):366.
[10] B. Britton, I. Holton, G. Meaden and D. Dingley. High angular resolution electron backscatter diffraction: measurement of strain in functional and structural materials. Microscopy and Analysis (2013), 27(4):8.

IOP Conf. Series: Materials Science and Engineering **426** (2018) 012023 doi:10.1088/1757-899X/426/1/012023

Examination of the effect of homogenization processes through compression tests in aluminum alloys

V Karpati[1], T Miko[1], P Barkoczy[1], D Angel[1], Z Gacsi[1], M Benke[1]

[1] University of Miskolc, Institute of Physical Metallurgy, Metalforming and Nanotechnology, Miskolc-Egyetemváros, Hungary

E-mail: cirkonium2000@gmail.com

Abstract. Formability characterises the behaviour of the metals during shaping, more precisely, the accumulated deformation until rupture or fracture. Formability can be influenced by several factors and can be characterized with several methods. In this paper, the effect of the processes occurring during the pre-heating and homogenization of formable aluminium alloys was studied. Three commercial aluminum alloys, namely, 3003, 5005 and 5052 were involved into the investigation to reveal the different behaviours resulting from composition differences. The characterization of hot formability must be characterized by a rapid and simple method in cases when large number of sample series must be handled. In parallel, reliable results of the examination method is required. To fulfil both requirements, cold and hot compression tests were carried out with an Instron 5982 universal material tester equipment. The rapid heating of the samples during hot compression tests was realized with a unique own-made inductive heating device. Thus, the effect of dynamic recrystallization during hot compression, while static recrystallization during cold compression tests and subsequent annealing could be studied. The performed experiments were used to simulate the effects of the cold-rolling and hot-rolling. The resulting grain structures were studied with Barker etching technique.

1. Introduction

The Faculty of Material Science at University of Miskolc and the Arconic-Köfém Ltd. with several other Universities in Hungary started a research project together. For the Arconic-Köfém Ltd., as the manufacturer of aluminum alloys, formability of the alloys is a key issue. Thus, the goal is to develop and carry out a series of experiments which provide a rapid, precise and reliable measurement method to study the formability of large numbered sample series in a short time.

The fundamental difference between cold and hot forming is that recrystallization process occurs at the temperature of forming or not. Recrystallization is a physical metallurgical process consisting of nucleation and grain growth, of which the driving force is the increased stored energy of the metal accumulated during deformation. During cold forming, recrystallization doesn't take place, thus dislocation density is increased. Increasing the density of dislocations will cause the material to become hardened and its formability will be depleted. During hot forming, the recrystallization takes place at the temperature of forming. Thus, dynamic recrystallization and static recrystallization must be distinguished [1-3]. The aim of this manuscript is to study the effect of recrystallization process (both static and dynamic) on the formability of three commercial aluminium alloys, namely: 3003, 5005 and 5052 subjected to different durations of homogenization [4,5].

2. Experimental

Three different formable alloys were investigated, namely the 3003, the 5005 and 5052. The composition of the examined alloys is shown in Table 1. The main alloying element of the 3003 alloy is Mn around 1% content. This content is mostly in solution and different precipitations, which greatly influences the formability. The 5xxx alloys are also formable, although they belong to the non-precipitation hardening alloy. The main alloying element of this group is the Mg around 0.7-2.45% content.

Table 1. Composition of the examined alloys

Alloy	Al	Si	Mn	Fe	Mg	Zn	Pb	Cu	Cr	Ti	Cd	Ni	Zr
3003	97,99	0,23	1,1	0,53	0,016	0,023	0,001	0,057	0,004	0,022	0,0005	-	-
5005	98,48	0,14	0,054	0,47	0,77	0,008	0,001	0,012	0,014	0,012	-	-	-
5052	96,89	0,12	0,08	0,26	2,41	0,01	0,002	0,044	0,16	0,012	-	-	-

One of the industrial homogenizations (shown in Fig 1.) used by the Arconic-Köfém Ltd. was applied. The peak temperature of the homogenization is 510°C, and the holding times are 1, 5 and 10 hours. Heat treatments were performed in an air furnace. The different homogenization durations modelled cases where different sections of the ingot (beginning, middle, end part) for different reasons (the non-planned execution of the heat treatment, etc.) spends different time intervals in the homogenization furnace.

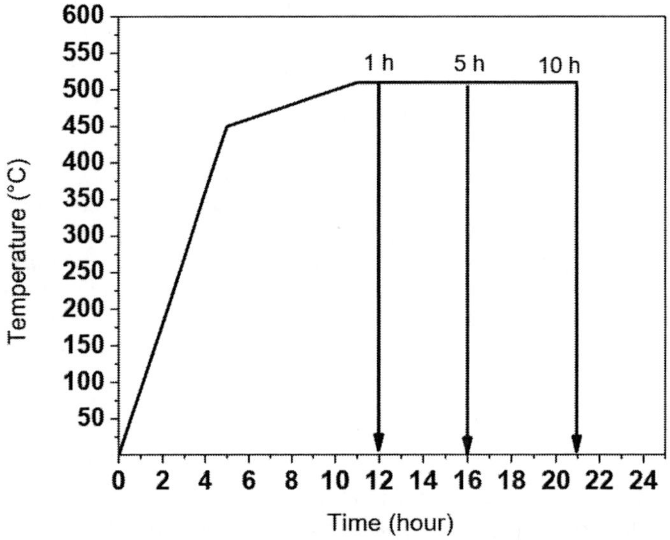

Figure 1. The applied homogenization programs

Cylyndrical samples were cut from the ingots and cold and hot compression tests were performed on them with an Instron 5982 type universal material tester equipment. To prevent barrelling [6,7], a small edge was left on the two flat sides of the compression samples in which a Teflon film was inserted to reduce friction between sample and sample holder. Cold compressions were performed at 23°C and hot deformations were carried out at 400°C with 1/s deformation rate up to 60% deformation. Sample heating was realized with an own-built inductive heating device providing rapid heating and cooling to prevent any recrystallization process after the compression tests.

The cold compression was followed by an annealing treatment at 500°C for 1 hour during which static recrystallization was expected to occur. The tested samples were cut in the plane of pressing (longitudinal section) embedded in resin and prepared with Barker etchant to reveal the grain structure (Fig. 2.).

Cold compression

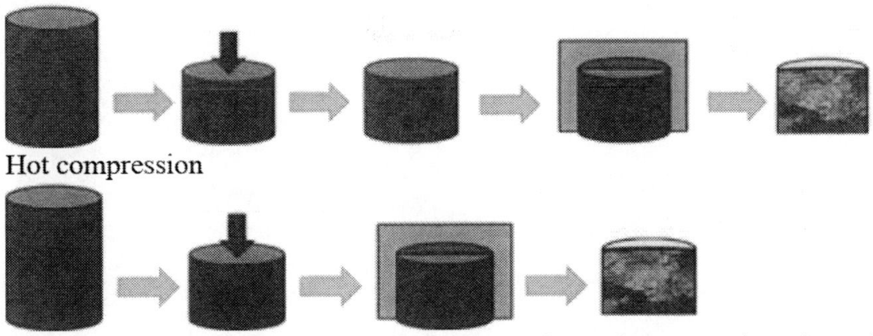

Figure 2. Schematic of compression tests subsequent and sample preparations for grain structure investigations. Cold compression tests (blue) were followed by annealing (red), and sectioning, Hot compression tests (red) were followed by sectioning.

3. Results

Figures 3-5 show the true strain-true stress of the examined alloys homogenized for 1, 5 and 10 hours during cold and hot compression tests; and the grain structure after the tests and subsequent annealing in case of cold compression and after tests in case of hot compressions.

IOP Conf. Series: Materials Science and Engineering **426** (2018) 012023 doi:10.1088/1757-899X/426/1/012023

Alloy 3003

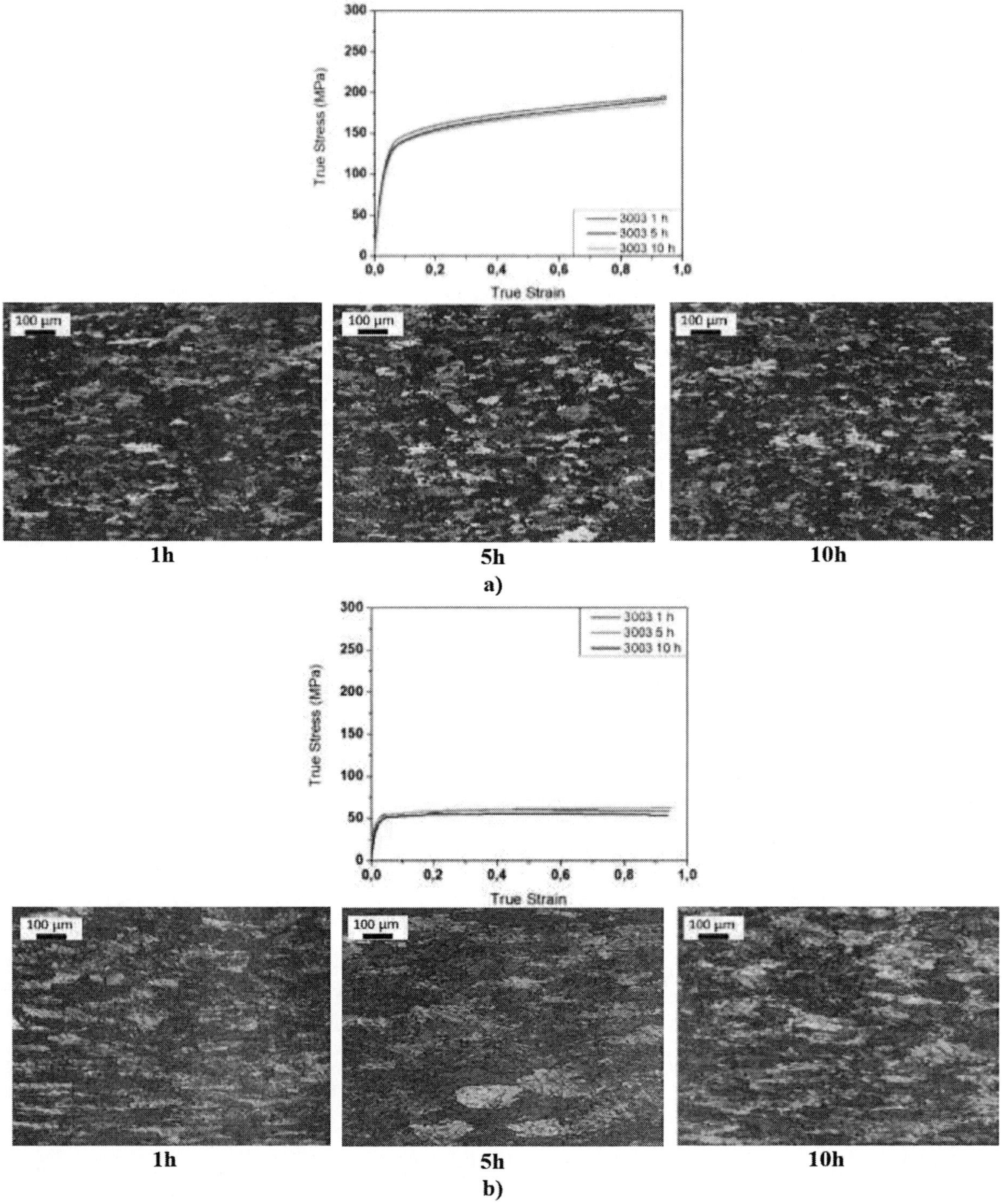

Figure 3. True strain-true stress curves of alloy 3003 during a) cold compression tests and grain structure after cold compression and annealing b) hot compression tests and grain structure after hot compression

Alloy 5005

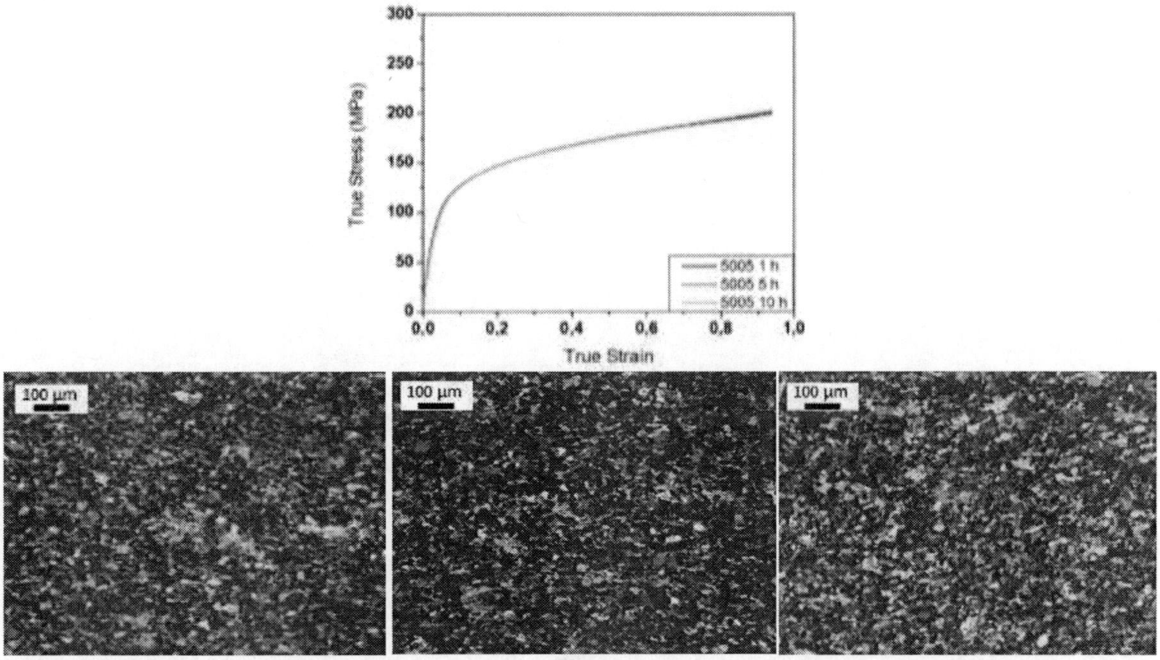

1h **5h** **10h**

a)

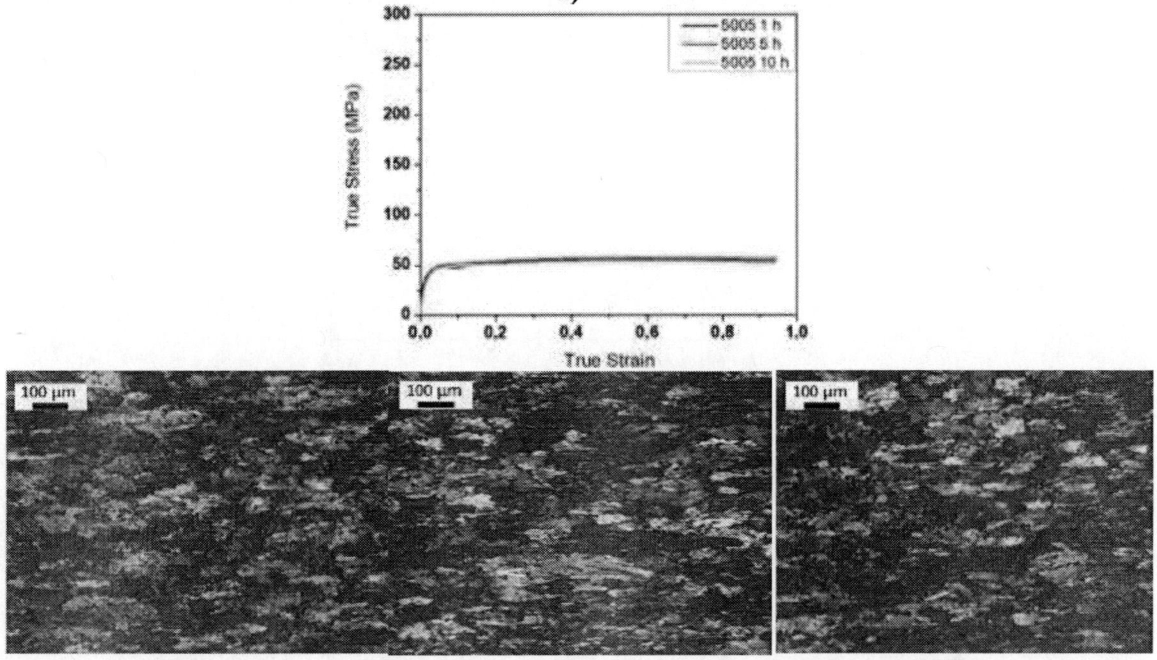

1h **5h** **10h**

b)

Figure 4. True strain-true stress curves of alloy 5005 during a) cold compression tests and grain structure after cold compression and annealing b) hot compression tests and grain structure after hot compression

IOP Conf. Series: Materials Science and Engineering **426** (2018) 012023 doi:10.1088/1757-899X/426/1/012023

Alloy 5052

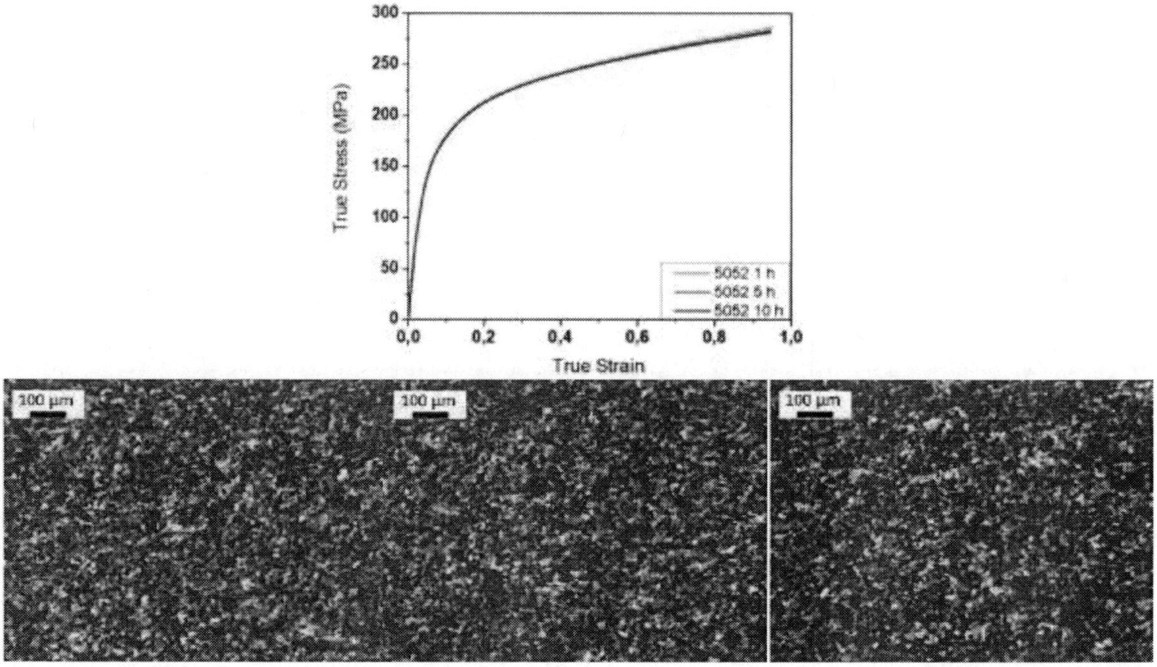

1h 5h 10h

a)

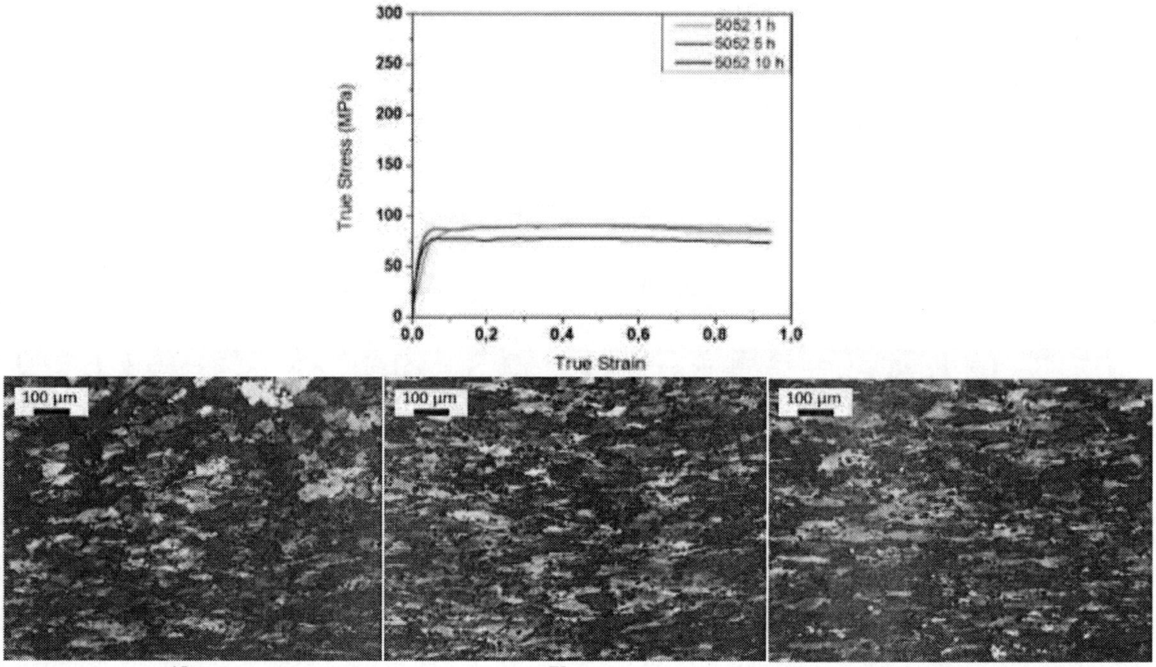

1h 5h 10h

b)

Figure 5. True strain-true stress curves of alloy 5052 during a) cold compression tests and grain structure after cold compression and annealing b) hot compression tests and grain structure after hot compression

In Fig. 3. it can be seen that the true strain-true stress curves of alloy 3003 obtained during cold and hot compression tests are very similar in case of different homogenizing durations. According to the micrographs, after cold compression tests and annealing, most of the grains are elongated suggesting that recrystallization partially occurred during the annealing. It is true for all three alloys that after hot compression tests, the grain structures are mixtures of grains that are severely deformed and others that are mildly deformed. The less deformed grains were formed during the dynamic recrystallization occurring during the compression. Their slightly elongated shape is the result of the combination of dynamic recrystallization and compression [8].

In Fig. 4. the true strain-true stress curves of alloy 5005 obtained during cold and hot compression tests are very similar in case of different homogenizing durations. After cold compression and annealing, equiaxial grains are present which means that the classic, high-angle rain boundary involved recrystallization occurred during the annealing treatment [8].

In Fig. 5. the true strain-true stress curves of alloy 5052 obtained at different homogenization durations during cold compressions completely match with each other. As for hot compression tests, a slight difference can be seen: the sample homogenized for 10 hours has lower stress values compared to the other two. After cold compression and annealing, the grain structure is equiaxial and the grains are finer compared to the alloy 5005 after cold compression and annealing. This suggests that the classic, high-angle grain boundary movement assisted recrystallization occurred with higher nucleation rate compared to the 5005 alloy.

Fig. 6. shows the maximal values of true stress of the three alloys as a function of homogenization time. As can be seen, the maximal stresses of cold compression tests do not depend upon the homogenization time. The values of alloys 3003 (Mn content 1.1 m/m%) and 5005 (0.054 m/m% Mn and 0,77 m/m% Mg), are close, but that of 3003 are slightly smaller. Alloy 5052 has the highest maximal stress value. This is probably due to the high (2.41 m/m%) Mg content [8].

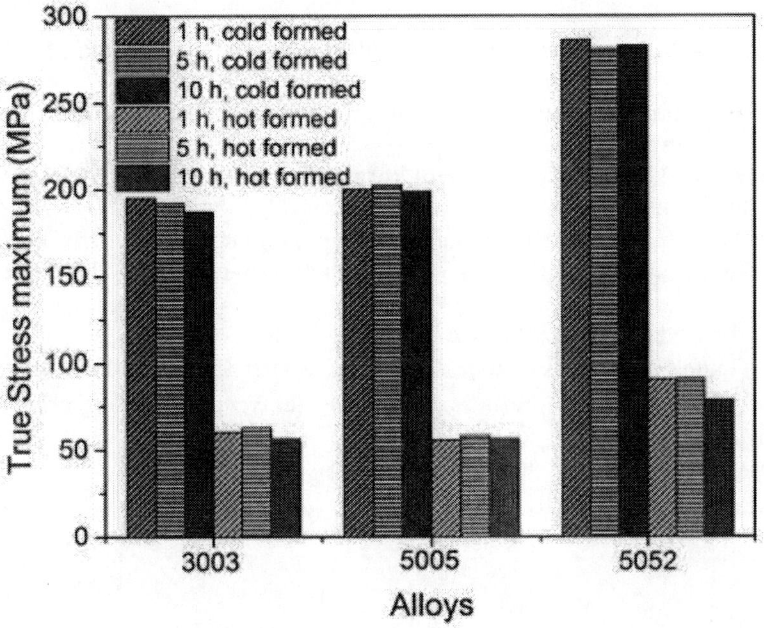

Figure 6. Maximal true stress values as a function of homogenization time of the 3003, 5005 and 5052 alloys

4. Conclusions

The hardening behaviour was examined with cold and hot compression tests of 3003, 5005 and 5052 alloys subjected to different durations of previous homogenization heat treatments. In addition, grain structures formed after cold compression and subsequent annealing, and hot compression was examined. It was found that the variation in the homogenization time between the 1 … 10 h interval does not have a major effect on the obtained true strain-true stress curves of all three alloys or resulting grain structure. Alloys 3003 and 5005 showed similar maximal true stress values, while alloy 5052 had higher maximal stress values after both cold and hot compressions. The maximal compression stresses are due to the high Mg content of alloy 5052 (2.41 wt%). For alloy 3003, the high angle boundary involved classic recrystallization occurred partially, while for alloys 5005 and 5052 it took place completely. The domination of high angle grain boundary involved recrystallization is probably owing to the higher Mg content of alloys 5005 and 5052 (0.77 and 2.41 wt%, respectively). For alloy 5052, the classic recrystallization occurred with higher nucleation rate. Again, this is due to the high Mg content, which supports secondary phase precipitation, which in turn promotes nucleation during classic recrystallization. Thus, it was shown that the increasing Mg content increase compression stresses and promotes high angle grain boundary recrystallization in Al alloys. The effect of Mg on mechanical properties are in agreement with observations on AlCuMg alloys [9] and on AlMgSi alloys [10]. Thus, the applied method is suitable to reveal the effect of Mg content on mechanical properties of formable Al alloys through simple and rapid examinations. It was observed for all three alloys that after hot compression tests, the grain structures were mixtures of grains that are severely deformed and others that are mildly deformed. This is characteristic of the dynamic recrystallization that occurred during hot compression.

Acknowledgments

The described article was carried out as part of the GINOP-2.2.1-15-2016-00018 project implemented in the framework of the Szechenyi 2020 program. The realization of this project is supported by the European Union.

References

[1] Dr. Gácsi Z and Dr. Mertinger V Fémtan, *Educatio Társadalmi Szolgáltató Nonprofit Kft. (in Hungarian)*
[2] Mikó T 2016 Alumínium-magnézium ötvözetek folyási viselkedése zömítés során *doktori (PhD thesis) (in Hungarian)*
[3] Altan T, Ngaile G and Shen G Cold and Hot Forging Fundamentals and Applications, Materials Park Ohio 44073-0002
[4] Dr. Gillemot L 1967 Anyagszerkezettan és anyagvizsgálat *Nemzeti Tankönyvkiadó Budapest*
[5] Szmirnov-Aljajev G Sz 1961 Szoprot'ivlenije materialov plaszticseszkomu deformirovaniju *Masgiz*
[6] Aluko O and Adeyemi M B 1998 *JMEPEG* **7** 474-478
[7] Williams J G and Gamonpilas C 2008 *Int J Solids Struct* **45** 4448–4459
[8] Dr. Zorkóczy B 1968 Metallográfia és anyagvizsgálat *Nemzeti Tankönyvkiadó Budapest (in Hungarian)*
[9] Girisha H N and Sharma K V 2012 *Int J Sci & Eng Res* **3** 2 1-4
[10] Yildirim M and Ozyurek D 2013 *Mat Design* **51** 767–774

IOP Conf. Series: Materials Science and Engineering **426** (2018) 012024 doi:10.1088/1757-899X/426/1/012024

Development of a shear-tensile testing method to qualify metal spraying technologies

B Katona[1] and K Bödök[1]

[1] Corweld Kft., Diósd, Hungary

E-mail: katona@corweld.hu

Abstract. The metal spraying is widely used process to extend the resistance against to wearing, erosion or corrosion. The related standard defines these surface protective methods as special processes because these cannot be qualify by a non-destructive testing method which clearly states the fulfilment of the requirements of a given work. Therefore the quality of the given surface protective method can be determined only during the later use. The main goal of this work is to define a control method which can be used for qualify the given metal spraying method. In our work the used base material as carrier material was P235GH steel and we investigated 6 different metal sprayed layers. To avoid the stress concentrator areas (small cracks) on the surface before the test the surface was grinded and EDM was used to cut out the samples from the manufactured work pieces. To investigate the properties of the sprayed layers tensile tests were used. From the data the strength of the composites and the layers can be calculated. Based on the shear-tensile test we created a calculation method which can be used to qualify the metal spraying methods before its use.

1. The metal spraying

The thermal spraying technologies like metal spraying is widely used for manufacturing or repair/maintenance works in the industry. The main goal of the metal spraying is to create such a layer which is improve the resistance against to wearing, erosion, cavitation, corrosion or the combination of the mentioned damage methods. Therefore the metal spraying technology extend the surface properties and reconstruct the geometrical dimensions of the given part.

The concerning standards (MSZ EN ISO 9000, MSZ EN ISO 14922 [1]) define the surface protective processes as special processes because these cannot be qualifying by a non-destructive testing method during the manufacturing which clearly states the fulfilment of the requirements of a given work. Therefore the quality and the possible mistakes of the given metal spraying method can be determined only during the later use. Because of this it is heavily necessary a method especially during the manufacturing which properly qualify the given work.

There are some researches research about the shear testing of the sprayed coatings and the bonding between the carrier material and the layer which gives the bond stress and the shear strength of the tested material as a result. The results are very promising in every case but the main disadvantage of these methods that they needs a special testing equipment therefore they can not be executed on a classical universal material testing machine [2-4].

A wide range of the thermal spraying method is known and used in the industry which is summarized by the MSZ EN 657:2005 [5] but 5 processes (Powder flame spraying (PFS), Wire flame spraying

IOP Conf. Series: Materials Science and Engineering **426** (2018) 012024 doi:10.1088/1757-899X/426/1/012024

(WFS), Arc spraying (AS), Plasma spraying in atmospheric condition (APS), High velocity oxygen fuel (HVOF)) are current. In this study we investigated only two of them which we discuss in detail. In the case of some technology the melting of the wire or powder executed by the same method as the arc welding [6] (in the case of AS) or as the plasma nitriding [7] (in the case of APS).

Powder flame spraying (PFS)

The PFS (Figure 1.) is the most common metal spraying technology worldwide by using oxygen-acetylene or propane heating gas. During the process the sprayed powder is being heated in plastic or melted condition and shot on the surface of the carrier material by the heating gas. Two different technologies are distinguished: warm and cold spraying. In the case of warm spraying the carrier material is preheated to 200-250°C and therefore no primer layer is needed in front of cold spraying where a nickel-base primer layer is used. The velocity of the powder is relative small (40-100 m/s) therefore the manufactured layer can be porous.

Arc spraying (AS)

In the case of AS (Figure 2.) two wires are dosed simultaneously and between them an electrical arc is created to melt the top of them like in the case of arc welding. The used wires can be similar or different materials as well. The melted drops are shot by gas on the surface by the velocity of 100-130 m/s (max. 200 m/s) which is similar to the WFS.

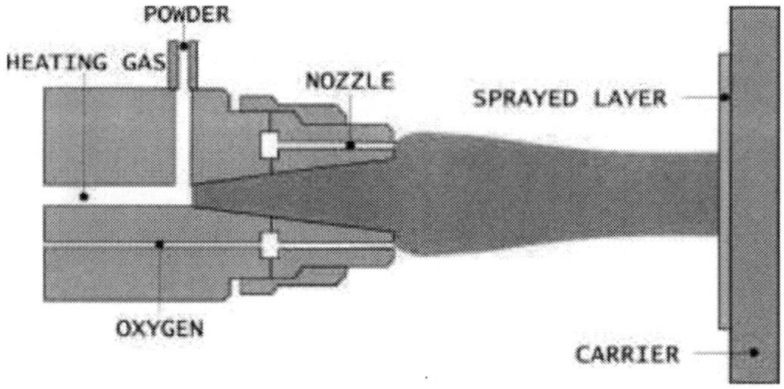

Figure 1. Schematic illustration of the powder flame spraying (PFS)

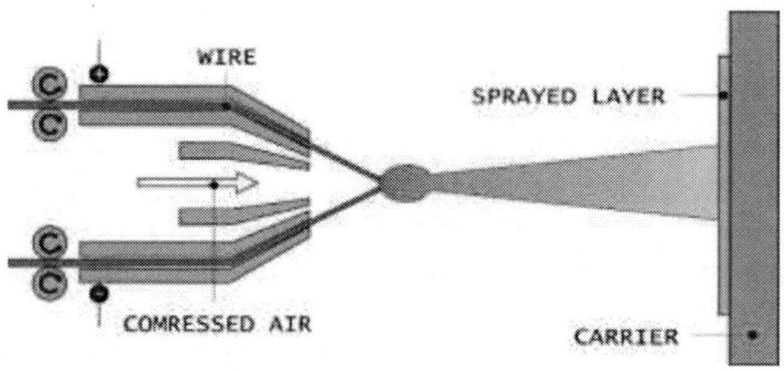

Figure 2. Schematic illustration of the arc spraying (AS)

2. Materials and methods

In our work P235GH micro alloyed steel was used as carrier material (Figure 3.) with the thickness of 5 mm. As metal spraying we used 6 different technology/material combinations (Figure 4.). Table1. shows the used technologies, materials and the thickness of the sprayed layers.

Table 1. The technologies, the materials and the thickness of the sprayed layers.

Sample	Technology	Primer layer	Layer	Layer thickness (mm)	Comment
A	AS	Kasamas KLD 75	Kasamas KLD 60	0.45-0.55	-
B	AS	Kasamas KLD 75	Kasamas KLD 60	1.45-1.95	-
C	AS	-	Soudokay 848M	1.65-1.75	-
D	PFS	-	Böhler UB5- 2545	1.35-1.45	free deformation
E	PFS	-	Böhler UB5- 2545	1.35-1.45	blocked deformation
F	PFS	UTP EB-1003	Metco 130	1.65-1.75	-

The main difference between the applied coating is the coating material and the thickness of the sprayed layer. In the case of sample A, B and F a primer layer was used which helps the bonding between the carrier material and the coating. In the case of sample C, D and E it was not necessary to use this primer layer. By the samples A-B and D-E the spraying technology and the applied materials were same the only difference is the thickness of the coatings. In the case of sample D and E because of the nature of the applied technology (Powder flame spraying (PFS)) the carrier heated and deformed. By the sample D this deformation was not blocked but in the case of sample E we fixed the carrier to avoid that.

To avoid the stress concentrator areas (small cracks) on the surface before the test the surface was grinded and EDM (Electrical discharge Machining) was used to cut out the samples from the manufactured work pieces (Figure 5.). As testing method a shear-tensile testing was used this was performed on a MTS 810 hydraulic materials testing machine at 1 mm/min speed rate. The elongation of the samples was monitored by Mintron video extensometer. To avoid the damage of the layer during the holding special sample geometry was used (Figure 6.). Before the test the main dimensions of the samples were measured.

To determine the properties of the carrier material a non-sprayed sample was investigated by the mentioned shear-tensile method. The test itself is a tensile test but because the plane of the applied load is out of the plane of the carrier-layer bonding the tensile load will result shear load in the boundary layer as well. Table 2. shows the measured and calculated values of the carrier material.

IOP Conf. Series: Materials Science and Engineering **426** (2018) 012024 doi:10.1088/1757-899X/426/1/012024

Figure 3. Plate from the carrier material

Figure 4. Metal spraying on the carrier surface

Figure 5. Grinded and EDM cut sample.

Figure 6. Prepared sample for the shear-tensile test with the holding accessories.

Table 2. The properties of the P235GH carrier material.

Upper yield stress (MPa)	$R_{p0.2}$ (MPa)	Tensile strength (MPa)	Elongation, A (%)	Contration, Z (%)	Young's modulus, E (GPa)
294	291	407	39	60	2,18x105

The first cracks in the sprayed layers appear at the first maximum peak of the tensile diagram which is the border between the elastic-plastic behaviour. At this point the measured force consist of two parts: the force of the carrier material and the force of the sprayed layer (Figure 7-8.).

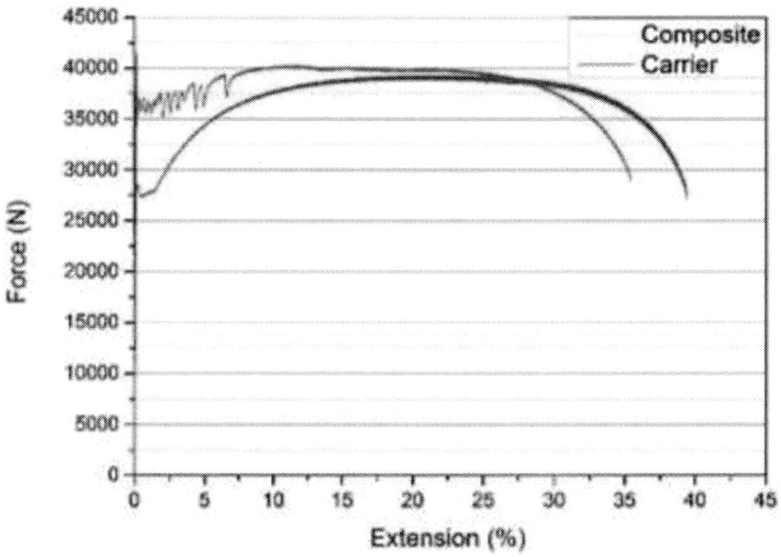

Figure 7. The force-elongation diagram of the carrier material and the carrier-sprayed layer composite by the shear-tensile test.

IOP Conf. Series: Materials Science and Engineering **426** (2018) 012024 doi:10.1088/1757-899X/426/1/012024

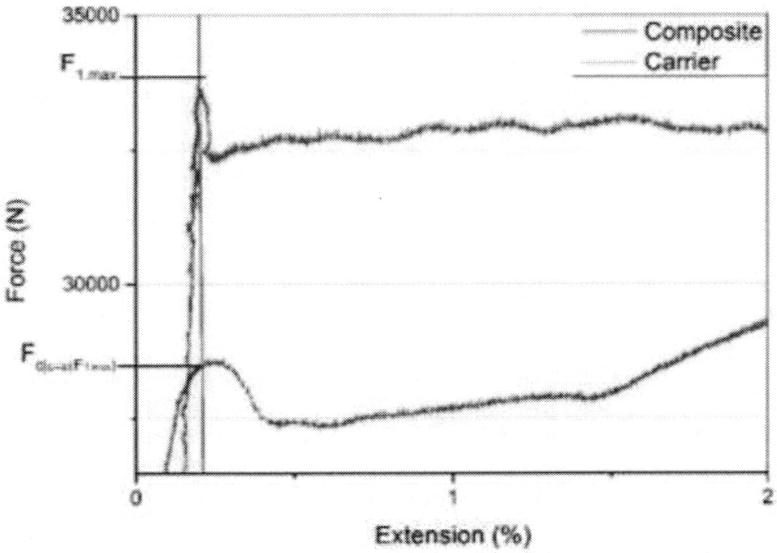

Figure 8. The beginning of the force-elongation diagram with the values for the calculations.

Due the pre-tests the mechanical properties of the carrier material are known therefore the strength of the sprayed layer can be calculated:

$$\sigma_{layer} = \frac{F_{1.max} - F_c|_{\varepsilon=\varepsilon(F_{1.\max})}}{S_{0,layer}}$$

where $F_{1.max}$ is the first maximal peak where the first cracks appear on the layer, $F_{c|\varepsilon=\varepsilon(F1.max)}$, is the force of the non-sprayed carrier material at the same elongation which was at $F_{1.max}$, $S_{o,layer}$ is the cross section of the sprayed layer.

For the comparison of the samples, the average stress in the carrier-sprayed layer composites can be determined at the first cracking in the sprayed layer using the following formula:

$$\sigma_{composite} = \frac{F_{1.max}}{S_{0,composite}}$$

where $F_{1.max}$ is the first maximal peak where the first cracks appear on the layer and $S_{o,composite}$ is the cross section of the composite.

3. Results

The mentioned testing method was performed on the samples which results shows a sizeable difference between the variant spraying materials and technologies.

Sample A (AS, Kasamas KLD 75+Kasamas KLD 60, thin layer)

The sprayed layer showed rigid behaviour, the cracks was appeared in the whole volume of the layer. After the first cracks the layer did not added strength to the composite. Before the contraction some larger piece of the sprayed layer removed from the carrier surface which means bad connection between the carrier and the sprayed layer. At the contraction the layer fully removed (Figure 9.).

Sample B (AS, Kasamas KLD 75+Kasamas KLD 60, thick layer)

In the case of sample B the sprayed layer started to peel from the carrier at the plastic zone therefore it was fully removed at the end of the shear-tensile test. It can be concluded that the applied materials and spraying parameters do not meet the requirements (Figure 10.).

Sample C (AS, Soudokay 848M)

The sprayed layer showed rigid behaviour, the cracks was appeared in the whole volume of the layer. After the first cracks the layer did not added strength to the composite. Before the contraction some larger piece of the sprayed layer removed from the carrier surface which means bad connection between the carrier and the sprayed layer. At the contraction the layer fully removed (Figure 11.).

Sample D and E (PFS, Böhler UB5- 2545)

The sprayed layer resulted significant increasing in the measured force in the elastic and the plastic zone as well. When the first crack was appeared a remarkable drop was seen on the tensile diagram. The connection between the sprayed layer and the carrier material is appropriate which is certified by the secondary cracks in the layer which was perpendicular to the direction of the tensile (Figure 12-13).

Sample F (PFS, UTP EB-1003+ Metco 130)

The sprayed layer showed rigid behaviour, the cracks was appeared in the whole volume of the layer. During the test the layer was removed from the carrier. The sprayed layer was very porous (Figure 14.).

Figure 9. Sample A after the shear-tensile test.

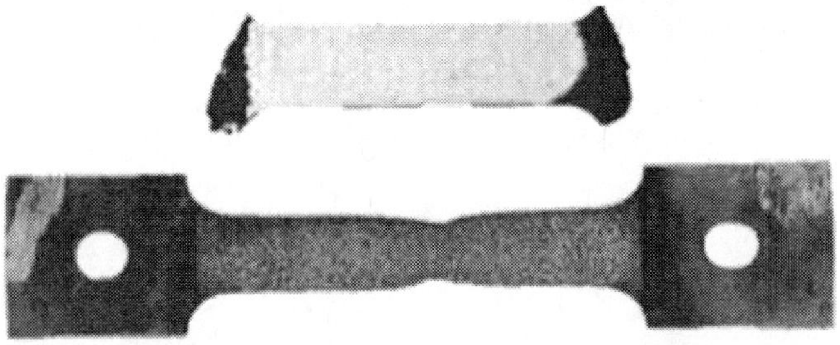

Figure 10. Sample B after the shear-tensile test.

Figure 11. Sample C after the shear-tensile test.

Figure 12. Sample D after the shear-tensile test.

Figure 13. Sample E after the shear-tensile test.

Figure 14. Sample F after the shear-tensile test.

4. Summary

The main advantage of the studied shear-tensile test method is it can be performed on a universal testing machine therefore it does not need any special testing equipment. Based on our results it can be concluded the presented testing-calculation method is appropriate to investigate and qualify the different metal spraying technologies right after the manufacturing process. With the discussed calculation method the different sprayed layer materials can be measured indirectly.

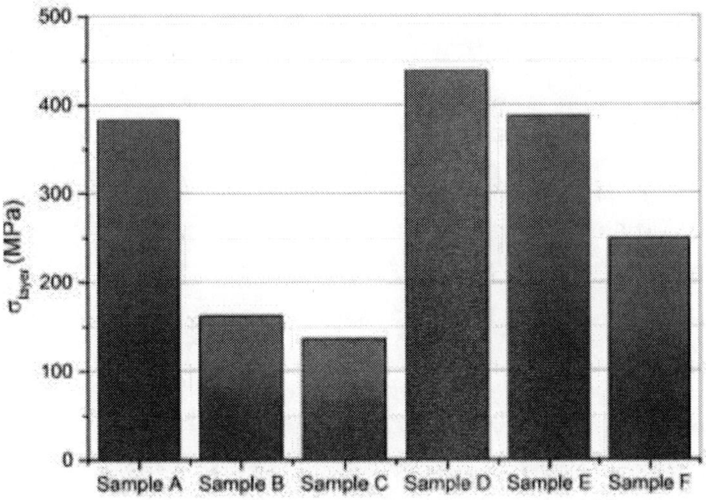

Figure 15. The calculated stresses in the sprayed layers.

IOP Conf. Series: Materials Science and Engineering **426** (2018) 012024 doi:10.1088/1757-899X/426/1/012024

In the case of the investigated spraying technologies with different powder or wire material can be seen that the PFS technology resulted more strength and more adhesive layers (Figure 15-16.).

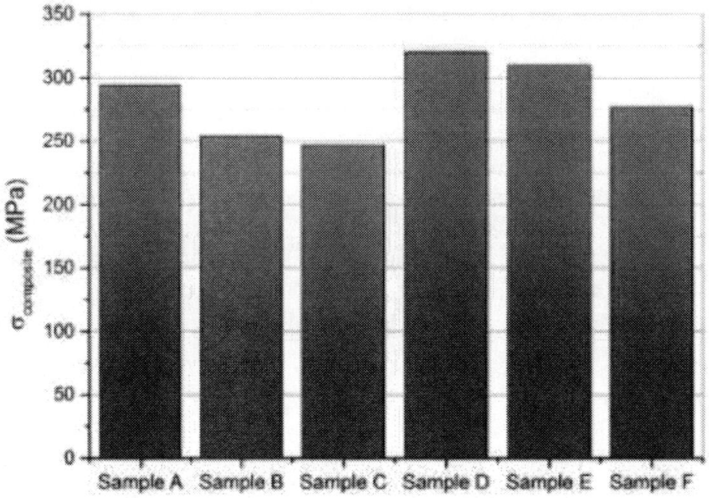

Figure 16. The calculated stresses in the carrier-sprayed layer composites.

References

[1] ISO 14922-4:1999: Thermal spraying -- Quality requirements of thermally sprayed structures -- Part 4: Elementary quality requirements

[2] S. Siegmann, M. Dvorak, H. Grützner et al.: Shear testing for characterizing the adhesive and cohesive coating strength without the need of adhesives, Proceedings of ITSC 2005 Thermal Spray connects: Explore its surfacing potential! (2005) 823-829

[3] C. C. Berndt: Tensile adhesion testing methodology for thermally sprayed coatings, Materials Engineering 12 (1990) 151-158.

[4] M. Kašparová, J.Volák, F.Zahálka et al.: Shear Strength of Thermally Sprayed Coatings, Chem. Listy 105 (2011) 818-819

[5] MSZ EN 657:2005: Thermal spraying. Definitions, grouping.

[6] Varbai B, Kormos R, Májlinger K 2017 Effects of Active Fluxes in Gas Metal Arc Welding Periodica Polytechnica Mechanical Engineering vol 61(1) pp 68-73

[7] D. Kovács, J. Dobránszky: Surface properties of plasma nitrided tempered steel, in Francesca Cosmi (ed.) "34th Danubia-Adria Symposium on Advances in Experimental Mechanics, University of Trieste, Italy, 2017", Trieste, EUT Edizioni Università di Trieste, 2017

IOP Conf. Series: Materials Science and Engineering **426** (2018) 012025 doi:10.1088/1757-899X/426/1/012025

EBSD examination of Ti-6Al-4V samples produced with additive technology

Z Keresztes[1], D Pammer[1], D Karoly[1] and J P Szabo[1]

[1]Budapest University of Technology and Economics, Faculty of Mechanical Engineering, Department of Material Science and Engineering, Budapest, Hungary

E-mail: kzoltan@eik.bme.hu

Abstract. Rapid Prototyping (RP) is a special technology which has many types included Direct Metal Laser Sintering (DMLS). The process starts by applying a very thin layer of material onto the construction platform, then fusing the powder at the exact points specified by the CAD model. After that, the platform is lowered and another layer of the substance is laid restarting these steps till the desire model is ready. Nowadays Additive Manufacturing (AM) expression is used instead of RP, because it is characterizing much better the process. This paper focuses on the microstructure, and EBSD examination of Ti-6Al-4V (Grade 5) samples produced with DMLS technology. Our main aim was to compare the properties of AM produced samples to the properties of traditional Grade 5 titanium alloy.

1. Introduction

Rapid Prototyping (RP) is a term used to describe fabrication processes responsible for creating representations and 3D models from digital information. This means that RP techniques are mostly used to materialize ideas and improve initiatives that are not finished yet, spending less money and time than traditional manufacturing procedures. RP is used for building final products as prototypes or can be found in different fields of art and hobbies. Nowadays the continuous improvement of this technology surprisingly increases the range of usage. Specialist have realized that instead of "Rapid Prototyping" a much more adequate name would be better in the scenario, so The American Society for Testing and Materials (ASTM) adopted the expression "Additive Manufacturing". This refers to the process of fabrications of parts using an additive approach [1, 2]. Of course we can find other synonyms such as: additive fabrication, additive process, etc. [3].

At the beginning of the 20th century scientists investigated new technologies to replace subtractive manufacturing processes and improved casting methods. During these researches prototypes were produced by human hand, but parts were not precise and repeatable with same geometry. These mistakes could be solve using machine-controlled technologies [4]. Despite of the fact, that AM just starts to spread around 1980, it is based on two technologies from the 19th century. The artist, Francois Willéme made 3D models taking photos of an object from 24 different angels. This is called photo-sculpture [5]. The other basic is from Joshep E. Blanther who prepared 3D maps placing layers onto layers [6]. In 1956 Jhon Munz successfully made 3D models using photosensitive raw material [7]. The very first real model related to Charles W. Hull, who used computer to build layer-by-layer parts from polymers in 1986 [8]. 2005 brings a great innovation, when an amateur team made available 3D printing worldwide for everyone using Scott Crump's thread-melting technology [9].

AM technology requires the raw material to be in the form of a liquid, powder or sheet, and it is usually ceramic, metal or polymer [10]. According to these there are numerous variations on AM processes separated into seven main categories.

- VAT Photopolymerization,
- Material Jetting
- Binder Jetting,
- Material Extrusion,
- Sheet Lamination,
- Direct Energy Deposition (DED),
- Powder Bed Fusion (PBF) [11].

As previously mentioned, DMLS is included in the Powder Bed Fusion category. Figure 1. shows how the method works schematically [12].

Raw material must be in a special form, called bended powder and nowadays has a wide range of choice such as, powder like stainless steels, titanium alloys, aluminum alloys, copper-based alloys, metal

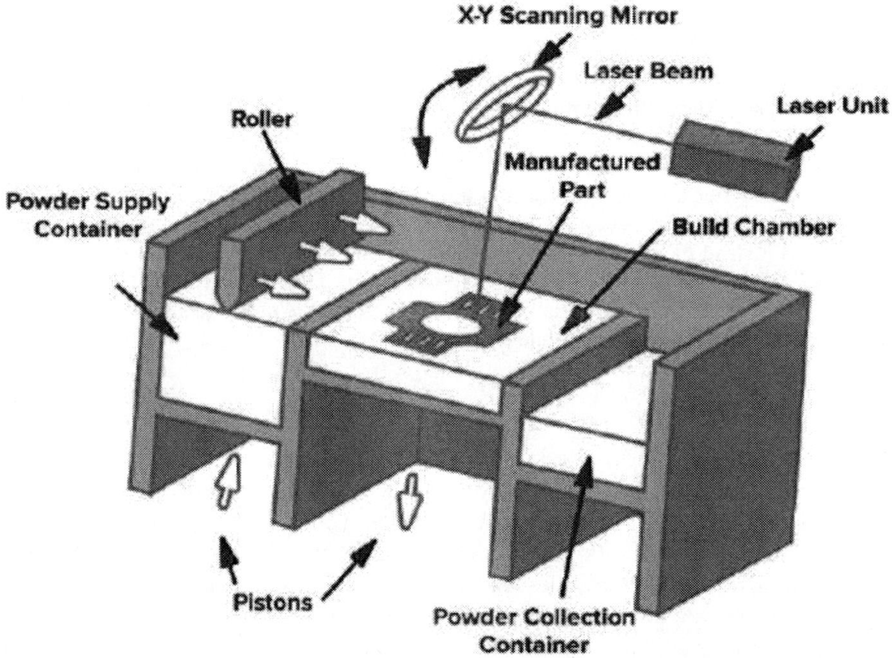

Figure 1. Schematic illustration of DMLS process

matrix composites [13]. In medical researches special composite materials are improved that can be used as implants and prosthesis [14].Another advantage of DMLS technology is the complexity of produced parts [15], which can be built in one production process, reducing post-processes and wasting less material [16]. The parts also have improves mechanical properties due to overlapping of laser paths, which create melt pools that solidify with characteristic fine microstructure [17].

The number of researches about AM grew ten times between 2011 and 2012, while the world market observed and increased from US$ 1.7 billion in 2011 to US$ 2.2 billion in 2012. It is clear that AM technologies are shaping the future of product development and manufacturing [18]. Current study's main aim is to examine the microstructure of DMLS produced Ti-6Al-4V titanium alloy samples.

IOP Conf. Series: Materials Science and Engineering **426** (2018) 012025 doi:10.1088/1757-899X/426/1/012025

2. Material

During tests we examined two variant materials to compare the microstructure of them. The difference was the way of production, one of them was a traditionally made titanium alloy, and the other one was produced with DMLS technology.

2.1. Chemical composition of Ti-6Al-4V material

The commercial available Ti-6Al-4V alloy does not have exact percentage of elements rather has intervals. Table 1. shows the table of chemical composition intervals for traditionally produced Ti-6Al-4V alloys.

Table 1. Chemical composition intervals of traditionally produced Ti-6Al-4V

Elements	Percentage (%)
Titanium (Ti)	88,1 – 91
Aluminum (Al)	5,5 – 6,5
Vanadium (V)	3,5 – 4,5
Iron	< 0,25
Oxygen	< 0,13
Carbon	< 0,08
Nitrogen	<0,03
Hydrogen	< 0,125

3. Metallography

For metallography study first of all the samples must be fixed in resin to make the grinding, polishing and etching effective. For etching we used Keller's reagent (190 ml H_2O, 5 ml HNO_3, 3 ml HCL, 2 ml HF), which is optimal for titanium alloys. Figure 2. shows the microstructure of a traditionally produced and an additive manufactured Ti-6Al-4V sample.

After the investigation using light microscope we used S50 30 kV scanning electron microscope (SEM) for more precise information and examine the chemical composition – especially focusing onto the main alloys - of DMLS produced samples comparing to Table 1.

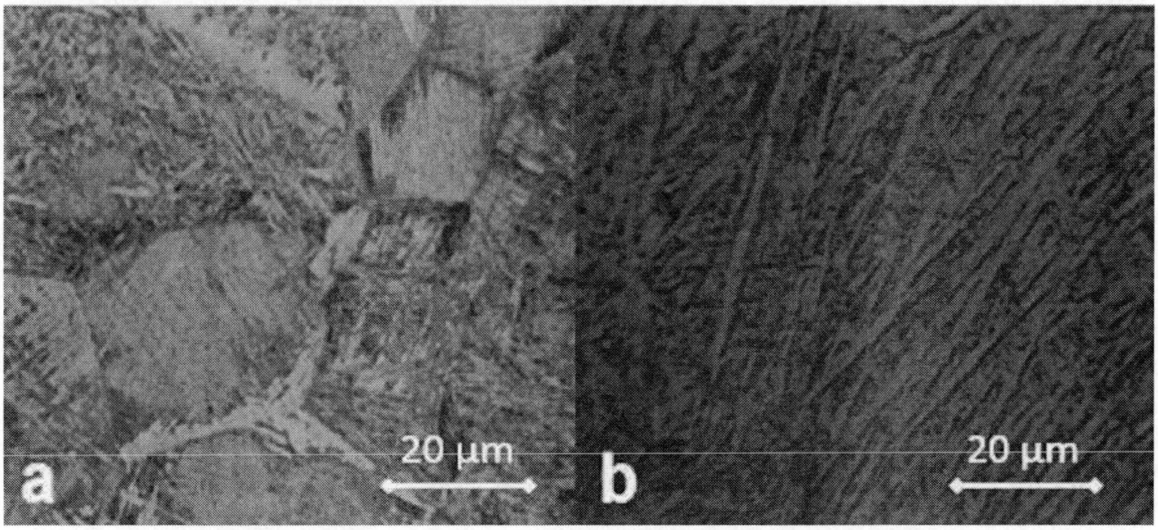

Figure 2. (a) Microstructure of traditionally produced titanium alloy,
(b) Microstructure of DMLS produced titanium alloy

IOP Conf. Series: Materials Science and Engineering **426** (2018) 012025 doi:10.1088/1757-899X/426/1/012025

Table. 2. Percentage of main elements of DMLS produced Ti-6Al-4V

Component	Element interval of traditional Ti-6Al-4V	Element percentage of DMLS Ti-6Al-4V
Titanium (Ti)	88,1 – 91	86,86
Aluminum (Al)	5,5 – 6,4	10,58
Vanadium (V)	3,5 – 4,5	2,56

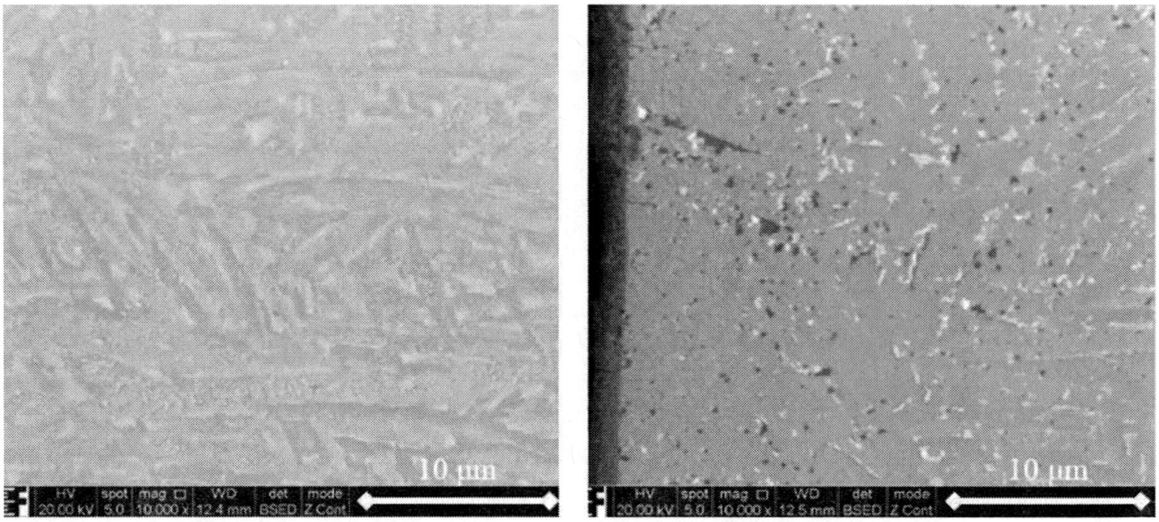

Figure 3. (left) SEM picture of traditionally produced titanium alloy, (right) SEM picture of DMLS produced titanium alloy

3.1. EBSD

Electron Back Scattering Diffraction (EBSD) is a special function of SEM and usually used to examine grain orientation and crystal structure of different materials. A short introduction of EBSD can be found in [19] reference. Our main aim was to obtain information even if DMLS produced titanium alloy has orientation or not. Furthermore we examined the traditionally produced sample's microstructure and compared the results with each other. Figure 4. shows information about traditionally processed Ti-6Al-4V.

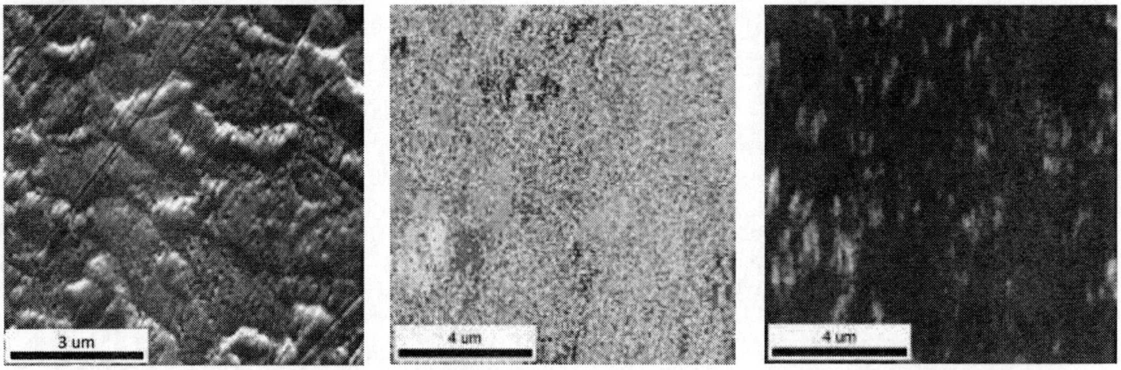

Figure 4. Traditional samples: (left) Test area, (middle) IPF picture with grains, (right) image quality picture

Figure 5. shows the same information about DMLS produced Ti-6Al-4V alloy.

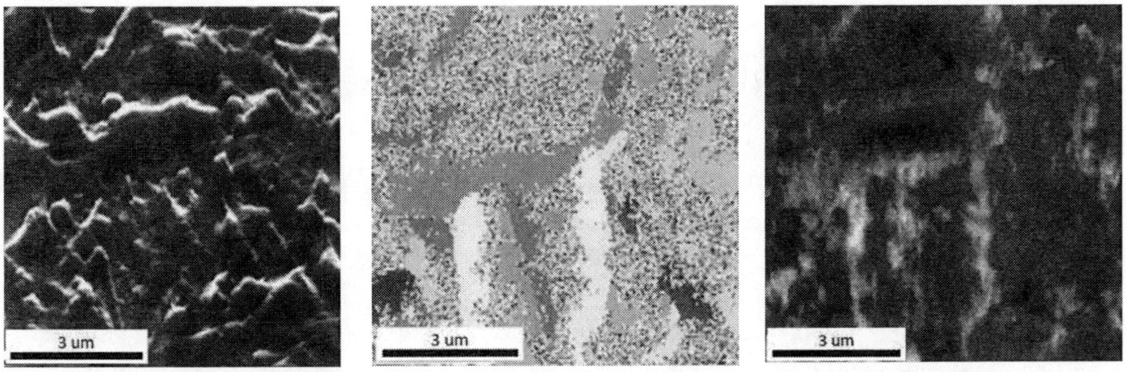

Figure 5. DMLS samples: (left) Test area, (middle) IPF picture with grains, (right) image quality picture

4. Conclusion

DMLS technology shows a great innovation to the economy during the past century, and now it has been gaining space in manufacturing processes. It is very important to continue this improvement in the future, and this is possible because of the standards created by ISO since 2009. DMLS technology is still a new method, so all the tests were performed to show the efficiency of it. It is possible to realize that the grains are much more visible in case of traditionally produced sample, but both have quite the same structure. As chemical composition, additive manufactured sample's elements are nearly in the interval. As the results of EBSD tests grains were shown with nearly equal size and orientation is also possible to examine.

Acknowledgements

The project is funded by the National Research, Development and Innovation (NKFIH) Fund. Project title "Developing a new generation of customized medical implants and medical aids for additive technologies". The application ID number: NVKP-16-1-2016-0022. The developers are grateful for the support.

The research reported in this paper was supported by the Higher Education Excellence Program of the Ministry of Human Capacities in the frame of Nanotechnologyresearch area of Budapest University of Technology and Economics (BME FIKP-NANO).

References

[1] Gibson Ian, Rosen David, Stucker Brent 2009 *Additive Manufacturing Technologies 2nd ed.* (New York)
[2] Bence Rochlitz, David Pammer 2017 Designe and Analysis of 3D Printable Foot Prothesis *Periodica Polytehcnica Mechanical Engineering* **61** Budapest 282-287
[3] Frazier William E 2014 Metal Additive Manifacturing: A Review *Journal Of Materials Engineering And Performance* **23** 1917-1928
[4] Singh R 2006 *Introduction To Basic Manufacturing Processes And Workshop Technology* (New Delhi)
[5] Szobieszek R A 1980 Sculture as the Sum of Its Profiles: Francois Willeme and Photosculpture in France *The Art Bulletin* **4**. 617–630
[6] Blanther J E 1892 *Manufacture of Contour Relief Maps* (USA)
[7] Munz O J 1951 *Photo-Glyph Recording* (USA)
[8] 3D Systems About Us, 3D Systems. Online 2013 *http://www.3dsystems.com/30-years-innovation*
[9] Hopkinson N, Hague R J M, Dickens P M 2006 *Rapid Manufacturing: An Industrial Revolution for the Digital Age* (USA)

IOP Conf. Series: Materials Science and Engineering **426** (2018) 012025 doi:10.1088/1757-899X/426/1/012025

[10] Doré Sylvie 2014 Additive Manufacturing: An Introductory Overview *Research article: Materials and Fabrication* **1**

[11] Loughborough University 2016 *The 7 Categories of Additive Manufacturing* (UK: Loughborough University)

[12] ACF Group 2016 *Laser Sintering* (USA: Florida)

[13] Shellabear M, Nyrhilla O 2004 *DMLS – Development history and state of the art In: Lane Conf.* Erlangen 1 - 12

[14] Barnatt Ch 2014 *3D Printing 2nd ed. CreateSpace Independent Publishing Platform* (UK:Nottingham)

[15] GU Dongdong, SHEN Yufi 2007 Direct laser sintered WC-10CoCu nanocomposites *Applied Surface Science* 3971-3978

[16] Mengucci P et al. 2016 Effects of thermal treatments on microstructure and mechanical properties of a Co-Cr-Mo-W biomedical alloy produced by laser sintering *Journal of Mechanical Behavior of Biomedical Materials* 106-117

[17] Cabrini M et al. 2016 Effect of heat treatment on corrosion resistance of DMLS AlSi10Mg alloy *Electrochimica Acta* **206** 346-355

[18] Ford Sharon L N 2014 Additive Manufacturing Technology: Potential Implications for U.S. Manufacturing Competitiveness Journal Of International Commerce And Economics 1-35

[19] Keith Dicks 2003 Introduction to EBSD (England: Oxford)

IOP Conf. Series: Materials Science and Engineering **426** (2018) 012026 doi:10.1088/1757-899X/426/1/012026

The effects of red mud on clay compounds

I Kocserha[1] , A Hamza[1] and R Géber[1]

[1] University of Miskolc, Institute of Ceramic and Polymer Engineering, Miskolc-Egyetemváros, Hungary

E-mail: fempityu@uni-miskolc.hu

Abstract. Recent study focused on the mixing properties of the clay and red mud compounds and the effects of red mud addition on clays with and without earth alkali carbonate content. Raw materials were added to compounds in natural form. Raw materials were applied as received in order to examine how mixing is possible in natural form without drying and grinding the red mud into powder. Highest amount of red mud content (50 wt%) was determined by mixing experiment. Compounds were prepared in pan mill and samples were extruded and analysed. Results indicated that red mud behaved in different way in carbonate and carbonate free clay but it is applicable for manufacturing of heavy clay products. It also revealed that the adequate mixing of red mud into the clay body is one of the most important question of the industrial application.

1. Introduction

Utilization of industrial by-products (e.g. fly ash, red mud etc.) and handle them as secondary raw material becomes more relevant as their volume grow constantly causing increasing dumping cost. Beside economical side they also may cause environmental problems. During Bayer-process in aluminium production, highly alkaline red mud is generated from the reaction of mineral bauxite and sodium-hydroxide. The production of 1 ton aluminium results 0.5-1.5 tons of red mud as waste material. There are several patents and studies dealing with the re-use of red mud in several industrial branch [1-3]. Among them the heavy-clay ceramic industry seems to be one of the possible field where huge volume of it may be consumed. In the production of fired ceramic building blocks beside red mud other organic waste material or by-product are also applied such as rice hull, paper sludge or marble powder [4-8]. In recent years, researchers studied the red mud in clay matrix, for example the change of microstructure in composition during heat treatment, physical and chemical properties of the manufactured samples were also analysed [9-17]. Based on researches above it can be said that the reuse of the red mud in brick clay resulted suitable products which fulfil the technological requirements. In these earlier studies [9-17] there is a uniformity: the raw materials were dried and ground in order to set the proper water content and to help the better mixing of the materials. However, in the heavy clay industrial practise it does not have the technical background for handling a huge volume additive material (i.e. red mud) at the present time. Beside drying, the grinding of red mud would demand also large energy consumption and introduction of the red mud into the production also requires to build up a technological line for raw materials preparation (i.e. special dryers and mills with powder technology system). Maybe that is why the one of the reasons why the red mud consumption is reported mainly in scientific articles. On the other hand, the utilization of red mud consumption in the heavy-clay industry has another aspect as well. Red mud has significant amount of natural radionuclide content of bauxite as residual from the Bayer-process. These materials are the naturally occurring radioactive materials (NORM) like ^{226}Ra, ^{232}Th or ^{40}K [18,19,22]. A study on raw materials found that the addition of Hungarian red mud is not proposed over a limit in order to fulfil EU recommendations on activity concentration index (I) of the mix (i.e. clays + red mud). Analysing 27 of hungarian brick clays [20], the maximal mixing ratio of red mud was calculated between 12-39 wt%. However, heat-treatment of the samples can reduce the risk originating from exhaled radon [21] which may elevate the volume of used secondary materials.

Based on the above considerations, we applied the raw materials as received in order to examine how mixing is possible in natural form without drying and grinding the red mud into powder. For brick and roof tiles, the earth alkali carbonate content of the clay affects the colour, water absorption and porosity of the fired products determining the frost-resistance. Clays with and without earth alkali carbonate content were selected to examine the effects of red mud addition on the properties of the products.

2. Materials and methods

2.1. Materials

For the purpose of examining clay and red mud compounds, three clays were obtained from two Hungarian brickworks. Two of them, which will be mixed later, consist considerable weight percentage of carbonates while third clay is carbonate free. The clays were named as follows: Grey clay – carbonate clay: GCC; Yellow clay – carbonate clay: YCC and the carbonate free clay: CFC. The red mud (RM) was obtained from the upper surface of an old Hungarian dump. Before the preparation of the compounds, the mineral composition of the raw materials was analysed by X-ray Powder Diffraction (Rigaku Miniflex II, Cu Kα, 2θ range from 3 to 90°). The quantitative results were calculated by Rietveld full-profile refinement analysis. The XRD patterns and the results are presented in Figure 1. and Table 1. respectively. After sampling, the water content of the raw materials was measured by RADWAG MA50 moisture analyser. Results are also included in Table 1.

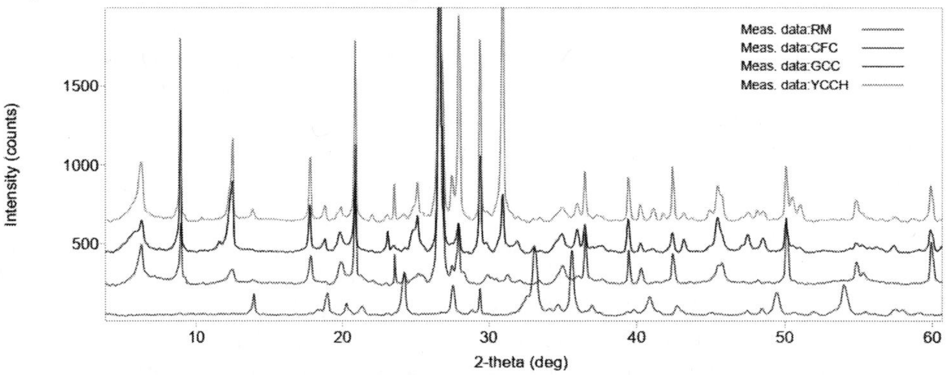

Figure 1. XRD patters of raw materials

Table 1. Mineral composition of raw materials

Raw mat.	Q	I	M	A	SM	G	K	M	C	D	CH	AM	WC
GCC	43.2	13.7	5.3	6.7	4.5	1.1	0.3	6.9	6.3	11.1	0.8	8	12.7
YCC	29.7	25.1	6.8	3.2	7.1	0.5	6.1	1.9	13.4	3.6	2.7	0	14.3
CFC	41.1	24.2	5.0	5.5	6.1	3.3	6.7	4.4	-	-	2.3	10	12.5
	Q	GI	C	H	CA	G	KI						
RM	1.0	7.2	5.5	44.2	32.9	4.5	4.7					0	33.1

(Q: quartz; I: illite; M: muscovite; SM: smectite; A: albite; G: goethite; K: kaolinite; M: microcline; C: calcite; D: dolomite; CH: chabazite; AM: amorphous content; WC: water content – H: hematite; GI: gibbsite; CA: cancrinite; KI: kimzeyite;)

GCC and YCC contain high amount of calcite and dolomite. The quantity of clay minerals was similar for YCC and CFC while the lowest was at GCC. In red mud hematite and cancrinite were the main phases.

2.2. Methods

2.2.1 Mixing and sample preparation

As raw materials were added to compounds in natural form without drying and grinding, it was necessary to take the nature of clays into account. Adequate mixing of two plastic materials having different water content required more experiment, as paste with higher plasticity could flow around the less plastic material. Red mud was received in soil wet form with around 33 % water content, so it behaved as a paste. The YCC and CFC were soft clays. GCC had very stiff nature so according to industrial practice a pre-grinding was applied. GCC and YCC clays were blended together in the ratio 1:3 and 2:3 to form a carbonate clay mix.

During preliminary mixing experiments, taking the original moisture content of the raw materials (Table 1.) into account, moisture content of the compounds was set to 22 wt% which was most suitable to produce samples by extrusion.This determined the maximal amount of RM that is incorporated into the compound. Both carbonate (CC) and carbonate free clay (CFC) were mixed with different amounts of red mud ranging from 0 to 50 wt% (Table 2.). The masses were homogenized by laboratory pan mill (Figure 2.) and the technical water was also added at this stage gradually. Mixing period was 5 minutes for each compounds. For CFC, increasing amount of red mud, the mass showed an improved consistency. Homogenized mass was illustrated in Figure 3.

Table 2. Composition of prepared compounds in wt%

Carbonate clay/ RM	100/0	90/10	80/20	70/30	60/40	50/50
Carbonate free clay / RM	100/0	90/10	80/20	70/30	60/40	50/50

Figure 2. Laboratory pan mill

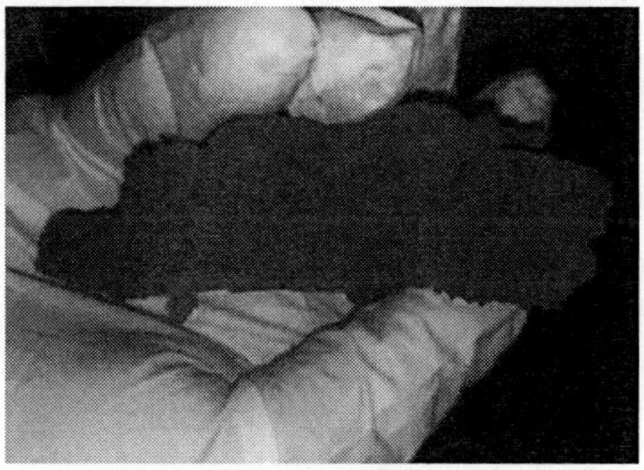

Figure 3. Homogenized mass

After preparation, masses were sealed and rested for 48 hours. KEMA, PVP5/s type laboratory vacuum extruder with 50mm screw diameter was used to produce 12 pieces of cylindrical samples for each compounds. The diameter of extruded specimens was 24 mm and the length was 50 and 150mm. After extrusion, samples were dried for 24 hours long at room temperature and for 48 hours long in drying chamber at 90 °C. Sintering was performed in a laboratory furnace at 860 °C, with the heating rate of 150 °C.

2.2.2 Testing methods

For discover cracks, colour differences or other faults, the steps of preparation of the samples were followed by a visual inspection. Considering that, the red mud is strongly alkaline (pH=13) the pH value was also measured in some steps of production. In order to follow the effects of red mud on the most important properties of the produced brick samples, typical brick industry tests were performed on the samples according to the related standards, such as drying and firing shrinkage, compressive strength, bending strength, density, and water absorption. For measuring flexural strength, samples with 150 mm length were applied. Average results obtained from the examination of compounds containing RM are illustrated in Figures 5-10.

3. Results and discussion

3.1 Mixing

During mixing experiments, it was observed that the type and the moisture content of clays determine the way of the mixing RM to clay. It was necessary to find a water content to set the proper consistency of the clays. It was not possible to mix the materials together at any moisture content. Larger pieces of the harder material were flowed around by RM having lower consistency without mixing. If the clay was softer, it was also not possible to press the RM among the clay particles as the two material slipped away on each other.

3.2 PH measurement

The original RM samples was strongly alkaline with the pH value of 13. That is why the condensate from the clay strip after extrusion and the evaporated moisture after drying were analysed by pH measurement. According to this, the condensates were no harmful as it showed a neutral pH. This is beneficial for industrial application of RM.

3.3 Colours

The change in colours can be seen in Figure 4 after drying and firing. The impact of red mud on colours was stronger visibly after drying. Firing caused a decrease in the colours difference. RM balanced the yellow colour of fired CC samples to "brick red" owing to its high iron-oxide content. On surface of CC samples there were lot of lime concretion.

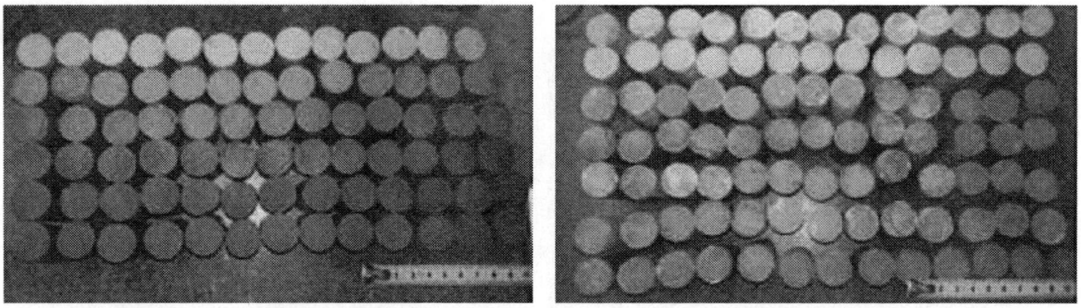

Figure 4. Colour difference after drying and firing

3.4 Drying losses and drying shrinkage

The drying loss in both clay was the same as the percentage of water added. The effect of RM addition on drying shrinkage is illustrated in Figure 5. The shrinkage of carbonate clay (CC) with additional RM slightly increased in all cases while the shrinkage of carbonate free clay (CFC) decreased from 20.41 % to 12.21 %. Overall, CFC samples which content RM showed less shrinkage. The pure CC had 4 % higher shrinkage value as CFC.

3.5 Burning shrinkage and density

The numerical values of shrinkage and density can be seen in Figure 6 and Figure 7, respectively. The CC had around 29 % of shrinkage value which is relatively high related to CFC. The amount of red mud had no significant effect on the firing shrinkage of the CC. At CFC samples with additional 10 % RM, the shrinkage decreased from 21.2 % to 18.3 %. Lager amounts of RM in CFC had but not tendentious effect on shrinkage value. The maximum is 23.82 % (at 30-40 wt%) RM, which is 10 % change related to pure clay. Compare to CC to CFC the shrinkage was always higher at CC.

In all cases, samples of CFC had higher density. For the two types of clays, RM behaved in similar way. Density of CC without RM was increased from 1,81 g/cm^3 to 1,91 g/cm^3 up to 30 wt% RM addition. Over this value density began to lower. The effect of RM on CFC was parallel in grow to CC but in this case the density increased up to 40 wt%. For both CC and CFC, the pure clay samples had the lowest density.

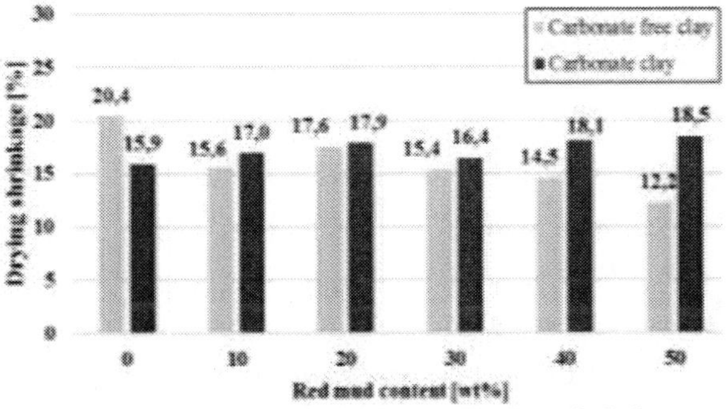

Figure 5. The effect of RM on drying shrinkage

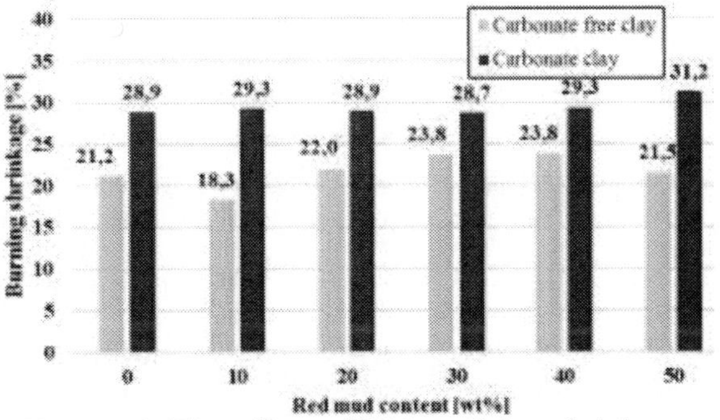

Figure 6. The effect of RM on firing shrinkage

For CC and CFC, RM affected the water absorption similar in tendency (Figure 8). Up 30 wt% RM content in CFC, there was no considerable change, the CFC held its closer structure. Over 30 wt% of RM, absorption capacity raised from about 14.4 % to 21.6 %. Correspondingly, adding over 30 wt% RM to CC, RM facilitated higher water absorption by opening more pores on the surface of the samples.

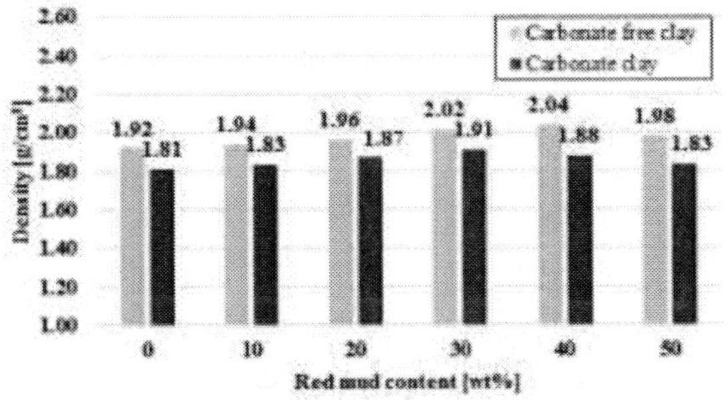

Figure 7. The effect of RM on density

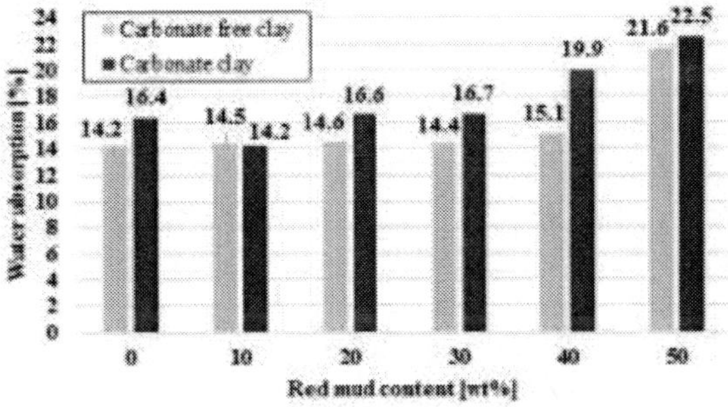

Figure 8. The effect of RM on water absorption

Adding RM to CFC had good effect on compressive strength (Figure 9.) since pure clay has the lowest strength value. Up to 40 wt% RM content the strength increased constantly from 23.8 to 36.8 MPa. It represented an 54% increase in strength. Other researchers experienced the same but at higher sintering temperature [12]. For CC, the use of RM caused lower compressive strength than the pure clay except at 30 wt% RM addition where it was higher. The more RM added to clays, the more it contained calcite which resulted in more open pores. It caused higher water absorption and lower strength.

For CFC samples, RM addition up to 20 wt% had no significant effect on flexural strength (Figure 10.). Similar to compressive strength at 30-40 wt% RM content in CFC resulted the highest strength value. Adding RM to CC, flexural strength decreased in tendency. Higher strength value at 10 wt% of RM content was only obtained.

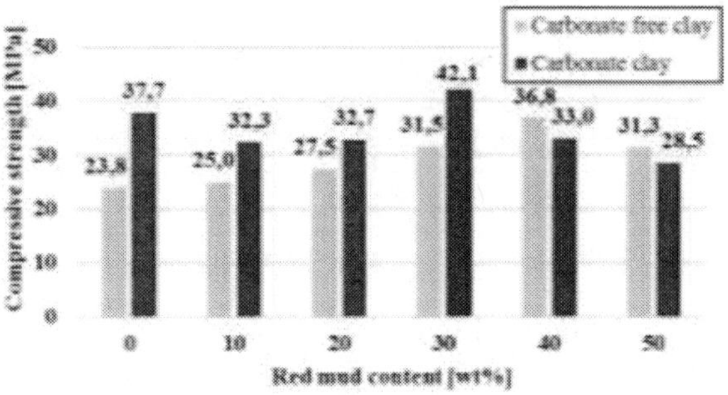

Figure 9. The effect of RM on compressive strength

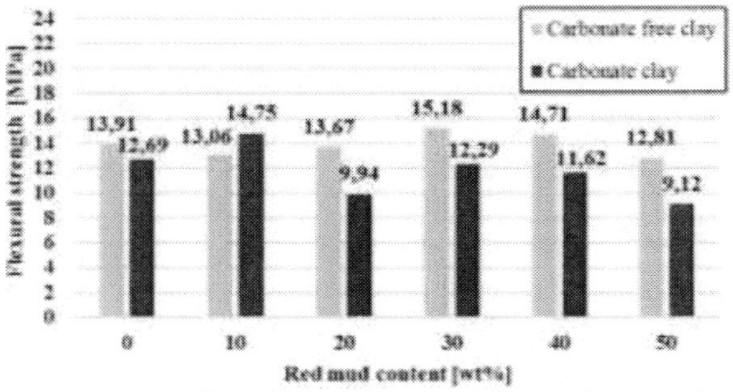

Figure 10. The effect of RM on flexural strength

4. Conclusions

Clays and RM were added to compounds in natural form without drying and grinding. The moisture content of clays determined the maximal amount of RM (50 wt%) that can be mixed into compounds. Adequate mixing of two plastic materials having different water content required more experiment, as paste with higher plasticity could flow around the less plastic material. It was observed that if the red mud content was higher than 50 wt%, than the compound became more plastic and the extrusion was not possible. The type and the moisture content of clays determined the way of the mixing RM to clay. It was not possible to mix the materials together at any moisture content. We controlled and set the proper water content for mixing to had the adequate plasticity of the compounds.

The test results of extruded and fired products showed that firing shrinkage of CC and RM compounds were always higher than that for CFC. In all cases, samples of CC had lower density owing to the high carbonate content.

The RM content in CFC and CC up 30 wt% had no considerable change in water absorption. On the contrary, adding over 30 wt% RM to CC, RM facilitated higher water absorption by opening more pores on the surface of the samples.

Adding RM to CFC had good effect on compressive strength. For CC, the use of RM caused lower compressive strength than the pure clay. It raised the compressive strength for CFC since up to 40 wt% RM content the strength increased constantly from 23.8 to 36.8 MPa as but it was decreased for CC. RM affected the values of compressive and flexural strength inordinately.

According to these results it can be said that usage of RM in heavy clay industry is possible without expansive drying and grinding the RM into powder.

5. Acknowledgement

The described study was carried out as part of the EFOP-3.6.1-16-2016-00011 "Younger and Renewing University – Innovative Knowledge City – institutional development of the University of Miskolc aiming at intelligent specialisation" project implemented in the framework of the Szechenyi 2020 program. The realization of this project is supported by the European Union, co-financed by the European Social Fund.

IOP Conf. Series: Materials Science and Engineering **426** (2018) 012026 doi:10.1088/1757-899X/426/1/012026

6. References

[1] C Klauber, M Gräfe and G. Power 2011 *Hydrometallurgy* 108 pp 11–32

[2] R K Paramguru, P C Rath and V N Misra 2005 *Mineral Processing & Extractive Metall. Rev.* **26** pp 1-29

[3] W.Liu, J. Yang and B. Xiao 2009 *Int. J. Miner. Process.* **93** pp 220–231

[4] K Junge 2000 *Ziegelindustrie International* **12** pp.25-39

[5] Z Xuanye 2003 *Ziegelindustrie International* **4** pp.22-27

[6] V Ducman and T Kopar 2007 *Materials and technology* **41/6** pp. 289–293

[7] I Kocserha and L A Gömze 2010 *Applied Clay Science* **48/3** pp 425-430

[8] F Kristaly, L A Gömze and I Papp 2010 *Materials Science Forum* **659** pp 37-42

[9] V M Sglavo, S Maurina, A Conci, A Salviati, G Carturan and G Cocco 1999 *Journal of the European Ceramic Society,* **20** pp 245-252

[10] Y Pontikes, P Nikolopoulos and G N Angelopoulos 2007 *Journal of the European Ceramic Society,* **27** pp 1645-1649

[11] H He, Q Yue, Y Su, B Gao, Y Gao, J Wang and H Yu 2011 *Journal of Hazardous Materials* **213-214** pp 53-61.

[12] L. Pérez-Villarejo, F.A. Corpas-Iglesias , S. Martínez-Martínez, R. Artiaga, J. Pascual-Cosp 2012 *Construction and Building Materials* **35** pp 656-665.

[13] H He, Q Yue, Y Qi, B Gao, Y Zhao, H Yu, J Li, Q Li and Y Wang 2012 *Applied Clay Science* **70** pp 67-73

[14] S Pişkin, A K Figen, E Özkan and Ü Özçay 2013 *IFAC Proceedings* **46** pp 484-487.

[15] M P Babisk, T P Altoé, H J O Lopes, U S Prado, M C B Gadioli, S N Monteiro and C M F Vieira 2014 *Materials, Science Forum* ***798-799*** pp 509-513

[16] S Liu, X Guan, S Zhang, C Xu, H Li and J Zhang 2016 *Materials Letters* **191** pp 222-224.

[17] C Scribot, W Maherzi, M Benzerzour, Y Mamindy-Pajany and N E Abriak 2018 *Construction and Building Materials* **163** pp 21–31

[18] J Somlai, V Jobbágy, J Kovács, S Tarján and T Kovács 2007 *Journal of Hazardous Materials* **150** pp 541–545

[19] S Landsberger, A Sharp, S Wang , Y Pontikes and A H Tkaczyk 2016 *Journal of Environmental Radioactivity* **173** pp 97-101

[20] Z Sas, J Somlai, G Szeiler and T Kovács 2015 *Journal of Radioanalytical and Nuclear Chemistry* **306** pp 271–275

[21] Z Sas, J Szántó, J Kovács, J Somlai and T Kovács *Journal of Environmental Radioactivity* **148** pp27-32

[22] S S Nenadovic, G Mucsi, L M Kljajevic, M M Mirkovic, M T Nenadovic, F Kristály and I S Vukanac 2017 *Nuclear Technology & Radiation Protection* **32/3** pp 261-266

IOP Conf. Series: Materials Science and Engineering **426** (2018) 012027 doi:10.1088/1757-899X/426/1/012027

Effects of plasma nitriding on tempered steel

D Kovács[1], A Kemény[1], J Dobránszky[2] and I Quintana[3]

[1] Budapest University of Technology and Economics, Faculty of Mechanical Engineering, Department of Materials Science and Technology, Budapest, Hungary
[2] MTA–BME Research Group for Composite Science and Technology, Budapest, Hungary
[3] IK4-Tekniker, Parke Teknologikoa, Goenaga, Spain

E-mail: dorina@eik.bme.hu

Abstract. Conventional plasma nitriding is highly widespread in the industry, but it has essential problems such as edge effect and hollow cathode effect. These problems induced the development of active screen plasma nitriding (ASPN) where the plasma forms on the active screen instead of the sample, which is completely insulated. However, this arrangement does not work with large furnaces, so the usage of bias voltage is necessary in industrial circumstances. In this experiment, the effects of different biases were examined, specifically the hardness, layer thickness, formed phases and the diffusion of nitrogen. In conclusion, the 20 % bias had the best effects on the sample with the defined nitriding parameters in this research.

1. Introduction

Plasma nitriding is a thermochemical surface treatment which can improve the hardness and the wear resistance of the steel. When the nitrogen – hydrogen gas mixture enters the process chamber, the gas becomes ionized between the anode (furnace wall) and the cathode (workpiece). The positively charged ions bombard the surface of the piece which heats up [1–3]. This issue a glow discharge when the particles absorb into the surface of the material. The three most important factors of the treatment are the pressure, the high voltage and the ratio of the used gas mixture. The temperature and time of the treatment also influences the properties of the nitride layer.

The direct current plasma nitriding (DCPN) is a traditional surface treatment over 40 years. The DCPN technology has different shortcomings which cause a damage on the parts, such as 'edge effect' or 'hollow cathode effect' [3–5]. The active screen plasma nitriding was developed to solve the previous problems. In this technology the plasma is produced on the screen, not directly on the workpiece therefore it causes a more homogenous surface [6–8]. In this case the worktable is isolated from the place of the voltage, so heating of the ASPN is slow. To increase this process the components are in a floating potential or a relative lower bias voltage [8, 9].

2. Materials and methods

The materials used in this study are tempered 42CrMo4 type low alloy steel and annealed C45 type carbon steel, their chemical compositions are shown in table 1. The sample disks were 20 mm in diameter and 6 mm in thickness. The sample's surface was mechanically ground with 80 to 2500-grit SiC emery paper, finely polished with 3 μm diamond suspension. They were cleaned in acetone and dried before placement to the vacuum chamber.

Table 1. Chemical composition of materials in wt.%

Steel	C	Si	Mn	P	S	Cr	Mo
42CrMo4	0.38–0.45	≤ 0.4	0.6–0.9	≤ 0.025	≤ 0.035	0.9–1.2	1.5–0.3
C45	0.42–0.5	≤ 0.4	0.5–0.8	≤ 0.045	≤ 0.045	≤ 0.4	≤ 0.1

The plasma nitriding system was built by own plan. Parameters easily set to the different methods. Figure 1 shows the schematic workplace of the chamber. The active screen was made of unalloyed 1.0330 type steel. The dimension is ∅100×85 mm with ∅ 5×8 mm holes and centers. All treatments were carried out using the same parameters as shown in Table 2 and following the same procedures. Bias voltage was adapting to the main voltage. The bias is 10 %, 20 % of the main voltage at the same parameters of the process. After the nitriding, free cooling started to the turning of the plasma. The workpiece was cooling down from the process temperature to the room temperature under the nitriding pressure.

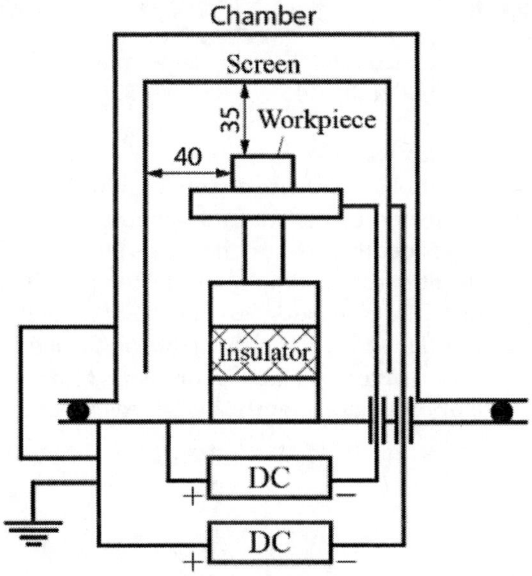

Figure 1. Schematic model of the nitriding chamber (biased arrangement)

Table 2. Parameters of plasma nitriding

Voltage (V)	Current (A)	Gas mixture N_2:H_2 (%)	Temperature (°C)	Pressure (torr)	Time (h)
490 – 540	0,9 – 1,5	25:75	490	2,8	4

Optical microscope and scanning electron microscope were used to examine morphology and measure thickness of nitrided layer. The hardness profile was determined by micro Vickers indenter with a load of 50g. The phase composition of the compound layer on the surface was determined by X-ray diffractometer (XRD) using CuKα radiation [10–15] .

3. Results and discussion

3.1. Surface analysis by optical microscope

First, DCPN and ASPN treated C45 samples were compared. The edge effect was observed at the treatment of DCPN as it can be seen in Figure 2a) and b). The well-visible erosion ring has reduced the hardness toward the centre of the sample, while the microhardness value is much higher in the edge area, due to thermal gradient, as it was presented in another work; this causes unequal microhardness values on the surface [6]. The nitride layer depth is unequal in the erosion zone compared to the centre.

a) b) c)

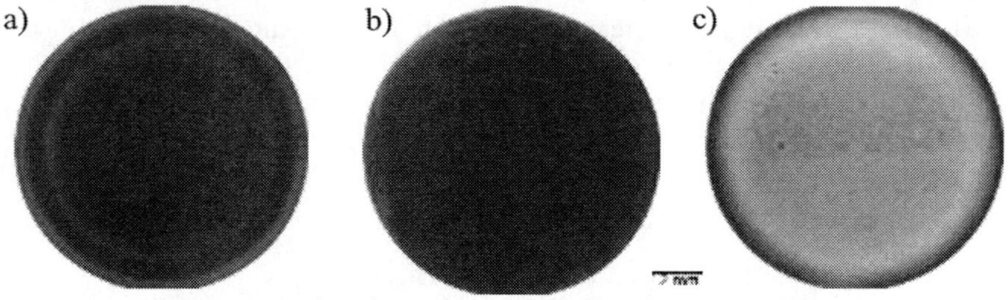

Figure 2. Discoloured surface after plasma nitriding
a) C45 – DCPN, b) C45 – ASPN, c) 42CrMo4 - biased

3.2. Hardness and thickness measurements

Figures 3 and 4 show the cross-sectional hardness distribution of the different samples. The compound layer thickness was measured by using the distance measurement tool of the scanning electron microscope. The hardness decreased considerably toward the core of the material in each condition. The samples of steel C45 show a significant hardness drop after the compound layer while the samples of steel 42CrMo4 have smooth transition to the core hardness. This means that the diffusion zone is substantially larger with the alloyed steel. The effect of bias is also visible on the hardness profile. As higher percentage of the voltage was used as bias, higher surface hardness and thicker compound layer was formed. In proportion to DCPN, the use of bias cause significant increasing of hardness.

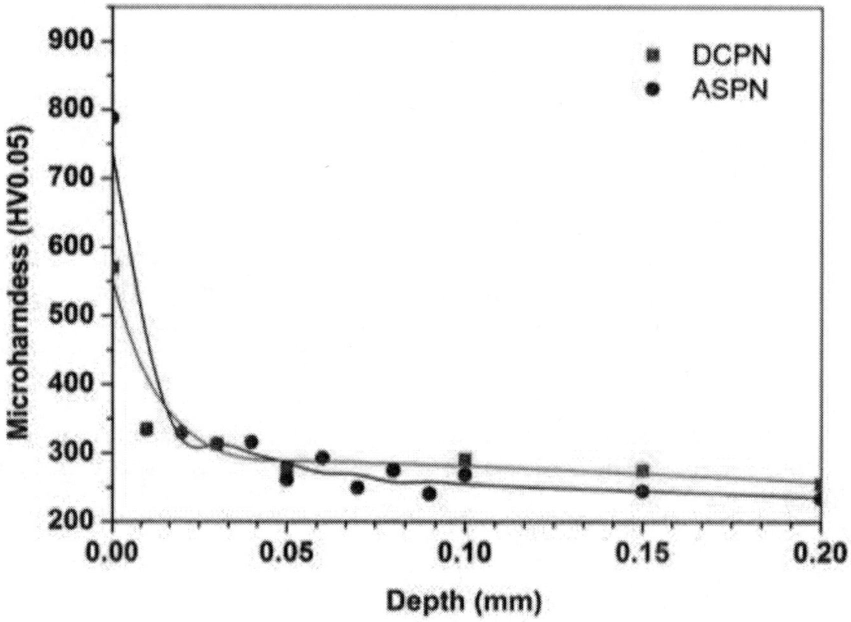

Figure 3. Two hardness profiles of steel C45

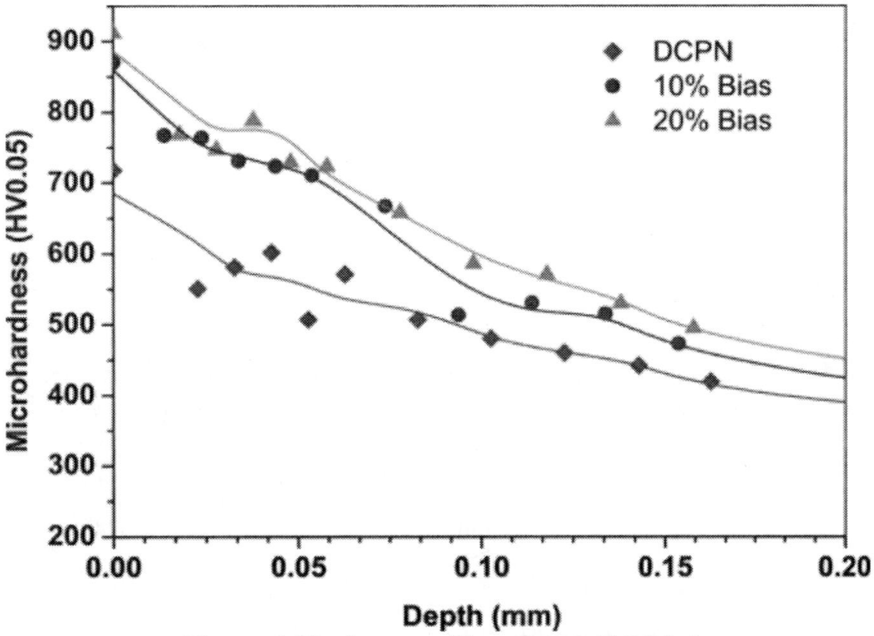

Figure 4. Hardness profiles of steel 42CrMo4

Scanning electron microscope was used to determine the depth of the compound layer. Figure 5 shows the cross section of two C45 steel samples. The layer thickness was 8.6 μm with DCPN and 5.5 μm with ASPN, respectively.

Figure 6 shows the cross section of the 42CrMo4 type low alloy steel samples. The layer thickness was 15 μm with DCPN, 11.2 μm with 10% bias and 15.5 μm with 20% bias.

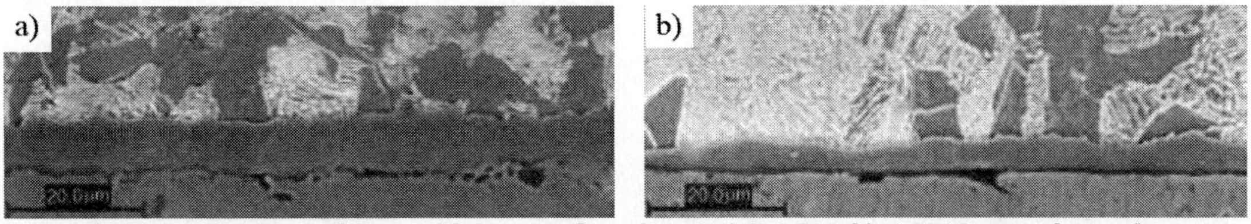

Figure 5. Compound layer thickness of steel C45 a) DCPN, b) ASPN treated samples

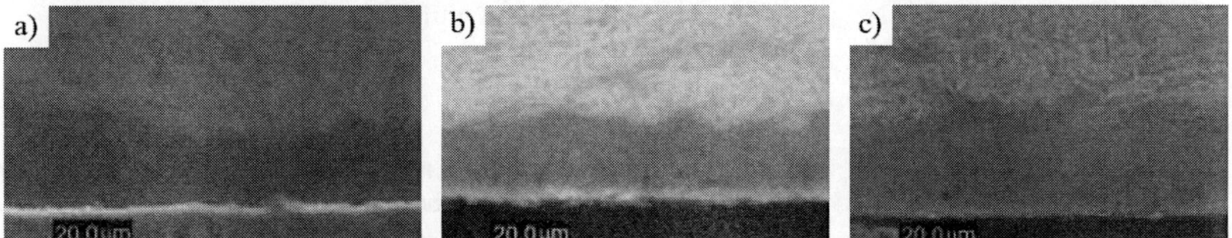

Figure 6. Compound layer thickness of steel 42CrMo4 a) DCPN, b) using 10% Bias, c) using 20% Bias treated samples

3.3. X-Ray diffraction analysis

DC plasma nitriding samples were investigated by XRD analysis. Figure 7-8 contain the diffraction pattern of C45 and 42CrMo4. γ' phase, as Fe_4N, was formed on both sample with the same intensity [15]. As observed in Figure 8, magnetite is also formed on the surface. It means some oxygen is present in the vacuum chamber, but this rate is minimal, it didn't influence the compound layer.

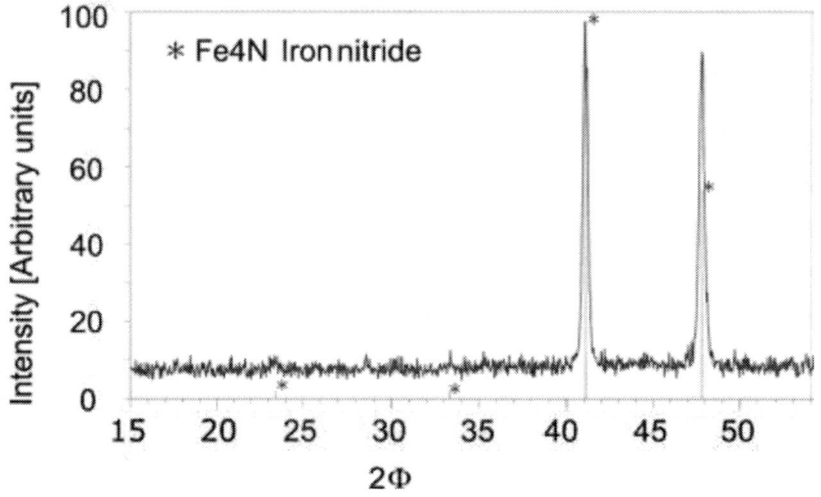

Figure 7. X-ray diffractogram of DC plasma nitrided steel C45

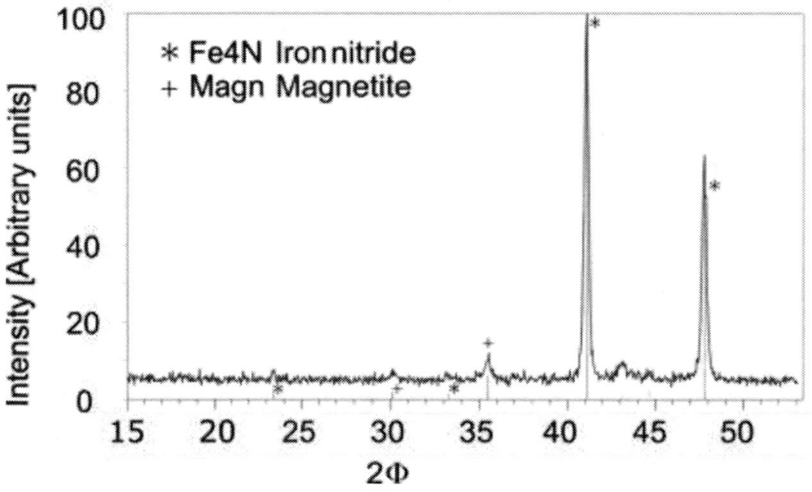

Figure 8. X-ray diffractogram of DC plasma nitrided steel 42CrMo4

4. Conclusions

The annealed C45 type carbon steel is not a typical grade for plasma nitriding. The compound layer thickness is formed on the surface like as other tempered steel, but the depth of the diffusion zone is lower than the other case.

The effect of bias on steel 42CrMo4 was analysed with different methods. The results show that higher hardness can be reached with higher bias voltage. The hardness profile also shows that the alloyed steel has a much thicker diffusion zone than the unalloyed. With higher bias, slightly thicker compound layer is formed which can be examined well using SEM.

Acknowledgments
The authors would like to acknowledge the IK4-Tekniker for their support and access to their facilities. This work was supported by foundation of Richter Gedeon Talentum.

IOP Conf. Series: Materials Science and Engineering **426** (2018) 012027 doi:10.1088/1757-899X/426/1/012027

References

[1] Pye D 2003 *Practical Nitriding and Ferritic Nitrocarburizing* (Ohio: ASM International)
[2] Hossein A and Behrangi S 1988 *Plasma nitriding of steels* vol 17
[3] Janosi S, Kolozsvary Z and Kis A 2004 Controlled hollow cathode effect: New possibilities for heating low-pressure furnaces *Met. Sci. Heat Treat.* **46** 310–6
[4] Li Y, Wang L, Shen L, Zhang D and Wang C 2010 Plasma nitriding of 42CrMo low alloy steels at anodic or cathodic potentials *Surf. Coatings Technol.* **204** 2337–42
[5] Alves C, de Araújo F O, Ribeiro K J B, da Costa J A P, Sousa R R M and de Sousa R S 2006 Use of cathodic cage in plasma nitriding *Surf. Coatings Technol.* **201** 2450–4
[6] Corujeira Gallo S and Dong H 2009 On the fundamental mechanisms of active screen plasma nitriding *Vacuum* **84** 321–5
[7] Zhao C, Li C X, Dong H and Bell T 2006 Study on the active screen plasma nitriding and its nitriding mechanism *Surf. Coatings Technol.* **201** 2320–5
[8] Hubbard P, Dowey S J, Partridge J G, Doyle E D and McCulloch D G 2010 Investigation of nitrogen mass transfer within an industrial plasma nitriding system II: Application of a biased screen *Surf. Coatings Technol.* **204** 1151–7
[9] Bonyár A and Szabó P J 2015 A method for the determination of ferrite grains with a surface normal close to the (111) orientation in cold rolled steel samples with color etching and optical microscopy *Mater. Sci. Forum* **812** 297–302
[10] Han L, Dai J T, Huang X R and Zhao C 2013 Study on the fast nitriding process of active screen plasma nitriding *Phys. Procedia* **50** 94–102
[11] Wang L, Li Y and Wu X 2008 Plasma nitriding of low alloy steels at floating and cathodic potentials *Appl. Surf. Sci.* **254** 6595–600
[12] Sun Y, Li X Y and Bell T 1999 X-ray diffraction characterisation of low temperature plasma nitrided austenitic stainless steels *J. Mater. Sci.* **34** 4793–802
[13] Olzon-Dionysio M, Campos M, Kapp M, de Souza S and de Souza S D 2010 Influences of plasma nitriding edge effect on properties of 316L stainless steel *Surf. Coatings Technol.* **204** 3623–8
[14] Li Y, Wang L, Xu J and Zhang D 2012 Plasma nitriding of AISI 316L austenitic stainless steels at anodic potential *Surf. Coatings Technol.* **206** 2430–7
[15] Hirsch T, Clarke T G R and da Silva Rocha A 2007 An in-situ study of plasma nitriding *Surf. Coatings Technol.* **201** 6380–6

IOP Conf. Series: Materials Science and Engineering **426** (2018) 012028 doi:10.1088/1757-899X/426/1/012028

Investigation of clinch joints made of similar and dissimilar materials

P Kovács[1] and M Tisza[2]

[1] Associate professor, University of Miskolc, Institute of Materials Science and Technology, Miskolc-Egyetemváros, Hungary
[2] Head of department, professor, University of Miskolc, Institute of Materials Science and Technology, Miskolc-Egyetemváros, Hungary

E-mail: metkpz@uni-miskolc.hu, tisza.miklos@uni-miskolc.hu

Abstract. Clinching is a mechanical cold forming method to join two or more overlapping sheets without using of rivet. This technology is already in use for production of aluminum and steel sheet car body panels. This process is carried out without heat effect on the material structure, thus the clinching is one of the technologies which may replace traditional processes as resistance spot welding or friction stir welding. The clinching technology has been well known for many years, but only during the last decades become more and more widely applied particularly in the automotive industry. This paper presents the results of experimental research work in the field of clinching of different strength sheets developed for automotive application.
Keywords: clinching, joining processes

1 Introductions

The variety of sheet materials applied in the vehicle industry is very versatile and the manufacturers can choose the most suitable ones for the given application according to their needs. It often happens that they apply dissimilar materials within one structural unit: it may mean different sheet thicknesses of the same material grade, but it may also mean the application of dissimilar material grades – e.g. aluminium and steel – with different material properties. However, they often have to join these sheets to assemble the suitable final product. There are various methods suitable for joining sheet metal parts. Among the many processes several groups of welding may be regarded as the most widely applied joining techniques, e.g. resistance spot welding, soldering, and besides them adhesive bonding, as well as various mechanical joining, or their combinations are also applied. The manufacturers choice is affected by many parameters, e.g. which process is more suitable for preparing the proper joint between the sheets, what is the cost for purchasing the right equipment, the raw material, and to what extent the given process can be mechanized, since all these can greatly affect the manufacturing time, the needed human resources quantity, and in this respect the final product price. These points of view altogether determine the process to be chosen. In this paper we will mainly focus on mechanical joining of similar and dissimilar sheet materials by clinching [1, 2].

IOP Conf. Series: Materials Science and Engineering **426** (2018) 012028 doi:10.1088/1757-899X/426/1/012028

2 Mechanical joining by clinching

Clinching nowadays is a widely used joining process belonging to the group of mechanical joining. Clinching means pressing two or more sheets into each other with a punch and die pair with the help of a special or universal pressing machine.

2.1 Main types of clinching

Clinching is more and more often applied in the vehicle industry for joining coated or uncoated thin sheets of similar or dissimilar materials. Clinching of dissimilar materials either mean different material grades or different thicknesses. In clinching processes, two or up to three sheets are overlapped on each other and the sheets are joined by stamping. Clinching has several variants, and among them the round clinching is regarded as the most general one. In Figure 1, the round clinching process can be seen together with the main elements of a clinching tool.

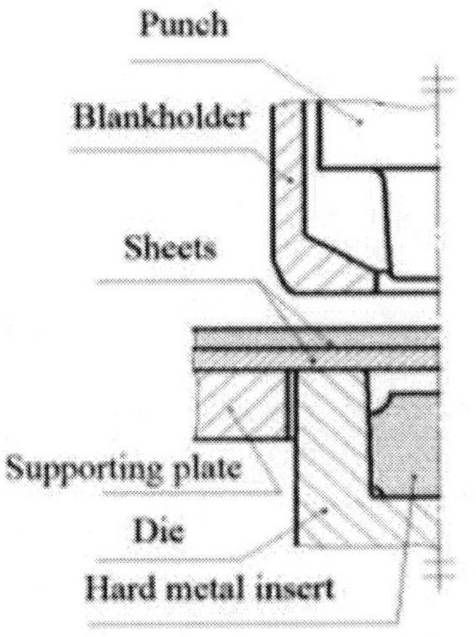

Figure 1. Main tool elements in a round clinching process.

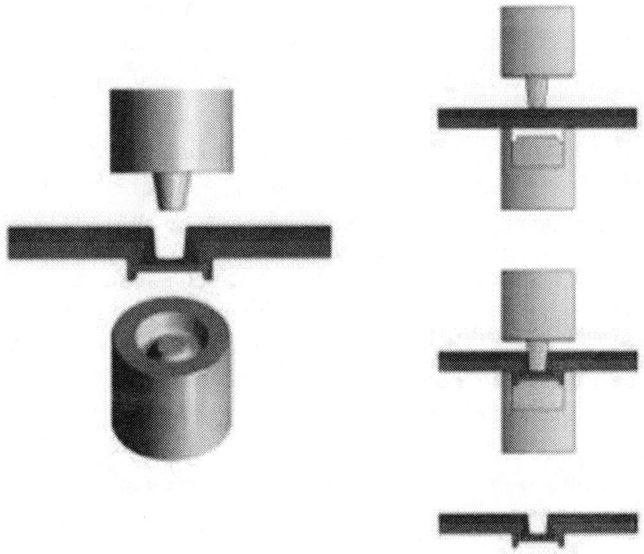

Figure 2. Process sequence in round joint clinching.

The process sequence and the material flow can be seen in Figure 2. The process sequence follows from up to down. As it can be seen in this Figure, a simple round punch presses the materials into the die cavity. As the force increases, the punch side material is forced to spread outwards within the die side material. The result is an aesthetically good looking round button at the end of this clinching process: the sheets are mechanically joined, clearly without any burrs or sharp edges.

Due to its versatility and its many advantages (summarised later) there are many different types of joints made by clinching. The TOX®-Twin Point [3] process provides two similar buttons with almost double strength of a single point. A special advantage of this process is that it prevents the rotation of two joined layers against each other. It may happen that the button normally formed by a round joint clinching is not acceptable: in this case flat joint clinching should be applied.

2.2 Characteristics, main advantages and disadvantages of clinched joints

As it was mentioned, round joint by clinching may be regarded as the most often used type of clinching processes. This is the simplest and the clearest mechanical joining of sheets. There is no need for preliminary works. The tools do not contain cutting or moving elements causing problem.

The most important advantages of clinching [3] may be listed as follows:

- 30 – 60% cost saving compared to resistance spot welding,
- the dynamic load resistance of clinched joints is usually better than in resistance spot welding,
- the necessary preparation for clinching can be well automated,
- simple, non-destructive quality inspection is possible by using a simple instrument to measure the remaining thickness at the bottom of the joint,
- during clinching there is no any heat generation thus no microstructural changes occur,
- there are no damages at different surface protecting layers (e.g. galvanised, pre-painted, etc.) since the layer flows with the base material together,
- parts joined by clinching have good electrical conductivity,
- since there is no temperature effect, clinching can be applied together with adhesive bonding,
- sheet thickness can vary in large interval (usual range between 0.3 mm and 6 mm),
- intermediate layers (e.g. paper or glue) can also be used,
- the process can be excellently monitored,
- joints made by clinching may have long life duration,
- since there is no contamination during the process, it does not require supplementary work,
- the process itself is extraordinarily environment friendly,
- clinching is significantly cheaper than laser welding.

Besides the many advantages, obviously there are some disadvantages, too. As one of the few disadvantages, we have to mention that the strength properties of joints made by clinching usually are weaker than that of made by resistance spot welding, however, in many cases it is even not always necessary. We could also mention that the strength parameters made by the new TOX technology already reach about 70% of the values of resistance spot welding. Furthermore, it is also worth mentioning that in spite of weaker mechanical properties, clinching is more and more often applied instead of resistance spot welding, since the clinching process often better fits into the manufacturing process route.

2.3 Quality factors of clinching

In a general round joint clinching process [4], the strength of the joints is determined by the neck thickness and the under cutting: both parameters are affected by the shape and parameters of tools as the diameter of the punch and the die, as well as the die deepness. The main geometric parameters of a round joint clinching tool are shown in Figure 3.

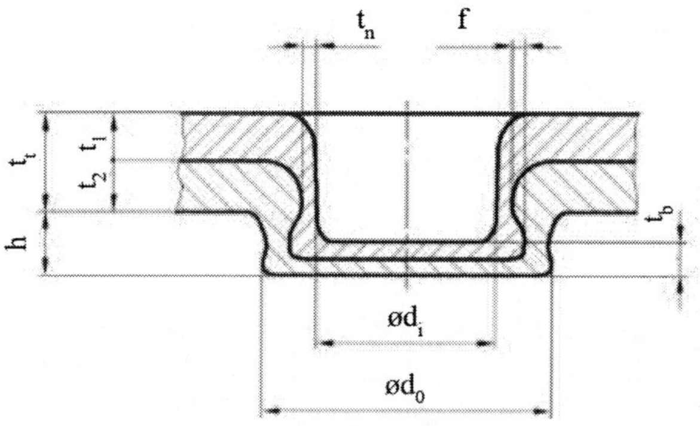

$ød_0$: outside diameter
$ød_i$: inside diameter
t_1: top sheet thickness
t_2: lower sheet thickness
t_t: total thickness
t_b: bottom thickness
t_n: neck thickness
h: point height
f: under cutting

Figure 3. Main technological and geometric parameters in clinching.

2.4 *Investigation of clinched joints*

To provide good quality clinched joints it is strongly advised to perform experimental investigations to define the appropriate process parameters and to choose the proper tool set for a given application. It is also desirable to perform mechanical tests applying static and dynamic loads.

The usually applied tests are the same as they are in case of spot welded joints, where the shearing and the pulling breakup tests are widely used as shown in Figure 4. During these tests, the neck width is also measured to investigate the effect of the geometric parameters on the strength of the joints.

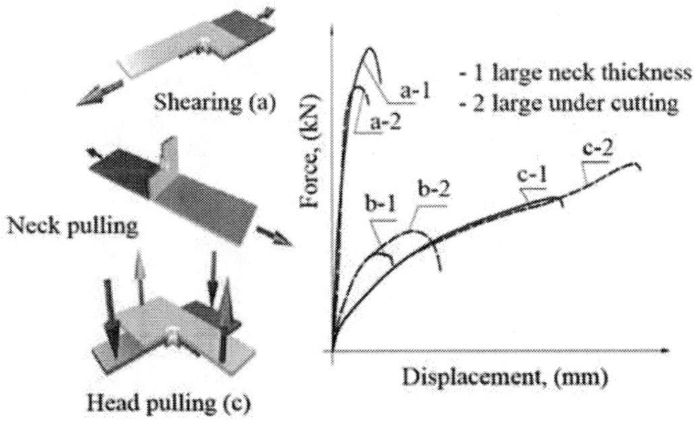

Figure 4. Illustration of shearing-, neck- and head pulling tests with the main types of failures [5].

2.5 *Experimental investigations*

During the experimental investigations a Round Tox tool was applied for testing clinched joints. The tool was specially designed for producing round joints. With this standard tool setup, it is possible to create mechanical joints between high strength sheets of 1 mm thickness. According to the tool manufacturer [3], the best properties of the joints can be reached when the bottom thickness is about 0.45–0.5 mm.

The tests were performed in the Material Testing Laboratory of the Institute of Materials Science and Technology. The testing machine was an MTS-801 type, electro-hydraulic, computer controlled, universal material testing equipment. The nominal force of this material testing system: $F_{max} = 250$ kN, The testing machine with a clinching tool mounted on it can be seen in Figure 5. On the right side of this Figure an enlarged view of the clinching tool is shown.

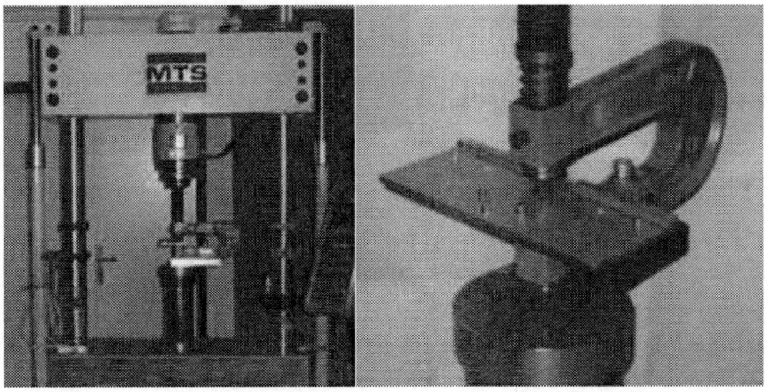

a) b)

Figure 5. The experimental setup; a) MTS-801 Testing System, b) the Clinching tool

In order to analyse the clinching process, first some preliminary clinching tests were performed with different punch displacements to produce different bottom thicknesses. After clinching, the parts are cut into two parts to create meridian sections as shown in Figure 6, for two different states (T0 and T6) of AA6082 material pairs. In the first case we analysed the joining in T6 (artificially aged) state then in T0 (solution heat treated) condition. From this figure it can be seen that successful joints could be made in solution heat treated condition, while in T6 condition cracks always appeared.

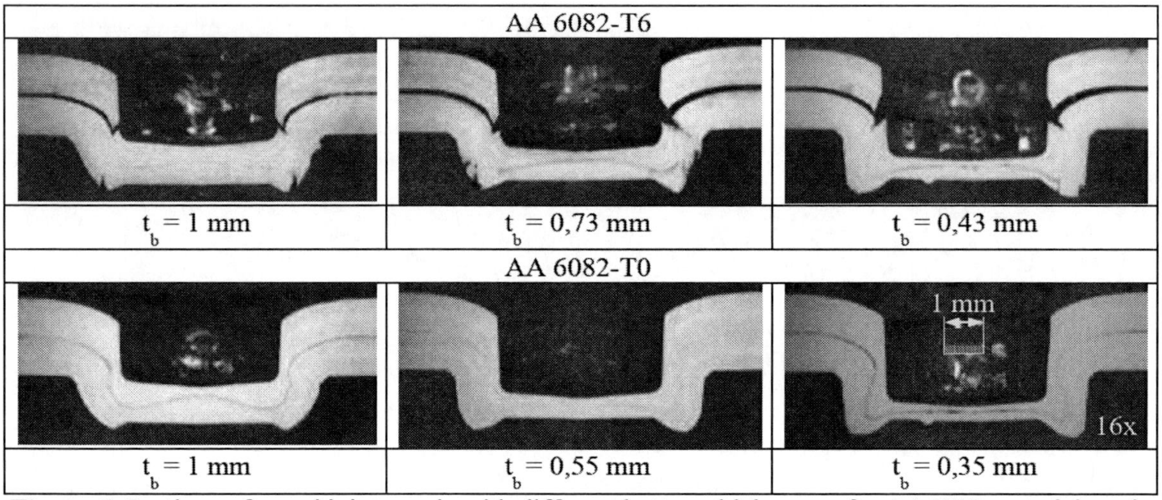

Figure 6. Sections of round joints made with different bottom thicknesses for AA6082 material grade in T6 and T0 condition (wall thickness 1mm).

On the microscopic samples micro-hardness measurements (HV 0.5) were also done. The micro-hardness values were measured in the meridian sections of the samples. Measurements started at the centre point of the joints, and the distance between the hardness measurement points was selected to 0.5 mm. The measured micro-hardness values of the specimen with 0.35 mm bottom thickness are shown

Then we investigated the effect of the bottom thickness (its optimum value is suggested by the tool manufacturer) on the strength properties of joints. The results of these investigations are shown in Table 1.

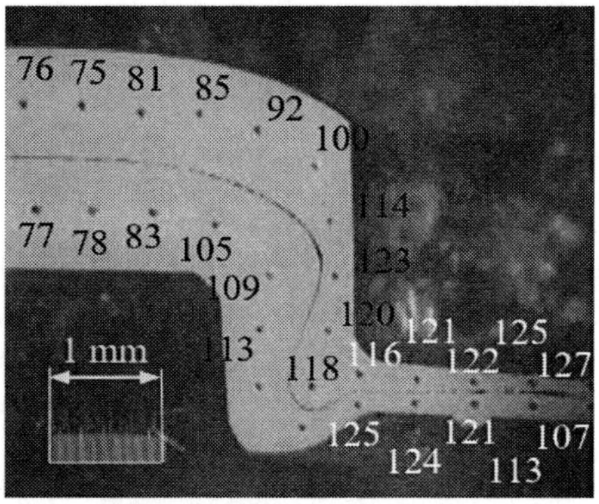

Figure 7. Results of HV 0.5 hardness measurement for 0.35 mm-bottom thickness on AA6082-T0 material quality (sheet thicknesses: 1 mm)

Besides the same AA6082 material quality clinched joints were made using dissimilar material grades with the same material thickness (t = 1 mm): in these tests AA5754 aluminium and DP600 steel materials were used. The optimum bottom thickness was also determined for these materials pairs, too. Those bottom thickness values were regarded as optimum when the highest tensile and neck pulling forces were measured (these bottom thickness values are shown with shaded colour in Table 1). Three tests types of samples were prepared: 1. Shearing – tensile test, 2. Head pulling test, 3. Neck pulling test. The samples were prepared by laser cutting. The sample geometry and dimensions are shown in Figure 8. The samples used for different tests are shown in Figure 9.

Table 1. Results of the determination of optimal bottom thickness

AA6082-T0 / AA6082-T0			AA6082-T0 / AA5754			AA5754 / AA6082-T0			AA6082-T0 / DP600			DP600 / AA6082 -T0		
t_b (mm)	Tensile test (kN)	Tensile test average (kN)	t_b (mm)	Tensile test (kN)	Tensile test average (kN)	t_b (mm)	Tensile test (kN)	Tensile test average (kN)	t_b (mm)	Tensile test (kN)	Tensile test average (kN)	t_b (mm)	Tensile test (kN)	Tensile test average (kN)
0,64	1,37	1,30	0,65	0,8	0,85	0,60	1,00	1,03	0,74	0,60	0,50	0,88	1,68	1,89
0,65	1,24		0,64	0,9		0,59	1,05		0,77	0,40		0,87	2,10	
0,55	1,40	1,29	0,59	1	1,03	0,55	1,48	1,24	0,67	0,58	0,42	0,75	2,40	2,38
0,55	1,17		0,6	1,05		0,55	1,00		0,68	0,25		0,77	2,35	
0,47	1,31	1,47	0,52	1,20	1,06	0,50	0,92	1,21	0,55	0,68	0,69	0,62	-	-
0,47	1,63		0,52	0,92		0,50	1,50		0,56	0,70		0,72	-	
0,43	1,60	1,38	0,48	1,70	1,50	0,47	1,46	1,53	0,47	0,72	0,72	0,55	-	-
0,43	1,16		0,47	1,30		0,46	1,60		0,46	0,72		0,56	-	
0,36	1,35	1,24	0,41	0,90	1,09	0,42	1,58	1,45	0,40	0,50	0,56	0,45	-	-
0,36	1,13		0,42	1,28		0,43	1,32		0,41	0,61		0,47	-	

The upper part of Figure 8, shows the original sample geometry and dimensions, while its lower part indicates the overlapping position of joined parts. The results of the tensile tests are summarised in Table *2*.

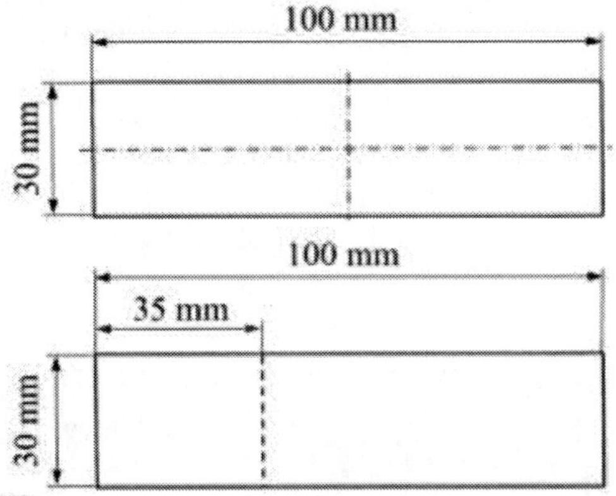

Figure 8. Geometry of test samples.

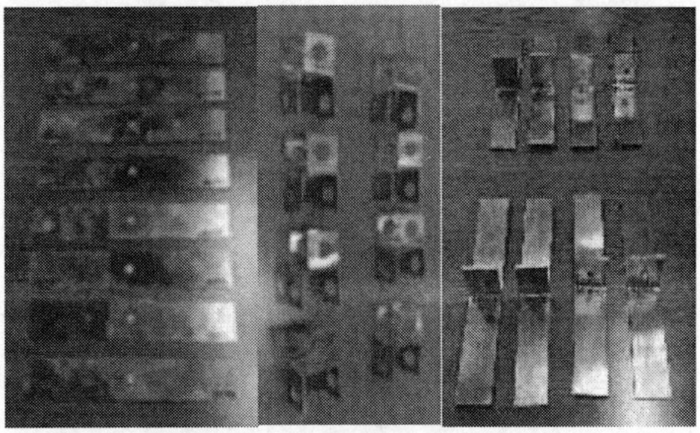

Figure 9. The samples used for tensile, head pulling and neck pulling tests.

Table 2. The results of the tensile, head and neck pulling tests for T6 and T0 condition of AA6082

	AA6082-T0 t_b = 0,47mm T = 525°C, t = 30 min			AA6082-T6 t_b = 0,47mm T = 525°C, 30 min + T = 190°C, t = 8 h		
No. of samples	Tensile test (N)	Head pulling test (N)	Neck pulling test (N)	Tensile test (N)	Head pulling test (N)	Neck pulling test (N)
1	1 171	709	254	1 755	803	307
2	1 662	737	279	1 738	816	295
3	1 506	692	310	1 751	872	290
4	1 329	737	277	1 717	883	304
average	1 417	719	280	1 740	843	299

Based on these results, it can be stated that the aluminium AA6082 alloy in T6 condition practically cannot be clinched successfully. After solution heat treatment at T=525°, with t=30 min holding time, the clinching can be done successfully. The best joints were obtained by using 0.45 and 0.5 mm bottom thicknesses. After ageing at T=190°C, t=8 h, the strength properties were significantly improved for all three types of tests.

As it was already mentioned, we also investigated clinched joints with dissimilar materials. The tested material pairs (thickness: 1 mm) were: AA6082-T0/AA5754, AA5754/AA6082-T0, AA6082-T0/DP600, DP600/AA6082-T0. In Figure 10, the clinched joints of AA6082 and AA5754 are shown. For clinched joints formed with Al 6082 / Al 5754 pairs, perfect joints were created (see Figure 10).

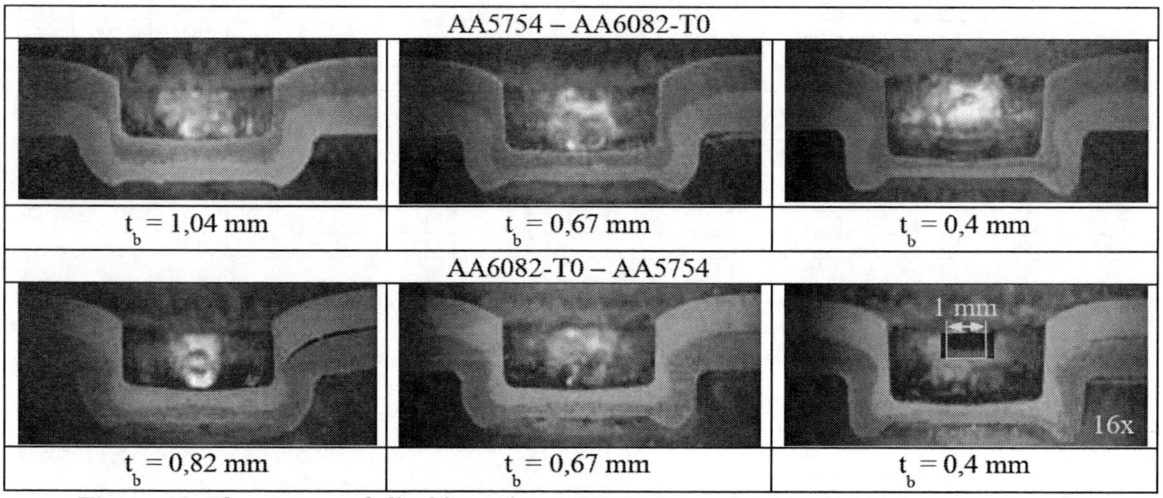

Figure 10. The process of clinching using AA6082-T0 and AA5754 (thickness t =1 mm).

In Figure 11, clinching of AA6082 and DP 600 material qualities are shown. In these experiments, we applied different joining configurations, i.e. punch side and the die side was differently applied to the materials joined.

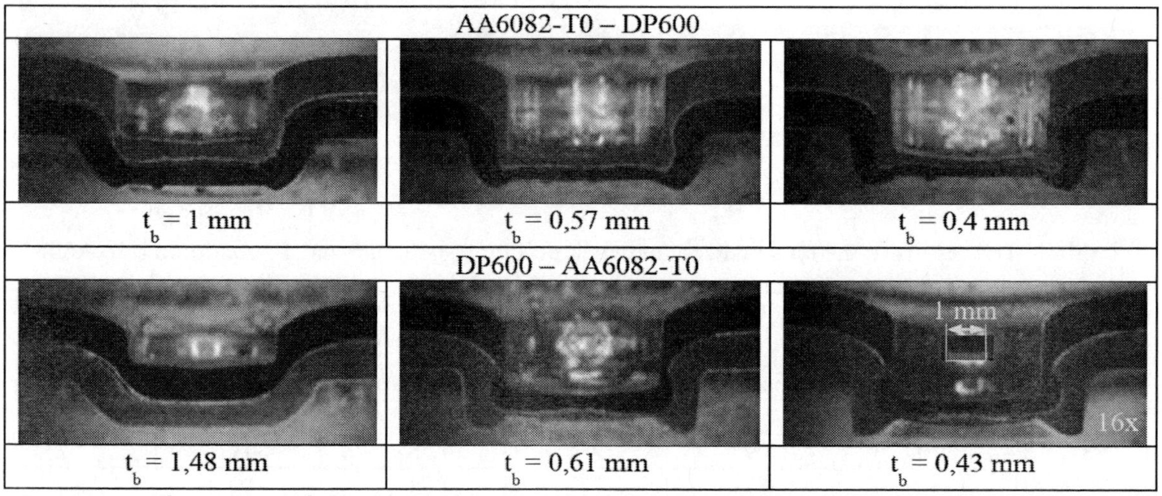

Figure 11. The process of clinching AA6082-T0 and DP600 material qualities (thickness t =1 mm).

For joints made with AA6082 / DP600 material qualities, it was not always possible to successfully create joints at different bottom thicknesses when the higher strength material, i.e. DP600 was on the punch side. The results of the tensile tests for dissimilar materials are summarised in Table 3.

Table 3. The results of the tensile tests for dissimilar materials

	AA6082-T0 / AA6082-T0		AA6082-T0 / AA5754		AA5754 / AA6082-T0		AA6082-T0 / DP600		DP600 / AA6082 -T0	
	T0 - (T = 525°C, t = 30 min)									
specimen	Tensile test (kN)	Head pulling test (kN)	Tensile test (kN)	Head pulling test (kN)	Tensile test (kN)	Head pulling test (kN)	Tensile test (kN)	Head pulling test (kN)	Tensile test (kN)	Head pulling test (kN)
1	1,17	0,71	1,35	0,70	1,25	0,55	0,85	0,58	1,15	0,30
2	1,66	0,74	1,30	0,65	0,92	0,60	0,30	0,40	2,50	0,55
3	1,51	0,69	1,38	0,66	1,40	0,62	0,45	0,43	0,70	0,37
4	1,33	0,74	1,30	0,62	1,20	0,53	0,50	0,52	1,40	0,51
average	**1,42**	**0,72**	**1,33**	**0,66**	**1,19**	**0,58**	**0,53**	**0,48**	**1,44**	**0,43**
	AA6082-T6 / AA6082-T6		AA6082-T6 / AA5754		AA5754 / AA6082-T6		AA6082-T6 / DP600		DP600 / AA6082 -T6	
	T6 - (T = 525°C, t = 30 min + T = 190°C, t = 8 h)									
specimen	Tensile test (kN)	Head pulling test (kN)	Tensile test (kN)	Head pulling test (kN)	Tensile test (kN)	Head pulling test (kN)	Tensile test (kN)	Head pulling test (kN)	Tensile test (kN)	Head pulling test (kN)
1	1,75	0,80	1,55	0,72	1,10	0,70	0,75	0,40	2,95	0,88
2	1,74	0,82	1,51	0,75	1,40	0,70	0,75	0,55	2,80	0,89
3	1,75	0,87	1,42	0,80	1,38	0,70	0,85	0,55	3,00	0,80
4	1,71	0,88	1,40	0,72	1,30	0,65	0,65	0,60	3,20	0,90
average	**1,74**	**0,84**	**1,47**	**0,75**	**1,30**	**0,69**	**0,75**	**0,53**	**2,99**	**0,87**

Based on these results, it can be stated that AA6082 solution heat-treated aluminium sheets can be successfully clinched with similar strengths material grade AA5754, while with other high strength sheets (DP600) cannot always be joined successfully.

3 Summary

In this paper, the mechanical joining of similar and dissimilar materials was analysed. Similar clinched joints were made from the same the material grades (e.g. AA6082) with the same sheet thicknesses. Then dissimilar clinched joints can be made either applying different material qualities (e.g. AA5754/AA6082, or AA6082/DP600) or different material thicknesses. The clinched joints were tested using tensile, head pulling or neck pulling tests. Several clinching parameters were also tested, e.g. the effect of bottom thickness and the heat treatment on the mechanical properties of joints. According to the results, it can be stated that the use of heat treatment allows to avoid fractures, improves the flow of material and consequently improves the mechanical behaviour of the joints.

Acknowledgements

The authors would like to express their greatest thanks to the European Commission for their support on the H2020 project entitled "*Low Cost Materials Processing Technologies for Mass Production of Lightweight Vehicles (LoCoMaTech)*", Grant No: H2020-NMBP-GV-2016 (723517).

References

[1] Tatsuya Sakiyama, Yasuaki Naito, Gen Murayama, Kenji Saita, Hatsuhiko Oikawa, Tetsuro Nose, Dissimilar Metal Joining Technologies for Steel Sheet and Aluminium Alloy Sheet in Auto Body, Nippon Steel Technical Report, No. 103, May 2013

[2] J. E. Gould, Joining Aluminium Sheet in the Automotive Industry – A 30 Year History, Welding research, January 2012, Vol. 91

[3] https://us.tox-pressotechnik.com

[4] Dr. Danyi József, Dr. Végvári Ferenc: Manufacturing and Maintenance in Mechanical Engineering (in Hungarian), Textbook, 2011.

[5] Gábor Béres, József Danyi, Ferenc Végvári: Clinching of steel sheets used in automotive industry, Modern Technologies in Manufacturing. 394 p., Cluj-Napoca, Románia, 2015.10.14-2015.10.16

IOP Conf. Series: Materials Science and Engineering **426** (2018) 012029 doi:10.1088/1757-899X/426/1/012029

Structural analysis of titanium alloys

K Kulcsár[1], J Kónya[2] and I Zsoldos[3]

[1] Development engineer, Dent-Art-Technik Kft., Győr, Hungary
[1] Head development engineer, Dent-Art-Technik Kft., Győr, Hungary
[1] Head of Department, Szechenyi Istvan University, Győr, Hungary

E-mail: kulcsar.klaudia@dentarttechnik.hu

Abstract. The material of choice for dental implant devices is titanium alloy (Ti-6Al-4V). In this study, the two fundamental manufacturing technologies of dental implant prostheses are compared. Titanium alloy specimens and models were created with conventional precision casting and with modern, innovative additive manufacturing technologies. Hereupon, the microscopic analysis of specimen cut-offs was carried out. We were focusing on fundamental microstructural differences. The thus-created specimen models were subjected to tensile tests for comparative examination.

1. Introduction

In the field of dental applications, there are several different materials available such as metals, polymers, and ceramics. Metallic materials are even more widespread due to their high strength and good processability. Titanium and its alloys bear with outstanding corrosion resistance and biocompatiblity [1]. The most common titanium alloy for prostheses, implants, and medical devices is Ti-6Al-4V alloy, which is the standard material for bone replacements [2]. Manufacturing technologies were classified into two groups until the end of the 1980s. One group was metal forming, during which adequate external forces were exerted to obtain the proper geometry of the workpiece. During the processing technology, material continuity was not broken, and the mass of the workpiece did not change. The other processing technology was subtractive manufacturing, during which the required geometry was created with cutting.

Because of the development in digital sciences, a new manufacturing technology emerged by the end of the 20th century. This method utilizes additive manufacturing instead of subtractive technologies to build up the required workpiece geometry. During this new process, the workpiece is built up layer-by-layer, and prototypes can be manufactured using digital data [3]. Nowadays, these processes are the so-called Additive Manufacturing (AM) technologies [4].

Different synonims are in use, such as additive procesess, additive manufacturing, etc. [5]. By the 1980s, computer-aided design and manufacturing (CAD/CAM) became widespread in dental fields [6].

Additive manufacturing enables a high-degree freedom of design of implants and prosthetic frames which meet all the mechanical and physical requirements [7]. 3D printing makes it able to create complex, hierarchic structures directly from a computer model, thus structured, porous dental implants can be manufactured [8]. More and more studies prove that porous structure helps tissue connection, thus enhances osseointegration between the implant and bone tissue [9-11].

Here we can mention that the corrosion resistance of the metal structure comes from the high affinity of oxygen to titanium. This helps osseointegration process by the formation of a passive oxide layer [12]. The oxide layer formed on titanium alloy is made up of TiO_2 (titanium-dioxide) [13]. The thickness of the formed oxide layer is approximately 10nm, and provides high resistance against several chemical effects. Thus, titanium-based prosthetic implant structure frames are highly corrosion resistant also in oral environments [14]. Conventional manufacturing technology of titanium structures is casting, which can create the required geometric contours. Shaping and coating of titanium implant surfaces raise important questions [15].

The use of titanium casting machine developed for dental applications increased the widespread use of cast titanium. Titanium is heated in a vacuum atmosphere, so oxidation rate is kept low during the melting of the metal [16-19]. In everyday routine, professionals encounter several problems during lost-wax casting technology. Possible emerging problems can be the distortion of wax, casting defects that lead to imprecision of the product [20]. Together with casting technology, modern additive technologies also show an ever increasing development.

2. Materials and Methods

2.1. Casting of specimens

Cast specimens were created with a Dentaurum Universal 230 Autocast casting unit (Fig. 1). The first step of the manufacturing process was 3D-printing of specimens from polymeric material. A pre-designed model was made during manufacturing, then the wax pattern was embedded. With the help of silicone casting material, a wax copy was made. The necessary organic wax for lost-wax casting had to be gassing-free and had to be suitable for burning without sudge formation. Afterwards, spruing and ceramic embedding of the models took place. The open-pore embedding material and the pattern inside is burned out in a furnace. Then titanium is casted into the remaining places of the pattern with a vacuum-pressure casting machine. This process takes place in an argon atmosphere. After this, embedding material is removed from the cast and the testing of geometrical sizes starts (Fig. 1).

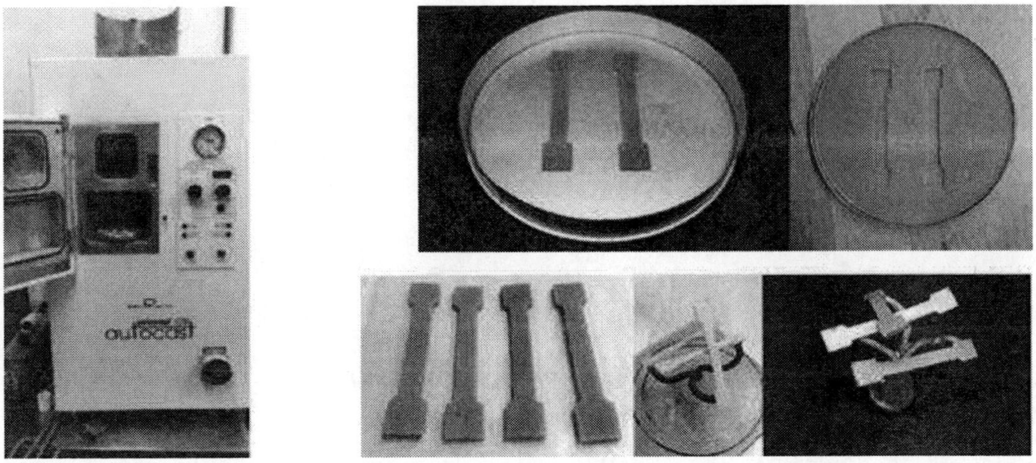

Figure 1. Manufacturing steps of titanium alloy specimens at casting processes

2.2. 3D printing of specimens

For the comparison of casting technology and 3D printing, specimens were also created with a SismaMysint 3D printer (Fig. 2). Hereby, printing was carried out with Laser Metal Fusion (LMF) technology. A programmable furnace was deployed for post-production heat treatments considering specific printing strategy and heat-diverting support materials. Different software and milling machine are available for precision post-processing. Additive manufacturing process is preceded by digital model design. 3D printing is an additive manufacturing process that builds up parts by layering and melting together metal powder particles.

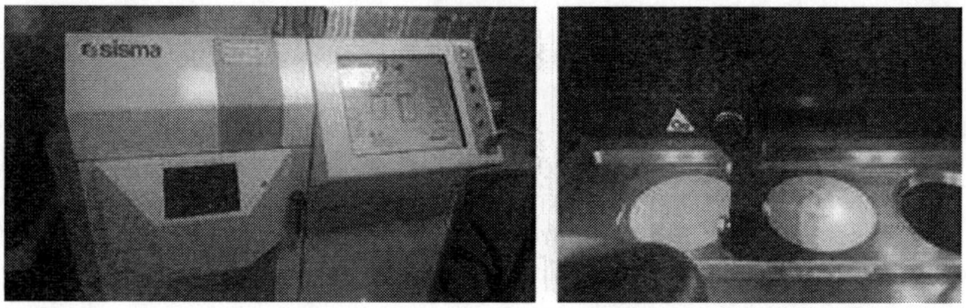

Figure 2. 3D printing unit and an image showing the printing process

2.3. *Heat treatment*

After the preparation of models, the necessary time interwall for 3D printing was approximately 4 hours. The specimens were separated from the building platform. The disc and the support material connecting the specimens were removed. Next, the workpieces were subjected to heat treatment with parameters determined by the manufacturer. Fig. 3 shows the process diagram for heat treatment.

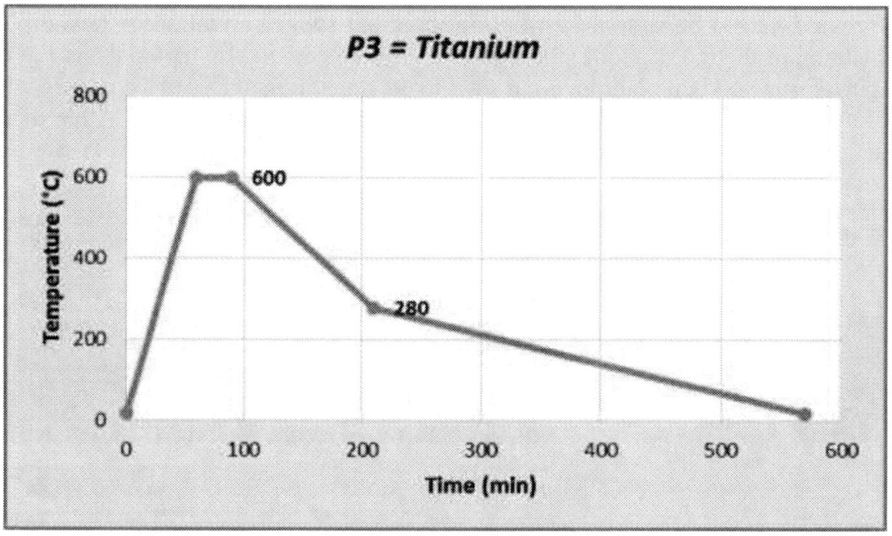

Figure 3. Heat treatment process diagram

2.4. *Materials of choice*

For the experiments, Ti-6Al-4V titanium alloy, and commercially pure titanium materials were chosen. Grade 5 ELI (marked "A") for casting, and Grade 23 (marked "B") for 3D printing were used for metallographic analysis. Grade 1 (marked "C") titanium was examined during tensile tests. The chemical composition of different titanium and titanium alloys are shown in Table 1.

Table 1. Chemical composition of different titanium alloys

Chemical composition			
Component	Percentage (%)		
	„A"	„B"	„C"
Oxygen (O)	0.20	0.13	0.18
Nitrogen (N)	0.05	0.03	0.03
Carbon (C)	0.08	0.08	0.08
Hydrogen (H)	0.015	0.0125	0.015
Iron (Fe)	0.4	0.25	0.2
Aluminum (Al)	5.5-6.75	5.5-6.5	0
Vanadium (V)	3.5-4.5	3.5-4.5	0

3. Testing of specimens

Printed and casted specimens were subjected to microscopic metallographic analysis and mechanical tensile tests.

3.1. Metallographic analysis

Microscopic images were made in the material testing laboratory at the Department of Material Science and Technology in Szechenyi Istvan University. Microscope: multifunctional optical microscope; type: ZEISS Axio Imager M 1. The first test was the comparison of Ti-6Al-4V samples manufactured with casting and 3D printing.

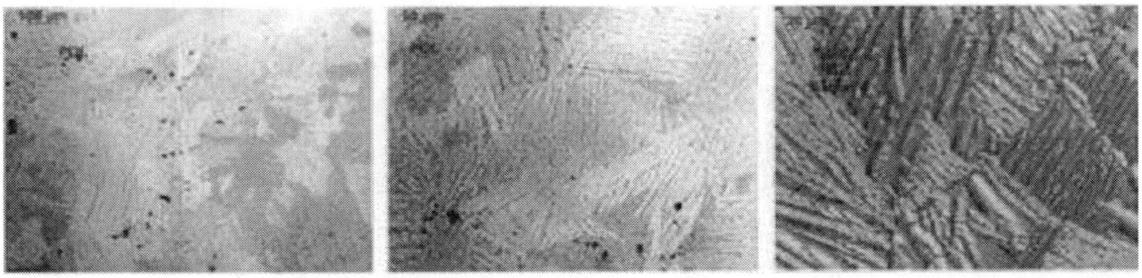

Figure 4. Grain structure of cast titanium samples

Fig. 4 shows the grain structure of cast titanium samples at different magnifications. No anisotropy can be discovered in the grain structure, typical grain size is relatively large exceeding 50-100 μm. At higher magnification (middle and right images), the characteristic pattern of α-β titanium alloys can be observed. Alpha-phases (light-colored parts) have a relatively regular geometry represented in plate-like shapes. Between alpha phases, thin areas of beta-phase (dark-colored parts) appear. In the left and middle images, small black grains can be detected, which prove that a certain extent of porosity remain in the structure after the casting process.

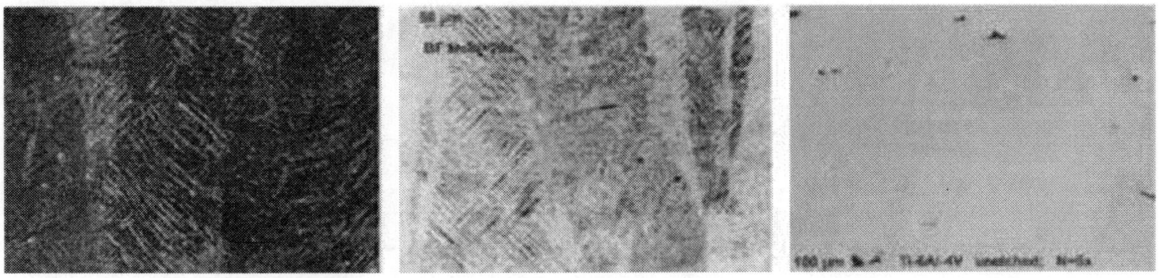

Figure 5. Grain structure of 3D-printed titanium samples

Images in Fig. 5 show the grain structure after 3D printing. The left and middle images were shot with different expositions after etching. The right image shows the sample before etching. Grain structure can be characterized with the left and middle images. Grain structure is anisotropic. Grains are elongated, and the longitudinal direction is parallel with the direction of laser scanning. The typical thickness of columnar grains lies between 50-100 μm, which is in the same order of magnitude than that of cast specimens. Black grains observed in the right image are small porosities. Generally, the same extent of porosity is present both in cast and 3D-printed specimens based on visual inspection.

IOP Conf. Series: Materials Science and Engineering **426** (2018) 012029 doi:10.1088/1757-899X/426/1/012029

3.2. Tensile tests

During the testing procedure, the previously introduced material at 3D printing was compared to commercially pure titanium. Commercially pure titanium is widely used in dental practice to create cast parts. Its chemical composition is shown in the following.

The test-specimens were fixed into a tensile testing machine, and then were subjected to controlled tension until failure. Fig. 6 shows the state of the specimens after testing. In the right image, the fracture site can be observed, which shows signs of brittle fracture.

Figure 6. Photos of a broken specimen after loading test.

Fig. 7 shows load-extension curves of specimens. Numbers from 1 to 3 mark 0°, 45° and 90° manufacturing orientations, respectively. Cast titanium specimen is marked no. 4.

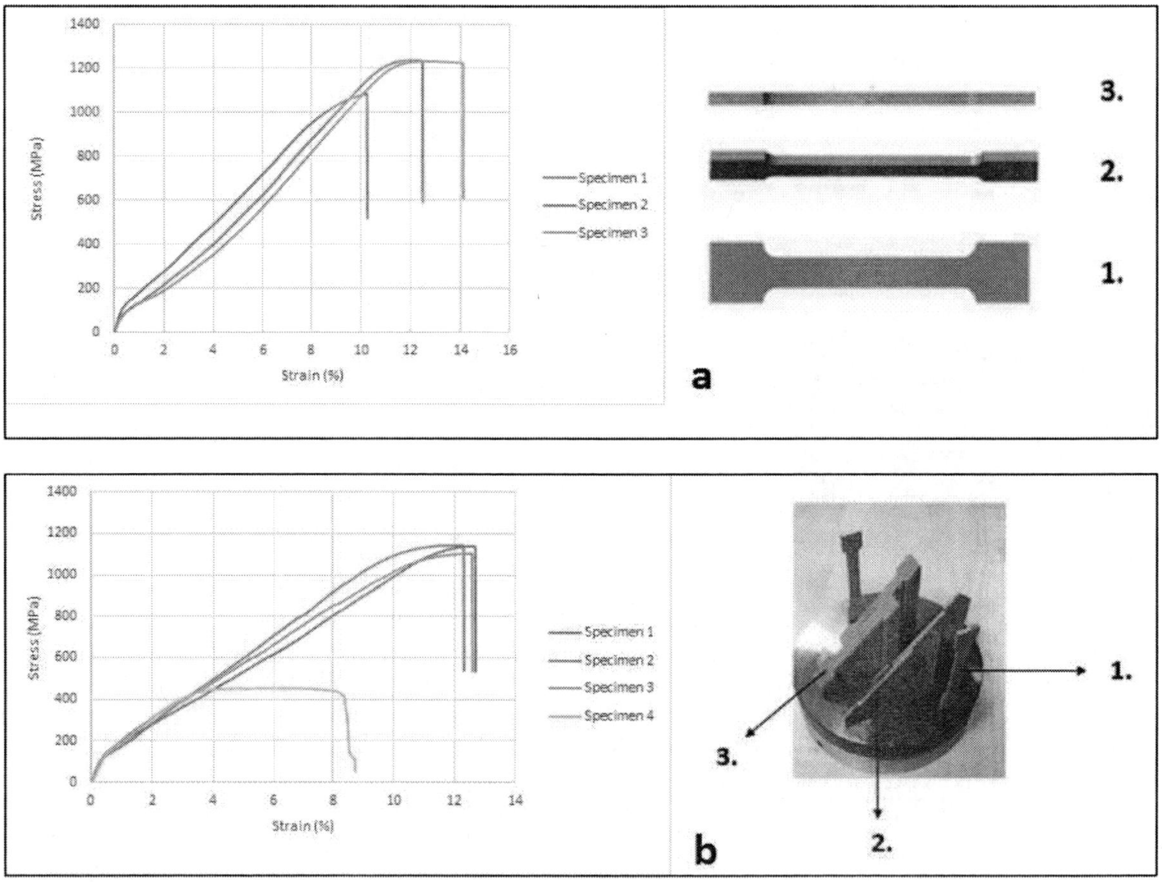

Figure 7. Load-extension curves and different printing orientations of specimens created with additive manufacturing. a: loading diagram for various orientation around the specimen axis. b: loading curves according to various specimen orientation angles in the workplace of the 3D printer

IOP Conf. Series: Materials Science and Engineering **426** (2018) 012029 doi:10.1088/1757-899X/426/1/012029

Table 2 and 3 detail the material properties determined by tensile testing.

Table 2. Material properties of specimens shown in Fig. 7a determined by tensile testing

	Tensile stress at Tensile Strength (MPa)	Modulus (E-modulus) (GPa)	Tensile strain at Tensile Strength (%)
1	1138	29.03	12.49
2	1142	31.48	12.06
3	1103	28.92	12.44
4	453	27.10	5.68

Table 3. Material properties of specimens shown in Fig. 7a determined by tensile testing

	Tensile stress at Tensile Strength (MPa)	Modulus (E-modulus) (GPa)	Tensile strain at Tensile Strength (%)
1	1081	29.68	10.12
2	1235	19.83	12.12
3	1231	15.79	12.64

The differences in strength values originate from anisotropic grain structure. The elongated grains in the laid specimen are parallel to the load, which is favorable in a mechanical point of view. However, these elongated grains in the standing specimen are directed perpendicularly to the load. There are more grain boundaries present in the direction of the stress, and this orientation is mechanically unfavorable. The orientation of 45 degrees is a transition between these two cases. It is noticeable that higher values are present in the second group. The distinctness of values raises other issues about the structured melting of layers and the orientation of the specimen.

Table 4 contains material properties for commercially pure titanium obtained from literature, and data determined from our cast specimens.

Table 4. Properties after the casting process

	Ultimate Strength (MPa)	Modulus of elasticity (Gpa)
Commercially pure titanium (literature)	345	103
Commercially pure titanium (cast)	453	27

Table 5 contains the comparison of mechanical properties of Ti-6Al-4V titanium alloy obtained from literature and from testing of specimens after additive manufacturing and heat treatment. The values from literature apply to rolled, grinded fiber material. It is not subjected to heat treatment before machining.

Table 5. Data for fiber and 3D-printed material

	Ultimate Strength (MPa)	Modulus of elasticity (Gpa)
Ti-6Al-4V (literature)	895	114
Ti-6Al-4V (3D-printed)	1138	29

IOP Conf. Series: Materials Science and Engineering **426** (2018) 012029 doi:10.1088/1757-899X/426/1/012029

4. Summary

In conclusion, assume that laser sintering – 3D printing technology – is going to replace the role of casting in terms of titanium processing in dental applications. The homogeneity of products created with additive manufacturing is more uniform. The geometry to be realized is going to be determined. Both in terms of microstructure and tensile tests, laser sintered specimens outperformed casted ones. During casting process, both internal and external inclusions are formed. A special and expensive casting material is needed. Human expertise is necessary for post-processing that guarantees the precision of the finished product. Laser sintering provides more precise, and faster processing. As a result of this, 3D printing of titanium is becoming even more widespread in medical technology.

References

[1] Brunette D M, Tengvall P, textor M and Thomsen P 2001 *Titanium in medicine* (Springer Berlin) pp 1-12

[2] Williams D F 2001 *Titanium for medical application)*In: Brunette D M, Tengvall P, Textor M and Thomsen P (eds.) Titanium in medicine (Springer-Verlag Berlin) pp 13-24

[3] Háber I E 2015 *3D adatfeldolgozás és gyártás* (II Pécsi Tudományegyetem Műszaki és Inforamtikai Kar Pécs) TÁMOP-4.1.1. F-14/1/KONV-2015-0009

[4] Gibson I, Rosen D and Stucker B 2009 *Additive Manufacturing Technologies* 2 ed. (New York Springer)

[5] Frazier W E 2014 *Metal Additive Manufacturing* (Journal Of Materials Engineering And Performance) Volume 23 Issue 6 pp 1917-1928

[6] Al-Mesmar H S, Morgano S M and Mark L E 1999 *Investigation of the effect of three sprue designs on the porosity and the completeness of titanium cast removable partial denture frameworks* (J. Prosthet Dent) 82 pp 15-21

[7] Arvidsson A, Sarve H and Johansson C B 2015 *Comparing and visualizing titanium implant integration in rat using 2D and 3D techniques* (J. Biomed Mater Res. B Appl Biomater) 103 pp 12-20

[8] Yang F et al 2017 *Laser beam melting 3D printing of Ti6Al4V based porous structured dental implants: fabrication, biocomplatibility analysis and photoelastic study* (Scientific reports) pp 1-10

[9] Maniatopoulos C, Pilliar R and Smith D 1986 *Threaded versus porous-surfaced design for implant stabilization in bone-endodontic implant model* (Journal of biomedical materials research) 20 pp 1309-1333

[10] Unger A S, Lewis R J and Gruen T 2005 *Evaluation of a porous tantalum uncemented acetabular cup in revision total hip arthroplasty: clinical and radiological result of 60 hips*. (The Journal of arthroplasty) 20 pp 1002-1009

[11] Wu S et al 2013 *Porous titanium -6 aluminium-4 vanadium cage has better osseointegration and less micromotion than a poly-ether-ether-ketone cage in sheep vertebral fusion* Artificial organs 37 pp 191-201

[12] Gittens R A, Scheideler L, Rupp F, Hyzy S L Geis-Gerstrfer J, Schwarz Z and Boyan B D 2014 *A review on the wettability of dental implant surfaces II: biological and clinical aspects* (Acta Biomater) 10 pp 2907-2918

[13] Khanlou H M, Ang B C, Barzani M M, Silakhori M and Talebian S 2015 *Prediction and characterization of surface roughness using sandblasting and acid etching process on new non-tonic titanium biomaterial: adaptive-network-based fuzzy inference system* (Neural Comput & applic) 26 pp 1751-1761

[14] Revathi A, Borrás A D, Munoz A, Richard C and Manivasagam G 2017 *Degradation mechanisms and future challenges of titanium and it alloys for dental implant applications in oral environment* (Materials Science and Engineering) C 76 pp 1354-1368

[15] Takács J and Ozsváth P 2004 *Development of human implants with plasma sprayed surfaces* First (Hungarian Conference on Biomechanics Budapest, Hungary June 11-12) pp 464-471

[16] Miyakawa O, Watanabe K, Okawa S, Nakano S, Kobayashi M and Shiokawa N *1989 Layered structure of cast titanium surface*. (Dent Mater J) 8 pp 175-185
[17] Tamaki Y and Miyazaki T 2002 *Casting technology of titanium in dentistry*. (J Jpn Soc Biomat) 20 pp 380-386
[18] Okuda R 1995 *Clinical application of titanium precision casting technique today*. (Jpn J Conserv Dent) 38 pp 9-10
[19] Doi H, Yoneyama T, Kotake M and Hamanaka H 1992 *Influence of surface reaction layer on mechanical properties of titanium castings*. (J J Dent Mater) 11 pp 817-822
[20] Anusavice K J 2003 *Casting Investments and Procedures philip~s* (Science of Dental Materials 11th ed. Saunders St. Luois Missouri) pp 295-300

IOP Conf. Series: Materials Science and Engineering **426** (2018) 012030 doi:10.1088/1757-899X/426/1/012030

Structure and properties of NR- and XNBR-toughened polypropylene produced with latex by melt compounding

L Lendvai[1] and J Karger-Kocsis[1]

[1] Budapest University of Technology and Economics, Faculty of Mechanical Engineering, Department of Polymer Engineering, Budapest, Hungary

E-mail: lendvai@pt.bme.hu,; karger@pt.bme.hu

Abstract. Binary blends composed of polypropylene (PP) and natural rubber (NR) or carboxylated acrylonitrile butadiene rubber (XNBR) were produced in batch-type melt compounding process. The rubbers used as toughening agents were introduced into PP in up to 20 wt% in their dry and latex forms. Structure-property relationships of the blends were studied. Both NR and XNBR were homogenously dispersed in the PP matrix: XNBR in microscale, NR in the submicronic scale. Fracture toughness and energy, determined on notched Charpy specimens, were improved by rubber toughening, however, at the cost of stiffness and strength, determined in tensile tests. NR toughened blends were superior to blends containing XNBR with respect to impact resistance. The introduction of rubbers in latex form resulted in slightly better mechanical performance compared to addition in dry form.

1. Introduction

Nowadays great efforts are made to improve the mechanical, thermal and other properties of polymers. Among those, the development of rubber-toughened polymers (RTP) has been a subject of active research in the last decades to extend their range of applications [1]. Polypropylene (PP) is a thermoplastic polymer, mostly used in the packaging and textile industries. It exhibits, however, a relatively poor impact resistance, especially at room temperature (close to its glass transition temperature, T_g) and below. Two methods are used to improve its toughness: copolymerization with suitable monomers [2] and blending with various elastomers [1, 3, 4]. The elastomer particles in both cases should be finely and homogenously dispersed and their interparticle distance (matrix ligament - controlled by the elastomer amount) should be adjusted to the matrix [5, 6]. These are the prerequisites of efficient toughening, the mechanism of which is as follows: stress concentration at the particles, followed by their cavitation/break-up/elongation, accompanied with massive shear yielding and/or crazing of the matrix [7]. It is noteworthy that rubber toughening – albeit via a somewhat different mechanism – also works in thermosets [8]. To get the suitable dispersion is, however, a challenging task. During melt blending the dispersion characteristics mainly depend on two controlling parameters: the viscosity ratio and the capillary or Weber number [9]. The closer the viscosity of the rubber modifier to that of the matrix under the melt blending conditions (temperature, shear rate), the finer the resulting dispersion is. The physical meaning of the Weber number is the balance between the shear stress acting locally on the particle and the deformation resistance of the particle. The elongated rubber particle breaks up when the Weber number exceeds a critical value, which also depends on the viscosity ratio [9]. Besides melt compounding and "reactor blending" (copolymerization involved), there is another possibility for the rubber modification of thermoplastics. Many rubbers are available in latex form,

IOP Conf. Series: Materials Science and Engineering **426** (2018) 012030 doi:10.1088/1757-899X/426/1/012030

which is an aqueous dispersion of the micron- or submicron-scaled elastomer particles. Therefore, it is reasonable to assume that by evaporating the water content of the latex during melt compounding, we may incorporate the elastomer particles for the level of dispersion required for toughening in the actual thermoplastic matrix [10]. It should be born in mind that a further requirement to this water-assisted melt compounding is the resistance of the corresponding matrix to hygrothermal decomposition. This concept has already been used for the toughening of polymers, such as polyamide [11, 12] and polystyrene [13]. Attempts have also been made to produce toughened and nanoparticle-reinforced thermoplastics in this way [10].

The objective of this paper was the toughening of polypropylene (PP) by blending it with uncured rubbers by latex-mediated and traditional melt compounding. The method followed would enable the preparation of PP-based materials with good impact resistance and thus broaden their application range. As modifiers, natural rubber (NR) and carboxylated acrylonitrile butadiene rubber (XNBR) were incorporated in PP. Their amount was varied between 0 and 20 wt.%. The morphology, tensile mechanical and dynamic mechanical properties were determined and the related structure-mechanical property relationships deduced.

2. Experimental

2.1. Materials

A polypropylene of medium molecular weight, PP H 543 F, supplied by MOL Petrolkémiai Zrt, Tiszaújváros, Hungary was used as matrix (MFR = 4 g/10 min at 230 °C and 2.16 kg). Note that PP does not degrade hygrothemally. Two different types of uncured rubbers were introduced as toughening agents: a natural rubber (NR) and a carboxylated acrylonitrile butadiene rubber (NBR). The NR latex (G-TEX LATZ) with 60 wt% dry content was obtained from Variachem Kft., Budapest, Hungary. XNBR latex (Chemigum Latex 550) with 41 wt% solid content was kindly supplied by OMNOVA Solutions Inc., Beachwood, OH, USA. A laser diffraction particle analyser Malvern 2000 (Malvern Instruments Ltd., Malvern, UK) was used to determine the initial particle size distribution of the rubber particles in these two latices.

2.2. Preparation of the samples

Binary blends were produced by melt compounding of PP with rubbers added in their dry and latex forms. The rubber content in the corresponding blends was set to 0, 5, 10 and 20 wt%. Table 1 shows the composition of each sample. Firstly, the PP pellets were masticated in a counter rotating internal batch mixer PL2000 Plasti-Corder (Brabender GmbH., Duisburg, Germany) at a temperature of 180 °C and 60 rpm. Subsequently, the rubber additives were introduced. When adding rubber in its latex form to the PP melt, we followed the fast evaporation mixing method described by Gao et al [14]. Therefore, the latex was dosaged dropwise. The compound was mixed for a further 5 min to achieve a fine dispersion and to eliminate the residual water. Then, the compound was dumped from the mixing chamber and allowed to cool to room temperature.

Afterwards, the prepared PP/rubber blends were compression molded to sheets of 2 mm thickness in a hot press (Teach-Line Platen Press 200E, Dr. Collin GmbH., Munich, Germany) for 3 mins at 200 °C. Specimens for testing were cut out of the compression molded sheets with a Mutronic Diadisk cutter (Mutronic GmbH, Rieden, Germany).

Table 1. Composition of NR- and XNBR-toughened polypropylene samples

Sample code	PP [wt%]	NR [wt%]	XNBR [wt%]	Rubber [latex/dry]
PP	100	-	-	-
5NR	95	5	-	latex
10NR	90	10	-	latex
20NR	80	20	-	latex
5XNBR	95	-	5	latex
10XNBR	90	-	10	latex
20XNBR	80	-	20	latex
10NR_DRY	90	10	-	dry
10XNBR_DRY	90	-	10	dry

2.3. Characterization and testing

The static mechanical properties were determined with tensile tests performed on a universal testing machine (Zwick Z005, Ulm, Germany) at a crosshead speed of 5 mm/min at room temperature. Tests were stopped at 150% elongation in the case of the very ductile specimens. The average value was calculated from five measurements.

Charpy tests were performed at room temperature (RT) on notched specimens. The notch length to width ratio of the Charpy specimens was constant, viz. 0.5. The notch was made with a V-type saw blade (Mutronic GmbH, Rieden, Germany). The impact test was performed on an instrumented pendulum (Ceast Spa, Pianezza, Italy), equipped with a data acquisition unit (DAS 8000 of Ceast). For the tests, the edge of the hammer (impact energy: 2 J) was cushioned with plasticine to suppress vibrations. The given test results represent average values of five tests.

Dynamic mechanical analysis (DMA) was performed in dual-cantilever mode at a frequency of 1 Hz with a DMA Q800 analyzer (TA Instruments, New Castle, USA). The storage modulus and the mechanical loss factor were determined as a function of the temperature (T = -50 °C ... + 75 °C). The amplitude was set to 0.02% of deformation. DMA curves were registered at a heating rate of 3 °C/min.

The morphology of the prepared samples was studied in a scanning electron microscope (SEM; JEOL JSM 6380LA, Tokyo, Japan). Fracture surfaces, produced in the impact tests, were sputter-coated with gold/palladium alloy prior to their inspection.

3. Results and discussion

3.1. Particle characteristics

The particle size distribution of the rubbers within the latices used are displayed in Figure 1. Figure 1 clearly demonstrated the large differences in both the mean particle sizes and size distributions of the latices. The distribution curves peaked at about 190 nm and 620 nm for the XNBR and NR latices, respectively. Particle size distribution also differs: NR particles have a considerably broader distribution than XNBR particles.

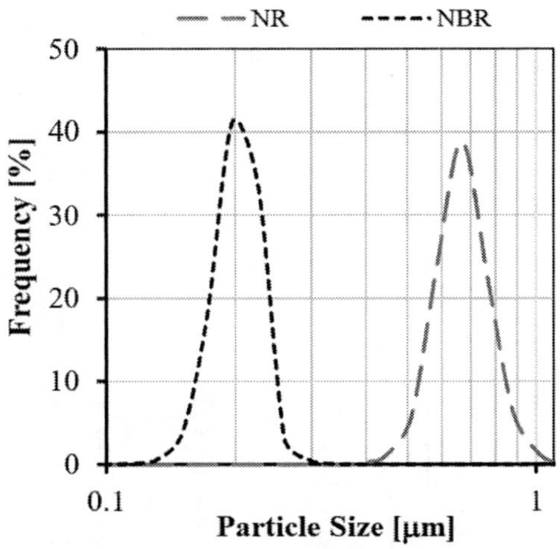

Figure 1. Particle size distribution of the rubber particles in the NR and XNBR latices used. Note: the abscissa in the diagram is in a logarithmic scale.

3.2. Tensile mechanical properties

The tensile mechanical properties (tensile strength, Young's modulus and elongations) are listed in Table 2. The addition of rubber particles markedly reduced both the strength and stiffness compared to neat PP. This is due to the "soft" rubbery character of the incorporated NR and XNBR. The incorporation of NR decreased the initial 29.9 MPa tensile strength of PP to 27.3, 24.8 and 19.7 MPa at 5, 10 and 20 wt% loading, respectively. The presence of XNBR had the same effect on strength. NR reduced the Young's modulus to a lesser extent than XNBR, suggesting a higher stiffness for NR compared to XNBR. This is supported by the elongation at maximum load (F_{max}) values, which were always below those of PP/XNBR in the case of NR toughening. Necking only appeared for neat PP and PP/NR blends. The possible reason behind this fact may be linked with the dispersion state of the rubber particles within the PP (see later in the morphology analysis).

The type of rubber introduction (latex/dry) makes no significant difference in strength, stiffness and elongation at F_{max}, although samples prepared through dry addition exhibited no necking (by breaking before it could occur), showing the advantage of using the aqueous media.

Table 2. Tensile mechanical properties of rubber-toughened polypropylene samples with the use of natural rubber and acrylonitrile butadiene rubber. Note: 150+ means higher than 150%

Sample	Tensile Strength [MPa]	Young's Modulus [GPa]	Elongation at F_{max} [%]	Elongation at break [%]
PP	29.9 ± 0.5	1.21 ± 0.04	11.8 ± 1.4	150+
5NR	27.3 ± 0.9	1.14 ± 0.03	10.3 ± 0.6	150+
10NR	24.8 ± 0.4	1.05 ± 0.01	9.6 ± 0.5	150+
20NR	19.7 ± 0.1	0.86 ± 0.01	7.7 ± 0.4	150+
5XNBR	27.2 ± 1.6	1.09 ± 0.06	11.8 ± 0.4	37.9 ± 7.9
10XNBR	22.9 ± 0.5	0.97 ± 0.02	11.8 ± 0.6	33.3 ± 7.5
20XNBR	18.4 ± 1.3	0.77 ± 0.07	12.8 ± 1.3	43.6 ± 9.1
10NR_DRY	23.5 ± 0.8	0.92 ± 0.03	11.2 ± 0.4	35.5 ± 8.3
10XNBR_DRY	22.7 ± 0.6	0.98 ± 0.02	10.5 ± 0.5	21.4 ± 4.1

IOP Conf. Series: Materials Science and Engineering **426** (2018) 012030 doi:10.1088/1757-899X/426/1/012030

3.3. Dynamic mechanical analysis

DMA results, i.e. storage modulus and loss factor versus temperature traces, are shown in Figure 2. It is well resolved that with increasing content of both NR and XNBR, stiffness is monotonously reduced. Storage modulus values of the PP/rubber blends above -10 °C are identical irrespective of rubber type. At lower temperatures, however, the moduli of the XNBR-toughened samples are markedly higher than those of samples containing NR because of the vitrification of XNBR (T_g at about -28 °C). The glassing of NR domains occurs at temperatures below the analysed temperature range.

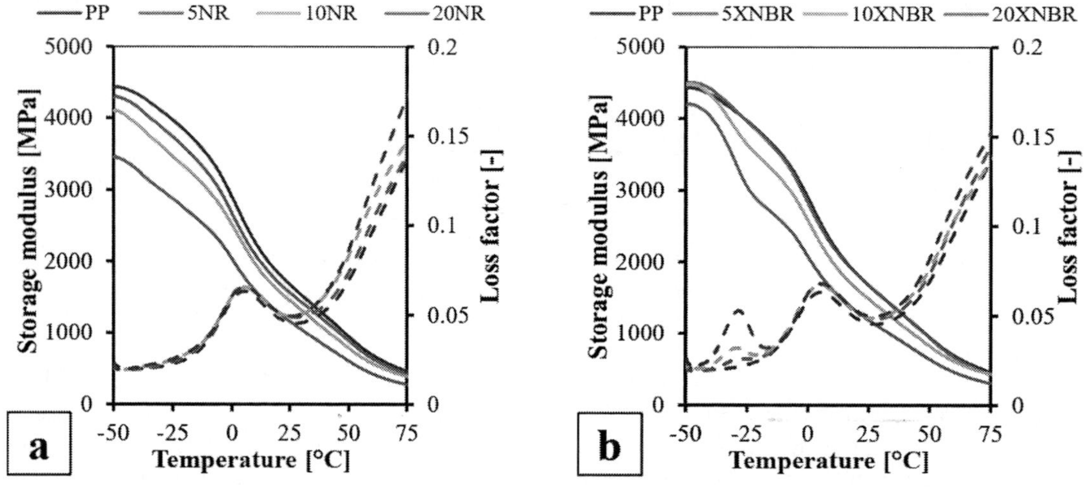

Figure 2. Storage modulus and loss factor as a function of temperature for the rubber-toughened PPs containing NR (a) and XNBR (b)

3.4. Impact performance

Figure 3 displays characteristic force-time traces of PP, PP/NR and PP/NBR blends registered at room temperature. The incorporation of rubber particles in PP resulted in a maximum force increment along with increased fracture time. This confirms that both rubbers acted as toughening agents for PP thereby increasing fracture energy. The largest fracture time was observed when NR was incorporated in a water-assisted way.

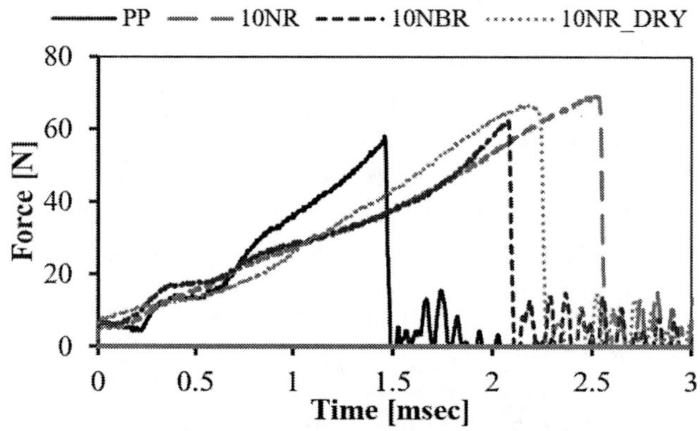

Figure 3. Characteristic force-time curves registered on PP/rubber (NR, XNBR) blends produced by latex and melt compounding. Note: comparison of the methods is given only for the PP with 10 wt% NR

The dependence of fracture toughness on rubber content is depicted in Figure 4 (a). It can be seen that the presence of both NR and XNBR leads to increased fracture energy values. For the PP/NR blends this increment is more significant than for PP/XNBR. For 5, 10 and 20 wt% NR loading the initial 5.2 kJ/m^2 fracture energy of PP increased to 8.1, 9.3 and 11.9 kJ/m^2, respectively. Samples containing XNBR exhibited slightly lower toughness values reaching a fracture energy of 8.8 kJ/m^2 at 20 wt% XNBR. Figure 4 (b) shows the differences in the fracture energy of the blends prepared by traditional and latex-mediated melt compounding methods. PP/rubber blends produced with latex dosage exhibited higher Charpy impact values than those with dry rubber, though the difference was practically within the scatter range.

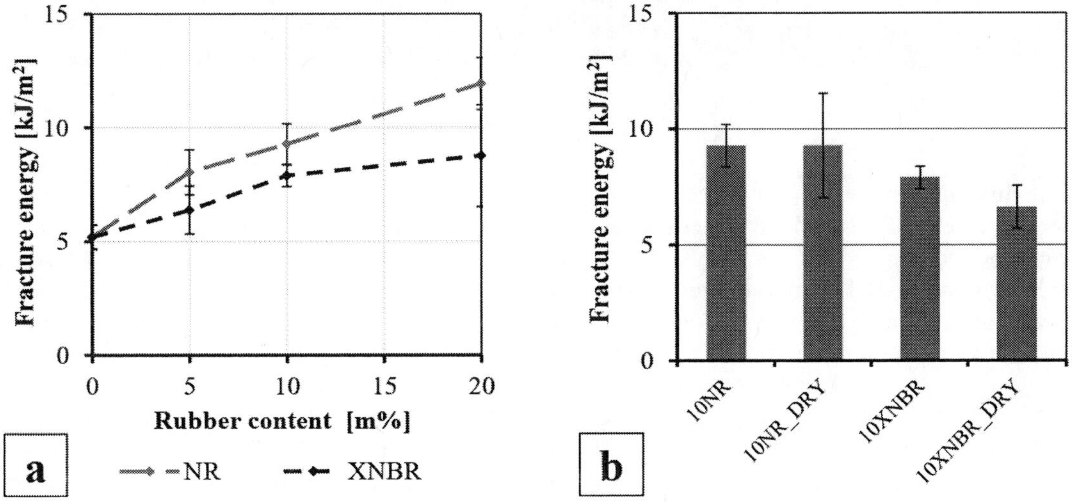

Figure 4. The fracture energy of rubber-toughened polypropylene samples with natural rubber and acrylonitrile butadiene rubber as a function of loading (a) and differences depending on the preparation method (b)

3.5. Morphology

SEM images taken of sample 10NR and 10XNBR are shown in Figure 5. Figure 5 (a) shows that NR particles within PP are in the size range of 500 nm to 2 µm. Compared to the initial particle size distribution of NR (cf. Figure 1), it means little to no increment. As a consequence, the original particle size of the NR latex also remained after compounding. By contrast, Figure 5 (b) shows that XNBR particles agglomerated to much larger domains compared to the original particle size in this latex. The initial size of XNBR increased from 190 nm to 3-8 µm, due to blending. It is well-known that in the presence of larger domains, resulting in larger matrix ligaments, there is no stress overlapping between the particles during deformation. This leads to a more brittle type of fracture in the case of PP/XNBR compared to the NR-toughened PP. This finding also supports our explanation when interpreting the measured mechanical properties (cf. Table 2).

Figure 5. SEM pictures from the fracture surface of rubber-toughened PP samples with NR (a) and XNBR (b) at a magnification of 2500x

4. Conclusions

Rubber-toughened PP-based blends, containing 0, 5, 10 and 20 wt% rubber were prepared in batch type melt compounding process. NR and XNBR rubbers were incorporated in their dry form by traditional or by latex-assisted melt compounding. Both NR and XNBR decreased the strength and stiffness of PP. This was in good agreement with a significant improvement in the notched Charpy impact energy. PP/NR blends outperformed the PP/XNBR blends with respect to Charpy toughness. Besides the better compatibility between PP and NR than between PP and XNBR, this was traced to the large differences in the dispersion characteristics of the rubber particles. XNBR particles were found in the range of 3 - 8 µm, by contrast to those of the NR particles, which were in the range of 0.5 - 2 µm. It is noteworthy that the particle size of the NR in the PP/NR blends agreed with that of the original latex. Opposed to that, the XNBR particles underwent a 15-40 fold size agglomeration upon blending compared to the parent latex. The static and dynamic mechanical properties of the PP/rubber blends prepared through dry rubber addition were inferior to those produced by latex mediation.

Acknowledgements

The work reported here was supported by the Hungarian Research Fund (OTKA) through the project K 109409. The authors thank Dr. J. Móczó for the particle size distribution measurements.

References

[1] Paul D R and Bucknall C B 2000 *Polymer Blends: Formulation and Performance* (Chichester, UK: John Wiley & Sons)
[2] Aarnio-Winterhof M, Doshev P, Seppala J and Gahleitner M 2017 Structure-property relations of heterophasic ethylene-propylene copolymers based on a single-site catalyst *Express Polymer Letters* **11** 152-61
[3] Utracki L A 1998 *Commercial Polymer Blends* (New York, USA: Springer US)
[4] Chow W S, Leu Y Y and Ishak Z A M 2016 Mechanical, Thermal and Morphological Properties of Injection Molded Poly(lactic acid)/Calcium Carbonate Nanocomposites *Periodica Polytechnica Mechanical Engineering* **60** 15-20
[5] Jang B Z, Uhlmann D R and Sande J B V 1985 Rubber-toughening in polypropylene *Journal of Applied Polymer Science* **30** 2485-504
[6] Liang J Z and Li R K Y 2000 Rubber toughening in polypropylene: A review *Journal of Applied Polymer Science* **77** 409-17
[7] Michler G H and Baltá-Calleja F J 2012 *Nano- and Micromechanics of Polymers* (Munich, Germany: Carl Hanser Verlag GmbH & Co. KG)

IOP Conf. Series: Materials Science and Engineering **426** (2018) 012030 doi:10.1088/1757-899X/426/1/012030

[8] Ozdemir N G, Zhang T, Aspin I, Scarpa F, Hadavinia H and Song Y 2016 Toughening of carbon fibre reinforced polymer composites with rubber nanoparticles for advanced industrial applications *Express Polymer Letters* **10** 394-407

[9] Karger-Kocsis J 1999 *Polymer Blends and Alloys,* ed G O Shonaike and G P Simon (New York, USA: Marcel Dekker Inc.) pp 125-54

[10] Karger-Kocsis J, Kmetty Á, Lendvai L, Drakopoulos S and Bárány T 2015 Water-Assisted Production of Thermoplastic Nanocomposites: A Review *Materials* **8** 72-95

[11] Siengchin S and Karger-Kocsis J 2009 Structure and creep response of toughened and nanoreinforced polyamides produced via the latex route: Effect of nanofiller type *Composites Science and Technology* **69** 677-83

[12] Ebrahimi Jahromi A, Ebrahimi Jahromi H R, Hemmati F, Saeb M R, Goodarzi V and Formela K 2016 Morphology and mechanical properties of polyamide/clay nanocomposites toughened with NBR/NBR-g-GMA: A comparative study *Composites Part B: Engineering* **90** 478-84

[13] Siengchin S and Karger-Kocsis J 2013 Binary and ternary composites of polystyrene, styrene–butadiene rubber and boehmite produced by water-mediated melt compounding: Morphology and mechanical properties *Composites Part B: Engineering* **45** 1458-63

[14] Gao W and Guo J 2017 A novel processing method namely fast evaporation mixing to prepare fluoroelastomer/montmorillonite composites *Composites Science and Technology* **139** 26-35

IOP Conf. Series: Materials Science and Engineering **426** (2018) 012031 doi:10.1088/1757-899X/426/1/012031

Development and characterization of innovative biopolymer foams

K Litauszki[1] and Á Kmetty[1, 2]

[1] Budapest University of Technology and Economics, Faculty of Mechanical Engineering, Department of Polymer Engineering, Budapest, Hungary
[2] MTA–BME Research Group for Composite Science and Technology, Budapest, Hungary

E-mail: litauszkik@pt.bme.hu, kmetty@pt.bme.hu

Abstract. Our development goal was to create foamed polymer systems based on biodegradable and renewable resource-based polymers. We used polylactic acid (PLA) and several types of chemical foaming agents (endothermic and exothermic), and produced foam samples by extrusion. We characterized and analyzed the foam samples with a wide range of tests (determining the cell structure, void fraction, compressive strength, etc.). Based on our tests, we showed the difference between the foaming agents and produced PLA-based foams with a void fraction of 57%.

1. Introduction

Today, with fast industrial development and rapidly increasing consumption, the need for packaging materials is increasing drastically. Most packaging is required by the food industry and the production of durable goods. The majority of these packaging materials, however, are one-way (single-use) packaging materials. The total production of the plastic industry was 322 million tonnes in 2015, and 58 million tonnes in the European Union, a major part of which – 39.9% – was packaging industry products. At present, these products are made from petroleum-based polymers, typically polypropylene, polyethylene and polystyrene. At the end of their lifetime, disposing of them presents a major problem since they are not biodegradable.

Based on the latest trends, poly(lactic acid) (PLA) is one of the strongest biopolymer candidates to replace petroleum-based polymers [1]. Bio-based and biodegradable PLA in a foamed structure can provide a possible marketable solution for engineering and costumer needs. There are various existing foaming techniques, like extrusion foaming [2], injection foaming, bead foaming [3] and syntactic foaming techniques are also [4,5]. Although the numerous article regarding PLA foams, there was no investigation regarding comparison the effect of endothermic and exothermic CBA effect on the PLA foaming behaviour in case of continuous extrusion foamed final product.

In this article we present the application of different types of Chemical Blowing Agents (CBA) to produce foamed PLA based biopolymer structures via extrusion to characterise the resulting foam structures in correlation of the CBAs endothermic or exothermic nature. The matrix polymer was PLA (foaming grade). We chose foaming agents according to the literature, and the manufacturer's information and recommendations. We used exothermic and endothermic CBAs. Extrusion was carried out on a twin-screw extruder. Each formulation contained 98m% PLA and 2m% blowing agent, and was produced with the same temperature profile, screw speed, and melt temperature. We investigated the resulting foamed products with a scanning electron microscope, and differential scanning calorimetry, cell-population and density measurements and compression tests to reveal and present the differences between endothermic and exothermic types of CBA.

2. Experimental

2.1. Materials

A commercial grade of linear polylactic acid (Ingeo™ Biopolymer 8052D, NatureWorks© LLC, USA) was used as polymer matrix, in granule form. The D-lactide content of this PLA was 4.5 mol% [6]. Its density was 1.24 g/cm^3, its glass transition temperature was 60.2 °C, its crystalline melting temperature was 153.3 °C, and its fraction of crystallinity before processing was 36.1%, which were measured by the authors. Two types of chemical blowing agents (CBA) were used, one exothermic (Tracel IM 3170 MS, Tramaco GmbH) and one endothermic (Tracel IMC 4200, Tramaco GmbH). Both foaming agents were in granule form and contained a polymeric carrier. Characteristic properties of foaming agents can be found in Table 1.

Table 1. Characteristic properties of foaming agents, according to product technical datasheet

Blowing agent	Type	Effective gases	Decomposition temperature [°C]	Yield gas at 220°C [mL/g]	Blowing agent content [%]	Recommended dosage [%]
Tracel IM 3170 MS	Chemical blowing agent, Exothermic	Azodicarbonamide, decomposes into: N_2, CO, CO_2, NH_3	170	50	30	1-2
Tracel IMC 4200 [a]	Chemical blowing agent + nucleating agent, Endothermic	CO_2, H_2O	135-220	60	40	1-2.5

[a] The active ingredients of the CBA are probably sodium bicarbonate and citric acid

Table 2. Notation for the experiment formulations

Notation	type of PLA	Name of CBA	Type of CBA	Dosage of CBA [m%]
8052	Ingeo 8052D	-	-	-
8052_3170	Ingeo 8052D	Tracel IM 3170 MS	(exothermic)	2
8052_4200	Ingeo 8052D	Tracel IMC 4200	(endothermic)	2

2.2. Material processing

Prior to manufacturing, the PLA granules were dried at 80°C for 6 hours in a WGL-45 drying oven. Before manufacturing, a fixed amount of chemical foaming agents was added by dry mixing to the poly(lactic-acid) at a rate of 2 m%, which resulted in a homogeneous mixture The extrusion chemical foaming process was performed with a Teach-line ZK25T twin-screw extruder produced by Collin GmbH (Figure 1.). The formulations and their notation can be found in Table 2. The temperature profile of the extruder was 155/165/175/190/190 °C; we used an increasing temperature profile, where zone 5 (Z5) is the temperature of the extruder die. The pressure and temperature of the melt were monitored and registered. The rotation speed of the screw was 10 rpm. The extrusion die was a cord type with a circular cross-section, and with a nominal diameter of 3 mm.

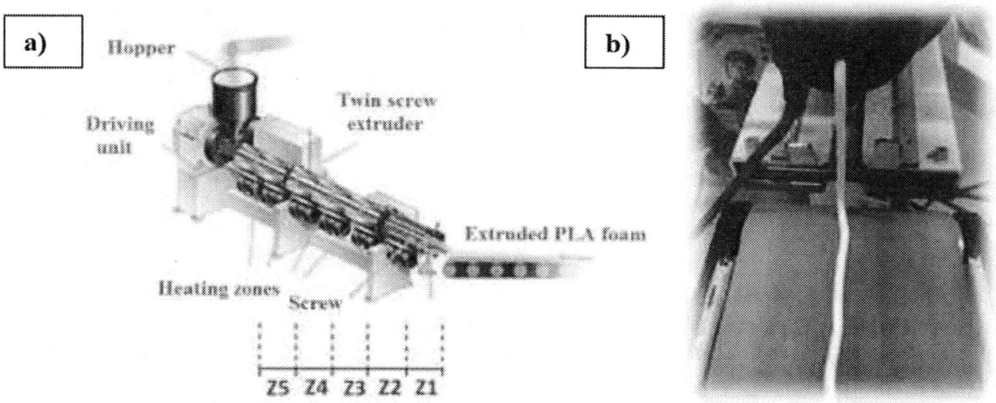

Figure 1. a) Extruder assembly used during the chemical foaming of poly(lactic-acid), b) foam extrusion

3. Methods of analysis

3.1. Thermal analysis

Differential scanning calorimetry (DSC) measurements were performed with a TA Instruments Q2000 automatic sampling device. The measurement temperature range was 0-200 °C, the heating rate was 5 °C/min, the mass of the samples was between 3 mg and 6 mg, and the tests were performed in nitrogen protective gas (20 ml/min) and with a nitrogen measuring atmosphere (20 ml/min). The degree of crystallinity (χ_c) was calculated according to Equation 1, where ΔH_m is the melt and ΔH_c is the crystallization enthalpies. $PLA_{100\%}$ is the theoretical melting enthalpy of 100% crystalline PLA, which is 93 J/g [7].

$$\chi_c = \frac{\Delta H_m - |\Delta H_c|}{PLA_{100\%}} * 100\ [\%] \tag{1}$$

3.2. Characterisation of biopolymer foam structures

The SEM images were produced with a JEOL JSM 6380LA scanning electron microscope with an acceleration voltage of 10 kV. The surfaces of the samples were gold plated in vacuum (Argon gas flush, 15 mA/Pa).

The volume of foam structures was measured with a 50 ml glass cylinder (accuracy 0.1 cm^3); the measuring medium was distilled water. Mass was measured with a Sartorius BP121S type balance. Its range is 120 g, its measuring accuracy is 0.1 mg and its resolution is 0.1 mg. Density was calculated according to Equation 2.

$$\rho_{foam} = \frac{m_{foam}}{V_{foam}} \tag{2}$$

Cell population density was calculated based on the SEM images, according to Eq. 3, where n [-] is the number of cells counted in the recorded image, A [mm^2] is the cross section area of the sample, M [-] is magnification and V_f [-] is the void fraction. [6, 8]

$$N_c = \left(\frac{n * M^2}{A}\right)^{\frac{3}{2}} * \frac{1}{1 - V_f} \tag{3}$$

This method not only takes into consideration the magnification applied, but corrects the density of the cell population with the expansion ratio, as well. The values thus calculated can be used to classify the relationships between processing and the foamed cell structure [6]. The expansion ratio can be calculated with Equation 4, where ER is the rate of expansion, ρ_{rel} is relative density and V_f is the void fraction [9].

$$ER = \frac{1}{\rho_{rel}} = \frac{1}{1-V_f} \quad (4)$$

3.3. Mechanical characterisation

Foam compressive strength tests were performed with a Zwick Z005 universal testing machine in compression mode. The load cell was a Mess & Regeltechnik KAP-TC type cell. Its range is 0-5000 N, the preload applied was 1 N. The measurement speed was 2 mm/min. The test piece was cylindrical, with a diameter of 3 mm and a height of 10 mm. The test was continued until a deformation of 10% was reached. Foam compressive strength was calculated with Eq. 5.

$$\text{compressive strength} = \frac{F_{10\%}}{A_{foam}} \quad (5)$$

4. Results and discussion

4.1. Foam processing

During foam extrusion, temperature was kept constant in the case of both formulations (195 °C ± 1 °C). Manufacturing parameters were also constant (within a range of ± 2 °C) except for melt pressure (Table 3.). The extrusion parameters can be found in Table 3. Melt pressure was 43 bar without a CBA. When a foaming agent was used, a drop in melt pressure can be observed, regardless of the type of CBA. The lowest melt pressure, 13 bar, was in the case of the 8052_4200 formulation with endothermic CBA, which is a 70% pressure drop (Δp_{melt}) compared to the reference PLA. A possible cause is that the decomposition products of Tracel IMC 4200 include CO_2 and H_2O, therefore the PLA was likely hydrolysed and its melt strength was reduced. In the case of the 8052_3170 formulation with the exothermic CBA, the pressure drop of the melt is 51%, which is a not drastic reduction.

Table 3. Manufacturing parameters registered during extrusion

Formulation	Extruder	Extruder			Adapter	Die	T_{melt}	p_{melt}	Δp_{melt}
-	RPM	Z1	Z2	Z3	Z4	Z5	T	p	Δp
PLA_CBA	1/min	°C	°C	°C	°C	°C	°C	bar	%
8052	10	155	162	176	189	191	195	43	-
8052_3170	10	155	161	174	190	190	195	21	51
8052_4200	10	155	159	175	191	190	196	13	70

4.2. Morphological structure of foams

We calculated the number of cells formed along the cross section, based on the SEM images (Figure 2. - Figure 3.), and calculated cell population density. Then density was measured and void fraction was calculated. The resulting morphological structure was characterised by differential scanning calorimetry (DSC), then foam compressive strength was measured. The equipment and methods used for the measurements are presented in Chapter 3.

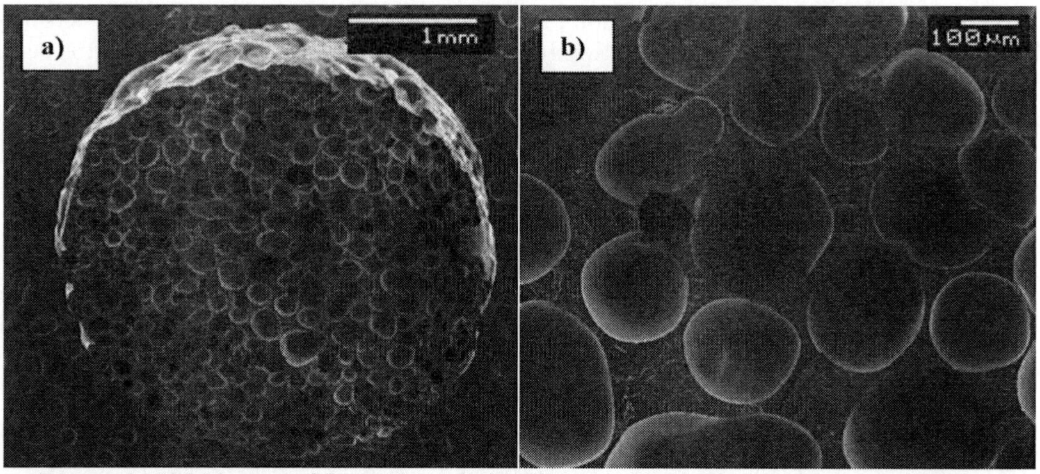

Figure 2. SEM image of foam, formulation 8052_3170, a) 20x b) 100x magnification

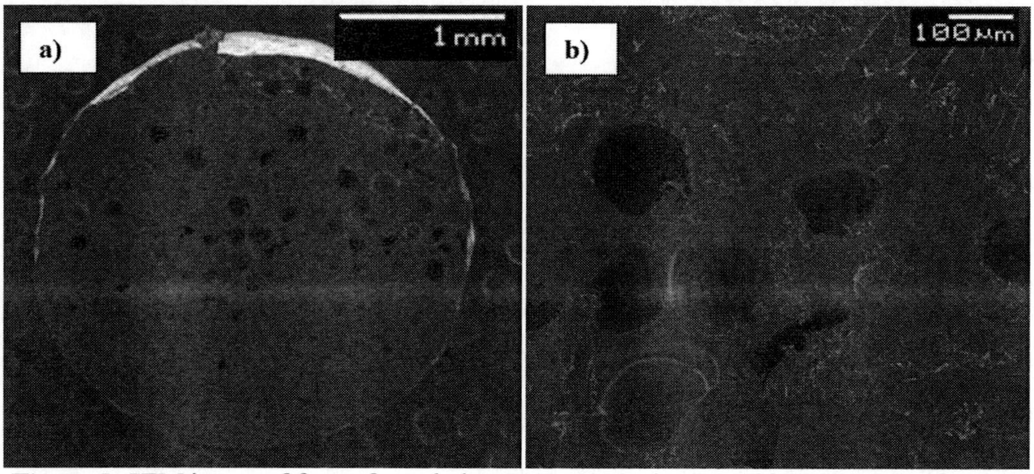

Figure 3. SEM image of foam, formulation 8052_4200, a) 20x and b) 100x magnification

Cell population density

Cell population density was calculated based on SEM images (Figure 2. - Figure 3.). Based on the results, the most promising foaming agent is Tracel IMC 3170. The unoptimized manufacturing production experiment resulted in a cell population density of 4.82 – 5.61x10^5/cm^3. These values are somewhat lower than the values of the foam structures produced by Julien et al. [10] (7.38 - 10.86x10^5), but they used an endothermic CBA (Hydrocerol OMAN698483), which product contains a nucleating agent. Mautana et al. [9] were able to approach 10^6 cells/cm^3 with an optimized temperature profile, extruder speed and using 0.5-1.5% of an endothermic CBA.

Density

The results of density measurements (Table 4.) correlate well with the results expected based on the SEM images. The neat, unfoamed material (Ingeo 8052D) has a density of 1.24 g/cm^3 and the lowest mass reduction was experienced with the 8052_4200 formulation. In the case of the 8052_3170 formulation, the decrease in density was 57%. The values are typical in the case of foaming thermoplastic polymers [10], but these values go beyond the values for the chemical foaming of PLA described in the literature [7–10].

Table 4. Density and void fraction, and cell population density (calculated from SEM images) of foam structures

Sample	Density [g/cm^3]	Void fraction [%]	Cell population density [cell/cm^3]
8052_3170	0.53	57.6	4.82E+05
8052_4200	1.00	19.1	5.58E+04

We investigated commercially available polylactic acid (in granular form) as well as the structure of the PLA after extrusion and the structure of foams formed by chemical foaming (Table 5.).

Table 5. DSC curves of different PLA samples and PLA foams, 1st heating up

PLA type	CBA type	CBA [wt%]	T_g [°C]	T_{cc} [°C]	ΔH_{cc} [J/g]	T_{m1} [°C]	T_{m2} [°C]	ΔH_m [J/g]	χ_c [%]	χ_{total} [%]
8052D	-	-	60.2	-	-	153.3		36.4	36.1	36.1
8052D (ext)	-	-	60.2	115.9	25.2	150.3	154.1	28.8	3.9	31.0
8052D	3170 (exo)	2.0	59.6	107.5	28.5	147.6	154.8	31.4	3.1	33.8
8052D	4200 (endo)	2.0	61.2	107.9	28.6	147.6	154.9	30.5	2.1	32.8

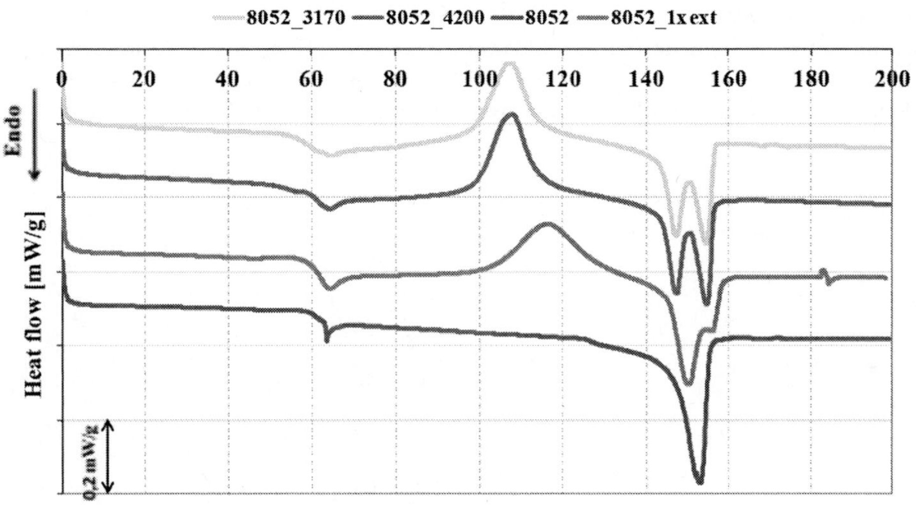

Figure 4. a) DSC curves of PLA granules, extruded PLA and the produced foam structures (1st heating up)

Figure 4. shows that that the commercially available (partially crystalline) PLA granule (8052) does not have an exothermic peak caused by cold crystallisation, but it appears in the case of the processed PLA (8052_1x ext) and chemically foamed samples. This is due to the fact that after processing the crystalline fraction cannot be formed because rapid (air) cooling leaves no time for the molecular chains to form ordered, crystalline regions. During the DSC measurement, the molecular chains are able to become more ordered, which is accompanied by heat release. These crystalline regions later melt at a higher temperature and the polymer melts. In case of the extruded (unfoamed and foamed) samples, the end of the cold crystallization process and the beginning of the melting process of the crystalline segments occur simultaneously and there is a sloping baseline between the two phenomena instead of a horizontal baseline.

It is interesting that there are two crystalline melting peaks. Unlike the original granules, a second crystalline phase was formed with a lower crystal melting temperature after processing (8052D_ext) and in foam structures. Frackowiak et al. [13] explained the phenomenon with the fact that the α' crystals of the lower temperature are smaller and were formed during the cold crystallization process, whereas the αcrystal types having a higher melt temperature were formed during the primary crystallization process, therefore they can be connected to the technology in the present case [14]. Based on this, it can be stated that foaming agents affect the crystalline structure of the polymer.

It can be noticed that after processing (compared to the reference PLA) only 3.9% crystalline part can form after rapid cooling. The unprocessed granules, however, have a crystalline fraction of 36%. Foaming further reduced the crystalline fraction, although only to a small extent.

4.3. Microscopic structure of foams

The comparison of 20x magnification SEM images (Figure 2.-Figure 3.) clearly shows that the foam structure produced by the Tracel IMC 4200 is not suitable either from the point of view of cell nucleation or cell growth. This is a visible consequence of the hydrolysis process described in section 4.1., which occurs during production. The best performing foaming agent is Tracel IM 3170. The foam produced with this CBA contains a considerable amount of cells and the cells formed were able to grow, thereby enhancing the rate of expansion. Although the shape and location of the cells do not make the structure completely homogeneous, but still, Tracel IM 3170 MS is a promising foaming agent in contrast to Tracel IMC 4200.

4.4. Mechanical characterisation

Figure 5. a) shows that a foam structure of a compressive strength of nearly 25 MPa was created with a significant reduction of density (57%), compared to the original PLA. The figure also shows two variants (XPS and EPS100) of one of the most widely used foams, polystyrene foam.

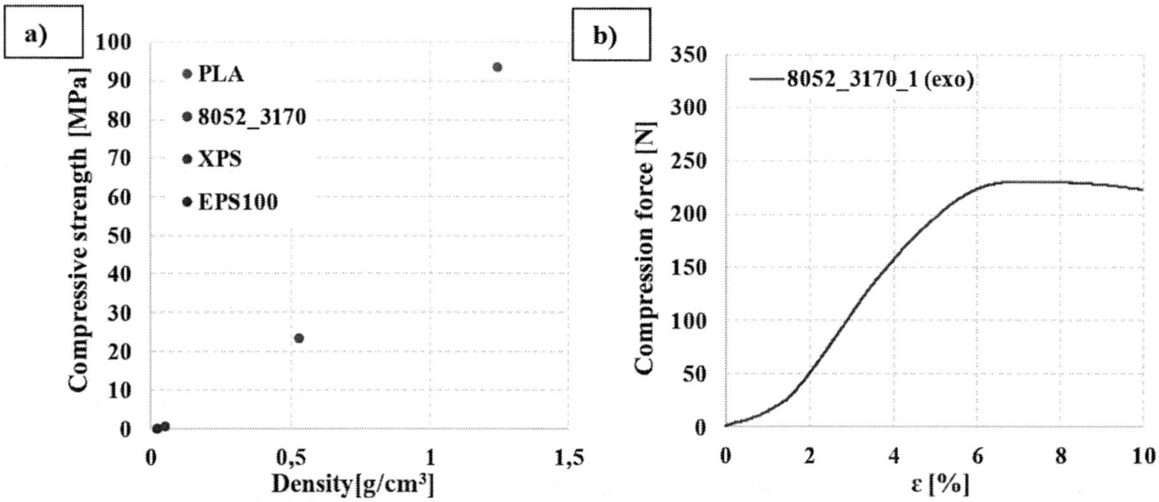

Figure 5. a) Compressive strength of the manufactured foam structure as a function of density (EPS - Expanded polystyrene foam. XPS-Extruded polystyrene foam). b) Compression force-deformation (ε) diagram in case of formulation 8052_3170

5. Conclusions

We performed several tests on the foam products in order to characterize them. Their density and void fraction were measured, then their structure and cell population density characterized based on SEM

images, and their morphological characteristics were measured by DSC. Finally, their compression strength was measured with a universal testing machine.
In the case of the exothermic Tracel IM 3170 MS, a foam was formed and the reduction of its density was significant (57%). A proper cell structure developed without any technological optimization. In the case of the endodermic Tracel IMC 4200 foaming agent, the foam structure was not adequate. During the extrusion process, water was also formed as a decomposition product of the CBA, which very probably hydrolysed the polymer chains, thereby reducing the strength of the polymer melt. Its density decreased slightly compared to the neat, unfoamed polylactic acid and relatively few cells developed.

Acknowledgements
SUPPORTED BY THE ÚNKP-17-4-III NEW NATIONAL EXCELLENCE PROGRAM OF THE MINISTRY OF HUMAN CAPACITIES. "PROJECT FINANCED FROM THE NRDI FUND THE NATIONAL RESEARCH, DEVELOPMENT AND INNOVATION OFFICE (NVKP_16-1-2016-0012)."

References
[1] Avérous L 2013 *Handbook of Biopolymers and Biodegradable Plastics* ed S Ebnesajjad (Oxford: Plastics Design Library)
[2] Bocz K, Tábi T, Vadas D, Sauceau M, Fages J and Marosi G 2016 Characterisation of natural fibre reinforced PLA foams prepared by supercritical CO_2 assisted extrusion *Express Polym. Lett.* **10** 3144
[3] Nofar M and Park C B 2014 Poly (lactic acid) foaming *Prog. Polym. Sci.* **39** 1721–41
[4] Katona B and Orbulov I N 2017 Structural Damages in Syntactic Metal Foams Caused by Monotone or Cyclic Compression *Period. Polytech. Mech. Eng.* **61** 146–52
[5] Al F, Huang Y, Naguib H E and Lo J 2016 Relation of impact strength to the microstructure of functionally graded porous structures of acrylonitrile butadiene styrene (ABS) foamed by thermally activated microspheres *Polymer (Guildf).* **98** 270–81
[6] Ding W, Chu R K M, Mark L H, Park C B and Sain M 2015 Non-isothermal crystallization behaviors of poly(lactic acid)/cellulose nanofiber composites in the presence of CO_2 *Eur. Polym. J.* **71** 231–47
[7] Fisher E, Sterzel H and Wegner G 1973 Investigation of the structure of solution growth crystals of lactide copolymers by means of chemical reactions *Kolloid- Zug Polym.* **251** 980–90
[8] Park C B 1993 *Ther role of polyimer/gas solution in continuous processing of microcellular polymers* (Massachusetts Institute of Technology)
[9] Matuana L M, Faruk O and Diaz C A 2009 Cell morphology of extrusion foamed poly(lactic acid) using endothermic chemical foaming agent *Bioresour. Technol.* **100** 5947–54
[10] Julien J, Bénézet J, Lafranche E, Quantin J, Bergeret A and Lacrampe M 2012 Development of poly (lactic acid) cellular materials : Physical and morphological characterizations *Polymer (Guildf).* **53** 5885–95
[11] Julien J M, Quantin J C, Bénézet J C, Bergeret A, Lacrampe M F and Krawczak P 2015 Chemical foaming extrusion of poly(lactic acid) with chain-extenders: Physical and morphological characterizations *Eur. Polym. J.* **67** 40–9
[12] Ludwiczak J and Kozlowski M 2015 Foaming of Polylactide in the Presence of Chain Extender *J. Polym. Environ.* **23** 137–42
[13] Frackowiak S, Ludwiczak J, Leluk K, Orzechowski K and Kozlowski M 2015 Foamed poly (lactic acid) composites with carbonaceous fillers for electromagnetic shielding *J. Mater.* **65** 749–56
[14] Tábi T, Hajba S and Kovács J G 2016 Effect of crystalline forms (α' and α) of poly(lactic acid) on its mechanical, thermo-mechanical, heat deflection temperature and creep properties *Eur. Polym. J.* **82** 232–43

IOP Conf. Series: Materials Science and Engineering **426** (2018) 012032 doi:10.1088/1757-899X/426/1/012032

Numerical modelling of hot forming of high strength Al-alloys

Zs Lukács[1] M Tisza[2]

[1] Associate professor, University of Miskolc, Institute of MS&T, Miskolc-Egyetemváros, Hungary
[2] Head of department, professor, University of Miskolc, Institute of MS&T, Miskolc-Egyetemváros, Hungary

E-mail: lzsolt@kugli.met.uni-miskolc.hu

Abstract. In this paper, we present the possibility of numerical modeling of high strength aluminum alloys gaining an increasing application in the vehicle industry. Our investigations are based on a comparison of real industrial part produced by Hot Forming and Quenching process (HFQ ™) and its numerical modeling results obtained by the AutoForm program. We will also show what material parameters are needed to successful forming of high-strength aluminum alloy grade AA2060.

1. Introduction of HFQ™ process

The automotive industry has been facing more and more challenges over the last few decades due to the tightening of environmental and safety standards. Mass reduction of vehicles may be regarded as one of the many potential research areas to meet these challenges.

The basic challenge for this is to guarantee greater security while reducing the mass. Therefore, nowadays an increasing development of high-strength metallic alloys can be observed. Among metallic alloys, high-strength aluminum alloys may be regarded as one of the best alternatives due to their low density compared to steel as one of the most potential alternatives for weight reduction [1]. However, their wider application has a serious limitation due to their lower formability properties.

It is well-known that the formability of metallic materials can be improved by increasing the deformation temperature; therefore, it is a plausible solution to apply hot forming processes for manufacturing high-strength aluminum alloys. A number of research projects have been based on this principle in the recent years [2]. As a result, the Hot Forming and Quenching (HFQ™) developed jointly by the University of Birmingham and Imperial College London (ICL) and patented by Impression Technologies Ltd. (ITL). Its basic principle is illustrated in Figure 1. The precipitates present in the alloy in disperse distribution will be totally solved by a Solution Heat Treatment (SHT) process forming a homogeneous solid solution and possessing good formability properties. Then, the sheet is formed in a cold die directly after this heat treatment process. The disadvantage of this process and at the same time the main reason is that it has not gain yet wider application in the industry that the aluminum alloy loses its high strength due to the solution heat treatment. Therefore, after the forming an artificial ageing has to be applied, to restore the high strength properties to the pre-heat-treatment condition.

The everyday work of design and process engineers working in the automotive industry nowadays cannot be imagined without the application of up-to-date numerical modeling software focusing on this target area. Therefore, one of the focal points of any technology to be introduced that processes can be examined using this software prior to the physical production of tools. Therefore, we have to examine

IOP Conf. Series: Materials Science and Engineering **426** (2018) 012032 doi:10.1088/1757-899X/426/1/012032

whether the forming process is suitable for numerical modelling with the required technological accuracy. Therefore, in this paper, we are aiming to demonstrate how accurately the HFQ™ process can be modeled with one of the industry-leading numerical applications (AutoForm R7^{plus}).

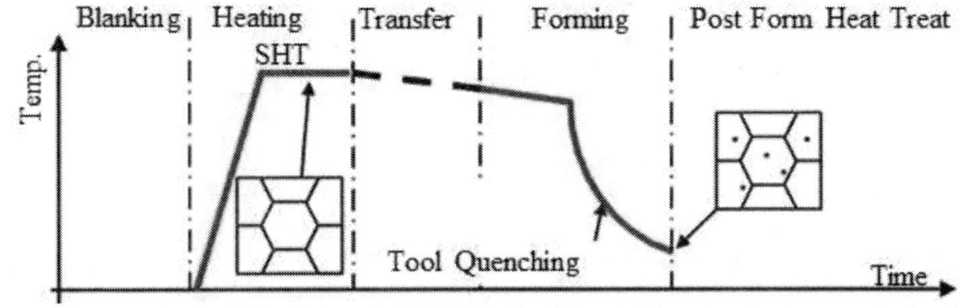

Figure 1. Principal diagram of HFQ™ technology in time vs. temperature coordinates

2. Experimental investigation of an industrial application

For the comparison of experimental investigation and numerical modelling a wing stiffener component will be studied with the geometry shown in Fig. 2 [3]. The raw material of the component is the AA2060 third generation Al-Li alloy not only providing outstanding strength, toughness, rigidity and high temperature performance but also provides superior corrosion resistance and weight saving [4]. The chemical composition of the material is shown in Table 1 [5].

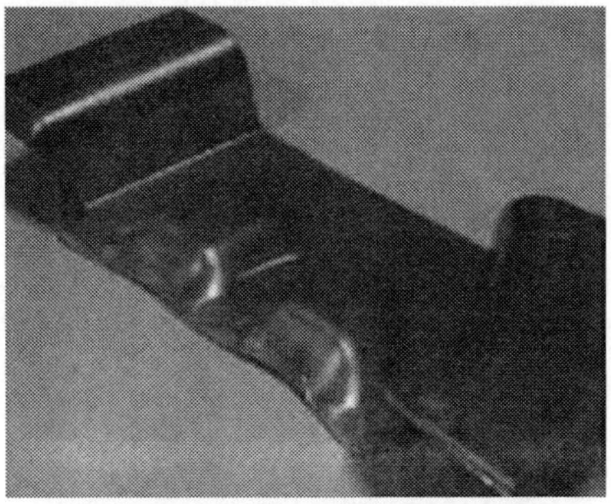

Figure 2. Industrial part: wing stiffener applied for comparison of experimental and numerical modelling results

Table 1. Chemical composition of AA2060 blank

Composition	Li	Mg	Cu	Ag	Mn	Zn	Al
(wt%)	0.6-0.9	0.6-1.1	3.4-3.5	0.05-0.5	0.1-0.5	0.3-05	Bal.

The physical forming process was published in several papers by Gao et al (2015), Mohamed and Szegda (2016), with the exception of the finite element software used for numerical modeling. The basics of these numerical modeling studies, in relation to the material model used, were reported by El Fakir et al (2014) [3], [5], [7].

The numerical modeling of aluminum alloys at high temperatures, which is well-aligned with the experimental results, can be accomplished applying a coupled thermo-mechanical simulation and with a viscoplastic material model [6].

Numerical modeling of the hot forming processes of the 22MnB5 Boron alloyed Manganese steel material (the well-known Press Hardening process) which is widely used in the automotive industry is already available at mass production in industrial circumstances. Therefore, this has provided a good basis for modeling the HFQ™ process that is very similar to the Press Hardening processes, except the technological process parameters. Furthermore, fundamental differences can be observed in the behaviour of the base materials. These differences are strongly connected to the phase transformations in the two types of materials: the phase transformation in the 22MnB5 steel during the forming and quenching cycle provides the high strength properties, while the effects of dissolving dispersed phases in the high strength aluminum alloys behaves differently. These different material parameters must be formulated with mathematical relationships in the numerical modelling system.

2.1. Material model for AA2060 in AutoForm

In sheet metal forming, the material models used in finite element software packages should focus on the followings:

- What changes will be caused in the strength properties by the forming?
- What stress state will result in plastic deformation?
- When the limits of the formed material will be reached?

For the first question, the material models should know the strength vs. deformation functions, which are referred to as flow curves. The second question is answered by the yield condition, while the last question is answered by the various forming limit theories.

For high-strength aluminum alloys, the development of material models has been a major field of research for many years besides many other applications in AutoForm. As a result, AutoForm has a material model for hot forming of sheet material AA5457 [8]. Its material parameters can be changed by using the AutoForm Material Generator according to the material properties of the AA2060 considered.

2.1.1. Flow curves

The peculiarity of hot forming processes is that the hardening curve depends on the strain-rate and the forming temperature. In Figure 3.a. the flow curves are shown at T = 350 °C for various strain-rates, while Figure 3.b. shows the flow curves at 1 s^{-1} strain rate for various temperatures. The applied flow curves are based on [3], [5].

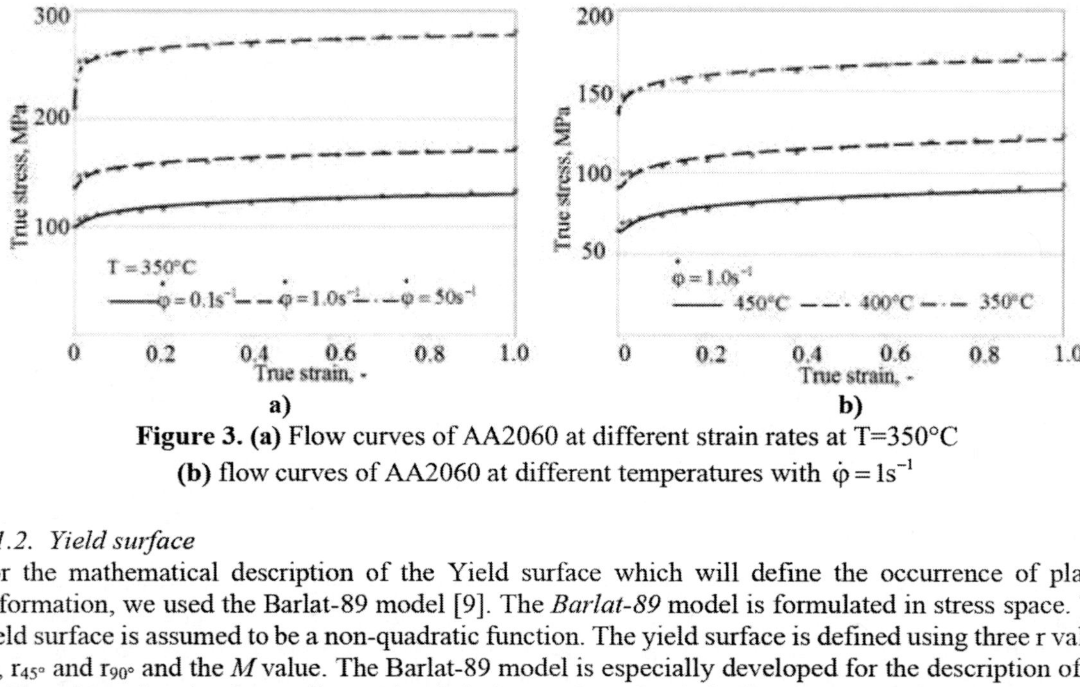

Figure 3. (a) Flow curves of AA2060 at different strain rates at T=350°C
(b) flow curves of AA2060 at different temperatures with $\dot{\varphi} = 1s^{-1}$

2.1.2. Yield surface

For the mathematical description of the Yield surface which will define the occurrence of plastic deformation, we used the Barlat-89 model [9]. The *Barlat-89* model is formulated in stress space. The yield surface is assumed to be a non-quadratic function. The yield surface is defined using three r values $r_{0°}$, $r_{45°}$ and $r_{90°}$ and the *M* value. The Barlat-89 model is especially developed for the description of the yield surfaces for aluminum alloys. The default *M* value of 8 typically gives a good representation of the aluminum behaviour.

The temperature dependence of Yield surface can be taken into consideration by specifying the temperature dependence of anisotropic coefficients. To get these values, the n-r tests must be performed at the given temperatures. In Table 2. the temperature dependence of normal anisotropic coefficients and biaxial anisotropic coefficient are given in the different directions relative to the rolling direction [8]. Using these parameters, the Yield surfaces for each temperature can be determined (Figure 4.a).

Table 2. Temperature dependence of r-values in different directions

Temperature	r_0	r_{45}	r_{90}	r_b
350°C	0.710	1.080	0.730	0.937
400°C	0.630	0.970	0.660	0.893
450°C	0.630	0.970	0.660	0.893

2.1.3. Process parameters

In the forming process, the tool elements and their position are shown in Figure 4.b. Accordingly, the forming process performed with the following schedule using the markings of the figure:

- First, the *Blank* (1) with the temperature corresponding to the SHT temperature (450 °C) of the material to be formed positioned on the *Bottom die* (4). The *Bottom die* is on room temperature.
- Subsequently, when the Top Blankholder (2) and the Top punch (3) together with the Blankholder (2) are closed on the Bottom die (4), thus the Blank to be formed is strongly hold by the Blankholder (2).
- The so-called *Top punch* (3) forms the characteristic U-shape of the component, while during its movement the forming of the U-shape is done by the *Bottom blankholder* (5).
- At the end of the forming process, the *Bottom punches* (6) will form the square and triangular ribs (shown in Figure 2.) moving into the blank in backward direction.

The further technological settings of the forming process are shown in Table 3.

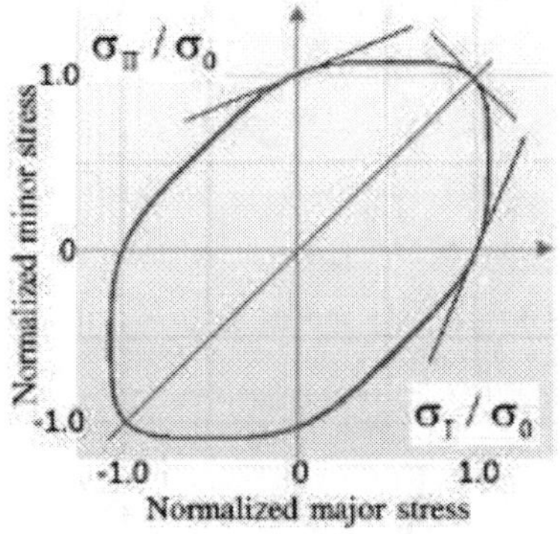

a)

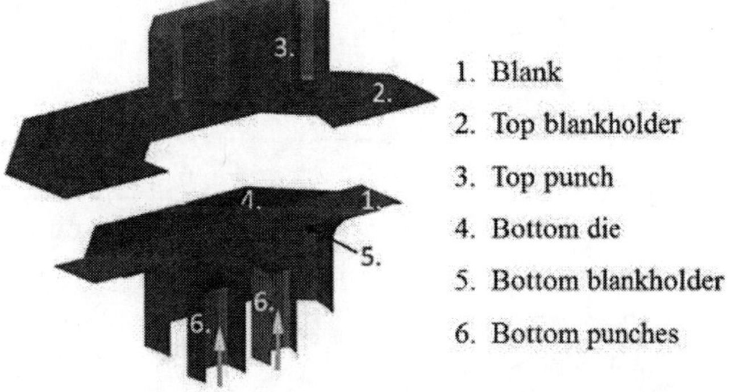

b)

Figure 4. (a) Temperature dependence of yield surface **(b)** Schematic drawing of tool elements

Table 3. Process parameter of forming the wing stiffener component

Process parameter	Value
Initial blank temperature	450°C
Initial tooling temperature	20°C
Forming speed	250 mm/s
Blankholder force	15kN
Gas spring stiffness	1kN/mm
Friction coefficient	0.2

3. The comparison of the result of the HFQ™ modelling

The results of the numerical modeling were examined by measuring the thickness distribution along defined sections. The reason for this is that a relatively large number of result variables of mathematical modeling, therefore it is advisable to choose such a measurable parameter that can be measured on the physical workpieces. Furthermore, the results of the chosen forming process were compared with the results of the modelling done by the PAM STAMP system and in this case the result variable was also the thinning. Therefore, the results of the AutoForm modeling were compared with this result. On the physical workpiece, the degree of thinning was measured on the basis of distorted grids using optical measuring principle applying the GOM ARGUS system [5].

In Figure 5, the thinning distribution of the workpiece is shown for a relevant section (X-Y) of formed component. It can be seen that both numerical simulation follow the measurement points with an acceptable error relative to the physically measured thinning. It is also common to say that the differences between the results of two software are not significant concerning both their nature and their actual values. The deviations do not exceed the pre-defined range of thinning applying optical measuring principle. Similar statements can be made for the section V-Z shown in Fig. 6.

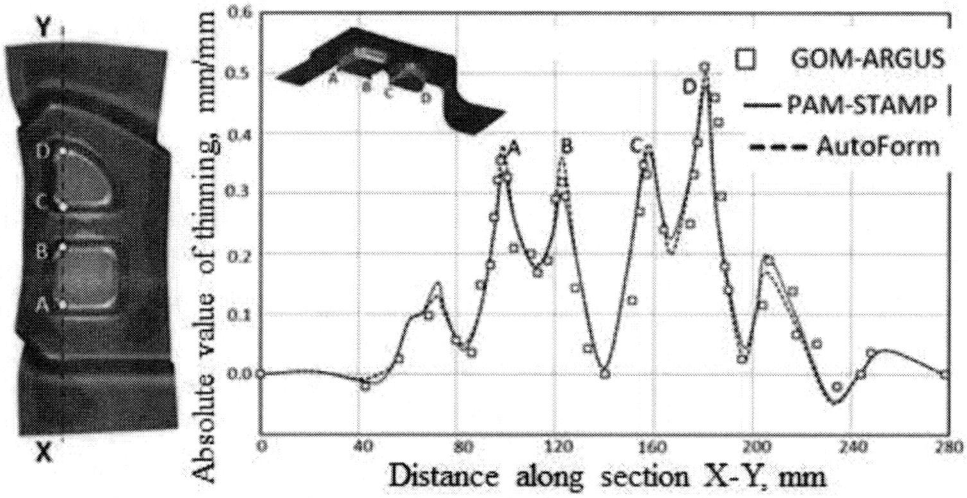

Figure 5. Comparison of thinning distribution through X-Y section

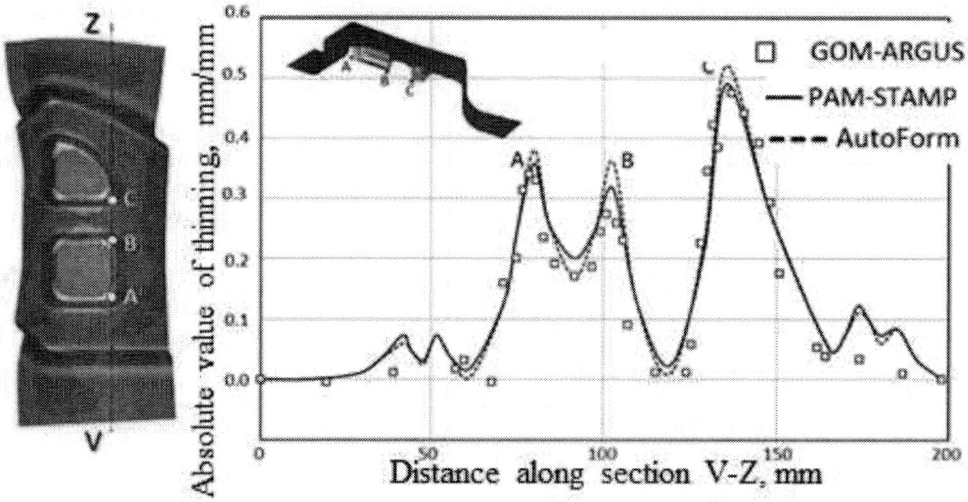

Figure 6. Comparison of thinning distribution through V-Z section

IOP Conf. Series: Materials Science and Engineering **426** (2018) 012032 doi:10.1088/1757-899X/426/1/012032

4. Conclusion

In this paper, we have presented a relevant field of research for the production of sheet metal parts made of high strength aluminum alloy (AA2060) for weight reduction of vehicles (HFQ™). The technological and tool design of automotive sheet parts is nowadays unimaginable without the use of integrated software packages capable of designing the full process cycle.

Therefore, with the help of a market leader software package in this field, we have examined the modeling ability of AutoForm software through a numerical modeling of a specific industrial part and process. As an overall result we can state that the process presented is acceptable for both process and design engineers in terms of the thinning as a parameter essential to the success of forming.

Acknowledgment

The authors express their grateful thanks to the European Commission for their support on the H2020 project „Low Cost Materials Processing Technologies for Mass Production of Lightweight Vehicles (LoCoMaTech)" , Grant No: H2020-NMBP-GV-2016 (723517)

References

[1] Mike Farish (2016) *Aluminium with form* Automotive Manufacturing Solution 2016 nov-dec. pp. 39-41

[2] Horizont 2020 EU FP7 NPM LoCoLite 604240 *Low Cost forming of Lightweight structures for transportation industries* (2013-2016)

[3] M. Mohamed, D. Szegda 2017 *An Industrial application of the continuum damage mechanics (CDM) model for predicting failure of AA6082 under HFQ process* Procedia Engineeering Vol 183. pp 277-282.

[4] N.E. Prasad, A.A. Gokhale, R.J.H. Wanhill B. (2014) *Aluminium lithium alloys processing, properties, and applications*, 63

[5] Haoxiang Gao at al (2015) *Hot Stamping of an Al-Li alloy: A feasibility study* MATEC Web of Conferences Volume 21 4th International Conference on New Forming Technology (ICNFT), article number:05007

[6] R. P. Garet at al (2005) An investigation of the effects of solution heat treatment

[7] Omer El Fakir at al 2014 *Numerical study of the solution heat treatment, forming, an in-die quenching (HFQ) process on AA5457* International Journal of Machine Tools&Manufacture Vol 87. pp 39-48.

[8] P-Y. Manach, J. Coer, A. Jégat, H. Laurent, J. W. Yoon, Benchmark 3 – Springback of an Al-Mg alloy in warm forming conditions, NumiSheet 2016 Conference, Bristol, 4-6 September 2016.

[9] F. Barlat J. Lian (1989) *Pastic behavior and stretchability of sheet metals: A Yield function for orthotropic sheet under plane stress conditions* International Journal of Plasticity Vol 5. pp 51-66.

IOP Conf. Series: Materials Science and Engineering **426** (2018) 012033 doi:10.1088/1757-899X/426/1/012033

The effect of gas permeability on the pressure of artificial resin-bonded core gases

L Mádi[1] and L Varga[1]

[1] University of Miskolc, Institute of Foundry, Miskolc-Egyetemváros, Hungary

E-mail: ontlaura@uni-miskolc.hu

Abstract. In case of the decomposition of foundry artificial resin-bonded moulding materials, the amount and the pressure of the produced gases can have significant effect on the quality of the casting surface during the early phases of the solidification. The produced gases can cause casting defects if the casting skin is not solidified before the gas could pass through. The cores were made out of commercially available multi-element (SH33-34) and sieved SH33 sands with phenol no-bake process, with the same resin content. The gas permeability and gas pressure of the released gases were examined. Furthermore, the changes in the gas pressures of the gases released from various resin quality cores at different casting temperatures were also examined.

1. Introduction

In case of the decomposition of foundry artificial resin-bonded moulding materials, the amount and the pressure of the produced gases can have significant effect on the quality of the casting surface during the early phases of the solidification. The produced gases can cause casting defects if the casting skin is not solidified before the gas could pass through. The examination of the gas and its pressure generated from sand cores is a matter of priority. The composition of the produced gases is an important environmental issue, as the regulations are getting stricter. The intense gas release from the core can result in blow holes, while the metal can also penetrate the sand. The relation of the speed of the produced and released gas determined the gas pressure in the core.

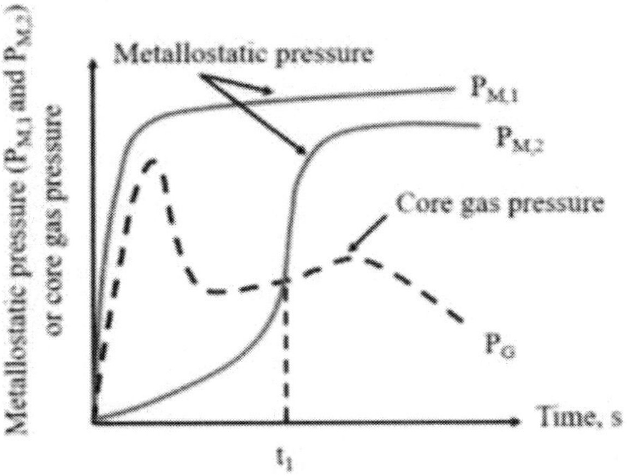

Figure 1. The pressures influencing the production of blow holes [1]

The metallostatic pressure can be observed in Figure 1. In case of $P_{M,1}$, the metallostatic pressure is higher during the solidification than the pressure of the released gases. Thus, products with blowholes are not produced. In case of $P_{M,2}$, the core gas pressure is higher than the metallostatic pressure until t_1, so during this time interval the released gas from the core can enter the melt [1]. In the case of different core recipes, the time course of the gas pressure curves may differ significantly. The aim of the research work is to compare the pressure of gasses developing from various core wit different grain size and quality.

2. Objectives

The changes of compactness and gas permeability of the form are reverse in relation to the compaction. Gas permeability is primarily dependent on the granulometry and the geometry of the pore structure during core shooting [2]. There is an interdependence between the pressure of the core gas and gas permeability. Gas permeability test were carried out with two sands on the market, SH33 and SH34, sands with fine and medium fineness that were separated to various fractions under laboratory conditions. Using the same amount of resin, Ø50×50 mm standard test bars were produced with phenol no-bake process from the sands. The pressures of the gases produced from cores with various gas permeabilities and resin qualities (no-bake, HB-phenol and HB-furan) were examined. The released gases from the different resin quality cores casted at 720 and 1400 °C were also examined

3. Experimental

The fire-resistant matrix of the core was washed, separated silica sand from Sajdikovó. The effect of gas permeability was examined using silica sand with various grain sizes (SH33-34) and sieved to fractions: 110-220, 220-300 μm. The dominating variable included in this experiment was sand grain size, so sieve analysis was performed. The particle size distributions of the used sands can be seen in Table 1. Based on the particle size distribution diagram it can be concluded that SH33 sand is ternary and SH34 sand is binary. The sand grain sieve analysis implied that there was no issue of grain distribution. This grain distribution together with the grain surface (specific surface) determines the indispensable amount of resin and catalyst. Grain fineness number (GFN) can be determined based on the quantity of sieve residues. To produce high quality sand moulds, good grain size distribution is needed. Grain fineness is defined by the number of mesh-openings per col, in which sand with consistent particle size distribution would pass through. It is basically based on the calculation of weighted mean value. The retained sand is multiplied with the so-called sieve constant and the results are cumulated. This number is divided by the full quantity of the sand used for the analysis [5].

$$GFN = \frac{\sum G_n K_n}{G_n}$$

where,

GFN= number of fineness,
G_n= weight percent of material for each sieve,
K_n= sieve constant for each sieve.

Table 1. The granulometric properties of the sands

Granulometric properties	Sand type, mm			
	SH34	SH33	0.11-0.22	0.22-0.3
Average particle size, mm	0.19	0.23	0.19	0.26
Number of fineness	97.73	74.93	105	55
Theoretical specific surface, g/cm^3	121.74	100.68	126.5	88
Measured specific surface, g/cm^3	140	105	155	115
Equality ratio, %	34	74	100	100

Three resin qualities were examined during the experiments. The compositions of the mixtures can be seen in Table 2. The effect of the granular structure was examined only in case of phenol no-bake cores. The cores for the resin quality and casting temperature tests were prepared from commercially available SH33 sand.

Table 2. The composition of the sand mixtures

Resin quality	Resin content, wt%	Catalyst content, wt%	Curing temperature, °C	Curing time, min
No-bake	1.2	0.5	no-bake	
HB-phenol	1.5	0.3	220	15
HB-furan	1.33	0.33	200	10

The cores were prepared with constant compacting energy and approximately the same bulk density (1.6 g/cm^3) under laboratory conditions. The setting time of the no-bake sand mixtures was 3 hours. The geometry of the test bars was Ø28×134.3 mm. The steel pipes used for collecting the produced were (Ø6×300mm) placed 5 mm away from the base of the test bars during core making. The form was prepared using a bentonite mixture containing 8% bentonite and 4% water. The fire-resistant matrix of the form was silica sand. The lower moulding box was Ø330×200 mm and the upper moulding box was Ø330×140 mm. Casting was carried out simultaneously using a tangential access channel in a three-cavity form. Approximately 45 kg bentonite mixture was necessary for each forms which were produced with a Multiserw Morek low speed mixers. The mould cavities were made with cone-shaped wood test bars (Ø70×150mm, draft: 4°). 720 °C foundry aluminium alloy was used for casting. The preparation steps of the waste mould can be observed in Figure 2.

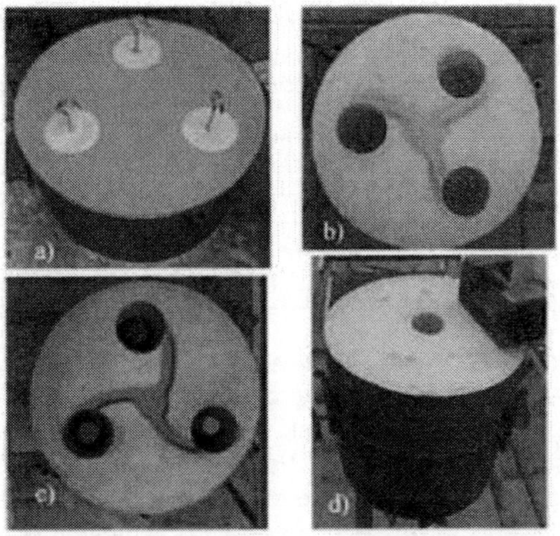

Figure 2. The experimental moulding box, a) lower part with the cores, b) tangential access channel without cores and c) with core d) lower and upper moulding box with weight on

The produced gases passed through a silicone tube to a pressure transducer. Three Baumer-type pressure transducers were used for the experiments. A pressure transducer is a front membrane device for low pressure measurement. The device can measure 0-0.1 bar pressure with ±0,5% accuracy. The output signal is 4-20 mA and the reaction time is ≤ 0,5 ms. A HBM QuantumX-MX840B universal 8 channel data acquisition module was used, which is connected to a computer with a HBM Catman software. The sampling rate was 30 Hz. The experimental three-core aluminium casting with the cores and the feeding system can be seen in Figure 3.

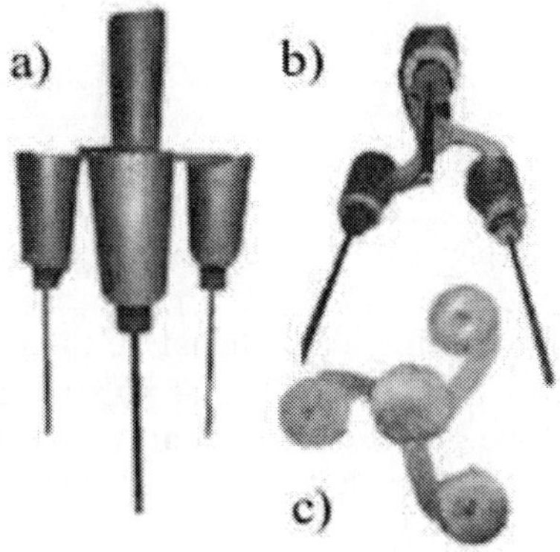

Figure 3. The three-core aluminium casting with the cores and the feeding system: a) cross-sectional view b) bottom view c) top view

4. The effect of granular structure on the pressure of the released gases

The gas permeability of test bars from the sieved, one-fraction sand and the multiple-fraction sand differ from one another. Significant difference can be observed between the gas permeability of fine and medium sands. The gas permeability of cores made out of medium sand was around 200, and around 100 for fine sand cores. The exact data on gas permeability can be found in Table 3.

Table 3. The gas permeability and bulk density of cores with various grain sizes

Sand	G_k	$\delta_{average}$, g/cm³
SH33	210	1.6031
SH34	100	1.6104
Medium	180	1.6057
Fine	87	1.5943

Based on the time course of the curves in Figure 7, two maximums can be observed. The increase in gas pressure is uneven. The first main pressure increase was at 5-10 seconds because the heating up and the expansion of the air trapped in the pores is fast. Furthermore, the amount of gas produced due to the high temperature gradient of the mould wall fills the pores and the gas release is started. More and more moulding mixture is getting heated up and the velocity of gas production is faster. Reaching the maximum pressure value at 30-40 seconds, a steady state is achieved [6, 7].

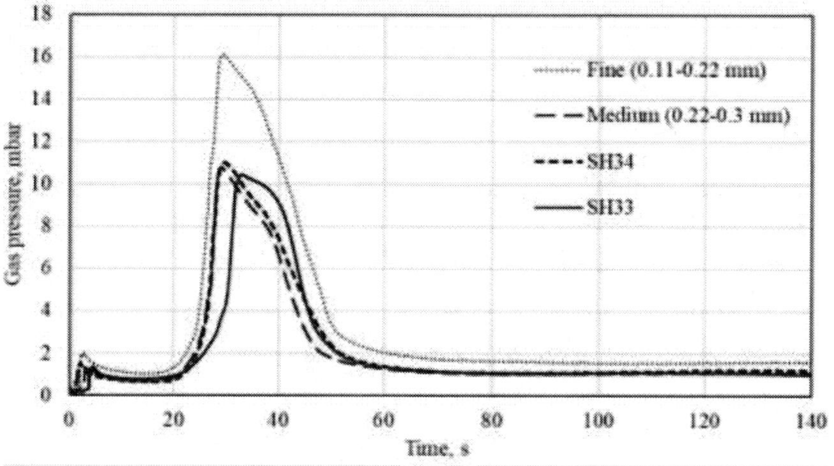

Figure 4. Pressures of the gases produced from cores with various particle sizes

As it can be seen on the diagram, the time dependency of the gas pressure curves of SH34 and the medium sand (0.22-0.3 mm) are almost identical. The gas pressure is higher in cores made from sieved sand due to the low gas discharge. The gas pressure of cores from fine one-fraction sand can reach up to 16 mbar as the gas permeability is very low. The probability of casting defects caused by gas penetration is also increased.

5. The effects of resin quality and casting temperature on the pressure of the released core gases

The effects of resin quality and casting temperature on the pressure of the released core gases can be observed in Figures 5 and 6.

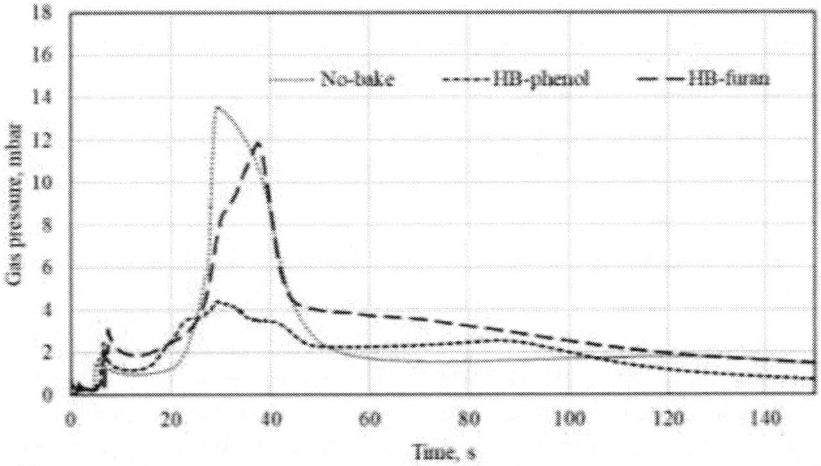

Figure 5. The effect of resin quality on the pressure of the released gases in case of casting aluminium alloy ($T_{cast.}$= 720°C)

The gas pressure curves are strongly influenced by the resin quality. In case of 720 °C casting temperature, the first pressure peak can be observed less than 10 s after casting. The p_{max} values vary between 4-14 mbar in case of the different resin qualities. Intermittent resin burning can be observed in the gas pressure curves of the HB-phenol resin cores. Rapid increase in the gas pressure can be observed in the pores of the cores made with no-bake process in 20 s after casting, which increases the possibility of the development of blow-holes.

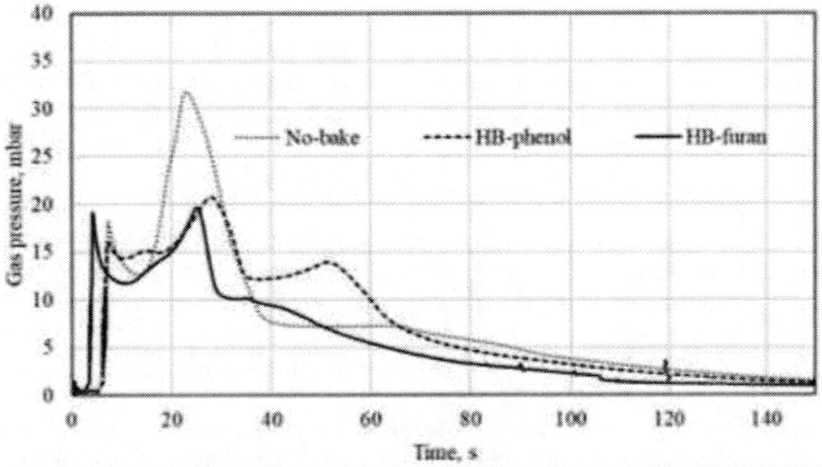

Figure 6. The effect of resin quality on the pressure of the released gases in case of cast iron ($T_{cast.}$= 1400±20°C)

The maximum gas pressure was 20-32 mbar in case of 1400 °C casting temperature. Two similar pressure maximums can be observed in the gas pressure curves of HB-furan resin cores which can be the result of the high moisture content of the binder. The cores made with no-bake process had the highest released gas pressure.

IOP Conf. Series: Materials Science and Engineering **426** (2018) 012033 doi:10.1088/1757-899X/426/1/012033

6. Summary

Casting defects caused by gas penetration are formed if the maximum gas pressure occurs seconds after casting, before the development of a solid sin.

There is an interdependence between the pressure of the core gas and gas permeability. The gas pressure of the fine-grained cores is higher and developed earlier than in case of medium grained cores. The reason behind is the low gas discharge. The possibility of casting defects due to gas penetration is higher if fine one-fraction sands are used.

The characteristics of the gas pressure curves are significantly influenced by the resin quality. The characteristic curves of the gas pressures of the various resin quality cores casted at different temperatures (720-1400°C) show similar tendencies. In case of 720 °C casting temperature, the first pressure peak can be observed less than 10 s after casting. The p_{max} values vary between 4-14 mbar in case of the different resin qualities. The maximum gas pressure was 20-32 mbar in case of 1400 °C casting temperature. Intermittent resin burning can be observed in the gas pressure curves of the HB-phenol resin cores. The gas pressure increases in the first 20 s after casting in case of both casting temperatures (720 and 1400°C), then an intermittent decrease can be observed. The difference in the gas pressures at various casting temperatures can be 15 mbar.

Acknowledgements

The research was also Supported BY the ÚNKP-17-3 New National Excellence Program of the Ministry of Human Capacities. The authors would like to Mária Ambrus for her linguistic advice.

References

[1] L. Winardi, H.E. Littleton, R.D. Griffin J.A. Griffin 2008 Gas from Green Sand Molds and Vapor Transport Zone, *AFS Transactions* p.381-399.

[2] L. Winardi, H. Littleton, C.E. Bates New 2005 Technique for Measuring Permeability of Cores Made from Various Sand, Binders, Additives and Coating, *American Foundry Society, Paper 395-396., AFS Transactions American Foundry Society, Schausburg, IL USA*

[3] P.F. Ambidge, P.D.E. Biggins 1984 Environmental Problems Arising From the Use of Chemicals in Molding *Materials Developments for Future Foundry Prosperity* p. 219.

[4] Mancuso Chemicals Limited, Foundry Products 2005 Technical Data Sheet *http://www.mancusochemicals.com/wp-content/ uploads/2018/01/Furan-Binder-Use.pdf*

[5] Shailee G. Acharya, Jeetendra A. Vadher, Manoj Sheladiya 2016 A furan no-bake binder system analysis for improved casting quality, *Copyright American Foundry Society, DOI 10.1007/s40962-016-0059-x*

[6] A.Chojecki, J. Mocek 1/2011, 9-14 Gas pressure in sand mould poured with cast iron, *Archives of Foundry Engineering Volume 11, Issue*

[7] Tóth L. 1997 Gázfejlődés az öntödei formázó és maghomok keverékekben. A bentonitos formázókeverékek aktív bentonit és víztartalmának összefüggése a plaszticitással és a repedési szilárdsággal, *PhD thesis (in Hungarian)*

IOP Conf. Series: Materials Science and Engineering **426** (2018) 012034 doi:10.1088/1757-899X/426/1/012034

Microstructural and magnetic investigations of duplex steel

I Mészáros[1] and B Bögre[1]

[1] Budapest University of Technology and Economics, Department of Materials Science and Engineering, Budapest, Hungary

E-mail: bbalint1113@gmail.com

Abstract. In this work, the eutectoid decomposition of δ-ferrite was investigated in 2507 type super duplex stainless steel. Samples were cold rolled in different extents and then heat treated at different temperatures. The ferrite contents of the specimens were determined by magnetic measurements and electron backscatter diffraction. The microstructure was examined by optical microscope and electron backscatter diffraction. The results showed that the cold rolling before the heat treatment significantly increases the rate of eutectoid decomposition and decreases the starting temperature of the ferrite decomposition. The measured ferrite contents with various methods showed diverse results because of the different principles of the measuring methods.

1. Introduction

DSS combines the advantages of the austenitic and ferritic steels thereby the microstructure of this corrosion resistant steel contains about 50-50% austenite and δ-ferrite. This coupling effects that the strength of the steel significantly increases and the corrosion resistance raises mainly in chloride ion containing environments. The yield strength of duplex stainless steel (DSS) can be two or three times higher than the yield strength of the traditional austenite steel.

The carbon content is limited strongly to avoid the formation of harmful precipitations. The aim of the chromium alloying is to increase the corrosion resistance, the nickel makes the steel tougher. Nitrogen can raise the strength and revises the resistance against the pitting corrosion. DSS is usually alloyed by molybdenum, tungsten and copper too [1, 2].

The austenite ensures good ductility, toughness and weldability until the δ-ferrite increases the resistance against the pitting, crevice and stress corrosion. Corrosion resistance of DSS can be determined by the pitting resistance equivalent number (PREN) which depends on the chemical composition. In case of super-duplex stainless steel (SDSS) the PREN index exceeds 40 [1, 2].

The upper application temperature of DSS is about 280-325°C. If the temperature is higher than this value, precipitations (intermetallic phases) can appear which cause the embrittlement and the reduction of the corrosion resistance of the material. The damageability of steels was simulated earlier [3]. The most important phase transformation in DSS is the eutectoid decomposition. During this process, the δ-ferrite transforms into σ-phase and secondary austenite ($\delta \rightarrow \sigma + \gamma_2$). The σ-phase can emerge between 600-1000°C and the chance of its appearance increases if the amount of the alloying elements is higher. It shows up mainly in SDSS. The χ-phase can turn up at 700-850°C. The $M_{23}C_6$ and M_7C_3 are high chromium content complex carbides. The Cr_2N has a hexagonal lattice and it settles down at the grain boundary of the δ-ferrite. The R-phase and π-phase appear between 550-650°C and these can cause significant embrittlement [4-8].

There are several publications which deal with the eutectoid decomposition and the formation of the σ-phase in SDSS but there is just limited information about how these transformations influenced by previous plastic deformation. The aim of this study was to investigate the changing of the process of the eutectoid decomposition in 2507 type SDSS if the samples were plastic deformed in different extents before the heat treatment.

2. Sample preparation

Samples were prepared from the basic material which sizes were 10×15×100mm. The specimens were cold rolled in 6 different extents in longitudinal direction at room temperature. Values of the rolling reductions were 10, 20, 30, 40, 50, 60% and there was a sample series without deformation. 5-5 samples were rolled in case of every deformation rates.

After the cold rolling the specimens were heat treated during 30 minutes at 700°C, 750°C, 800°C, 850°C and there was a sample series without heat treatment. After rolling every specimen were milled at same geometry which was 3,4×10×100mm.

Table 1. shows the chemical composition of the 2507 type SDSS.

Table 1. Chemical composition of the 2507 type SDSS (%)

C	Mn	P	S	Si	Cu	Ni	Cr	Mo	Nb	Ti	N
0,021	0,822	0,023	0,0004	0,313	0,178	6,592	24,792	3,705	0,008	0,005	0,264

3. Experiments

The ferrite content of the samples was firstly determined by an alternating current (AC) magnetometer as it was reported in our previous publication [9].

In the present paper the results of the Stablein-Steinitz direct current (DC) magnetometer, optical microscopic and electron backscatter diffraction (EBSD) measurements were summarized.

3.1. DC magnetometer measurement

The samples were measured by the DC magnetometer which is an especially suitable method to determine the ferrite content of samples. The Stablein-Steinitz DC magnetometer requires bulk samples so the longish, rolled specimens which were measured by the AC magnetometer were cut into smaller pieces and fixed to cubic shape. Fig. 1. shows the sketch of the applied Stablein-Steinitz DC magnetometer. This set-up contains a symmetrical yoke which stands two U-shaped parts and a cross bridge. There are four excitation coils on the main arms of the yoke. The set-up consists two Hall sensors, a reference and a measuring air gap too. If there is no sample in the measuring air gap the arrangement is symmetrical therefore there is no flux through the cross bridge. If the sample is put in the measuring air gap the symmetry ends and some part of the flux closes across the cross bridge. The Hall senor in the cross bridge measures this magnetic field which is proportional with the magnetization of the sample.

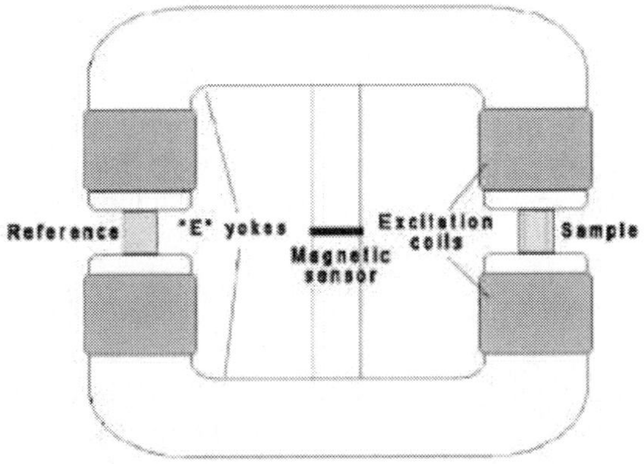

Figure 1. Sketch of the Stablein-Steinitz DC magnetometer

The set-up recorded the saturation magnetic hysteresis loop of the samples. The value of the saturation polarization which can be determined from the hysteresis loop are known to be proportional to the ferrite content of the sample [10]. Fig. 2. represents the calculated ferrite contents in function of the heat treatment temperature and the deformation rate. It can be seen that the ferrite content reduction is more moderate by the undeformed and less deformed samples. The ferrite content suddenly drops at 750°C by the strongly rolled samples. This means that the previous cold rolling before the heat treatment significantly influences the eutectoid decomposition. The stronger the deformation rate is the more active the eutectoid decomposition is. Much more paramagnetic σphase and secondary austenite begin to appear in effect of the heavy deformation which reduces the ferrite content appreciably. On the other hand, the previous cold rolling seems to decrease the starting temperature of the eutectoid decomposition as well.

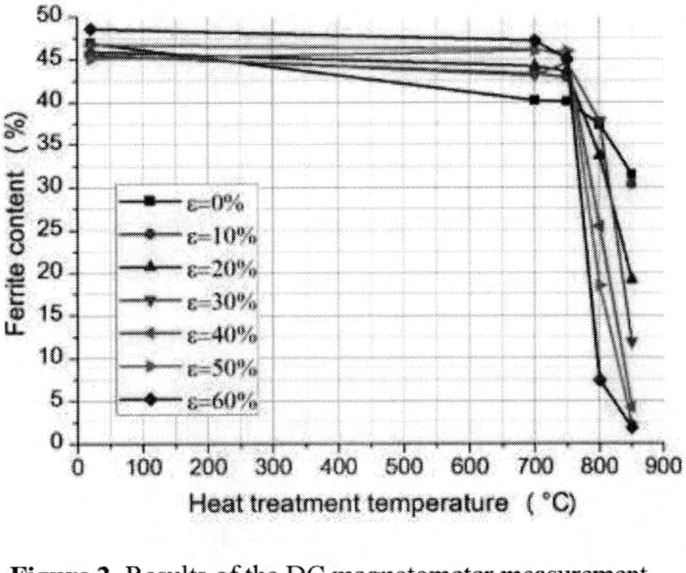

Figure 2. Results of the DC magnetometer measurement

The ferrite contents measured by the DC magnetometer were compared with the results of the AC magnetometer measurement. The details and results of the AC magnetometer measurement had already been published in a previous paper [9]. It was found that the AC magnetometer measured lower ferrite values than the DC magnetometer. This difference can be explained by the significantly different excitation levels. The DC magnetometer is able to excite the whole volume of the samples to their magnetic saturation in contrast to the AC magnetometer.

3.2. Metallographic investigation with optical microscope

All samples were investigated by metallography using optical microscope. The specimens were fixed in resin, grinded, polished and prepared by Beraha etch. Fig. 3. shows the microscope pictures in case of different deformation levels and different heat treatment temperatures. The magnification of the microscope was 1000×. The etching time of the samples were different for achieving the best image quality. It caused the diverse colour of the same phases on the different samples. The dark area (brown and green) demonstrates the ferrite (δ), the bright area is the austenite (γ). σ-phase, ferrite and austenite are indicated on the Fig. 3.

The pictures demonstrate obviously that the previous cold rolling can significantly influence the eutectoid decomposition. The heat-treated samples on 850°C show most expressively this phenomenon. The stronger the extent of the previous cold rolling is the more δ-ferrite transforms mainly into σ-phase and secondary austenite. It can be seen the decomposition of δ-ferrite appears slightly in case of the ε=20% deformed sample. However, the decomposition of the δ-ferrite is almost complete at the sample which belongs to the highest deformation level (ε=60% and heat treated at T=850°C) in good agreement with the DC magnetometer results.

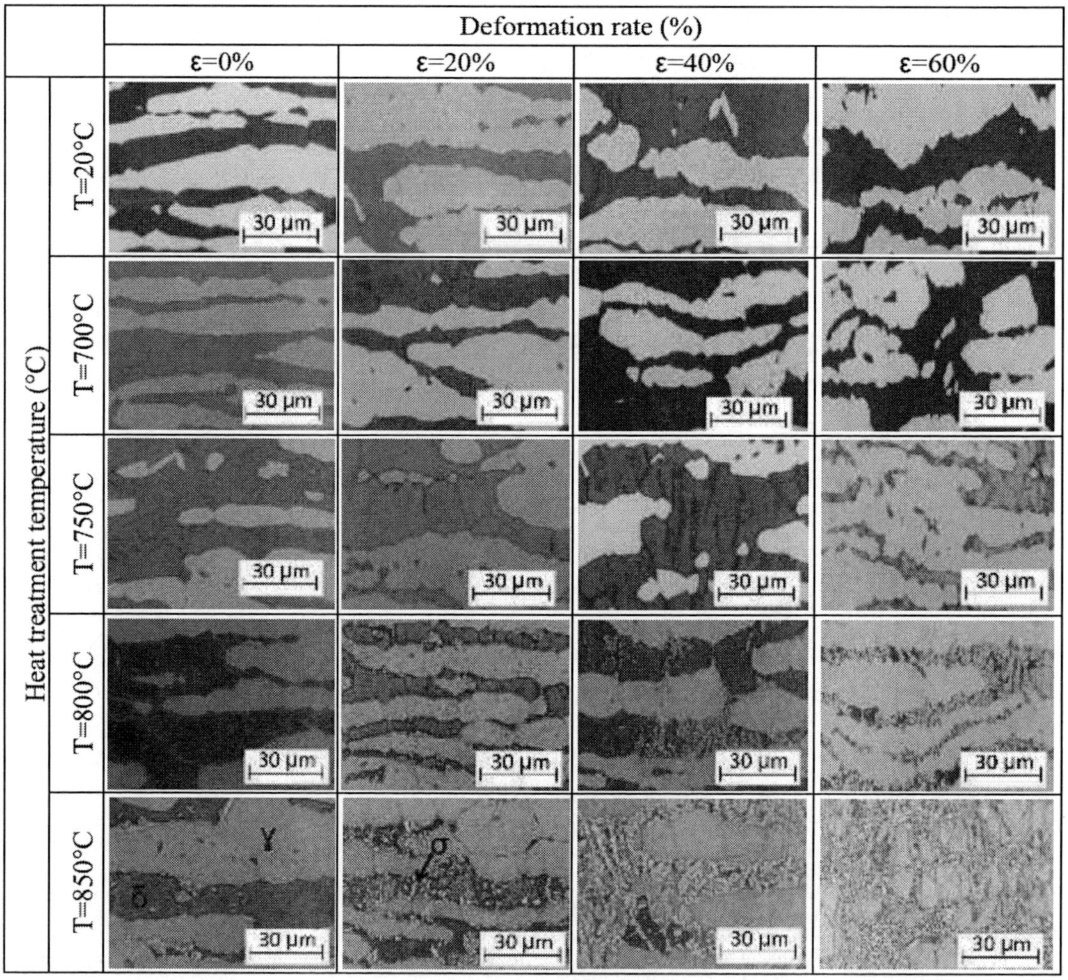

Figure 3. Microstructural images of 2507 type SDSS (magnification of the microscope was 1000×)

IOP Conf. Series: Materials Science and Engineering **426** (2018) 012034 doi:10.1088/1757-899X/426/1/012034

3.3. EBSD measurement

Samples were investigated by electron beam scattering diffraction technique (EBSD) as well. Fig. 4. represents the EBSD phase maps of four samples. All the four samples were heat treated at 850°C and the deformation rates are the following from left to right in the top line: ε=0%, ε=20% and in the bottom line: ε=40% and ε=60%. The red, green and yellow fields indicate the δ-ferrite, austenite and σ phase respectively.

The EBSD phase maps show similar results as the optical microscope pictures. If the deformation rate is stronger the amount of the σ-phase increases, because the previous cold rolling before the heat treatment decreases the starting temperature of the eutectoid decomposition.

The EBSD phase map is suitable also to determine the phase ratio. The EBSD showed different ferrite contents than the DC and the AC magnetometers. Fig. 5. compares the ferrite values of the EBSD, DC and AC magnetometer measurements. It can be seen that the difference is remarkable, the ferrite contents are far higher at the EBSD than at the DC and AC magnetometer measurements. The lowest ferrite values were showed by the AC magnetometer. The reason of this difference can be that the EBSD is a surface measurement which uses a two-dimension picture to calculate the ferrite values in contrast to the DC and AC magnetometers which are volumetric measurements. [11]

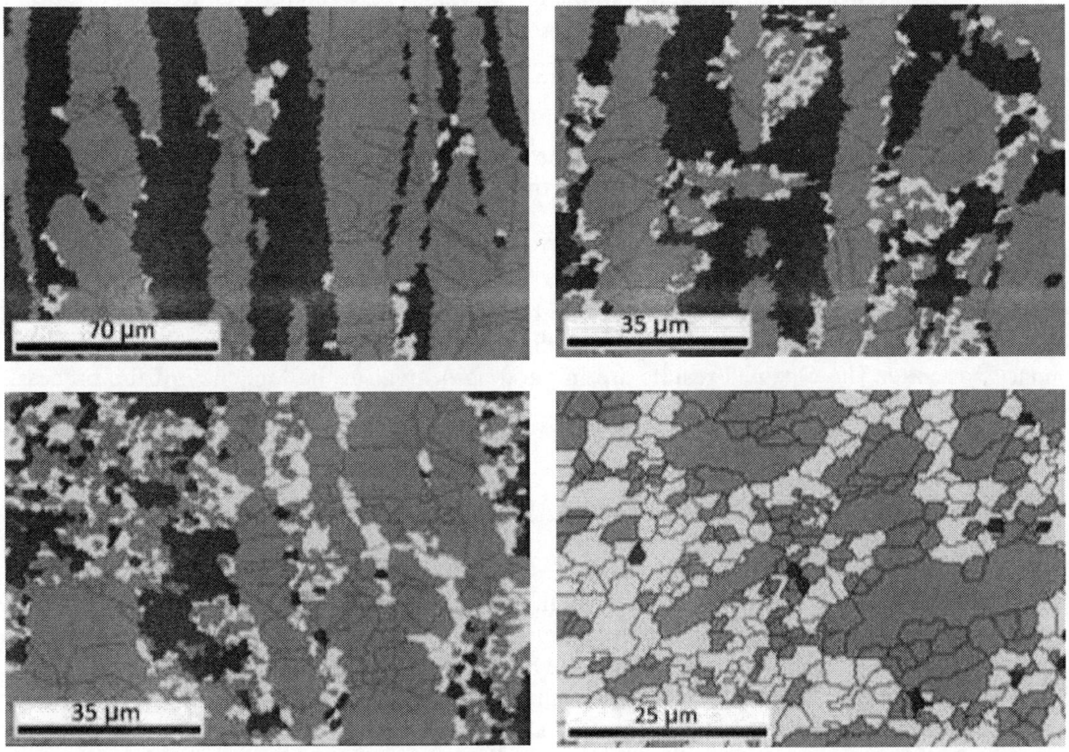

Figure 4. EBSD phase maps of cold rolled (upper line left side ε=0%, upper line right side ε=20%, lower line left side ε=40% and lower line right side ε=60%) and heat treated samples at 850°C

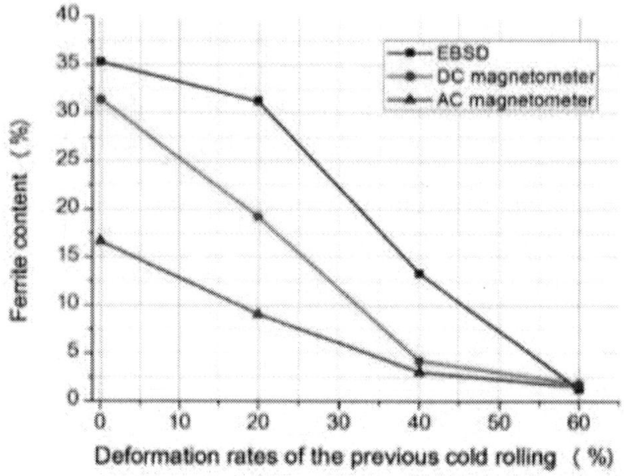

Figure 5. The difference of the ferrite values between the EBSD, DC and AC magnetometer measurement

4. Summary

The main goal of this study was that to investigate the eutectoid decomposition of ferrite phase in the 2507 type SDSS. The samples were plastic deformed in different extents before the heat treatment. The extents of the cold rolling were 0, 10, 20, 30, 40, 50 and 60% which was followed by an isothermal heat treatment at the temperatures of 20, 700, 750, 800 and 850°C.

Three methods were used to study the structural changes due to the decomposition of ferrite phase. Magnetic ferrite content measurements, metallography and EBSD were applied. The results of all these measurements demonstrated expressively the decomposition of the δ-ferrite into σphase and secondary austenite. The obtained results are in good agreement in the tendency of the process. The followings were proved by all the three methods;

- the previous cold rolling before the heat treatment significantly increases the rate of eutectoid decomposition,
- the previous cold rolling decreases the starting temperature of the eutectoid decomposition.

Naturally, some differences were also observed among the three applied measurements. The DC magnetometer detected higher ferrite values than the AC magnetometer measurement because only the DC magnetometer can excite the samples into magnetic saturation.

The ferrite contents determined by EBSD and DC magnetometer are different. The EBSD seems to be detected higher ferrite amount than the DC magnetometer. This difference is caused by the different principles of these measuring methods. The EBSD is a surface (2D) while the DC magnetometer is volumetric (3D) measurement. The EBSD derives the phase ratio from the tested surface area in contrast to the DC magnetic measurement which measures the total volume of the sample.

Finally, it must be stressed that magnetic measurements, the metallography and the EBSD supplied useful and coherent information about the details of the eutectoid decomposition of the ferrite in the studied duplex stainless steel.

References

[1] Bödök K 1997 Az ötvözetlen, gyengén és erősen ötvözött szerkezeti acélok korrózióállósága, különös tekintettel azok hegeszthetőségére *CorWeld* pp 225-254

[2] Guocai C and Pasi K 2016 Super and hyper duplex stainless steels: structures, properties and applications *Elsevier Procedia Structural Integrity* **2** pp 1755-1762

[3] Savkin A N, Sedov A A and Andronik A V 2014 The steel damageability simulation under random loading by the power, energetical and strain fracture criterions *Periodica Polytechnica Mechanical Engineering* **58(2)** pp 119-126

[4] Llorca-Isern N, López-Luque H, López-Jiménez I and Biezma V M 2016 Identification of sigma and chi phases in duplex stainless steels *Materials Characterization* **112** pp 20-29

[5] Sieurin H and Sandström R 2007 Sigma phase precipitation in duplex stainless steel 2205 *Materials Science and Engineering* **A 444** pp 271–276

[6] Dyja D, Stradomski Z, Kolan C and Stradomski G 2012 Eutectoid Decomposition of δ-Ferrite in Ferritic-Austenitic Duplex Cast Steel - Structural and Morphological Study *Materials Science Forum* **706-709** pp 2314-2319

[7] Breda M, Brunelli K, Grazzi F, Scherillo A and Calliari I 2015 Effects of Cold Rolling and Strain-Induced Martensite Formation in a SAF 2205 Duplex Stainless Steel *Metallurgical and Materials Transactions A-Physical Metallurgy and Materials Science* **46A** pp 577-586

[8] Calliari I, Bassani P, Brunelli K, Breda M and Ramous E 2013 Effect of Continuous Cooling on Secondary Phase Precipitation in the Super Duplex Stainless Steel *Zeron-100 Journal of Materials Engineering and Performance* **22** pp 3860-3866

[9] Mészáros I and Bögre B 2017 Plastic deformation and thermal induced phase transformation in 2507 type super duplex stainless steel *Materials Science Forum* **885** pp 184-189

[10] Cullity B D and Graham C D 2009 Introduction to magnetic materials *Wiley*

[11] Dobránszky J, Szabó PJ, Berecz T, Hrotkó V and Portkó M 2004 Energy dispersive spectroscopy and electron backscatter diffraction analysis of isothermally aged SAF 2507 type superduplex stainless steel *Spectrochimica Acta Part B: Atomic Spectroscopy* **59** pp 1781-1788

IOP Conf. Series: Materials Science and Engineering **426** (2018) 012035 doi:10.1088/1757-899X/426/1/012035

The effect of process control agents and milling atmosphere on the structural changes of $Ti_{50}Cu_{27,5}Ni_{10}Zr_{10}Co_{2,5}$ master alloy during short time milling

T Miko[1], F Kristaly[2], K Bohacs[c], M Sveda[3], A Sycheva[4] and D Janovszky[5]

[1] University of Miskolc, Institute of Physical Metallurgy, Metalforming and Nanotechnology, Miskolc-Egyetemváros, Hungary
[2] University of Miskolc, Institute of Mineralogy and Geology, Miskolc-Egyetemváros, Hungary
[3] University of Miskolc, Institute of Raw Material Preparation and Environmental Processing, Miskolc-Egyetemváros, Hungary
[4] MTA-ME Materials Science Research Group, Miskolc-Egyetemváros, Hungary

E-mail: femmiko@uni-miskolc.hu

Abstract. Properties of powders made from $Ti_{50}Cu_{27,5}Ni_{10}Zr_{10}Co_{2,5}$ alloy, milled under different conditions for only 3600 s in total were investigated. For this purpose two different milling atmospheres and two different Process Control Agents (PCA) were tested: argon without PCA, air without PCA, argon combined with stearic acid, argon combined with petroleum spirit. Results showed that the size and shape of particles, and the microstrucutre and hardness of partcicles were highly depended on the milling conditions. According to XRD analysis, the maximal amorphous fraction has been achieved in the air atmosphere withouth PCA milled sample (57 wt. %) powder. However the extent of Fe contamination and weight loss during milling powder resulted bigger values. As a result, the optimal combination of morphology, contamination amount and weight lost level has been observed in the case of powder milled in argon atmosphere using stearic acid.

1. Introduction

Recently, the research of Ti-Cu-Ni based alloys has attracted much attention due to its high glass forming ability [1–3]. Compared with the crystalline alloys, the amorphous or nanocrystalline bulk alloy has excellent mechanical properties, due to the special mechanism of deformation and fracture arising from its structure [4, 5]. The high energy ball milling is an often used method to change the crystalline structure of metallic powder to amorphous/nanocrystalline microstructure. The two most important mechanisms during this process are the cold welding (particle adhesion and sticking) and the fracturing of the powder particles. The quality, the purity and the yield of milled powder strongly depends on the dominancy of either cold welding or fracturing. Bay [6] investigated the cold-welding mechanism, and concluded that for cold welded bonding to occur, plastic deformation of the powder particles is necessary, but shear welding also occurs with just local micro plastic deformation. The extent of cold welding is influenced by the milled powder properties (ductile or brittle) as well as the milling conditions (atmosphere, presence of liquids or lubricants). The ductile particles tend to be cold welded, but in case of the brittle phases, notable cold welding cannot occur. Changing the milling atmosphere or adding Process Control Agent (PCA, also referred to as lubricant or surfactant) to the mixture can also reduce the effect of cold welding [7]. The PCA is adsorbed on the surface of the powder particles and minimizes cold welding [8]. One type of most commonly used PCA-s is solid (e.g. stearic acid), while others are liquid (e.g. hexane, methanol, or ethanol) [8]. Petroleum spirit however is not applied as PCA, one of the reasons to test its applicability, since it's easy to remove from the milled product and produces no contamination.

The atmosphere in which the milling process occurs is another important factor. Mechanical milling process is normally conducted under vacuum or in an inert atmosphere to prevent/minimize oxidation of the milled powder. The absence of oxide layer on the surface of broken particles causes that they can easily adhere to each other.

During ball milling, contamination is an additional process which is not easily prevented. In most cases, the aim is to decrease the extent of contamination also by using PCA-s. Contaminants will be nanoparticles of the milling equipment [8], which might be important regarding cold welding. Since ceramic particles will not be implied in mechanochemical reactions, metal contaminants (Fe, Cr or Mn) will tend to react with the milled alloys.

A reliable measure of cold welding activity is the powder yield [9] of milling, since the amount of unrecovered material has been adhered to the milling equipment. In the case of cold welding dominated milling, the powder yield is low, because big part of the powder attains to the balls and vial. If the powder yield is not high enough, then any PCA can be used to improve milling. It has been reported that after milling for 15h, only 50% of the powder was recovered if 2 wt% of polyethylene glycol was used, while almost 100% of the powder was recovered if stearic acid was used [9].

Four different milling conditions have been studied to highlight the high influence of ball milling parameters on the final products. The aim of this work is to compare the different milling atmospheres and some PCAs in terms of the particle size distribution, phase composition and microstructure.

2. Experimental conditions

Master alloy ingot with the compositions $Ti_{50}Cu_{27,5}Ni_{10}Zr_{10}Co_{2,5}$ (at. %) was prepared by vacuum induction melting of pure metal mixtures (min. 99.9 wt%). The ingot was grinded and fractioned to particle size below 320 μm for ball-milling. High energy ball milling was used with a Pulverisette 5 high-energy planetary ball-mill using a 250 ml hardened steel vial (63 HRC) and hardened steel balls (64 HRC) with a combination of Ø20 mm, Ø15 mm, Ø10 mm, Ø5 mm, and Ø2,5 mm, 10 pieces each. Ball to powder weight ratio and rotational speed were adjusted to 80:1 and 200 rpm, respectively. The angular velocity of the bowl and the supporting disc were calculated to be 231 and 200 rpm respectively. The initial weight of the powder was 6.4 g for each batch.

The micrographs of the powders were acquired with a Hitachi S-4800 scanning electron microscope (SEM) equipped with BRUKER AXS type energy-dispersive X-ray spectrometer (EDS) for elemental analysis. The particle size distribution of the ground material was determined by a Horiba LA-950 V2 type laser diffraction particle size analyser (LPSA) in distilled water (1 minute ultrasonic shaking, 1 ml of 50 g/l Na-pyrophosphate applied for dispersion). The as received and milled samples were examined by Bruker D8 Advance diffractometer (XRD) using Cu K□ radiation (40 kV, 40 mA), in parallel beam geometry obtained with Göbel mirror, equipped with Vantec-1 position sensitive detector (1° window opening), measured in the 2-100° (2θ) angular range, with 0.007° (2θ)/29 s speed. The specimen is rotated in sample plane during the measurement, to obtain data from the whole surface and to reduce in-plane preferred orientation effects. Quantitative results were obtained by combined use of Rietveld refinement and peak area calculation in TOPAS4, amorphous fraction was determined by fitting a broad peak (amorphous hump method). Microhardness measurements were performed by Instron Tukon 2100B equipment, applying a load of 10 g for 15 s for as milled powders.

3. Experimental results and discussion

3.1 Shape and size of the particles

SEM was used to visually follow the changes by milling. Coarse elongated-angular particles characterize the as received powder (Fig 1.a), with very different evolution depending on milling conditions (Fig 1.b,c,d,e). The volume distribution by LPSA of the as received powders particle shows that there are two modes x_{m1}= (72 µm), x_{m2}= (245 µm) (Fig.2), where the mode represents the most common particle size in a range. This bimodal character disappeared during milling except when using stearic acid PCA (Fig 1.c). In this case the higher mode decreased to x_{m2}= 78 µm, and the smaller mode decreased to x_{m1}= 15µm. The strangest behaviour was observed for Ar milled sample, resulting in large clustering (Fig 1.b) and strong adherence (Fig 2.) of fine particles.

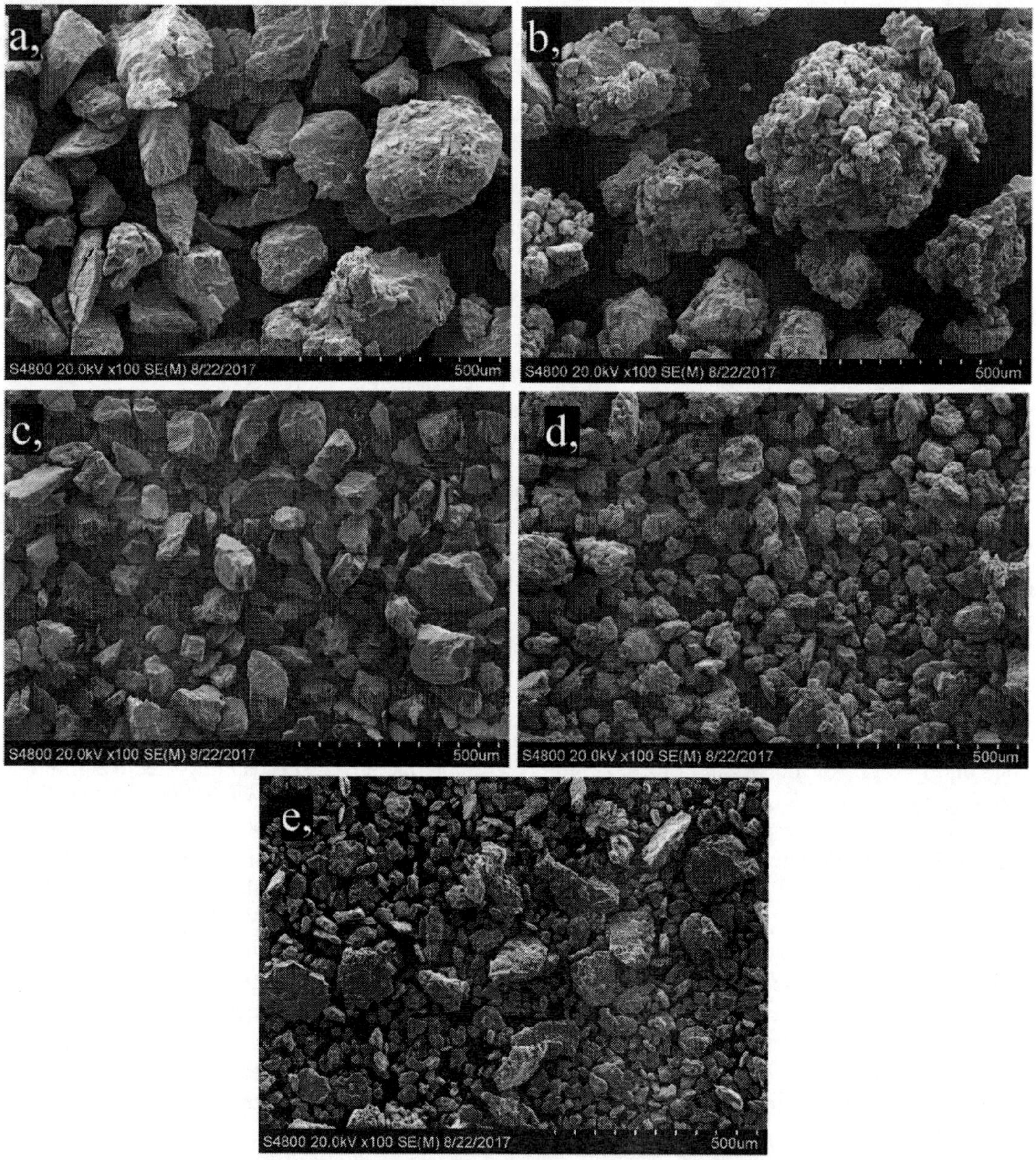

Figure 1. SEM images of as-received particles (a), and powders at 3600 s milling time: (b) argon (c) argon+ stearic acid, (d) petroleum spirit, (e) air.

Milling efficiency is reflected by LPSA values of D10, D50 and D90 (Fig 3.). The D50 value of Ar milled sample is practically the same as the stearic acid samples x_{m2}. However, in the argon milled sample the smaller particle fraction does not appear separately, due to clustering and surface adherence. Using stearate acid, small particles generated by milling are maintained as individual. As shown by D_{50}=80 μm of Ar milling, this batch produced the smallest particle size decrease.

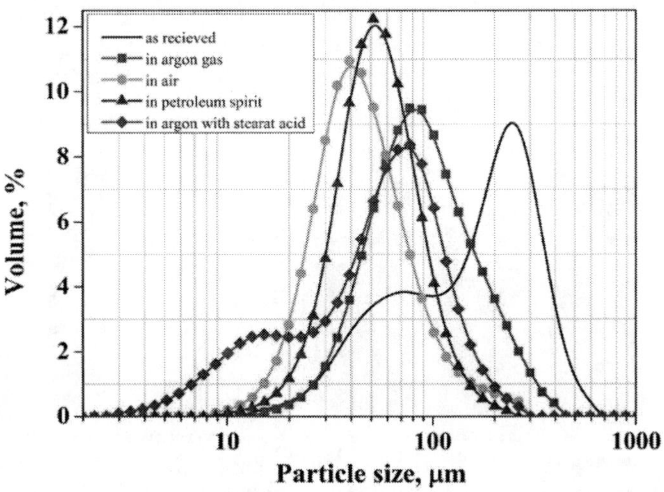

Figure 2. Volume distribution of as received and milled powders.

The smallest D_{50} value was measured in the air atmosphere milled sample (41 μm). This is because in absence of inert gas on the milling jar the cold welding was less significant. According to volume distribution of the air atmosphere and petroleum spirit milled sample, differences are minimal. However, based on SEM images, the shape and morphology is significantly different more platy and angular in air atmosphere milled, while more rounded and uniform in petroleum spirit milled sample (Fig 1 d, e).

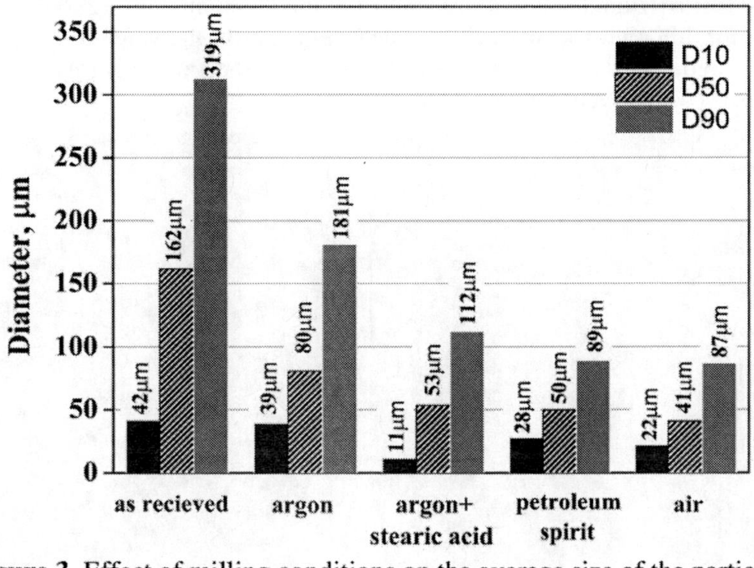

Figure 3. Effect of milling conditions on the average size of the particles

3.2. Contamination and yield properties

Fig. 4a shows the amount of powder before and after the milling processes, displaying also weight losses (wt %). The biggest difference between the weight of loaded and extracted mass is the case of dry milling with Ar atmosphere (-36 wt%). The reason for this is the significant degree of cold welding associated with adhering to the vial and ball surfaces. With a short (180 s) petroleum spirit milling step used after the 3600s milling time, significant quantity of milled powder was possible to recover from the surface of balls and vial (Fig 4a). Therefore, the total weight loss just (-9 wt%) Significantly smaller losses occurred when we used stearic acid as a PCA in argon atmosphere (-4 and -0.3 wt%) respectively. The PCA blocked the adherence mechanism, and as can see in Fig 5.b, the surface of vial and balls remained clean even after the milling, against the case of argon (Fig 5.a,). The two other investigated cases resulted in very similar end, with ca. -3.5 wt% losses.

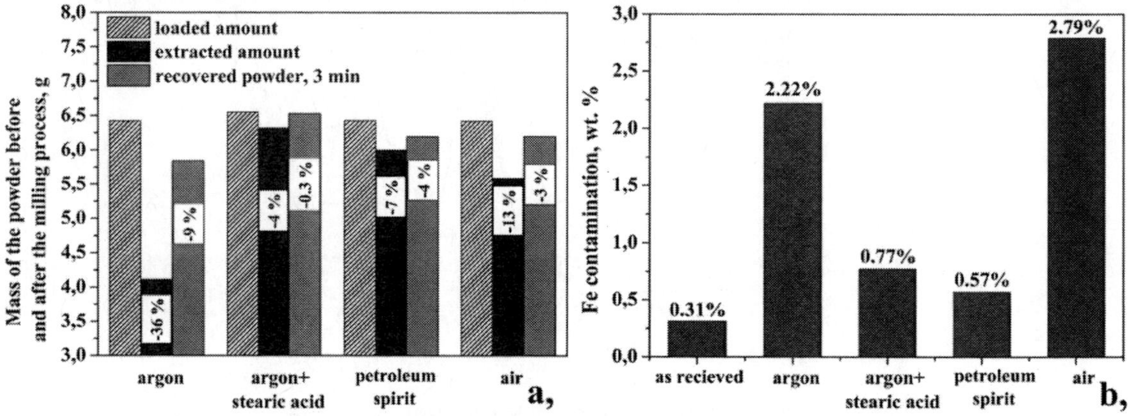

Figure 4. a, Weight loss during different milling conditions
b, Fe contamination in the case of different milling conditions (EDS)

Because we used steel vial and balls, the milled powder was contaminated with Fe during milling (Fig 4.b), highest value observed in air atmosphere without PCA (2.8 wt%), also significant in argon milled powder (2.2 wt%). The reason for this large quantity of Fe is the repeated cold welding and fracturing mechanism. The particles adhered to the balls and vial surface and then fractured from there by direct collision and relative friction, leading to transfer of atoms from the milling tool to the powder particles [7]. In the case of milling with stearic acid and petroleum spirit, the value of Fe contamination is much smaller, 0.8 wt% and 0.6wt% respectively. In these cases, the PCA-s act as lubricants between steel parts and alloy particles, preventing cold welding.

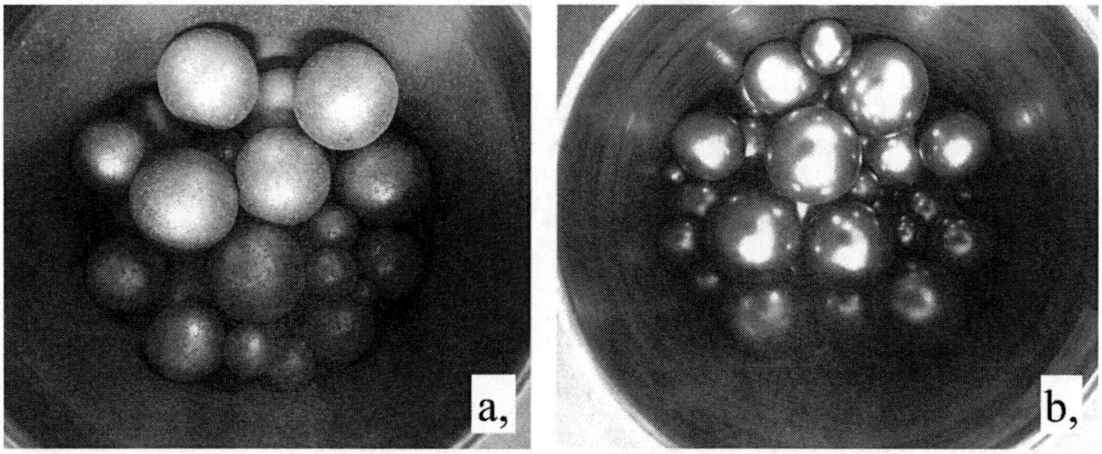

Figure 5. The surface of vial and balls after 3600 s milling a, in argon atmosphere without PCA b, argon atmosphere with 2 wt% stearate acid (PCA)

3.3 Microstructure and composition changes

Based on the XRD patterns eight different phases were identified in the as received $Ti_{50}Cu_{27.5}Ni_{10}Zr_{10}Co_{2.5}$ sample. In Fig 6.a, we have marked the main peaks of the major phases identified by Search/Match algorithm and Rietveld refinement. Transformations during milling can be followed based on the changes of these peaks (Fig.6.b). Milling in air produced similar peak broadening to milling in argon and petroleum spirit, but Ti peaks remain more visible. The cubic Ti has been destroyed by milling.

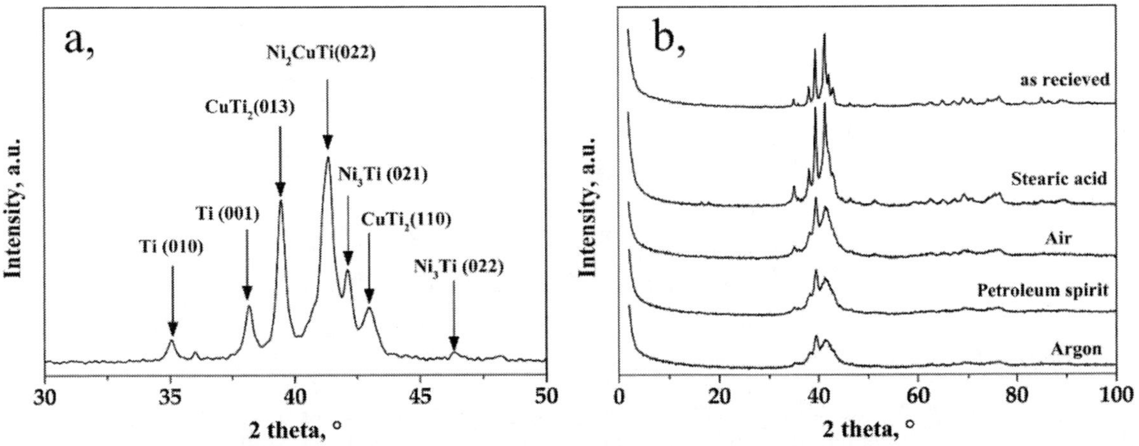

Figure 6. a, XRD pattern of $Ti_{50}Cu_{27,5}Ni_{10}Zr_{10}Co_{2,5}$ master alloy,
b, XRD patterns of powders with different milling agent

The change in the fraction of phases shows a very interesting trend as a function of the milling condition. The weight fractions of the different phases are summarized in Table 1. From the eight initial phases amount of five decreases or they disappear (Ti, $CuTi_2$, Ni_3Ti_3, Ni_2CuTi, Ti(Fm-3m)), while amount of three increases ($CuTi_3$, $Cu(Ni_{0.7}Cu_{0.3})Ti_2$, $Cu_{0.5}Ni_{0.5}Ti$). Amorphous phases arise in all samples, regardless the milling conditions, but their amount varies according to atmospshere and PCA-s.

Table 1. Fraction of crystalline and amorphous phase of the as received and milled powder

Phase name	space group	Crystalline size, nm				
		as recieved	Milling atmosphere/PCA			
			Air	Argon	Argon+stearic acid	Petroleum spirit
Ti	P63/mmc	11.5	3.7	4.3	6.4	8.4
$CuTi_3$	P4/mmm	2.8	16.7	13	19.4	10.7
$Cu(Ni_{0.7}Cu_{0.3})Ti_2$	Pmmb	1.2	2	1.2	5	2.1
$CuTi_2$	I4/mmm	35.3	5.5	6.9	3	7.5
Ti (Fm-3m)	Fm-3m	5.1	-	-	-	-
Ni_3Ti_3	P63/mmc	10	0.7	0.6	0.8	1.3
$Cu_{0.5}Ni_{0.5}Ti$	Pmmb	4.1	9.8	12.9	25.3	11.8
Ni_2CuTi	Fm-3m	30	4.6	6.3	2.1	6.2
amorphous	-	0	57	55	38	52

The phase evolution of powder caused by high energy ball-milling was plotted as function of different milling condition, which can be seen in Fig. 7. Compared to the four difference milled powders, the quantity of phases in argon+stearic acid sample significantly differs from the others. Using this milling condition the value of amorphous phase is the smallest (38 wt%), but the amount of $CuTi_3$ (19.4 wt%) and $Cu_{0.5}Ni_{0.5}Ti$ (25.3 wt%) are the biggest, compared with the other samples.

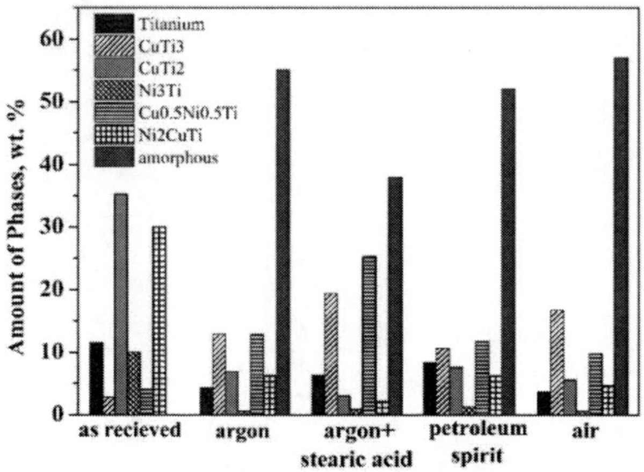

Figure 7. Fraction of crystalline and amorphous phases of the as received and milled powders

Measurement of the crystallite size in the powders is also very important since the phase constitution and transformation characteristics appear to be critically dependent on the crystallite size. The crystallite size decreases significantly due to the milling processes (Fig. 8.). The smallest amount of decreasing is in the case of stearic acid. This is because the low friction, the milling balls and the powder particles are mainly just rolling on the inner wall of the bowl. The plastic deformation of powder particles is low, so the surface of the flake shaped particles becomes very smooth (Fig. 1. c) compared to all other investigated cases.

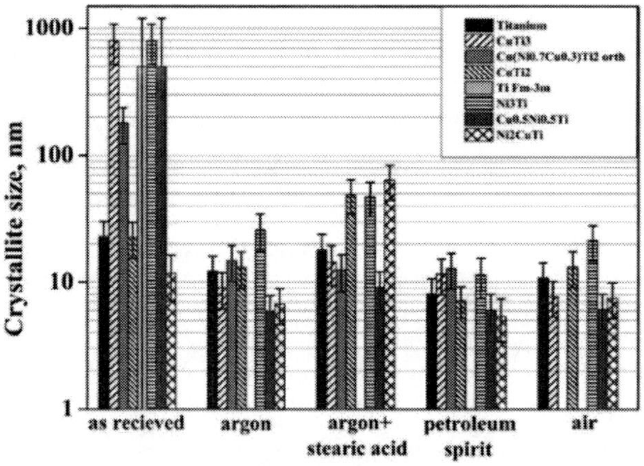

Figure 8. Crystallite size of as received and milled powders.

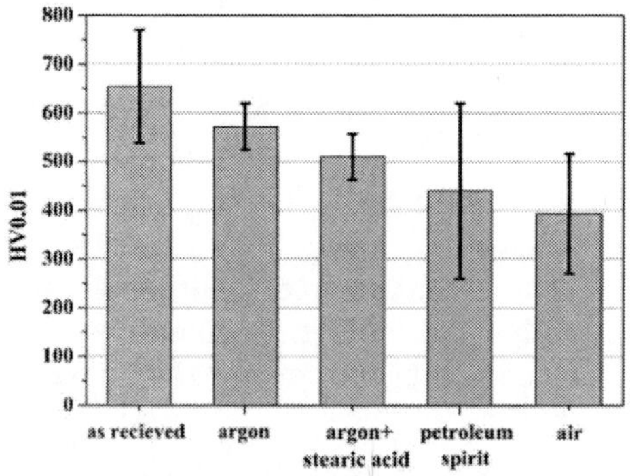

Figure 9. Hardness of the different condition milled powder particles

The measured hardness values of the different powder particles are shown in Fig. 9. The as received sample has the biggest mean hardness (655 $HV_{0.01}$) compared to the hardness of milled powders. However, the standard deviation is significant, because it contains many different phases besides the brittle intermetallic phases, contains also ductile parts, like Ti structures. The particle size of powder is large compared to the all other milled powder. Hardness of each milled powder samples are below to the as received sample. The tendency of hardness lowering is equal to the decreasing of mean particle size (Fig. 3.). The smallest hardness decrease was measured in the Ar milled sample (572 $HV_{0.01}$), while the biggest decrease is measured in the air atmosphere milled sample (393 $HV_{0.01}$).

4. Conclusion

In this study the effect of different milling atmosphere and PCA was studied on $Ti_{50}Cu_{27,5}Ni_{10}Zr_{10}Co_{2,5}$ alloy. During only 3600 s milling time, the particles size as well as the crystallite size and amorphous content went through a big change.

Use of petroleum spirit and stearic acid considerably reduces the Fe contamination compared to the case of milling without any PCA. The best yielding properties of the milled powder were measured using stearic acid (96%). The worst yielding properties of the milled powder measured milling in argon atmosphere without any PCA (64 wt %). Using a short (3 min) wet milling with petroleum spirit, this value increases to 91 wt%. The content of Fe and Cr impurities in the milled powder increased when the cold-welding process is dominant, Ar and air without PCA.

Milling with petroleum spirit and stearic acid the Fe contamination and the weight loss was much smaller than in the cases without any surfactant. The biggest amorphous content obtained milling without any PCA and inert atmosphere. Using stearic acid obtained the smallest amorphous content. The amorphization process of structure is not as rapid with use of stearic acid as PCA, then in the other investigated cases.

Acknowledgements

"The described article/presentation/study was carried out as part of the GINOP-2.3.2-15-2016-00027 "Sustainable operation of the workshop of excellence for the research and development of crystalline and amorphous nanostructured materials" project implemented in the framework of the Szechenyi 2020 program. The realization of this project is supported by the European Union."

References

[1] Pan Gong, Lei Deng, Junsong Jin, Sibo Wang, Xinyun Wang, Kefu Yao 2016 Review on the Research and Development of Ti-Based Bulk Metallic Glasses *Metals*; 264

[2] Tao Chang, Akihisa Inoue 1998 Thermal and Mechanical Properties of Ti-Ni-Cu-Sn Amorphous Alloys with a Wide Supercooled Liquid Region before Crystallization, *Materials Transaction*, 1001 to 1006.

[3] Jianli Wang, Baozhong Liu, Lidong Wang, Yaoming Wu, Limin Wang 2008 Glass formation ability and mechanical properties of Ti40Cu40Zr10Ni10 alloy, Journal of Non-Crystalline Solids *Journal of Non-Crystalline Solids* 3653–3658

[4] Christopher A. Schuh, Todd C. Hufnagel, Upadrasta Ramamurty 2007 Mechanical behavior of amorphous alloys Review on the Research and Development of Ti-Based Bulk Metallic Glasses *Acta Materialia* 4067–4109

[5] Y.C. Kim, W.T. Kim, D.H. Kim 2004 A development of Ti-based bulk metallic glass *Materials Science and Engineering* 127–135

[6] N. Bay 1986 Cold welding. Part 1: Characteristics, Bonding Mechanism, Bond Strength, *Metal Construction*, 369-372

[7] G. Cipolloni, M. Pellizzari, A. Molinari, M. Hebda, M. Zadra 2015, Contamination during the high-energy milling of atomized copper powder and its effects on spark plasma sintering, *Powder Technology* 51–59

[8] C. Suryanarayana 2001 Mechanical alloying and milling, *Progress in Materials Science* 1-184.

[9] Lai MO, Lu L 1998 Mechanical alloying Boston, MA: Kluwer Academic Publishers

Effect of hot rolling conditions on the nitride precipitation process in low carbon steel strips

A Mucsi[1]

[1] Óbuda University, Donát Bánki Faculty of Mechanical and Safety Engineering, Budapest, Hungary

E-mail: mucsi.andras@bgk.uni-obuda.hu

Abstract. The nitride precipitation process plays an important role in the final mechanical and aging behaviour of low carbon steels in hot rolled state as well as in cold rolled and annealed state. The nitride precipitation process is investigated in an indirect manner: instead of measuring the amount of nitrogen compounds, the free nitrogen content of the samples was measured using a special thermomechanical treatment and a thermoelectric power tester. It was found, that the free nitrogen content in a steel strip strongly depends on three main factors: on the position in strip transverse direction, on the coiling temperature and on the cooling method at the run-out table of the hot rolling mill.

1. Introduction

Low carbon steels are frequently used for cold forming applications in hot rolled as well as in cold rolled and annealed state [1]. Low carbon steel grades usually consist of elements in w.t. %: 0.02…0.05 C, 0.15…0.3 Mn, 0.01 Si, 0.03-0.06 Al, 0.004…0.006 N, 0.015…0.05 Cr, 0.01…0.06 Cu, and low amount of Ti, Mo, Ni, Nb. The basic production technology is shown in Fig. 1.

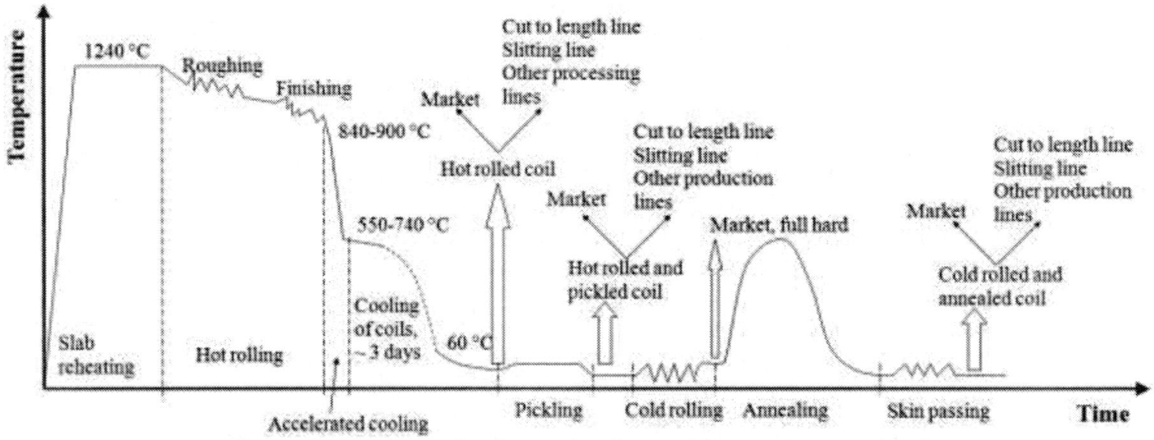

Figure 1. Basic production technology of low carbon steel strips

The dissolution of nitrogen containing precipitates (usually aluminium-nitride and aluminium-chromium nitride) occurs at the slab reheating temperature (~1240 °C) [1-3]. After reheating the slabs to the suitable temperature, the roughing mill rolls the material to an intermediate thickness, ranging usually between 20 and 30 mm. The precipitation of nitrides in this austenitic region is very slow compared to the duration of roughing [2]. The final strip thickness is reached on the finishing mill, where the most important parameters are the finish rolling temperature (FT), the cooling method after rolling and the so-called coiling temperature (CT, the temperature of the strip before reaching the coiling equipment). The hot rolled coils are stored for 3-4 days to reach the suitable temperature (~60 °C) for further processing. During the hot rolling process the 10-20 % of the nitrogen content could precipitate depending on the temperature history and alloying elements in the steel [4-6]. But during the slow cooling after coiling a large amount of nitrogen containing precipitates could form, since the temperature and the time available is large enough to precipitate nucleation and growth [1, 4, 6]. The amount of nitrides formed depends strongly on the coiling temperature [1] and on the cooling rate after coiling [4]. Since the sides of the coil (strip borders) cools down more rapidly than the center of the coil, it is expected to be a large difference in the amount of precipitated nitrides (and in the free nitrogen content) along the transverse direction. The aim of this study to estimate the free nitrogen content along the transverse direction and investigate the role of cooling method and coiling temperature.

2. Experimental

The material investigated was a low carbon steel containing 0.056 C, 0.3 Mn, 0.042 Al, 0.046 Cu, 0.072 Cr, 0.034 Ni, and 0.0050 N (expressed in wt%). In order to study the effect of hot rolling thermal history, three different hot rolling conditions (FT=900 °C and CT=570°C, FT=900 °C and CT=730 °C with rapid and delayed cooling, FT=811 °C and CT=667 °C without strip cooling) were applied. The final strip cross section dimensions were 3 x 1000 mm. The free nitrogen content was determined at 50 m from the head end and along the width of the strip.

The determination of the amount of precipitated nitrides is based on the change in the thermoelectric power (S) caused by the precipitation of interstitials from solid solution. The thermoelectric power of the specimens was measured according to the concept shown in Fig. 2 [4].

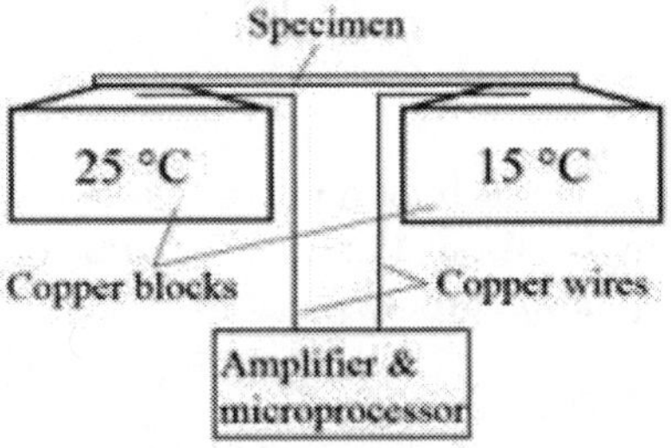

Figure 2. Basic assembly of the equipment used for measuring the thermoelectric power [4]

The methodology of Massardier et al. [7, 8] has been used for experimental investigation of nitride precipitation in hot rolled state. The method consists of an equalization treatment at 270 °C for 3 hours in order to set up the equilibrium carbon leveli n solid solution in each specimen. The cold rolling is intended to produce a high dislocation density, which allows the movement of all carbon and nitrogen atoms from solid solution (SS) to the vicinity of dislocations, see Fig. 3. The difference (ΔS_a) between the thermoelectric power before (S_ε) and after (S_a) the aging treatment is related to the amount of carbon and nitrogen atoms diffused to the dislocations. From that, using the Nordheim-Gorter law [9, 10], the amount of free nitrogen can be evaluated. The method is described in detail in Refs. [4-6].

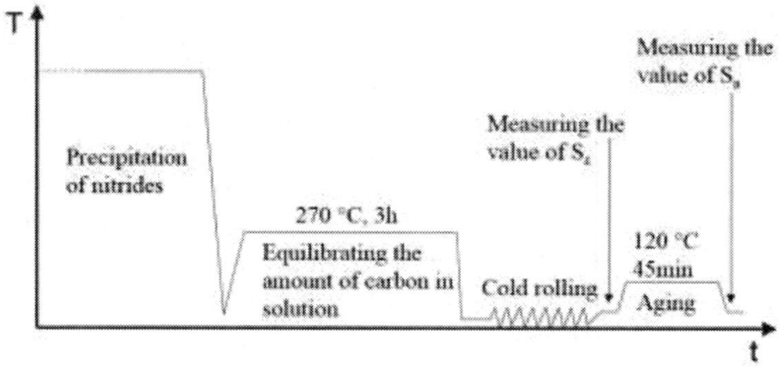

Figure 3. Methodology for determination the free nitrogen content [5]

The difference ΔS_a between S_a and S_ε carries the information about the amount of free interstitials at 270 °C [7]:

$$\Delta S_a = S_a - S_\varepsilon = K_C \cdot x_C + K_N \cdot x_N \tag{1}$$

where K_C and K_N are the Nordheim-Gorter coefficients for carbon and nitrogen (K_C=45 μV/K·wt%, K_N=24 μV/K·wt% [5,6]), x_C and x_N are the amount of carbon and nitrogen in solid solution (wt%). If all of the nitrogen are precipitated as nitrides (in case of high temperature and long treatment times), only the solute carbon determines the value of ΔS_a. So, firstly a long and large temperature treatment is necessary to determine the solubility of carbon in the given steel at 270 °C. After a long and high temperature treatment, no free nitrogen exist in the sample, so, the carbon in solid solution at 270 °C (called residual carbon) can be expressed as:

$$x_C = \Delta S_a / K_C \tag{2}$$

The value of x_C is forund to be 15±2 ppm by weight. Using this value the amount of free nitrogen in other samples can be calculated as:

$$x_N = (\Delta S_a - K_C \cdot x_C)/K_N \tag{3}$$

3. Results and discussion

The free nitrogen distribution in transverse direction is shown in Fig. 4. It can be seen, that the different hot rolling thermal histories and the local position has a marginal effect on the free nitrogen content (and, of course, on the nitride precipitation process).

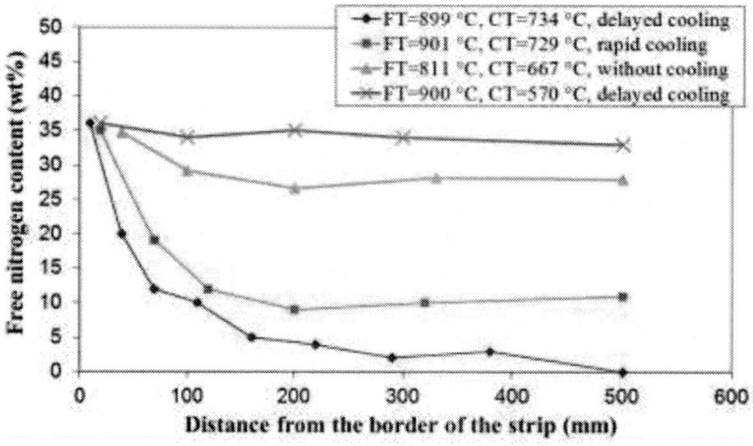

Figure 4. Distribution of free nitrogen content along the strip width

The major parameter, which influences the amount of nitride precipitation is the local position along the strip width. Quite close to the borders of the strip (<40 mm), the free nitrogen content is ranging between 34 and 36 ppm regardless to the hot rolling thermal history. The difference between the total nitrogen content (45 ppm) and this value is approx. 10 ppm (22% of the total nitrogen content). Since the cooling rate of the coil is the largest at this position, the nitride precipitation after coiling is negligible. This 10 ppm nitrogen is composed of two parts: a part of it was precipitated during hot rolling until the last finishing stand and another part is present as high stability precipitates (such as TiN) which were not dissolved during the reheating of the slabs. Titanium is present as a trace element in this steel (its content is lower than 10 ppm, and its more exact measurement with spectroscopy is difficult), but this amount is still large compared to the precipitated amount of free nitrogen (also ~ 10 ppm) at the borders of the strip. Considering a 10 ppm titanium level and taking the molar weights of titanium (M_{Ti}=47.9 g/mol) and nitrogen (M_N=14 g/mol), the nitrogen present as titanium-nitride ($x_{N,\text{ as TiN}}$) can be estimated as:

$$x_{N,\text{asTiN}} \approx x_{N,f,b} \cdot M_N / M_{Ti} = 10\text{ppm} \cdot 14/47{,}9 \approx 3\text{ppm} \quad (4)$$

where $x_{N,f,b}$ is the nitrogen present as precipitates at the border of the strip. Considering this simple calculation, the nitride precipitation process (altogether 10 ppm) until the beginning of strip coiling is composed of: 3ppm/45ppm≈ 6.7% could present as TiN, while 7ppm/45≈ 15.6% precipitates during hot rolling.

The free nitrogen content decreases strongly to the center of the strip, if the coiling temperature is high (670-730 °C). The large and the small CT at the borders of the strip decreases quickly after the coil is placed into the coil store and the time available and temperature are not large enough for aluminium (-chromium) nitride precipitation. In contrast to this, the regions of the strips located closer to the centerline cool much more slowly causing the gradual increase of the rate of nitride precipitation. When the strip is coiled at low temperature (CT=570 °C), negligible nitride precipitation (1-2 ppm nitrogen forms nitrides) takes place inside the coil.

The cooling method at the run-out table has a detectable effect of the nitride precipitation process. When the strip is cooled down quickly to the coiling temperature CT=730 °C just after the last stand of the finishing mill (rapid cooling), the free nitrogen content is found to be approx. 10 ppm in the centerline. When the strip is cooled down to the same temperature at the end of the run-out table (just before the coiler) the free nitrogen content is tending to zero at the centerline. The strip temperature in this case decreases gradually from FT=900 °C to an unknown value until it reaches the cooling unit at the end of the run-out table. During this travel of the strip (usually 6-8 s), 2-8 ppm nitrogen (the difference between the two lowest curves on Fig. 4.) precipitates.

With regard to the free nitrogen content of such mild steels, a simple observation connected to this topic should be mentioned. It is well known, that the free nitrogen content is in close relation with the elastic-plastic behavior of mild steels (Cottrell-theory [12]). Generally, it is believed, that the free nitrogen (and carbon) content of the steel is responsible for the discontinuous yielding phenomenon. The so-called coil break defects on low carbon steel strips are connected to the presence of pronounced yield point, they can form only if the steel exhibits discontinuous yielding [13-15]. Generally, according to the industrial investigations, coil break defects are not formed just at the borders of the strip where the free nitrogen content is high. This fact and other investigations [14, 15] suggest that the free interstitial content of a steel is not the individual reason of the discontinuous deformation.

4. Conclusions

On the basis of the experimental work outlined in this study, the following conclusions can be drawn:

1. The nitrogen distribution function in transverse direction of the hot rolled strip is determined by the coiling temperature and cooling method at the run-out table.
2. The free nitrogen content of the steel investigated is found to be 34-36 ppm (by weight) (approx. 22% of the total nitrogen of the steel) at the border of the strips regardless to the hot rolling temperature history.
3. The precipitated 22% nitrogen is composed of two parts: undissolved precipitates (in case of titanium-nitride it could not be more than 6-7%) and nitrides precipitated during hot rolling (in this case 15-16%).
4. From the 34-36 ppm level, the free nitrogen content decreases (while the amount of precipitated nitrides increases) gradually to the centerline of the strip. The decreasing rate depends on the coiling temperature and on the cooling method at the run-out table.
5. The strip cooling method cause a measurable difference in the free nitrogen distribution curves. When the strip is cooled quickly just after the hot rolling, the free nitrogen in the middle regions is 2-8 ppm larger than cooling at the front of the coiler.
6. The background of the uneven distribution of free nitrogen along the strip width is the uneven cooling of the coil after coiling. The difference between the border and the centerline became more pronounced, when the coiling temperature is high.

Acknowledgement

This work was supported by the ÚNKP-17-4-I. New National Excellence program of the Ministry of Human Capacities.

References

[1] Wilson F G and Gladman T 1988 *Int. Mater. Rev.* **33** 223
[2] Meyzaud Y and Parniere P 1974 *Mem. Etud. Sci. Rev. Met.* **71** 415
[3] Radis R and Kozeschnik E 2010 *Model. Simul. Mater. Sc.* **18** 1
[4] Mucsi A 2014 *J. of Mat. Proc. Tech.* **214** 1536
[5] Massardier V, Guétaz V, Merlin J and Soler M 2003 *Mat. Sci. Eng. A-Struct.* **355** 299
[6] Massardier V, Lavaire N, Soler M and Merlin J 2004 *Scripta Mater.* **50** 1435
[7] Rawlings R and Tambini J 1956 *J. Iron Steel Inst.* **184** 302
[8] Beeghly H F 1949 *Anal. Chem.* **21** 1513
[9] Merlin J, Merle P, Garnier S, Bouzekri M and Soler M 2004 *Metall. Mater. Trans. A* **35** 1655
[10] Nordheim L C and Gorter J 1935, *Physica* **2** 383
[11] Zolotorevsky N Y and Pletenev V P 1998 *Model Simul. Mater. Sc.* **6** 383
[12] Cottrell A H and Bilby B A 1949 *Proc. Phys. Soc. London Sect. A* **62** 49
[13] Thakur S K, Sarkar B, Ghosh B, Datta R 2014 *Case Stud. Eng. Fail. Anal.* **2** 76
[14] Mucsi A 2018 *Eng. Fail. Anal.* **83** 109
[15] Akama D, Nakada N, Tsuchiyama T, Takaki S and Hironaka A 2014 *Scripta Mat.* **82** 13

Following the precipitation process during homogenization by correlative microscopy in aluminium manganese alloy

E Nagy [1], Á Kovács[2], P Barkóczy[2], B Schweitzer[2], Z Gácsi[2], V Mertinger[2]

[1] MTA-ME Materials Research Group, H-3515 Miskolc-Egyetemvaros
[2] Institute of Physical metallurgy, Metalforming and Nanotechnology, University of Miskolc, Miskolc-Egyetemvaros, Hungary

E-mail: femzsofi@uni-miskolc.hu

Abstract. The correlative microscopy provides an opportunity for combining two visual impressions based on different principles. By this method, the taking of optical microscopic and electron microscopic images become possible about the same experimental area in an appropriate hardware-software environment.
In the aluminium-manganese alloys, precipitation form during homogenization which is one of the processes in the rolling technology sequence. The features, morphology and composition of developing precipitation depend on the homogenisation temperature and time, these characteristics change within wide limits.
In this study, the precipitates forming during homogenization were examined by correlative method. In 3003 manganese alloy, the appearance of particles at different homogenization temperature and time was followed. The microstructure of sample before and after homogenization was investigated, the phases were identified in the microstructure and the formation processes were followed.

1. Introduction

In the 3xxx aluminium alloys, the main alloying element is Mn with about 1w% (3003), besides it can contain maximum 0.6% Si, maximum 0.2 % Cu and maximum 0.7% Fe. The manganese in dissolved state or in different intermetallics/precipitates has a strong effect on formability; the formed phases have an influence on the dynamic recrystallization during hot rolling and recovery process after cold rolling and subsequent annealing. The nature, the composition, the morphology and the dispersion of precipitates can be changed in wide range by changing the homogenization time and temperature. In the manganese aluminium alloy, intermetallic phases appear as e.g. Al_6Mn, ($Al_6(Mn,Fe)$ as a function of the cooling rate during casting. Depending on the time and temperature of the homogenization, other processes can also take place in the course of the homogenization process, such as the destabilization of non-equilibrium phases or the spheroidization of equilibrium phases. The orthorhombic $Al_6(Fe,Mn)$ often transforms to cubic α-(Fe,Mn)-Si („6 to α" transformation). [1, 2, 4]

The manganese precipitation then the dissolution from the aluminium solid solution starts at about 300°C in the 3003 aluminium. The eutectic in the cast microstructure mostly consists of $Al_6(Fe,Mn)$, the fragmentation of which starts at about 400°C. The $Al_6(Fe,Mn)$ phase transforms to α-Al(Fe,Mn)Si and the amount of this phase increases by increasing the temperature. The fragmentation of eutectic phases is the strongest at 560°C in addition spheroidization and coarsening can also be observed. The volume fraction of primary phases does not change significantly during homogenization, the manganese has not any effect on it but the finest secondary precipitation has. The formation of $Al_6(Fe,Mn) \rightarrow \alpha$-$Al_{12}(Fe,Mn)_3Si$ eutectoid transformation and the fine dispersed precipitation begin at the same temperature (300°C) and their kinetic is the same up to 500°C. Above 500°C, the small precipitates dissolve, while the amount of the ones having big size increase.

The average size of precipitation grows by increasing the homogenization temperature, but the homogenization has an effect on the morphology of precipitation as well. In the eutectic structure formed during the crystallization, spheroidization takes place during homogenization in addition small manganese contented precipitates also appear within the homogenised grains [4, 6, 7]. During growing of the precipitates, particle-free zones (PFZ) form in the matrix which do not contain any dispersoids. The increasing of PFZ has a beneficial effect on the subsequent formability. The size of particle-free zone grows by increasing the heating time especially in case of the alloys with high Si and Mn content which can be explained by the coarsening of fine disperse and primary big size particles [3, 4].

As the character, composition, morphology and distribution of precipitations appearing during the homogenizing heat treatment applied in the course of the rolling technology of manganese aluminium alloys can widely change as a function of the time and temperature influencing by this the properties of final product. A method suitable for describing the structure-changes taken place during the homogenizing heat treatment and for following the progress of these changes in time was searched. The correlative microscopy seemed to be suitable for this purpose. Various heat treatments were performed at different homogenization temperatures and times by using 3003 manganese aluminium alloy. The formed microstructure of homogenized samples was investigated by a correlative microscopic method.

2. Experimental

2.1. Correlative microscopy

In correlative microscopy, the possibilities offered by visual impressions operating at different principles are combined. As the visual imaging is always based on the information obtained from the interaction between the sample and the radiating source, therefore light contrast is used in optical microscopy, while material contrast is applied in electron microscopy for imaging. By applying correlative microscopy, optical and electron microscopic images can be made sequentially in a special hardware and software environment, which is free from the anomalies of mapping techniques.

The Zeiss method uses a special sample holder and software to connect the two investigation methods (Figure 1.). The sample holder has a frame which can be attached both to the optical microscope and to the electron microscope as well.

The three markers placed at certain distances on the frame help recording the accurate measuring coordinates. During calibration, the coordinates of markers are recorded in the software. After calibration, the actual coordinates are recorded in each case near the image taken of the investigated area of sample. After completing the heat treatment or other treatment result a structural change, the sample is returned to the holder and the software finds the examined area on the basis of the recorded coordinates.

Figure 1. Zeiss sample holder with markers for correlative microscopy

The method can be applied for (1) the determination of morphology and composition on a given area by two different methods (optical and electron microscopic) and for (2) observing in high magnification the process taken place on the given area.
In the course of our research work, the results of processes taken place during homogenization were followed by using the electron microscopic method based on the correlative microscopy method.

2.2. *Examinations*
The experiments were performed on 3003 aluminium alloy. The composition of alloy can be seen in Table 1.

Table 1. Composition of 3003 aluminium alloy [w%]

	Al	Mn	Cu	Fe	Si	Zn	Residuals
3003	Balanced	1.0-1.5	0.05-0.20	max 0.7	max 0.6	max 0.1	max 0.15

Polished samples were used for the measurements. The investigations were performed by using Zeiss EVO MA 10 scanning electron microscope in the Institute of Physical Metallurgy, Metalforming and Nanotechnology, University of Miskolc. In case of each sample, the images were taken in by a magnification of 500 and 1000. Following the heat treatment, the phases developed on the areas investigated earlier could be observed on the basis of the recorded and saved coordinates. The homogenization heat treatment was performed at 500 and 600°C for 1, 5 and 10 hours period. In this study, the results obtained during the homogenization process of max. 10 hours are presented.

3. Results and Discussion
Due to the resolution limit, optical microscopy is not suitable for observing small precipitates with a size of few hundred nanometre developing during the homogenization process. In case of the sample homogenized at 500°C for 10 hours, the information content of images taken of the investigated area by optical and electron microscopy method is clearly visible on **Figure 2**. The small particles cannot be seen on the image taken by optical microscope. Since the purpose of our study is to observe the precipitates forming during homogenization, the changes will only be followed in the electron microscopy images.

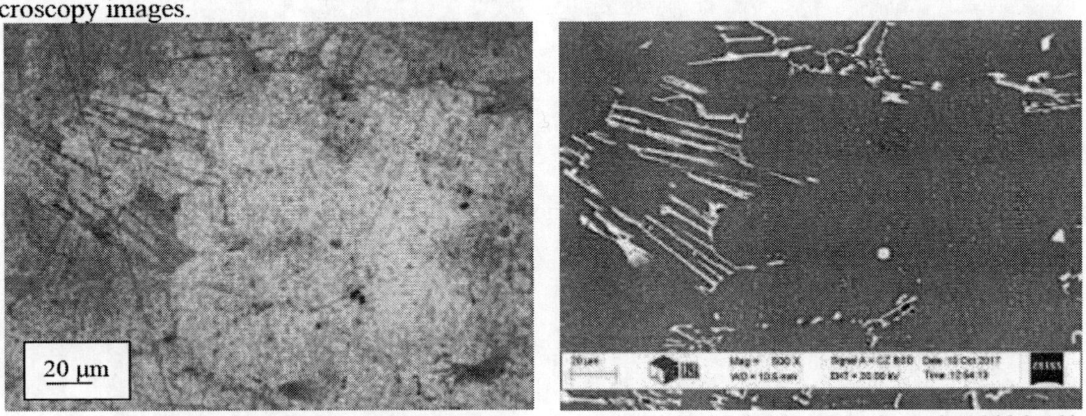

Figure 2. Optical microscopic (left) and electron microscopic (right) images; Sample 500°C 10h

3.1. Results of homogenization

The microstructure of a homogenized 3003 aluminium sample is shown in **Figure 3**.

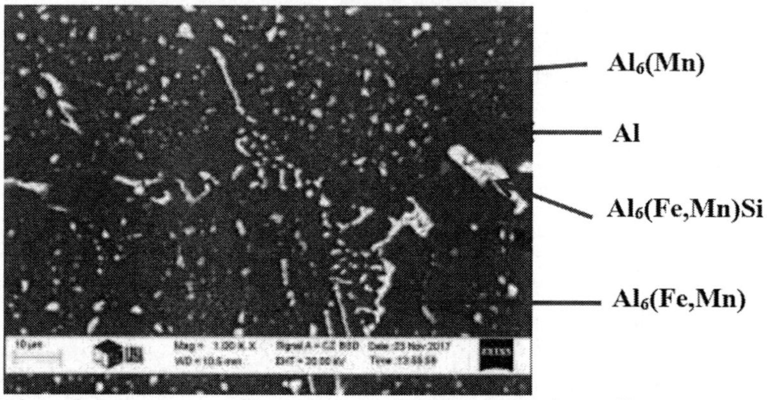

Figure 3. The microstructure of homogenized 3003 aluminium alloy

In the 3003 alloy, the phases having a larger size similar to the light colour Chinese characters are $Al_6(FeMn)$ primary intermetallics. The small particles appearing in aluminium solid solutions are Al_6Mn precipitates. The amount of precipitates increases as a function of the homogenization time (Figure 4.).

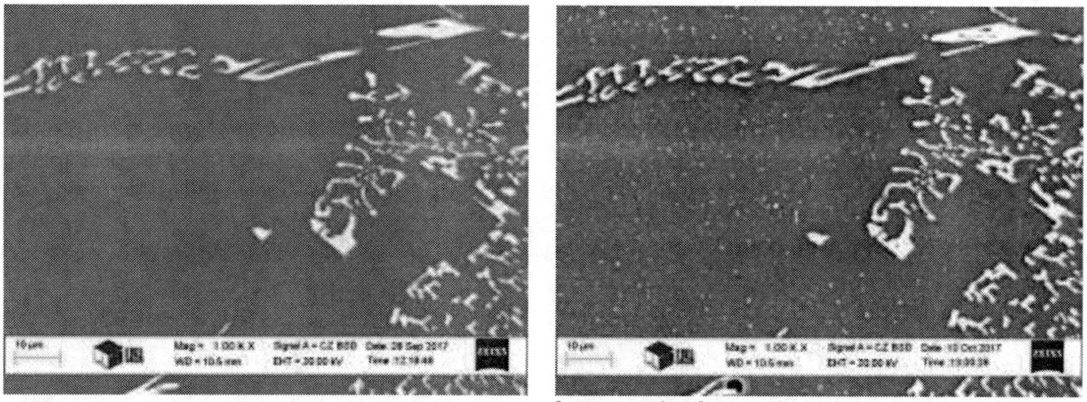

cast homogenised

Figure 4. Microstructure of sample after homogenisation (500°C, 10h)

The precipitates are coarsening by increasing the homogenization temperature (Figure 5.).

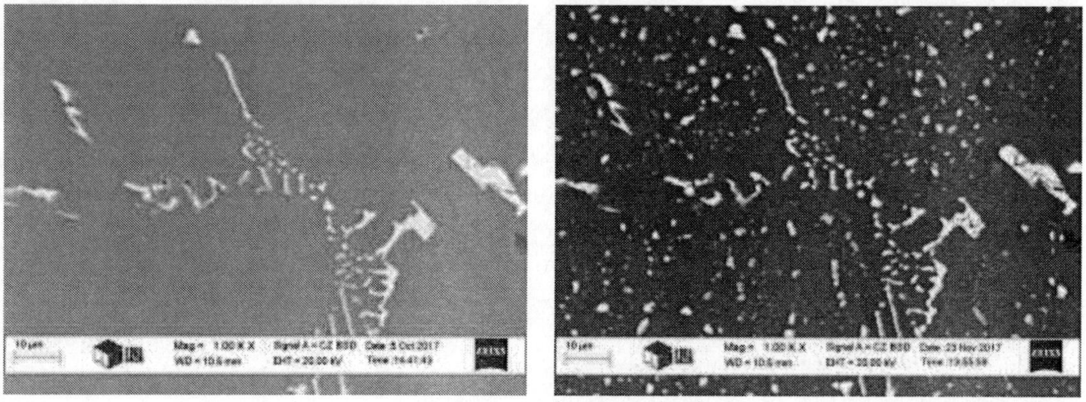

cast homogenised

Figure 5. Microstructure of Sample after homogenisation (600°C 10h)

3.2. Precipitates in the homogenised samples

High magnification images were performed for characterising the precipitation. In case of the sample homogenised at 500°C, many small particles of a size of about 200-300 nm and some larger particles with a size of about 0.7-0.8 μm next to the small particles can be observed. The distribution of small dispersed phases is uniform and it surrounds the previously formed particles having a larger size. (Figure 6.).

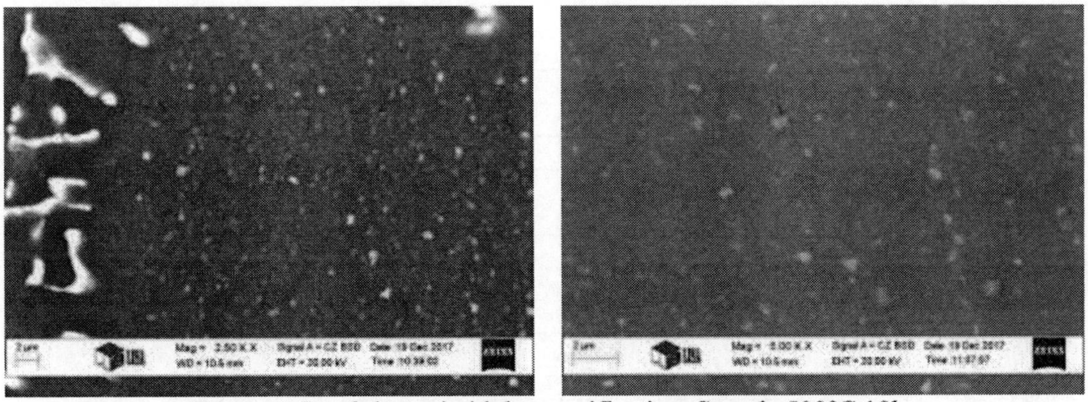

Figure 6. Precipitates in high magnification, Sample 500°C 10h

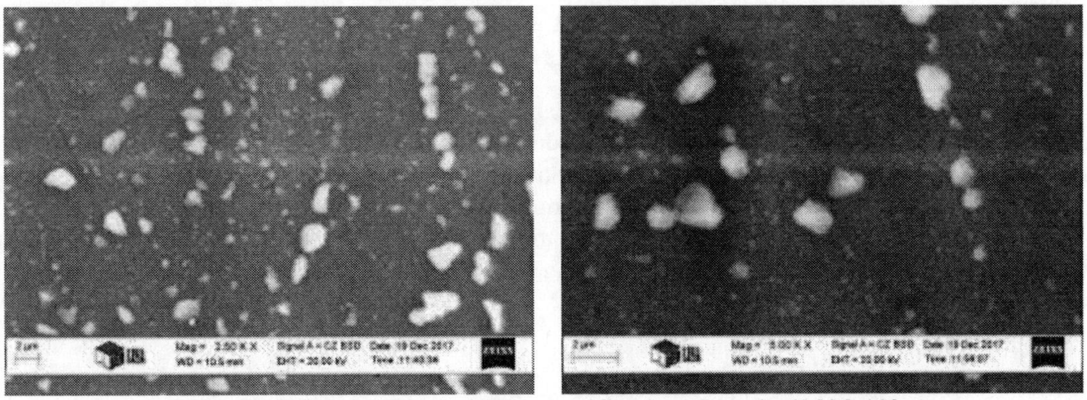

Figure 7. Precipitates in high magnification, Sample 600°C 10h

In the sample homogenised at 600°C, small precipitations of a size of about ~100-200 nm can be seen in an uneven distribution around the larger spherical particle of a size of ~1.5-2 μm but grains falling between the two sizes can hardly be observed. The signs of coarsening can obviously be seen in the microstructure: the bigger particles are grown, while the smaller precipitates are dissolved and the primary phases start to solve; it is shown by the change of curvature and spheroidization of the phase boundary surface. The presence of tiny particles shows the traces of precipitation process which proceeds parallel with the coarsening, but the coarsening process is dominant (Figure 7.).

The particle diameters measured in different areas in the homogenised samples are shown in Table 2.

Table 2. Particle diameters in the homogenised samples

Sample	Examined area	Particle diameter [μm]	Average particle diameter [μm]
500°C 10h	Field of view No. 1	0.826	0.890
		0.768	
	Field of view No. 2	0.945	
		0.903	
	Field of view No. 3	1.049	
		1.104	
	Field of view No. 4	0.744	
		0.784	
600°C 10h	Field of view No. 1	2.574	2.380
		2.727	
	Field of view No. 2	2.694	
		1.971	
	Field of view No. 3	2.437	
		2.916	
	Field of view No. 4	2.020	
		1.703	

4. Summary

It can be stated that the correlative microscopy is suitable for following and studying the processes taken place in aluminium alloys during homogenization. The correlative microscopy method provides an opportunity for detecting the precipitation, coarsening and dissolving processes too.
In the 3003 alloy, the precipitates appear at a homogenization temperature of 500°C, the number of particles increase by increasing the time of homogenization. The coarsening of precipitates starts at a homogenization temperature of 600°C, the size of particles increases by increasing the time of homogenization.

Acknowledgement
The study was carried out as part of the GINOP-2.2.1-15-2016-00018 project implemented in the framework of the Szechenyi 2020 program. The realization of this project is supported by the European Union.

References

[1] Merchant H.D, Morris J.G. and Hodgson D.S. 1990 *Mater. Character.* **25** 339
[2] Alexander D.T.L. and Geer A.L. 2002 *Acta Mater.* **50** 2571
[3] Muggerud A.M.F et al. 2013 *Mat. Sci. Eng. A* **567** 21
[4] Dehmas M. et al 2005 *J. Alloy Comp.* **400** 116
[5] Li T. 2011 Rok
[6] Huang H-W and Ou B-L 2009 *Mater. Design* **30** 2685

IOP Conf. Series: Materials Science and Engineering **426** (2018) 012038 doi:10.1088/1757-899X/426/1/012038

The variation of earing and texture components during annealing treatments of a 3003-type aluminium alloy

D Petho[1], A Hlavacs[1], P Barkoczy[1], M Benke[1]

1 University of Miskolc, Institute of Physical Metallurgy, Metalforming and Nanotechnology, Miskolc-Egyetemváros, Hungary

E-mail: femdani@uni-miskolc.hu

Abstract. In this research work the effect of manufacturing parameters on the relationship between the developed crystallographic texture and the anisotropy characteristic earing was studied for a 3003 aluminium alloy, which alloy is broadly applied worldwide. The parameters of the cold rolling subsequent annealing heat treatments were varied and the developed texture and earing were studied. The initial material was a hot rolled sheet, which was cold rolled to the final thickness ($\varepsilon \cong 1.94$), then the samples were heat treated at 190°C, 280°C and 380°C, for 1 to 5 hours. Afterwards XRD examinations were used to evaluate the crystallographic texture. The earing numbers were determined by an industrial partner. The low-temperature annealing resulted in increased rolling texture due to the recovery of the sheet. At temperatures of 280°C the time-dependence of the recrystallization can be seen. Thus, on these temperatures the recrystallization was PSN. At 380°C the classic high-angle boundary movement involved recrystallization occurred. It can be seen altogether that raising the temperature, or the time duration of the annealing generally causes the 45°-type earing index number to gradually decrease. The results provide a good base for the optimization of the given technology, moreover it provides data for the simulation based researches.

1. Introduction

In the most cases the aluminium sheets, these semi-products are deep-drawn. Therefore, the critical property of the sheets, is the deep-drawability. This is defined explicitly by the crystallographic texture. The crystallographic texture describes the crystallographic orientation of all grains in the material, thus it defines the degree of anisotropy. This is an important aspect, because currently, in most industrial environments, the effectivity of the annealing is characterized by the hardness, which does not completely describe the behaviour of the material.

The main alloying component in the 3003 aluminium alloy is the manganese. During heat-treatments, raising the temperature causes the dissolution of the manganese from the solid solution, forming compounds. These precipitations have effect on the recrystallization occurring during annealing, hence they are nucleating sites

This aim of this research is to characterize the kinetics of the physical metallurgical processes and to link it to the earing, the index number of the formability.

2. Examinations

During experiments a 7 mm thick hot-rolled sheet was cold-rolled to 1 mm thickness. This was followed by annealing heat treatments at three different temperatures (190°C, 280°C, 380°C) for five different time durations (1-5 h). On Figure 1. the route of the examinations can be seen. The heat treatments were carried out in an air atmosphere furnace: The sheets were put into the furnace, heated up to the annealing temperature and after the heat treatments the sheets were taken out and cooled down in room temperature air.

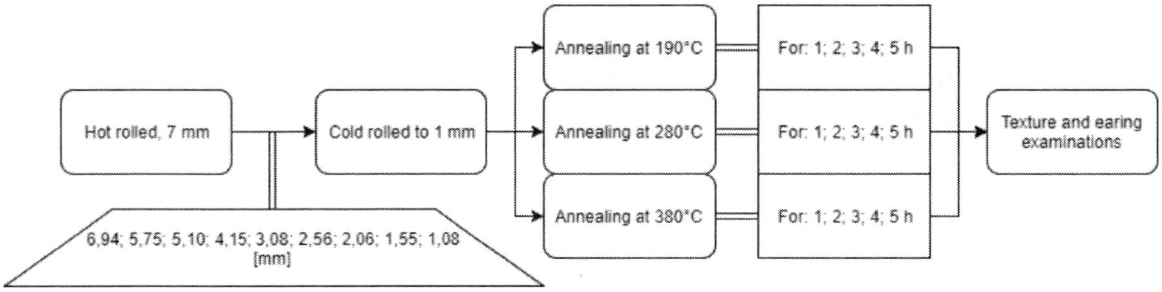

Figure 1. Route of the examinations.

The texture examinations were carried out on the Bruker D8 Advance X-ray diffractometer equipped with an Eulerian cradle. The parameters of the measurement were: 40mA, 40kV. The radiation is CoKα. The samples cut from the sheets were 30x30 mm in dimensions. During the examinations, the Bragg angles of the {111}, {200}, {220} lattice planes were measured and the corresponding pole figures were recorded. The texture evaluations were carried out with the TexEval program. The pole figures of the prementioned lattice planes were recalculated and provided the base for the Orientation Distribution Function calculation. ODF cuts were made at constant φ_2 values with 5 degree steps. The texture component volume fractions were defined by φ_1, Φ and φ_2 angles.

Earing examinations were carried out by the ARCONIC-Köfém Kft. The earing examinations were carried out at the laboratory at Szekesfehervar and results were sent back to evaluation. Besides the industrial measurement of earing, the earing tendency was also calculated with a method proposed by the authors and described in detail in Ref. [1]. Briefly, the method is based on defining the individual texture component poles as the vectorial combination of the (+) and (-) directions.

3. Results

On Figures 2-4. the volume fractions of the texture components are shown at constant annealing temperatures as the function of annealing time. The texture components resulting from the forming are: C, S, B; while the texture components resulting from the annealing are: G, Cube, R, P. The index numbers of the earing are also shown with an „x" on the same diagrams for comparison. The texture components of the rolled state can be seen at 0 h annealing.

Fig. 2. shows the samples annealed at 190°C. The texture components resulting from the forming unequivocally dominate the texture. It can be seen that the time duration of the heat treatment at this temperature causes no significant change in the texture or in the index number of the earing (rolled).

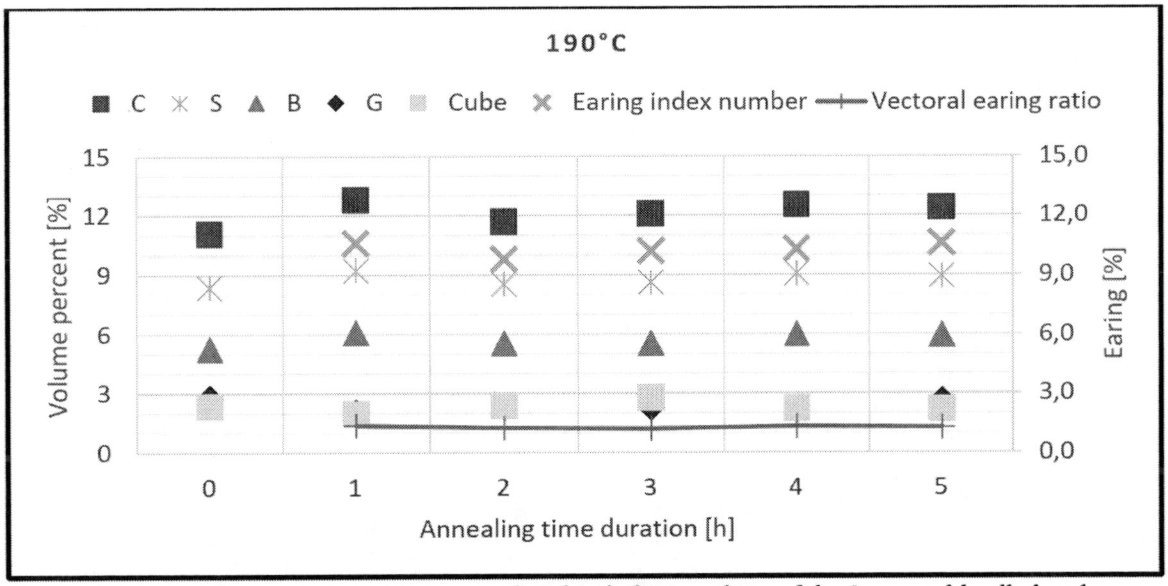

Figure 2. The texture components and earing index numbers of the 1 mm cold-rolled and 190°C annealed samples.

Fig. 3. shows the results of the samples annealed at 280°C. It can be seen that changing the duration of the annealing causes substantial change (contrary to Fig. 2.) in the texture of the samples. After one hour annealing, the rolling components dominate, furthermore, they are increased compared to the rolled state. Accordingly, the degree of earing (rolled) is high. The samples annealed for two and three hours show a considerable decrease in the rolling components and parallelly an increase of the recrystallization components. Proportionately, the index number of the earing decreases. After annealing of four and five hours compared to the three hours state no change is visible in the earing and in the texture components.

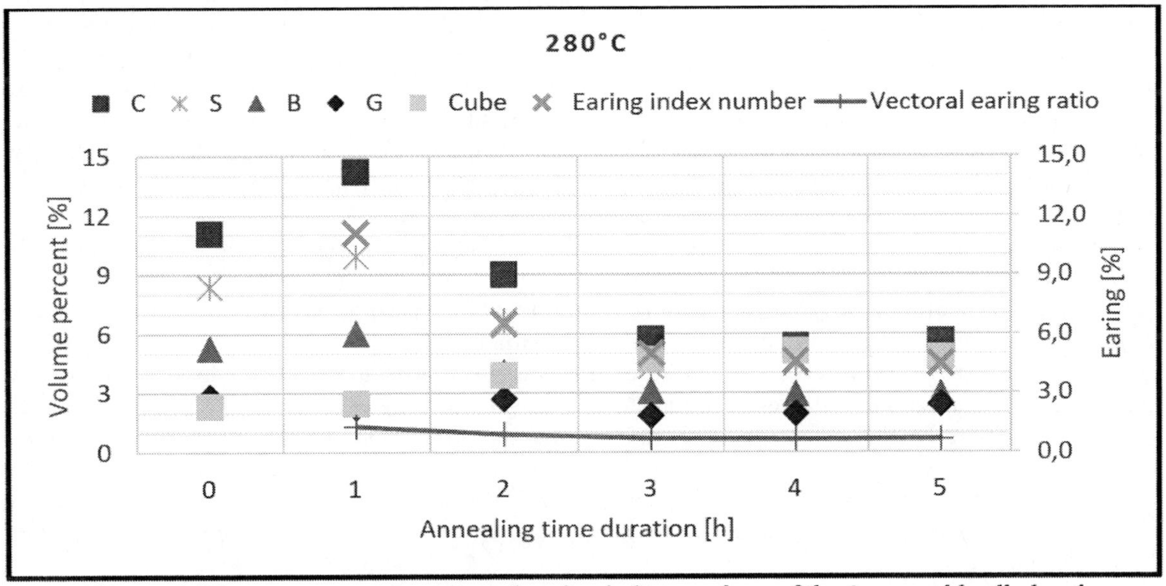

Figure 3. The texture components and earing index numbers of the 1 mm cold-rolled and 280°C annealed samples.

Fig. 4. shows the results of samples annealed at 380°C. It can be seen that at all annealing durations, the recrystallization texture-components have the highest volume percentage and the texture does not change significantly. The earing index number also does not show significant change in function of annealing time and reaches the lowest number (around 3%) throughout the study.

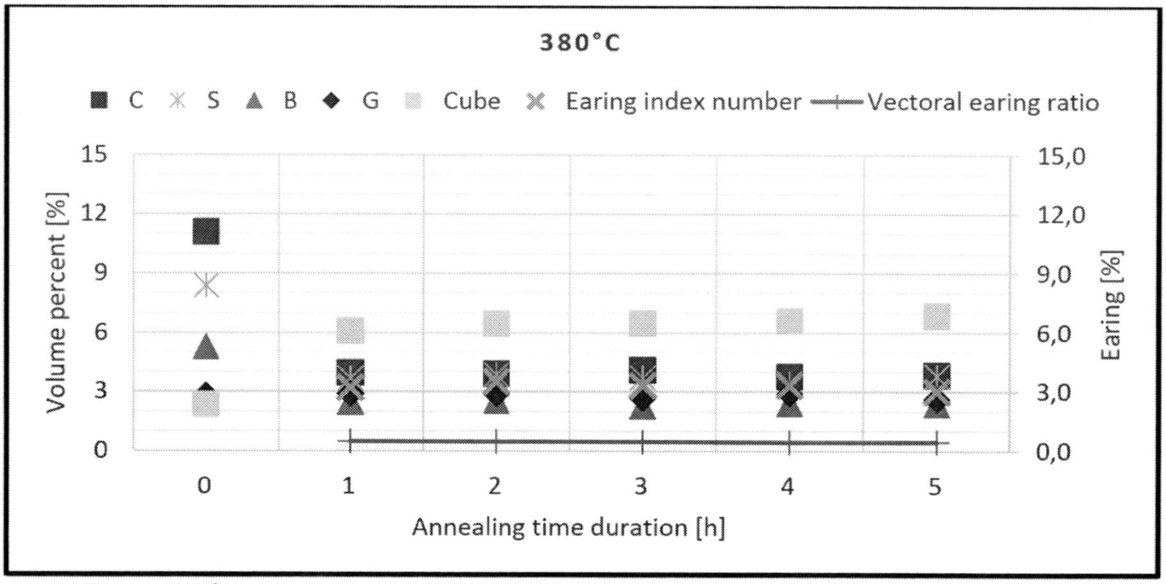

Figure 4. The texture components and earing index numbers of the 1 mm cold-rolled and 380°C annealed samples.

4. Discussion

It can be stated that during the annealing of the 3003 aluminium alloy at 190°C only recovery occurred. It is based on that even after five hours of annealing, the rolling texture-components remained the strongest and were larger compared to the rolled state, while the recrystallization texture-components remained at low values. The reason behind this is that during recovery, the dislocations are arranged to form tilt boundaries (polygonization) thus forming subgrains. This arrangement also yields a close to dislocation-free volume inside the formed subgrain, oriented more uniformly. It means that after recovery, the rolling texture-components have higher volume percentage than before annealing. Furthermore, it was observed that the earing values are in agreement with the texture-components and they were characteristic of rolling.

At 280°C, the rolling texture evolved into an almost isotropic texture during in annealing. The process can be followed in detail with the variation of the volume fraction of the texture-components as a function of annealing time: First, the rolling components increased after one hour annealing due to recovery, then they decreased and the recrystallization components slightly increased. Hence, even after five hours of annealing the recrystallization texture was not dominant. Thus, it can be stated that at 280°C the recrystallization mechanism produced an almost texture-free structure. This mechanism is the Particle Stimulated Nucleation (PSN). The analysis of the texture components results very detailed and quantitative outcome of this process. The volume fraction of the components in between three and five hours did not change notably. This data shows, that at 280°C the PSN type recrystallization peaks at three hours. The earing values decrease in between one and three hours of annealing, afterwards their value stagnates.

At 380°C annealing, the sheets recrystallized at all annealing time durations, even after one hour. The recrystallization texture-components dominate, which indicates that at this temperature, the classic high-angle boundary movement involved recrystallization was activated. Hence, the recrystallization components are not too high (~6-7%), which suggests that besides high angle recrystallization, the PSN mechanism was also activated. The texture-components show, that at 380°C after one hour of annealing, recrystallization takes place and no notable change can be observed in the texture at longer time durations (2-5 h). Comparing the sheet annealed for one hour with the sheet annealed at 280°C for five hours, it can be seen that the recrystallization texture components are higher and the rolling components are lower in the 380°C annealed sample. It can be concluded that the effect of annealing temperature cannot be compensated with longer annealing durations at a lower temperature. The earing values were the lowest in the samples annealed at 380°C. This corresponds well to that the rolling components were the lowest at the 380°C samples.

5. Conclusions

In this research, the evolution of texture and earing were characterized at three different temperatures in the function of annealing time in the 3003 aluminium alloys. The chosen temperatures are based on industrial heat treating temperatures. The conclusions of the texture and earing development, influenced by the annealing heat treatment are the following:

- At 190°C, at all the practical annealing time durations, only recovery occurs, which causes rolling texture and strong earing.
- At 280°C, recovery occurs after one hour annealing, which is followed by PSN recrystallization, which causes a texture-free structure and homogeneous formability after three hours of heat treatment.
- At 380°C annealing, even after one hour, classic recrystallization, involving high angle grain boundary movement is the dominant process, although PSN still occurred at a lesser extent. The texture and earing do not change with increasing annealing time. The result is a weak recrystallized texture and a marginally inhomogenity of the forming. Considering the deep-drawability, the annealing at 380°C results the most favourable structure and properties.

Acknowledgement

The described research work was supported through the National Research, Development and Innovation Office – NKFIH K119566 project. The described article was carried out as part of the GINOP-2.2.1-15-2016-00018 project implemented in the framework of the Szechenyi 2020 program. The realization of this project is supported by the European Union.

References:

[1] Benke M, Hlavacs A, Petho D, Angel D A, Sepsi M, Nagy E and Mertinger V 2018 A simple correlation between texture and earing *Institute of Physics (IOP) Conf. Series: Materials Science and Engineering*; preprint

IOP Conf. Series: Materials Science and Engineering **426** (2018) 012039 doi:10.1088/1757-899X/426/1/012039

On the possible generalization of JMAK-type kinetic functions

T Réti[1], A M Deus[2], I Felde[1] and M Réger[1]

[1] Óbuda University, Budapest, Hungary
[2] University of Lisbon, Lisbon, Portugal

E-mail: reti.tamas@bgk.uni-obuda.hu

Abstract. The Johnson-Mehl-Avrami-Kolmogorov (JMAK) kinetic equation is generally used for describing the progress of solid state diffusion-controlled transformations. The traditional JMAK model is applicable for modeling only monophase decomposition (precipitation) processes. This paper deals with possible generalizations of the classical JMAK kinetic model by taking into consideration the simultaneous occurrence of more parallel transformations. It is revealed that the Jones-Bhadeshia kinetic model belonging to the family of the strongly coupled multiphase models has only a limited practical application because it contains only few fitting parameters. Moreover, it has been shown and demonstrated that for predicting decomposition processes it is more advantageous to use a multiphase kinetic model including a finite set of ordinary autonomous differential equations.

1. Introduction, preliminary considerations

For the phenomenological description of diffusion-based transformation processes occurring in alloys various kinetic equations have been suggested in the literature [1-7]. The most popular, widely used kinetic model was established in the late 1930s, independently by Johnson and Mehl, Avrami and Kolmogorov [1]. It is referred to as the Johnson-Mehl-Avrami-Kolmogorov (JMAK) transformation model, and the corresponding kinetic equation is represented by

$$X(t) = 1 - \exp\left\{-[Kt]^n\right\} \tag{1}$$

where X=X(t) is the transformed fraction at time t, K is the reaction rate constant, and n is the Avrami exponent. In Eq.(1) the reaction rate constant K is given by

$$K = K(T,E) = A\exp\left(-\frac{E}{RT}\right) \tag{2}$$

where T is the absolute temperature in Kelvin, E is the activation energy, A is the frequency factor and R is the universal gas constant. For constant temperature, the transformation rate can be derived by differentiating Eq.(1) with respect to time

$$\frac{dX}{dt} = K^n(1-X)nt^{n-1}. \tag{3}$$

Several attempts have been made to generalize equations (1) and/or (3) to treat two or more transformation processes (phase decomposition processes) occurring simultaneously [8-14].

A detailed comparative analysis of various decomposition models devoted to the extension of the traditional JMAK kinetic equation to the prediction of simultaneous transformations is given in [7].

2. A possible extension of the Jones-Bhadeshia multiphase kinetic model

In what follows starting with the concept outlined in Refs. [9-12], we investigate some possible extensions of the isothermal kinetic model represented by Eq.(3). Let us assume that there exists a parent phase γ which decomposes simultaneously into J distinct phases (α_1, α_2,…,α_J) during isothermal transformation and this precipitation process is described by an ordinary differential equation system

$$\frac{dX_j}{dt} = f(X_1, X_2, \ldots X_J) g_j(t) . \tag{4}$$

The transformed fractions are denoted by X_j and $f(X_1, X_2,\ldots X_J)$ and $g_j(t)$ are appropriately selected non-negative real functions. To generate the solution of the above differential equation system, we have introduced some simplifying hypotheses. It is assumed that in Eq.(4)

$$f(X_1, X_2, \ldots X_J) = \left[1 - \sum_{j=1}^{J} X_j\right]^p \qquad \text{and } g_j(t) = K_j g(t)$$

by definition, where p and K_j are positive numbers, g(t) is a positive function of time if $t > 0$.

As a result of a simple transformation, we obtain a system of non-linear ordinary differential equations defined as

$$\frac{dX_j}{dt} = K_j \left\{1 - \sum_{j=1}^{J} X_j\right\}^p g(t) . \tag{5}$$

The set of ordinary differential equations (5) represents a strongly coupled multiphase kinetic model (SCMK model) because

$$\frac{dX_j}{dt} = \frac{K_j}{K_i}\frac{dX_i}{dt} \qquad \text{for any } i,j > 1.$$

This implies that functions $X_j(t)$ can be directly calculated by

$$X_j(t) = \frac{K_j}{K_1} X_1(t) \text{ for } j > 1. \tag{6}$$

Assuming that parameters p, K_j and function g(t) are appropriately defined, the differential equations (5) can be rewritten in the following alternative form:

$$\frac{dX_j}{dt} = K_j \left\{1 - \frac{X_j}{K_j} \sum_{i=1}^{J} K_i\right\}^p g(t) \tag{7}$$

Based on the concept outlined previously it is assumed that the solutions $X_j(t)$ of the differential equation system satisfy the following conditions:

i) $X_j(0)=0$, $\lim X_j(t)=Y_j > 0$ if $t \to \infty$ and $\sum_{J=1}^{J} Y_j = 1$,

ii) $\frac{dX_j}{dt} = \frac{K_j}{K_i}\frac{dX_i}{dt}$ and $X_j(t) = \frac{K_j}{K_i} X_i(t) = \frac{Y_j}{Y_i} X_i(t)$ for $1 \leq i, j \leq J$,

iii) $\sum_{j=1}^{J} X_j(t) = \frac{X_1(t)}{Y_1} \sum_{j=1}^{J} Y_j = \frac{X_1(t)}{Y_1}$,

iv) $\sum_{j=1}^{J} K_j(t) = \frac{K_1}{Y_1} = \frac{K_2}{Y_2}, \ldots, = \frac{K_J}{Y_J}$,

where Y_j stands for the maximum transformed fractions for which $\sum Y_j = 1$ holds.

If the above restrictions are fulfilled, the differential equations (7) can be transformed into a simplified form. As a particular case, the transformation rate for the phase α_1 can be calculated as

$$\frac{dX_1}{dt} = K_1\left\{1 - \frac{X_1}{K_1}\sum_{j=1}^{J} K_j\right\}^p g(t) = K_1\left\{1 - \frac{X_1}{Y_1}\right\}^p g(t) \tag{8}$$

It is important to emphasize that if $X_1(t)$ is known, then $X_j(t)$ for $1 < j \leq J$ can be simply generated by

$$X_j(t) = \frac{Y_j}{Y_1} X_1(t). \tag{9}$$

3. Investigation of the SCMK models

In what follows, starting with Eq. (8), we determine the solutions of differential equations included in the SCMK model for some particular cases.

a) ***The case of p=1.*** One obtains that

$$\frac{dX_1}{dt} = K_1\left\{1 - \frac{X_1}{Y_1}\right\} g(t) \tag{10}$$

where

$$g(t) = nt^{n-1} \tag{11}$$

and $n \geq 1$ is a positive number. Then we have

$$\int_0^{X_1} \frac{du}{1 - u/Y_1} = Y_1 \int_0^{X_1} \frac{du}{Y_1 - u} = K_1 \int_0^t g(\tau) d\tau = K_1 t^n.$$

It follows that

$$X_1(t) = Y_1\left[1 - \exp\left\{-\frac{K_1}{Y_1}\int_0^t g(\tau)d\tau\right\}\right] = Y_1\left[1 - \exp\left\{-\frac{K_1}{Y_1}t^n\right\}\right]$$

and

$$X_j(t) = \frac{Y_j}{Y_1} X_1(t) = Y_j\left[1 - \exp\left\{-\frac{K_1}{Y_1}t^n\right\}\right].$$

b) ***The case of p= 2*** In this case we get

$$\frac{dX_1}{dt} = K_1\left\{1 - \frac{X_1}{Y_1}\right\}^2 g(t) = K_1\left\{1 - \frac{X_1}{Y_1}\right\}^2 nt^{n-1}$$

and

$$\int_0^{X_1} \frac{du}{(1 - u/Y_1)^2} = Y_1^2 \int_0^{X_1} \frac{du}{(Y_1 - u)^2} = K_1 \int_0^t g(\tau) d\tau = K_1 t^n.$$

Because

$$Y_1^2 \int_0^{X_1} \frac{du}{(Y_1 - u)^2} = Y_1^2\left\{\frac{Y_1}{Y_1 - X_1} - 1\right\}$$

this implies that

$$\frac{Y_1 - X_1}{Y_1} = \left\{1 + \frac{K_1}{Y_1}\int_0^t g(\tau)d\tau\right\}^{-1} = \left\{1 + \frac{K_1}{Y_1}t^n\right\}^{-1}.$$

Consequently,

$$X_1(t) = Y_1\left[1 - \frac{Y_1}{Y_1 + K_1 t^n}\right].$$

It is easy to check that

$$X_j(t) = \frac{Y_j}{Y_1} X_1(t) = Y_j\left[1 - \frac{Y_1}{Y_1 + K_1 t^n}\right],$$

and lim $X_j(t) = Y_j$ holds if $t \to \infty$.

c) The case of p=1/2.

It is interesting to note that if p=1/2, then for $X_1(t)$ the condition lim $X_1(t) = Y_1$ is not fulfilled when t tends to infinity. Consider the kinetic differential equation given as

$$\frac{dX_1}{dt} = K_1\sqrt{1 - \frac{X_1}{Y_1}} n t^{n-1}$$

If p=1/2, it follows that

$$\int_0^{X_1} \frac{du}{\sqrt{1 - u/Y_1}} = \sqrt{Y_1}\int_0^{X_1} \frac{du}{\sqrt{Y_1 - u}} = K_1 \int_0^t n\tau^{n-1} d\tau = K_1 t^n.$$

But

$$\sqrt{Y_1}\int_0^{X_1} \frac{du}{\sqrt{Y_1 - u}} = 2\sqrt{Y_1}\left\{\sqrt{Y_1} - \sqrt{Y_1 - X_1}\right\} = K_1 t^n \quad (12)$$

consequently,

$$X_1(t) = Y_1 - \left\{\sqrt{Y_1} - \frac{K_1 t^n}{2\sqrt{Y_1}}\right\}^2 = Y_1\left\{1 - \left[1 - \frac{1}{2}\frac{K_1}{Y_1} t^n\right]^2\right\}. \quad (13)$$

By definition $0 \le X_1 \le Y_1$. From Eq.(12) and Eq.(13) it follows that if $X_1 = Y_1$ then $2Y_1 = K_1 t_1^n$ holds, consequently $t_1 = (2Y_1 / K_1)^{1/n}$. This observation implies that if t tends to infinity, then $\lim X_1(t) = -\infty$ holds.

4. Correspondence between the Jones-Bhadeshia model and the SCMK models

In what follows, the relations between the Jones-Bhadeshia kinetic model and the strongly coupled multiphase kinetic models are investigated. The Jones-Bhadeshia kinetic model describes the isothermal simultaneous precipitation of phases α_1 and α_2 from a parent phase [9]. The corresponding volume fractions of two product phases are denoted by X_1 and X_2. In the model the microstructural parameter χ plays a decisive role. It is defined by

$$\chi = \frac{G_2^3 I_2}{G_1^3 I_1} \quad (14)$$

where G_1, G_2, I_1 and I_2 are growth and nucleation rate of precipitating phases α_1 and α_2, respectively. Moreover it is assumed that both precipitating phases are represented by spherical particles, their volume fractions $V_1(t)$ and $V_2(t)$ are given by

$$V_i(t) = \frac{4\pi}{3} G_i^3 (t - \tau)^3$$

for time t if $t > \tau$. According to Jones and Bhadeshia the precipitated fractions of phases α_1 and α_2 can be calculated as

$$X_1(t) = \left(\frac{1}{1+\chi}\right)\left\{1 - \exp\left[-\frac{1}{3}(1+\chi)\pi G_1^3 I_1 t^4\right]\right\} \tag{15}$$

$$X_2(t) = \left(\frac{\chi}{1+\chi}\right)\left\{1 - \exp\left[-\frac{1}{3}(\frac{1+\chi}{\chi})\pi G_2^3 I_2 t^4\right]\right\} \tag{16}$$

Now, for comparison, consider the SCMK model represented by Eq.(8). Supposing that J=2, p=1 and n=4, then in this particular case, the corresponding set of kinetic differential equations will be

$$\frac{dX_1}{dt} = 4K_1\left\{1 - \frac{X_1}{Y_1}\right\}t^4 \tag{17}$$

$$\frac{dX_2}{dt} = 4K_2\left\{1 - \frac{X_2}{Y_2}\right\}t^4 \tag{18}$$

By solving the above system of differential equations one obtains that

$$X_1(t) = Y_1\left\{1 - \exp\left[-\frac{K_1}{Y_1}t^4\right]\right\} \tag{19}$$

$$X_2(t) = Y_2\left\{1 - \exp\left[-\frac{K_2}{Y_2}t^4\right]\right\} = Y_2\left\{1 - \exp\left[-\frac{K_1}{Y_1}t^4\right]\right\}. \tag{20}$$

Introduce the following parameters defined by

$$Y_1 = \frac{1}{1+\chi}, \qquad Y_2 = 1 + Y_1 = \frac{\chi}{1+\chi},$$

$$\frac{K_1}{Y_1} = \frac{1}{3}(1+\chi)\pi G_1^3 I_1, \text{ and}$$

$$\frac{K_2}{Y_2} = \frac{1}{3}\left(\frac{1+\chi}{\chi}\right)\pi G_2^3 I_2 = \frac{1}{3}(1+\chi)\pi G_1^3 I_1 = \frac{K_1}{Y_1}.$$

Based on the above formulas it is easy to see that the two kinetic models represented by Eqs.(15-16) and Eqs.(19-20) are equivalent, consequently, the Jones-Bhadeshia multiphase model can be considered as a special type of SCMK models.

Example 1 Consider the Jones-Bhadeshia kinetic model outlined in Ref. [9].
Let J=2, p=1, n=4, χ=8,

$$Y_1 = \frac{1}{1+\chi} = \frac{1}{9}, \qquad Y_2 = 1 - Y_1 = \frac{\chi}{1+\chi} = \frac{8}{9},$$

and C = K_1/Y_1 = K_2/Y_2=1/ 625. For the transformed fractions of phases α_1 and α_2 we get

$$X_1(t) = Y_1\left\{1 - \exp\left[-\frac{K_1}{Y_1}t^4\right]\right\} = \frac{1}{9}\left\{1 - \exp\left[-\frac{1}{625}t^4\right]\right\} \text{ and}$$

$$X_2(t) = \frac{Y_2}{Y_1}X_1(t) = 8X_1(t)$$

If the time is equal to $t_X = (Y_1/K_1)^{1/4} = 1/C^{1/4} = 5$, in this particular case, one obtains that

$$X_1(t_X) = Y_1\{1 - \exp[-1]\} = 0{,}63212(1/9) = 0{,}07024,$$
$$X_2(t_X) = Y_2\{1 - \exp[-1]\} = 0{,}63212(8/9) = 0{,}561885.$$

Example 2 Let J=3, Y_1=Y_2=1/4, Y_3 =1/2, p=1, n=4, K_j=CY_j and C=1/625. Then the solutions relating to the corresponding SCMK model are:

$$X_j(t) = Y_j\left\{1 - \exp\left[-\frac{K_j}{Y_j}t^4\right]\right\} = Y_j\{1 - \exp[-Ct^4]\}.$$

Example 3 Let J=2, Y_1=1/3, Y_2 =2/3, p=1/2, n=4, K_j=CY_j and C=1/648. Then

$$X_j(t) = Y_j - \left\{\sqrt{Y_j} - \frac{K_j t^4}{2\sqrt{Y_j}}\right\}^2 = Y_j\left\{1 - \left[1 - \frac{1}{2}Ct^4\right]^2\right\}.$$

Additionally, it follows that $t_1 = (2Y_j/K_j)^{1/4} = (2Y_1/K_1)^{1/4} = (2/C)^{1/4} = (2*648)^{1/4} = 6$, and $X_1(6)=Y_1=1/3$, $X_2(6)=Y_2=2/3$.

5. Autonomous multiphase kinetic models

In SCMK models the independent parameters are: $J \geq 2$, $Y_1 > 0$, $Y_2 > 0, \ldots, Y_{J-1} > 0$, $p > 0$, $n \geq 1$, and $K_1 > 0$. This means that if $J \geq 2$ is a fixed positive integer, then the number of freely selected model parameters is $J - 1 + 3 = J + 2$. From this observation it follows that the SCMK models contain only few free fitting parameters, and due to this fact, it is not expected that corresponding kinetic equations can be successfully applied to describe parallel multiphase decomposition processes. Based on our preliminary results obtained by some comparative tests, instead of the previously outlined SCMK models we suggest using another version of the isothermal multiphase decomposition models of JMAK type. This is called the autonomous multiphase kinetic model (*AMK model*) because it is represented by a finite set of autonomous ordinary differential equations:

$$\frac{dX_j}{dt} = m_j B_j^{1/m_j}(Y_j - X_j)\left\{\ln\frac{Y_j}{Y_j - X_j}\right\}^{1-1/m_j} \qquad (21)$$

where m_j, B_j and Y_j are positive numbers, and as usual, Y_j stand for the maximum transformed fractions for which $\sum Y_j = 1$ holds.

At constant temperature, if $X_j(0)=0$, the solutions of above kinetic differential equations are

$$X_j(t) = Y_j\{1 - \exp[-B_j t^{m_j}]\} \qquad (22)$$

for j= 1,2,…J. Investigating the characteristic features of AMK models including differential equations (21), the following conclusions can be drawn:

- The isothermal kinetic equations (22) represent a multiphase decomposition model where the simultaneous precipitation reactions are independent, and the progress of jth precipitation process does not depend on the degree of transformation of the others.

- In the particular case, if J=1 holds, the differential equation (21) is identical to the JMAK kinetic differential equation extended to the description of non-isothermal transformation processes [1,2].

- In the multiphase model represented by Eqs. (21 and 22) the independent parameters are: $J \geq 2$, $Y_1 > 0$, $Y_2 > 0, \ldots, Y_{J-1} > 0$, $n_1 > 0$, $n_2 > 0, \ldots, n_J > 0$, $B_1 > 0$, $B_2 > 0, \ldots, B_J > 0$. Consequently, if $J \geq 2$ is a fixed positive integer, then the number of freely selected model parameters is relatively large, i.e. $J - 1 + 2J = 3J - 1$.

Our preliminary tests have verified that the AMK model represented by Eq.(22) is successfully applicable for describing the isothermal austenite decomposition into ferrite and pearlite.

6. Using the AMK model for the prediction of isothermal austenite decomposition in a hypoeutectoid steel

Consider the isothermal decomposition of austenite in low alloy 34Cr4 steel [13]. The unknown model parameters (Y_j , n_j and B_j) of simultaneous ferrite and pearlite precipitations can be estimated from the measured isothermal transformation diagrams (IT diagrams), where times t_s and t_f characterize the start and finish of the isothermal austenite transformation as a function of temperature T plotted in a form of C-curves. According to the classical definition of C-curves, t_s and t_f stand for the times belonging to 1% and 99% of relative transformed fractions. For the computation of model parameters n_j and B_j characterizing the jth precipitation reaction, the following formulas are used [13]

$$n_j(T) = \frac{6{,}1273}{\ln(t_{fj}/t_{sj})}$$

and

$$B_j(T) = \frac{0{,}01005}{t_{sj}^{n_j(T)}}.$$

If T= 650 C, then in this particular case the kinetic parameters of ferrite precipitation are Y_1=0,25, t_{s1} = 2 sec, t_{f1}= 35 sec, and for the corresponding parameters of pearlite precipitation Y_2= 0,75, t_{s2}=15 sec, t_{f2}= 90 sec yield. From these data one obtains that n_1= 2,1408, B_1= 2,2789 $*10^{-3}$, n_2= 3,4197 and B_2=9,5559$*10^{-7}$, respectively.

The computed results relating to the simultaneous ferrite and pearlite precipitations at 650 °C are demonstrated in Fig. 1.

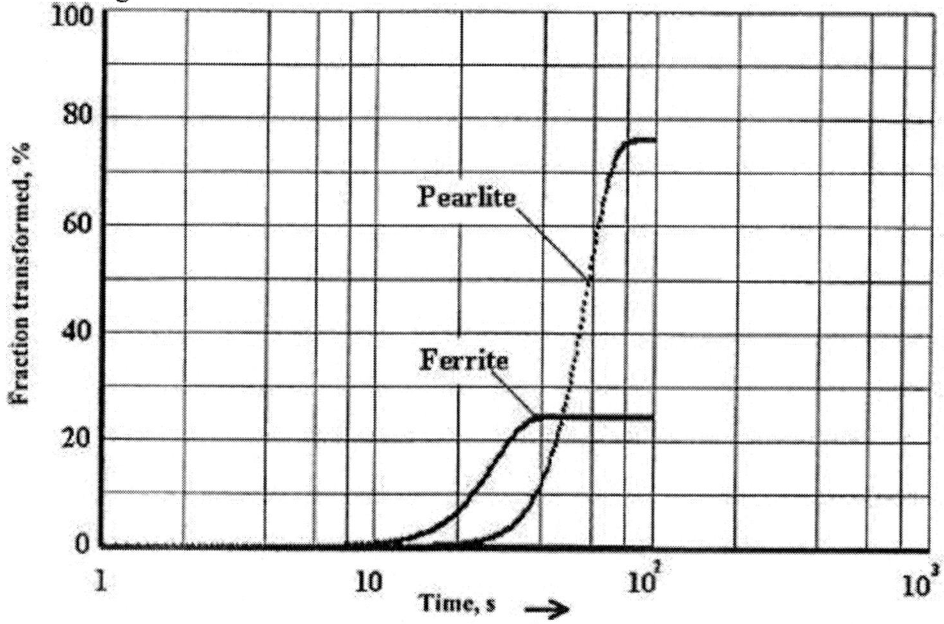

Figure 1 Predicted isothermal decomposition of austenite into ferrite and pearlite at 650 C

An important property of the autonomous decomposition model is that, disregarding the equality $\sum Y_j = 1$, the differential equations included in the kinetic model are considered to be independent. It should be noted that an advantage of using the autonomous decomposition model is that it can be applied to the prediction of isothermal and non-isothermal diffusion controlled transformation processes.

7. Conclusions

The multiphase SCMK kinetic model represented by Eq.(7) incorporates the classical JMAK kinetic model if equalities J=1, p=1 and $g(t)=nt^{n-1}$ are fulfilled. As a particular case, if J=1, p=1 and $g(t)=nt^{n-1}$ hold, the Austin-Rickett kinetic equation is obtained [4]. As it is known, the model parameter p represents the so-called impingement exponent related to the fraction untransformed [2].
Comparing the SCMK and AMK kinetic models we can conclude that the practical application of AMK models seems to be more advantageous because they contain more free fitting parameters.

Acknowledgements

We acknowledge the financial support of this work by the Hungarian State and the European Union under the EFOP-3.6.1-16-2016-00010 project and the Hungarian-Portuguese bilateral Scientific and Technological (TÉT_16-1-2016-0097) project.

References

[1] Christian J W 1975 The Theory of Transformations in Metals and Alloys, Part 1, 2nd edition, Pergamon Press, Oxford
[2] Lee E-S and Kim Y G 1990 A transformation kinetic model and its application to Cu-Zn-Al shape memory alloys-I. Isothermal conditions, Acta Metall. Mater. **38** 1669-1676.
[3] Hsiao B S et al 1991 Isothermal crystallization kinetics of poly(ether ketone ketone) and its carbon-fibre-reinforced composites, Polymer, **32** 2799-2806.
[4] Starink M J 1997 Kinetic equations for diffusion-controlled precipitation reactions, J. Mater. Sci., **32** 4061-4070.
[5] Bishnoi S 2013 Geometric limitations of nucleations and growth models: Revisiting the impingement assumption, Cem. Concr. Res. **46** 30-40.
[6] Blázquez J S et al 2012 Extension of the classical theory of crystallization to non-isothermal regimes: Application to nanocrystallization processes, J. Alloys Compd. **544** 73-81.
[7] Blázquez J S et al 2015 On the use of classical JMAK crystallization kinetic theory to describe simultaneous processes leading to the formation of different phases in metals, Int. J. Therm. Sci. **88** 1-6.
[8] Luiggi N J and Betancourt A E 1994 Multiphase precipitation of carbides in Fe-C systems: Part I. Model based upon simple kinetic reactions, Metall. Mater. Trans. **B 25** 917-925.
[9] Jones S J and Bhadeshia H K D H 1997 Kinetics of the simultaneous decomposition of austenite into several transformation products. Acta Mater. **45** 2911-2920.
[10] Robson J D and Bhadeshia H K D H 1997 Modelling precipitation sequences in power plant steels Part 1 – Kinetic theory, Mater. Sci. Technol. **13** 631-639.
[11] Kasuya T et al 1999 Real and extended volumes in simultaneous transformations, Mater. Sci. Technol. **15** 471-473.
[12] Bhadeshia H K D H 1999 Kinetic of Simultaneous Transformations, in Proceedings of Solid-Solid Phase Transformations '99, (JIMIC-3), Eds. M. Koiwa, K. Otsuka and T. Miyazaki, Japan Institute of Metals, Kyoto, Japan 1445-1453.
[13] Reti T et al 2001 Computer simulation of steel quenching process using a multi-phase transformation model, Comput. Mater. Sci. **22** 261-278.
[14] Wozniak T 2005 Modeling kinetics of the austenite simultaneous decomposition into two bainitic products in the range of swinging back, Mater. Sci. Eng. **A 408** 309-316.

IOP Conf. Series: Materials Science and Engineering **426** (2018) 012040 doi:10.1088/1757-899X/426/1/012040

Comparison the structure of PVC foams with different additives

K Roman[1] and G Zsoldos[1]

[1] University of Miskolc, Institute of Ceramics- and Polymer Engineering, Miskolc-Egyetemváros, Hungary

E-mail: polkrisz@uni-miskolc.hu

Abstract: The effects of various types of fillers in the PVC structure can be determined by using Scanning Electron Microscope. By the results of microscope, the modifying effects of fillers can be examined. Due to the most common failures which can occur in the production detected from the records. At different dose of fillers the structural changes were examined. The results of SEM analysis show that the most common structural failure was caused by the mixing preparation (especially for $CaCO_3$ blends). The types of fillers influence the shape-, orientation- and open-close of foam cells. Adding fillers of different percentage changes the thickness of the cells walls and the size and orientation of the cells. Based on these tests, it is possible to determine which filler is the most suitable for the formation of regular cellular structure.

1. Introduction

In the last 20 years, polymer foams are used in many applications, such as insulation systems, furniture or decorative elements. There are numerous reasons for use this, such as good mechanical and insulating properties and the low cost. Other advantages are low density and weight; therefore easy to transport and build it. The main disadvantage is difficulty to break down, because it is made from PVC raw materials [1]. PVC processes needs different types of additives. These additives are stabilizers and co-stabilizers, lubricants, polymeric auxiliaries, plasticizers, heat resistance, pigments and fillers [2]. Based on the literature review, the mechanical properties of blends change according to different fillers [3]. The mechanical and structural properties of the foam mainly depend on the cell's morphology, such as size-, shape of cells, the wall thickness and the density of the materials [4]. It is well known that calcium carbonate fillers modify the impact properties of the polymer. Some studies were dealing with the effect of rise husks fillers of the PVC foam. The measurements, determines that the density of the non-filled foam material is lower than the foam with filler. If the filler content in the mixture is higher than the original PVC, than cells of composite were deformations. The original PVC does not include fillers ($CaCO_3$ and marble). These fillers ($CaCO_3$, marble, rise husk) are the most important additives in the PVC formulation if we wanted to change the form of structure [5]. It was observed that using surface – treated fillers in the material causes higher resistant to the external influences [6].

2. Materials and sample preparations

The raw material was the PVC from the BorsodChem Zrt. company. The component the suspension grade PVC (K value=58) and the other components of the mixture were: solid Ca-Zn based stabilizer, paraffin based external- and partial glycerol ester type internal lubricants, acrylic based processing aid 1.5 phr azodicarbonamide. The azodicarbonamid [commercial name: Tramaco] was used as a blowing agent in all mixtures.

Types of fillers [commercial name]:

1. treated ultra fine calcium carbonate [OMYA EX-H1]
2. surface treated white marble [OMYACARB 1T-AV]
3. fine uncoated precipitated calcium carbonate [Socal P3]
4. ultrafine, white and odorless precipitated calcium carbonate [Socal 31]
5. ultrafine surface treated precipitated calcium carbonate [Socal U1S2]

6. ultrafine, white and odorless coated precipitated calcium carbonate [Socal 312]

Table 1. Particle size of fillers

	OMYA EX-H1	OMYACARB 1T-AV	Socal P3	Socal 31	Socal U1S2	Socal 312
Particle size (μm)	1.7	2.0	0.18-0.24	0.24-0.35	0.07-0.13	0.045-0.090

Table 2. Formulation of the blends

PVC	Stabilizer	Lubricants	Processing aid	Bowling agent	Fillers six each
100	2.4	2	7	1.5	5;10;20

In the blends the fillers were added 5; 10; 20 phr. Five types of $CaCO_3$ and one of marble filler were used in the materials. The differences between the different $CaCO_3$ is the surface treatment and the particle size. The mixtures were made using a laboratory dry-mixer. In the mixer the temperature increased to 110 °C. Extruded sheets were made from the mixtures by a twin screw laboratory extruder. The extruder parameters were: D=30mm, L/D= 20, compression rate 1:3 and the extruder temperature set at 165°C/170°C/170°C°/175°C/175°C/180°C from hopper to die. The extruded sheets were guided trough cooling cylinders to fix thickness cylinders. The space between the cylinders was 4 mm. The densities of final sheets were about 0.7027 g/cm³, there were only slight variations.

2.1. *SEM analysis*

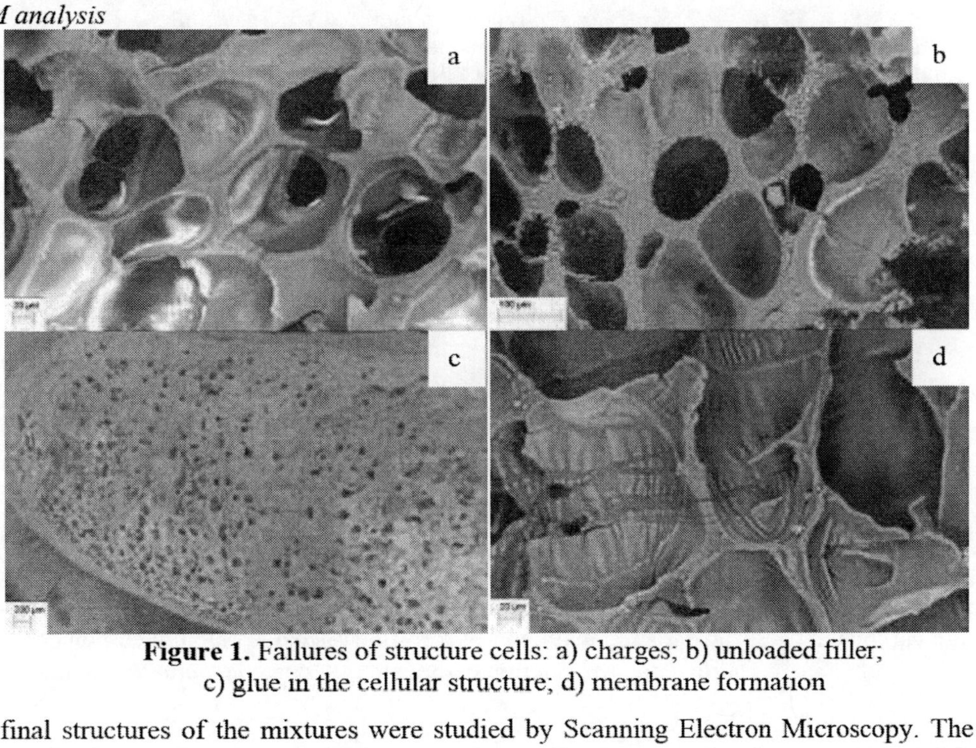

Figure 1. Failures of structure cells: a) charges; b) unloaded filler; c) glue in the cellular structure; d) membrane formation

The final structures of the mixtures were studied by Scanning Electron Microscopy. The SEM images show the fractured surface on different specimens. Furthermore the shape- and size of the cells can be tested.

Process of sample preparation [7]:

- Used liquid nitrogen to fractured the samples
- The samples must be electrically conductive. A conductive layer on the insulating materials has to be formed. The conductive layer was gold.
- The samples were fixed on a metal plate with universal glue.
- Used ZEISS SEM to observe samples, cells sizes and wall thicknesses were measured at 2 areas each

Figure 1. shows a little bright area on the surface, where the conductive layer was too thin (lach of gold), therefore the effect of reflecting electrons is visible. It was also observed that the particle sizes of $CaCO_3$ have poor dispersion in the PVC in Fig 1. b). The Fig 1. c) shows the open cells structure and the effect of glue infusion. The adhesive material migrated into the cellular structure thus destroying the whole foam structure. The phase of the specimen and the gilding formed a membranous material on the Fig 1. d). This determined that the gilding layer was too thin, so the surface charged.

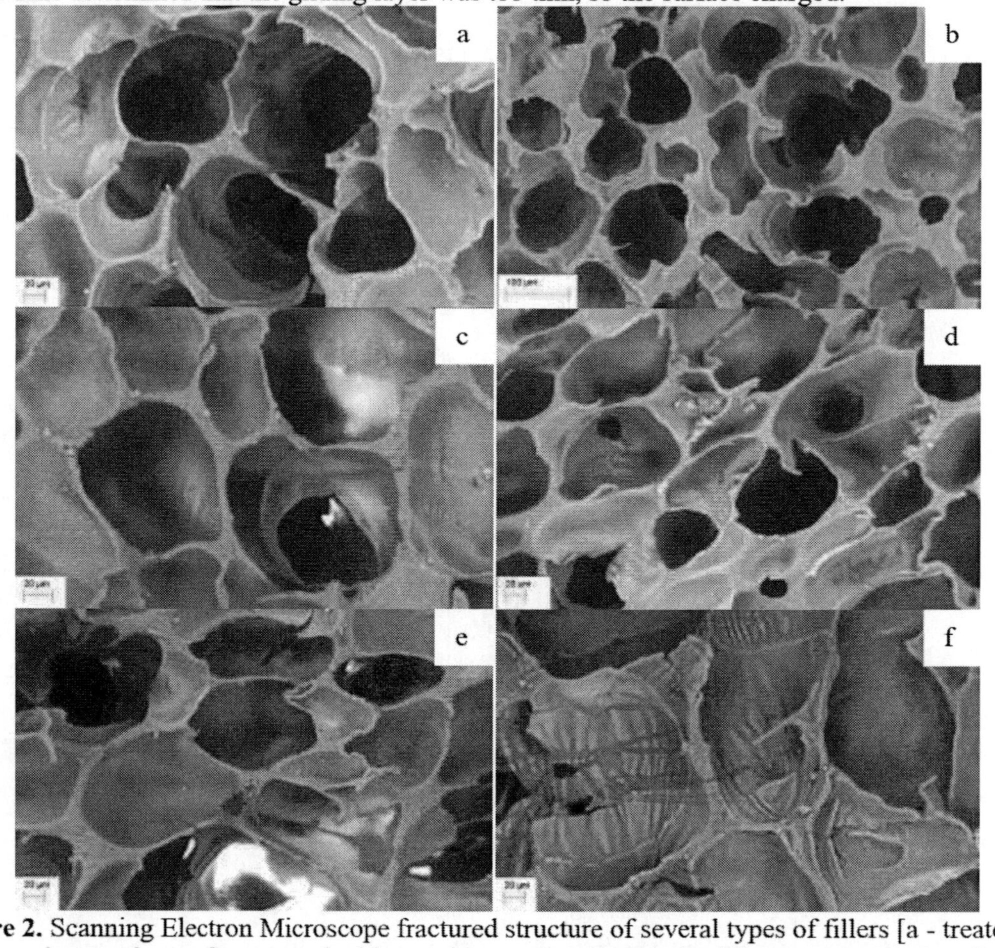

Figure 2. Scanning Electron Microscope fractured structure of several types of fillers [a - treated calcium carbonate; b - surface treated white marble; c - fine uncoated calcium carbonate; d –calcium carbonate; e - surface treated calcium carbonate ; f – coated calcium carbonate]

The Figure 2. illustrates the structural effect of various fillers. The cells shape (open and close cells) the changes in all samples. Fig 2. f) sample have much more closed cell structures than the other samples. Indicating that the fillers have modifier effect of the morphology. Based on Fig 2. it can be seen that the cells have seemingly similar density distribution in all samples. During the formation of the structure, the machine orientation has an effect on the shape of the cells. Fig 2. a) and 2. e) shows that the various fillers further orientate the foam. In Fig. 2. c) and 2. d) it is also noticed that the surface treatment, untreated surface and the uncoated filler does not have a major effect on the form of the final structure. Fig 2. c) has different magnification 350x vs. 250x this is why the cell sizes are larger here seemingly.

The mixtures differ in the 5, 10, 20 phr dosage of marble filler. Fig. 3. shows the changes of the wall thickness and cell size by added different marble concentration. The wall thickness and the size of the cells show irregularity distribution for mixtures of 5 and 20. The smallest size of cells and also unified structure is observed in case of the mixture with 10 phr. I show only one result of mixture 1, because the other measurements gave similar results.
In case of 10 phr wall structure was equable and the cells sizes were approximately the same in the whole sample. Based on Fig. 3. for the further production and examination the 10 phr samples were used.

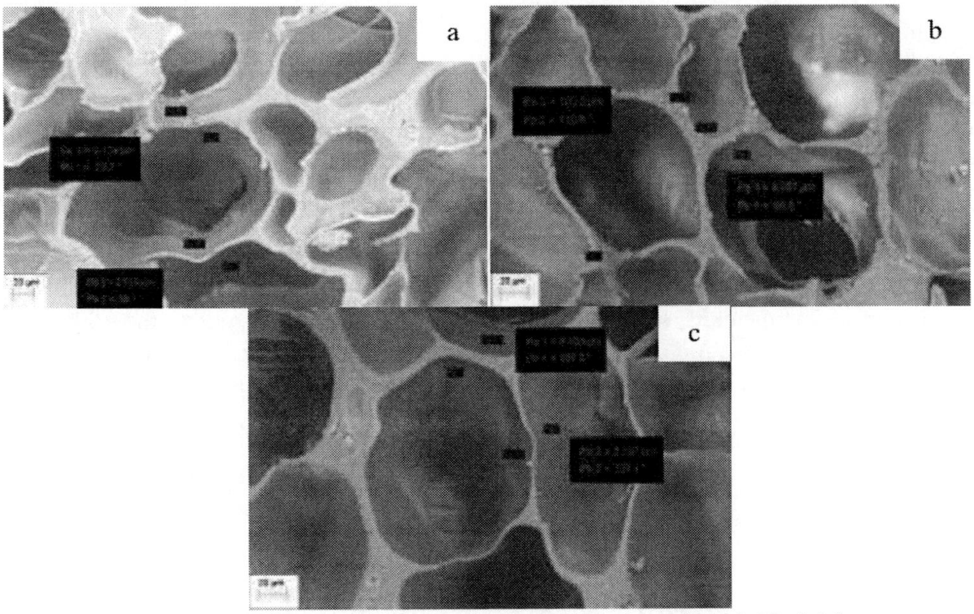

Figure 3. Effects of different dosage of marble: a) 5 b) 10 c) 20

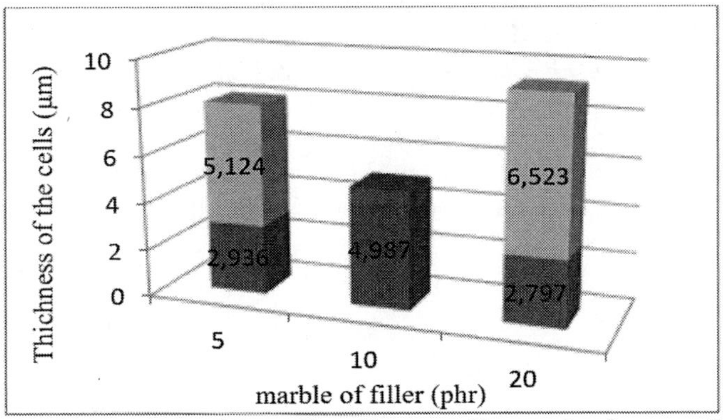

Figure 4. Thicknesses of the cell walls according to the dosage of marble

IOP Conf. Series: Materials Science and Engineering **426** (2018) 012040 doi:10.1088/1757-899X/426/1/012040

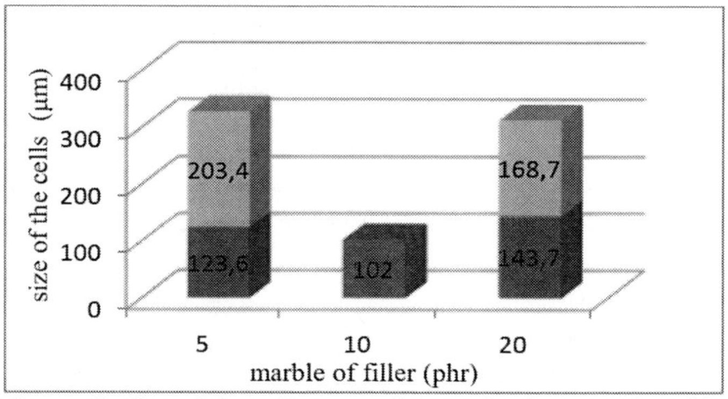

Figure 5. Size of the cells according to the dosage of marble

3. Conclusion

Several types of blends were produced with different types of fillers. In this paper the main goal was to determine the modifier effects of the $CaCO_3$ and the marble filler. Fillers greatly influence the morphology of the cellular structure. Accordingly, the prepared mixtures were tested by SEM analysis. From the SEM analysis the structure of the blend can be examined. The results of SEM examinations show that the surface treatment additive does not modify the structures. The different dose and type of fillers other changed the orientation and the shape of the cells in the structure. Furthermore, the size of the fillers more significantly affects the size and the wall thickness of the final cells. From the result of the six types of specimens I choose the marble filled for the future measurements, because despite of the small particle size of the cells. It has a more even distribution structure than the other samples. The aim of the research was formation of regular cells shape, and the apparent changes can be observed on the samples. Accordingly, certain structures can be formed, which are suitable for insulation foams.

References

[1] Garadimani, K R, Raju, G U & Kodancha, K G 2015 Study on Mechanical Properties of Corn Cob Particle and E-Glass Fiber Reinforced Hybrid Polymer Composites. **5,** 86–91.

[2] Vučak, M & Stöver, H-D 2005 Precipitated calcium carbonate in PVC *Gefälltes Calciumcarbonat PVC* **95,** 97–100

[3] Demir, H, Sipahioğlu, M, Balköse, D. & Ülkü, S 2008 Effect of additives on flexible PVC foam formation. *J. Mater. Process. Technol.* **195,** 144–153

[4] Roman, K, Zsoldos, G 2017 International Journal of Research - Granthaalayah, MEASUREMENTS OF RHEOLOGYCAL AND MECHANICAL- , THERMO-MECHANICAL PROPERTIES OF PVC / CORN COB FOAM. **5,** 313–321

[5] Hassan, A, Y J Wee, U L A Johari Impact strength of PVC-U-Effect of Calcium Carbonate particle size 2008 *J. Chem. Nat. Resour. Eng. 35-40*

[6] Chand, N, Fahim, M, Sharma, P & Bapat, M N 2012 Influence of foaming agent on wear and mechanical properties of surface modified rice husk filled polyvinylchloride. *Wear* **278–279,** 83–86

[7] H Károly Scanning Electron Microscope, Manuscript ELTE [In Hungary: *Pásztázó elektronmikroszkópia]*

IOP Conf. Series: Materials Science and Engineering **426** (2018) 012041 doi:10.1088/1757-899X/426/1/012041

The effect of silicon on the homogenization of CDC cast aluminium alloys

Zs Sályi[1], B Hatalák[1], V Kárpáti[1], D Angel[1], V Mertinger[1] M Benke[1], A Kovács[1], Z Gácsi[1], E Nagy[2] and P Barkóczy[1]

[1]University of Miskolc, Institute of Physical Metallurgy Metalforming and Nanotechnology, Miskolc-Egyetemváros, Hungary
[2]MTA-ME Material Science Research Group, Miskolc-Egyetemváros, Hungary

E-mail: femsalyi@uni-miskolc.hu

Abstract. The homogenization processes of commercial and modified Al 3003 alloys were investigated. The concentration of Si, Zn and the Mn were varied within the standard specification. Not only the temperature but the holding time of homogenization was also varied. The effects of alloying element on the precipitation process during the homogenization were investigated by optical and scanning electron microscopy. The correlative microscopy was also a very useful method to follow the evaluation of different precipitate. Hot and cold compression tests were used to detect the effect of different precipitation on the formability. It was experimentally showed that just a slight difference in the alloying elements resulted in significantly different microstructure and formability after the same homogenization process.

1. Introduction

Aluminum alloys have been widely used in the aerospace, automotive, and construction industries due to their high specific strength, low cost, and good corrosion resistance [1, 2]. The 3003 aluminum alloy is widely used in the container, packaging, and automobile industry, because of its excellent specific strength, corrosion resistance and formability. The alloy 3003 is typically Al-Mn alloys, but some further dissolved elements also used to improve strength and formability [1, 3, 4]. Such elements are, for example, silicon, iron, copper, magnesium, zinc or even zirconium. The effect of the alloying elements for the precipitations and the mechanical properties of the material are already known [5, 6]. For example, the effect of zinc, silicon and titanium elements for the microstructure and mechanical properties is well known [1, 7]. Small amount of zirconium addition helps to increase strength and provides grain coarsening. The addition of iron, silicon, copper, zirconium and titanium to the alloy has a strong influence on the formation of the precipitates. It is well known that iron and silicon reduce the solubility of manganese in the solid solution and accelerate the rate of precipitation of intermetallic compounds [2, 5, 6]. Fe promotes the precipitation of $Al_6(Mn,Fe)$ while Si promotes the precipitation of αphase $Al_{15}(Mn,Fe)_3Si_2$. Different equilibrium processes occur in the material during heat treatment after casting. In this regard the most important is the so-called 6 to αtransformation where $Al_6(Fe,Mn)$ transform to $Al_{15}(FeMn)_3Si_2$ [8-10]. In addition, further precipitations will be generated from the aluminium matrix during homogenization [11-13]. The type size and distribution of precipitation formed during homogenization have strong effect on the deformation behaviour, recrystallization and mechanical properties of alloys [14,15]. So the homogenization process is crucial to the optimization of the chemical composition and mechanical properties of the alloy. Controlling the state of precipitation is very important [16, 17]. In the present study the result of two different homogenization processes were compared for the commercial and the modified 3003. In case the modified alloy the concentration of Si, and Zn and the Mn were increased. Not only the temperature but the holding time was also varied.

2. Experimental

Table 1 present the compositions of the investigated ingots were CDC, (Closed Direct Chill) casted in the industrial environment. Comparing the two alloys the differences in Si, Zn and Cu contents are obvious. Sectioning was used to take samples from the cross section of the ingot at the ¼ of the with. The samples were homogenized under two conditions: condition A 510°C 1h, 5h or 10 h, and condition B 540°C 4h step 510°C 1h, 5h or 10h. The heat treatments were performed in laboratory environment. After the homogenization subsequent water quenching was applied. The as casted and the homogenized microstructure were observed by OM and TEM. Microstructure studies were performed using a Zeiss Imager M1m optical microscope. Barker etchant and Zeiss Axiovert M40 MAT optical microscope was used to visualize the grain size with polarized light filter. Further investigations were performed using a Zeiss EVO MA10 scanning electron microscope equipped with EDX. Correlative microscopy was used for direct monitoring the solid phase reaction during homogenization by the Zeiss EVO MA10 scanning electron microscope. The same part of the sample was investigated before and after the heat treatment. Specific sample holder provided the coordinates for the orientation on the sample. The as casted sample were prepared for the OM investigations and the quality of the surface was good enough after the heat treatment as well. The homogenization condition B was used for correlative sample.
For modeling hot or cold rolling, hot and cold compression tests were carried out on the cylindrical samples. The tests were performed by Instron 5982 material testing equipment. The cold and hot tests were carried out at 23 ° C and 400 ° C respectively. 1 / sec deformation rate with a 60% reduction were applied in both cases. The heating and holding time of pressing pieces was provided by an inductor. After the cold compression test, a subsequent heat treatment was carried out in 500 ° C in a furnace for 1 hour and then the pieces were cooled into water to control the driving force for the recrystallization. After the examinations the data were evaluated, namely comparison of true and engineering stresses and deformations, true stress and true deformation diagram, and the deformation strength chart. The microstructure was also investigated.

Table 1. Chemical composition of the specimens, wt%

identification	Si	Mg	Mn	Fe	Zn	Cu	Cr	Ti	Al
XH	0,79	0,01	1,61	0,25	1,49	0,04	0,002	0,01	bal.
X8	0,06	0,006	1,09	0,2	0,02	0,52	0,001	0,125	bal.

3. Results

The Figure 2 shows the initial as casted microstructure of the samples, no significant difference can be observed. Only the grain size is smaller in case of X8 sample. Typical coarse Al6(Fe, Mn) particles were formed during casting at the grain boundary of dendrites. Figure 2-3 present the LM image of the samples after the homogenization under 6 conditions. After performing the homogenizing heat treatments at lower temperatures, optical microscopic examinations suggest that the formation of the second phases is started after one hours in both alloys, which are increasing with time. The difference is observed in the distribution of the precipitations, as the second phase in the high silicon alloy grows around the eutectic, while the low silicon-containing alloys tend to appear along the grain boundary. Very fine precipitations can be observed in the matrix but precipitation free zones around the primary particles still exists. The effect of the temperature and time of the homogenization is different for the two different composition. Increasing the time and the temperature the number of fine particles was also increased. The dissolution of the primary particles also can be observed after 10 hours treatment of X8 alloy. The SDAS value seems to be also smaller which cannot be the result of the homogenization but the inhomogeneity of the initial casted sample, which was different for each case. So, correlative

IOP Conf. Series: Materials Science and Engineering **426** (2018) 012041 doi:10.1088/1757-899X/426/1/012041

investigations were performed to clarify this effect. The Figure 4-5 show the result after the heat treatment condition B. It is obvious that the primary particles were nearly unaffected in both alloy even after 10 hours treatment, just a slightly coarsening was observable. The existing of the precipitation free zone is not convincing. But zones with different type of particles is very clear especially at the alloy with lower Silicon concentration. The distribution of zones with very fine precipitates with size less than 1 μm associated with primary dendrites morphology. So it is clear that those formed from the supersaturated solid solution dendrites not from the dissolution of the primary $Al_6(Fe,Mn)$. Other zones with coarse precipitates and also needle like precipitates are located between the dendrites. The volume fraction of this zona is much higher after 10 hours. EDX analysis confirmed that the fine precipitates are Al-Mn while the needle like is Al-Fe rich. Because of the small size of the particle other methods with higher resolution are needed for more precise identification. These kinds of needles may have a harmful effect on the mechanical properties especially of the alloy because these sharp boundaries may be stress concentrators. The formability also depends on the morphology and distribution of precipitates. The hot press curves and the microstructures were find identical after the different treatments. But the cold curves behave quite different. The Figure 6 introduces the hardening curve after cold compression tests of the two alloys were homogenized under condition A. The Figure 7 shows the microstructure on the cross section parallel to the axis of the cylindrical sample after cold compression test and annealing heat treatment. The hardening curves of the samples were homogenized different time are similar in case of X8 alloy. Only a small part of the recrystallized volume was detected. Alloy XH where the concentration of alloying elements (Si, Mn, Zn) is higher but the precipitation was finer behaving a very different hardening, especially after 10 hours homogenization.

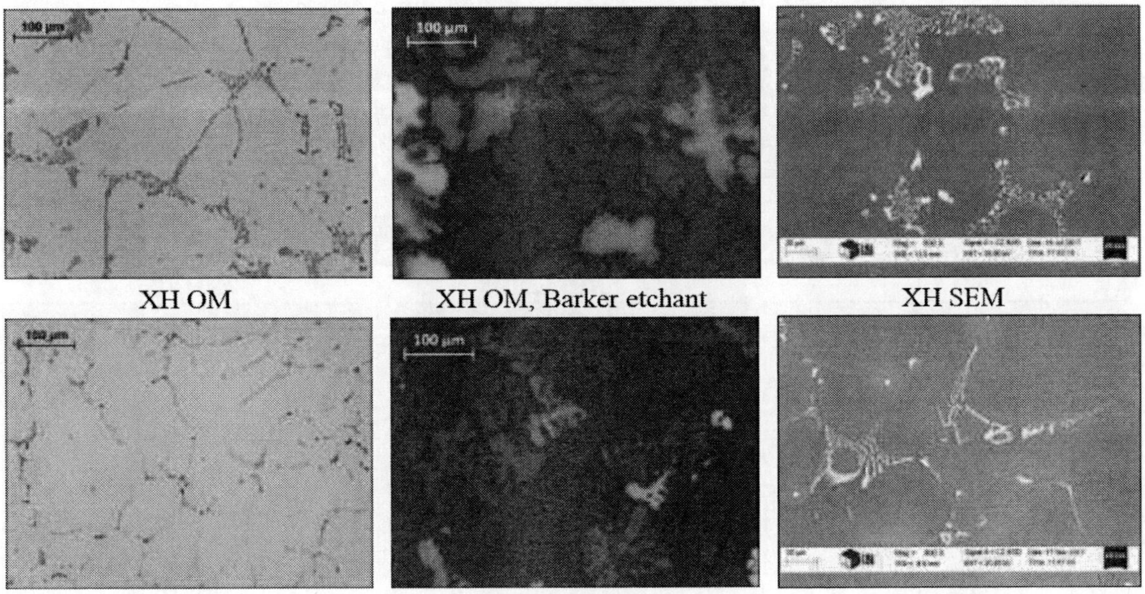

Figure 1. The as casted microstructure of the samples

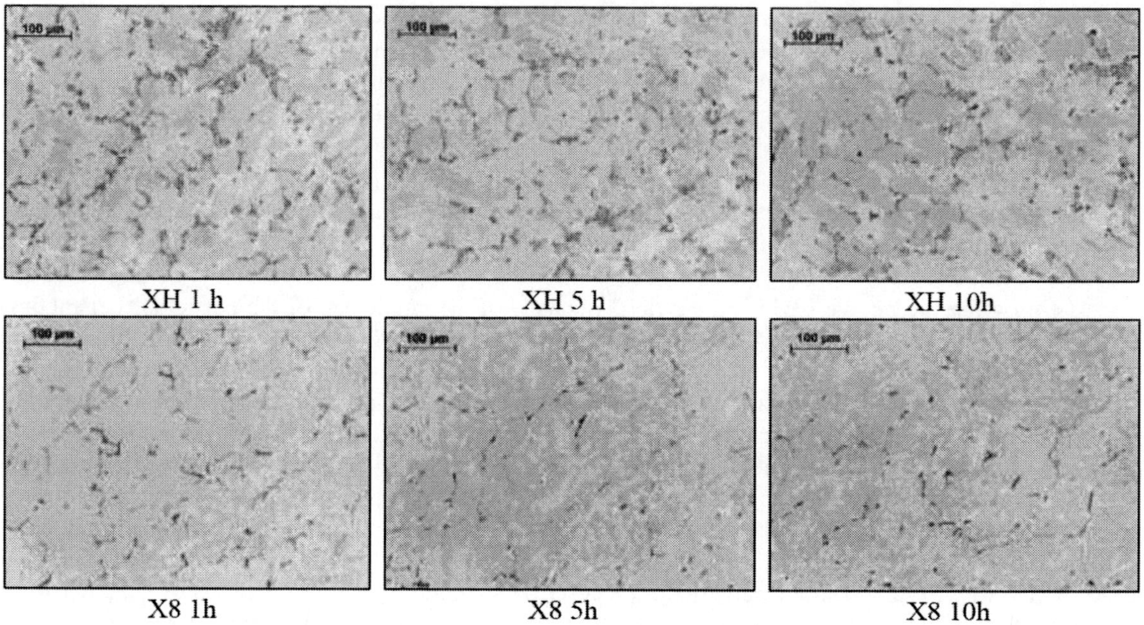

Figure 2. The light microscopy feature of homogenized sample at 510°C 1, 5 and 10 hours

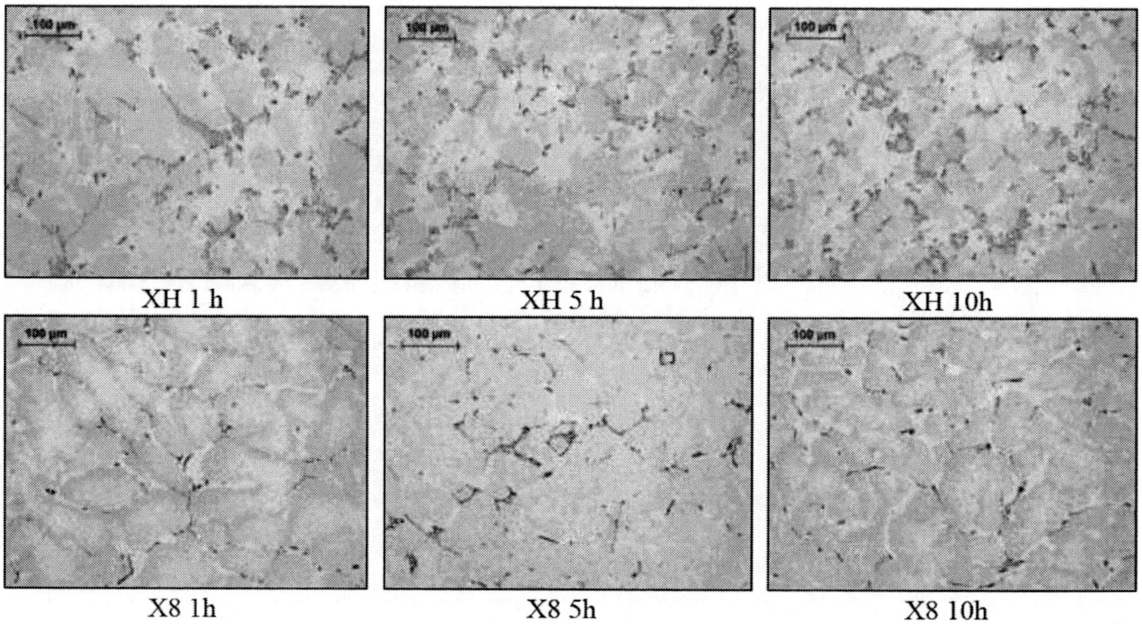

Figure 3. The light microscopy feature of homogenized sample at 540°C 4h step 510°C 1, 5 and 10 hours

IOP Conf. Series: Materials Science and Engineering **426** (2018) 012041 doi:10.1088/1757-899X/426/1/012041

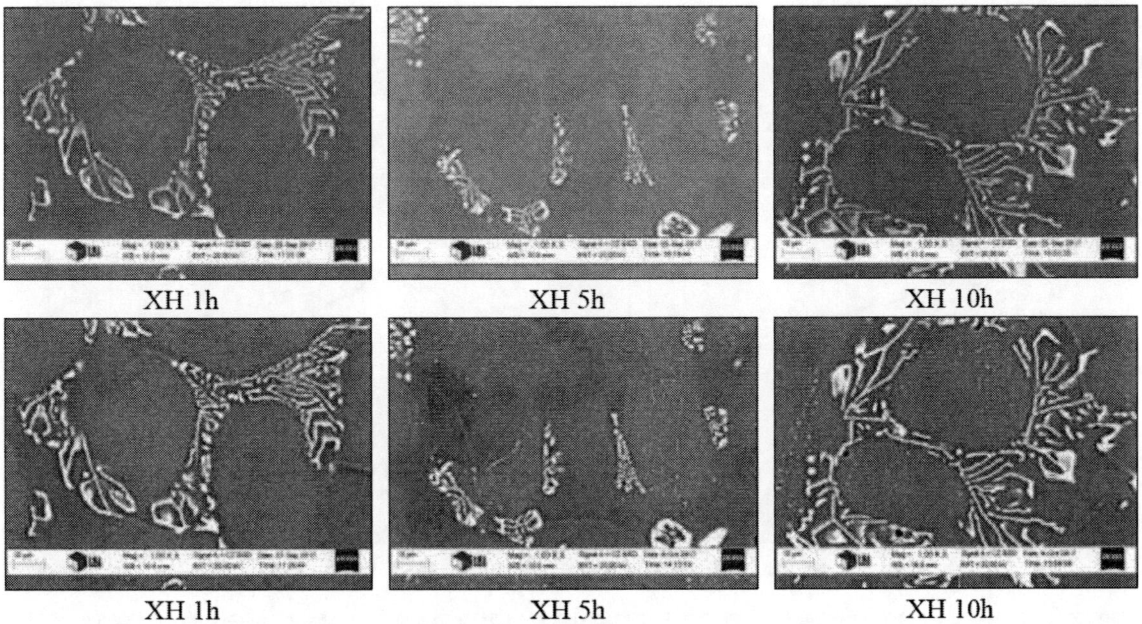

Figure 4. Correlative microscopy investigation of the XH samples before and after the homogenization

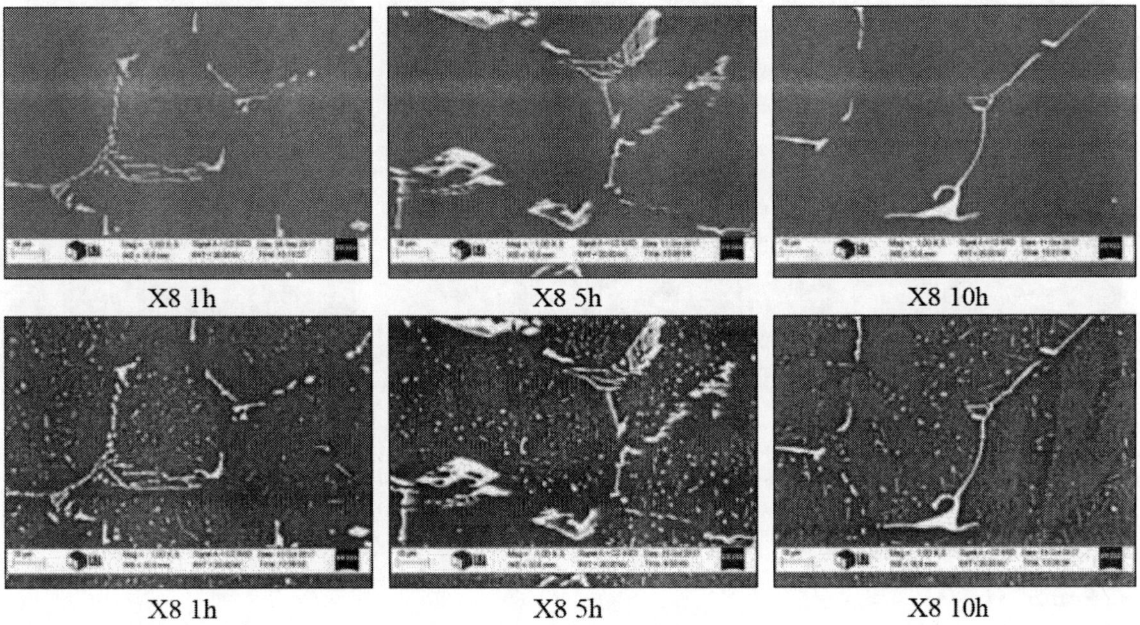

Figure 5. Correlative microscopy investigation of the X8 samples before and after the homogenization

IOP Conf. Series: Materials Science and Engineering **426** (2018) 012041 doi:10.1088/1757-899X/426/1/012041

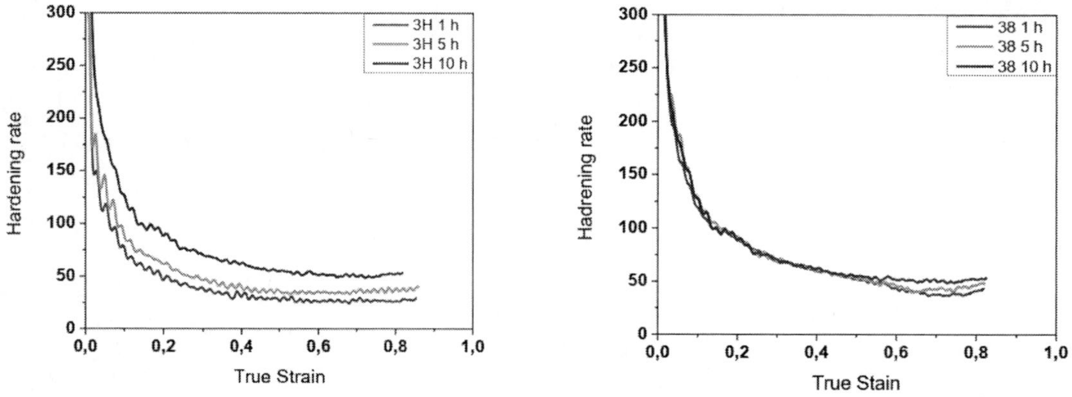

alloy XH alloy X8

Figure 6. Hardening curves determined during cold compression test of the homogenized sample under conditions A

alloy XH 1h alloy XH 5h alloy XH 10h

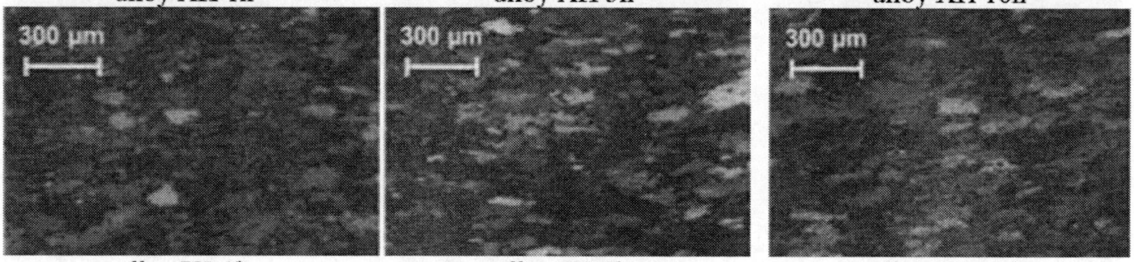

alloy X8 1h alloy X8 5h alloy X8 10 h

Figure 7. Microstructure of the two alloys were homogenized under condition A. After cold compression test and subsequent annealing. Barker etchant

4. Summary

Homogenization of commercial and modified 3003 aluminium alloys was investigated. The effect the increased Si, Mg, Zn were studied on the precipitation processes using different temperature and time for homogenization. Using the method of correlative microscopy, it has been clearly demonstrated that the small percentage of alloying variation causes a significant change in the microstructure. Smaller Si and Mn content causes iron-rich needle particles. In larger Si and Mn content, formation of needles is retarded and during cold deformation results in stronger hardening that increases the recrystallization tendency and may result in a finer grain size after the recrystallization process. So the dissolved elements caused hardening has stronger effect on the recrystallization than the particle stimulated process.

IOP Conf. Series: Materials Science and Engineering **426** (2018) 012041 doi:10.1088/1757-899X/426/1/012041

Acknowledgement

"The described article was carried out as part of the GINOP-2.2.1-15-2016-00018 project implemented in the framework of the Szechenyi 2020 program. The realization of this project is supported by the European Union."

References

[1] Majed M Jaradeh R and Carlberg T 2011 *J. Mater. Sci. Technol.* 27(7) pp 615-627
[2] Li Y J, Muggerud A M F, Olsen A and Fura T 2012 Precipitation of partially coherent α-Al(Mn,Fe)Si dispersoids and their strengthening effect in AA 3003 alloy *Acta Materialia* 60 pp 1004-1014
[3] Alexander D T L and Greer A L, 2002 Solid-state intermetallic phase transformations in 3XXX aluminium alloys, *Acta Materialia* 50 pp 2571-2583
[4] Fujii T, Watanabe C, Nomura Y, Tanaka N and Kato M, 2001 Microstructural evolution during low cycle fatigue of a 3003 aluminum alloy, *Materials Science and Engineering,* A319-321 pp 592-596
[5] MERCHANT H D, MORRIS J G, HODGSON D S ,1990 Characterization of intermetallics in aluminum alloy 3004, *Materials Characterization* 25 pp 339-373
[6] Li M, Tamura T, Omura N, Murakami Y, Tada S, Miwa K and Takahashi K 2014 Refinement of intermetallic compounds and aluminum martix in 3xxx aluminum alloys solidified by an electromagnetic vibration technique, *Journal of Alloys and Compounds* 610 pp 606-613
[7] Miroux A, Lok Z J, Marthinsen K and van der Zwaag S, 2007 Solute and second phase evolution during industrial processing of AA3103, *Materials Science Forum* Vols. 539-543 pp 281-286
[8] Huang H-W and Ou B-L, 2009 Evolution of precipitation during different homogenization treatments in 3003 aluminum alloy, *Material and Design* 30 pp 2685-2692
[9] Huang H-W, Ou B-L and Tsai C-T, 2008 Effect of homogenization on recrystallization and precipitation behavior of 3003 aluminum alloy, *Materials Transactions*, Vol. 49, No.2 pp. 250 to 259
[10] Marie A, Muggerud F, Mortsell E A, Li Y and Holmestad R, 2013 Dispersoid strengthening in AA3XXX alloys with varying Mn and Si content during annealing at low temperatures, *Materials Science & Engineering A* 567 pp 21-28
[11] Du Q, Poole W J, Wells M A and Parson N C, 2013 Microstructure evolution during homogenization of Al-Mn-Fe-Si alloys: Modeling and experimental results, *Acta Materialia* 61 pp 4961-4973
[12] Li Y J and Arnberg L, 2003 Quantitative study on the precipitation behavior of dispersoids in DC-cast AA3003 alloy during heating and homogenization, *Acta Materialia* 51 pp 3415-3428
[13] Sjolstad K, Engler O, Tangen S, Marthinsen K and Nes E 2002, Texture evolution of an AA3xxx alloy after different homogenisation treatments, *Materials Science Forum* Vols. pp 396-402 pp 463-468
[14] Liu Q, Yao Z, Godfrey A and Liu W 2009, Effect of particles on microstructural evolution during cold rollig of the aluminum alloy AA3104, *Journal of Alloys and Compounds* 482 pp 264-271
[15] Liu W C, Zhai T and Morris J G 2004, Texture evolution of continuous cast and direct chill cast AA 3003 aluminum alloys during cold rolling, *Scripta Materialia* 51 pp 83-88
[16] KWAG Y and MORRIS J G 1986, The effect of structure ont he mechanical behavior and stretch formability of constitutionally dynamic 3000 series aluminum alloys, *Materials Science and Engineering*, 77 pp 59-74
[17] Duan X and Sheppard T 2002, Three dimensional thermal mechanical coupled simulation during hot rolling of aluminium alloy 3003, *International Journal of Mechanical Sciences* 44 pp 2155-2172

IOP Conf. Series: Materials Science and Engineering **426** (2018) 012042 doi:10.1088/1757-899X/426/1/012042

Industrial application of a quick, non-destructive anisotropy characterisation method

M Sepsi, A Hlavacs, V Mertinger, M Benke

Institute of Physical Metallurgy, Metalforming and Nanotechnology; University of Miskolc; Miskolc-Egyetemvaros, Hungary 3515

E-mail: femsepsi@uni-miskolc.hu

Abstract. Centerless X-ray diffractometers are specially designed for non-destructive residual stress measurement. The residual stress, and the crystalline anisotropy are induced by elastic deformation and are stored within the crystal structure. The similarity between residual stress and pole figure measurement led to the idea of developing the texture test method for centreless diffractometers. After the determination of identical lattice plane series distribution that can be measured with centreless and conventional diffractometers, advanced anisotropy measurements techniques were invented. In some cases, to characterize the sample from the aspect of crystalline anisotropy it is enough to make measurements in the specific directions (e.g. χ cuts in the rolling, cross and diagonal direction). With our centreless X-ray diffractometer, we managed to measure a χ cuts in the rolling direction under two minutes without sample cutting. The implementation of this non-destructive technique into the production line can be possible, consequently monitoring of the texture evolution even on industrial-sized aluminium coils. In this paper we compare the conventional X-ray diffraction measuring technique with our developed method on the example of a cold rolled aluminium sheets at different reductions.

1. Introduction

The industrial significance of anisotropy is highly important in metal forming and shaping processes, for example, during the deep drawing, drawability. During this process the earring indicates the homogeneity of the formability. Of course, small earring and homogenous deformation is favoured, which can be achieved by a texture-free structure [1,2,3]. The pure deformation or recrystallization (annealing) texture cause large earring, which can be reduced by an appropriate combination of them. From the casting there are several reasons that influence the final texture after the last technological step in the case of aluminium alloys [4]. Obviously, the effect of annealing (rate of the last cold forming, annealing time, temperature) conditions are crucial, but the formation of anisotropic texture is also influenced by the presence of secondary phases formed during the casting or homogenisation. This is the so-called PSN (Particle Stimulated Nucleation) effect, when the number of dislocations increase in the surroundings of the rigid, not formable particles thus the stress field of these dislocations helps the nucleation of the recrystallization and the forming of isotropic microstructure [4,5,6]. This process highly depends on the size, shape and distributions of the precipitations determined by compositions, casting, preheating prior to hot forming and the conditions of homogenization. Due to this, it is possible to observe different earring rates at the start, middle, at the end and even in cross direction of the rolled aluminium coil. Therefore, very often, the edges of the coil are trimmed to separate the sections have different drawability. There is an industrially used method to characterise the drawability, called cup drawing. The results of the tests show the cause of earring (deformation or recrystallization), however, such results cannot give an answer with details. Furthermore, it is a destructive measuring technique. Fortunately, the X-ray diffraction can provide detailed information on texture. The only problem with conventional X-ray diffraction technique and the characterization by pole figure or ODF (Orientation Distribution Function) is that the measuring time is too long, and the sample size is limited which requires the need for sample cutting and preparation. To characterize the sample from the aspect of anisotropy, it is enough to make measurements in the preferred directions (e.g. χ cuts in the rolling, cross and diagonal direction). With our centreless X-ray diffractometer (Stresstech G3R), we managed to measure a χ cuts in the rolling direction under two minutes without sample cutting (Figure 1a). These centreless X-ray diffractometers are originally produced to measure residual stress and retained austenite volume fraction non-destructively and quickly even in industrial environment [7]. The implementation of this non-destructive technique into the production line can be possible, thus, monitoring the texture evolution even on industrial-sized aluminium coils. In this paper we compare the conventional X-ray diffraction measuring technique with our developed method on the example of a cold rolled aluminium sheets.

2. Experimental

The measurements were carried out on 3103 type aluminium cold rolled sheets that were rolled with small steps. The measurements were performed after each reduction at the following sample thickness: 5.74 mm; 5.36 mm; 4.97 mm; 4.49 mm; 4.01 mm; 3.52 mm; 3.04 mm; 2.54 mm; 2.05 mm; 1.58 mm; 1.16 mm; 0.85 mm; 0.68 mm and 0.45 mm. The χ cuts were measured on the {222} plane series. For the conventional pole-figure measurements, a Bruker D8 Advance diffractometer was used equipped with Eulerian cradle. The details of the conventional measurement were: CoK_α radiation, 40 kV acceleration voltage, 40 mA tube current. With this technique the {222} pole figures were measured, and the χ cuts in the rolling direction were selected from these pole figures. The non-destructive measurements were performed with the Stresstech G3R device in OMEGA mode, CrK_α radiation was used with 3 mm collimator and with 2 second exposure time. Figure 1 shows the measuring setup and the beam path.

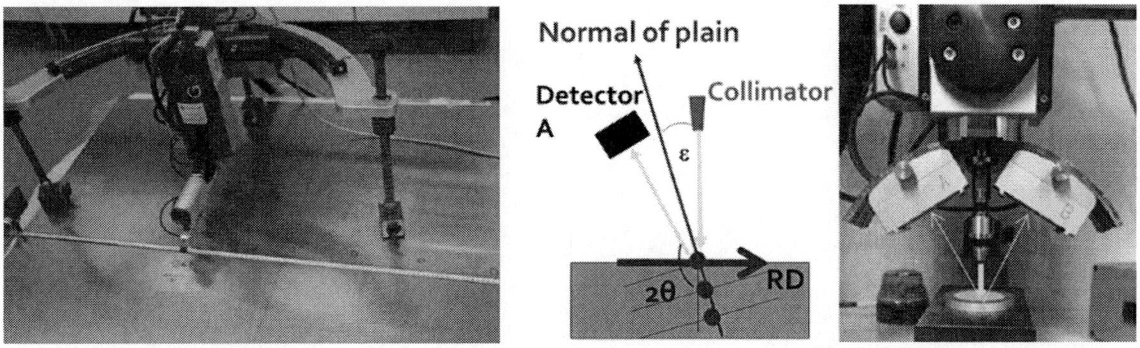

Figure 1. Non-destructive measuring with a Stresstech G3R device a) measuring of the aluminium sheet b) beam pass in OMEGA mode (schematic and photo - with A detector, Rd-rolling direction)

3. Results and discussions

Figure 2. shows the measured {222} pole figures on the specimens obtained with the Bruker D8 Advance. It is observable, how the rolling deformation texture builds up with higher reduction, the characteristic of the deformation texture getting stronger and stronger as it was expected. The last pole figure is rotated with 90° degree, and the red arrow shows the interpretation of the χ cuts in the rolling direction. Figures 3-4 represent the χ cuts, the blue cuts show the non-destructive technique measured directly with the Stresstech G3R, and the red cuts measured with the Bruker D8 Advance, calculated from the pole figures. The reason of the asymmetric characteristic of the individual curves is that asymmetric tribological conditions were built up in some cases. Comparing the conventional and the non-destructive methods, the same tendency is observable. Even the above-mentioned asymmetry is revealed and fits well for both equipment, for example, at the 5.75 or 0.45 mm thickness. As the specimens suffered higher and higher deformation, the results measured with the different techniques fit better and better to each other, since the texture is increased in the examined volume, and the texture gradient getting smaller. This agrees with the results of the pole figures. On the y axes the absolute intensity scales are not the same because they were measured by different devices and parameters, but the tendency is absolute convincing.

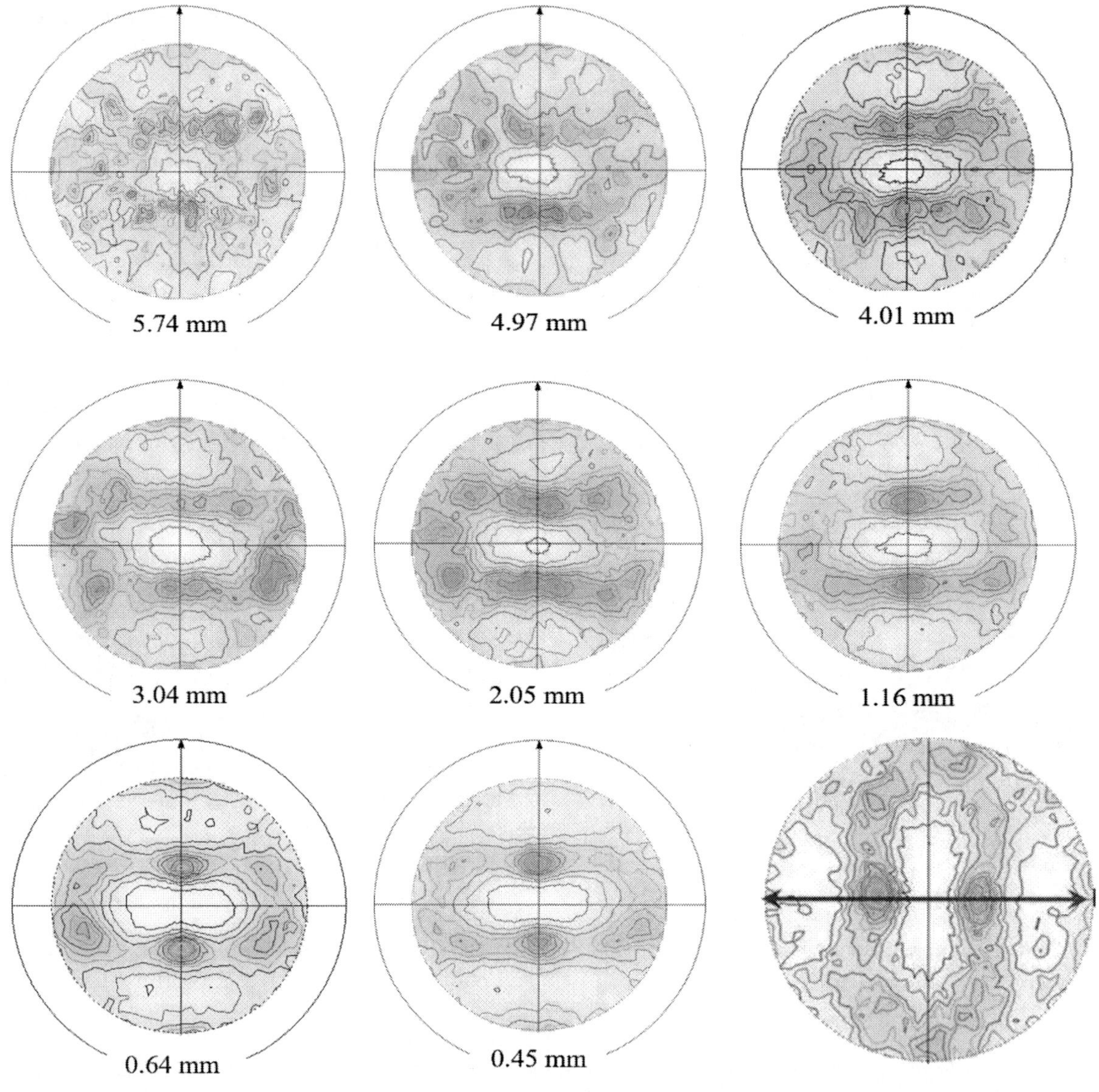

Figure 2. Rolled 3103 type aluminium {222} pole figures after different reduction (5.74 mm-0.45 mm), red arrow shows the interpretation of the χ cuts in the rolling direction

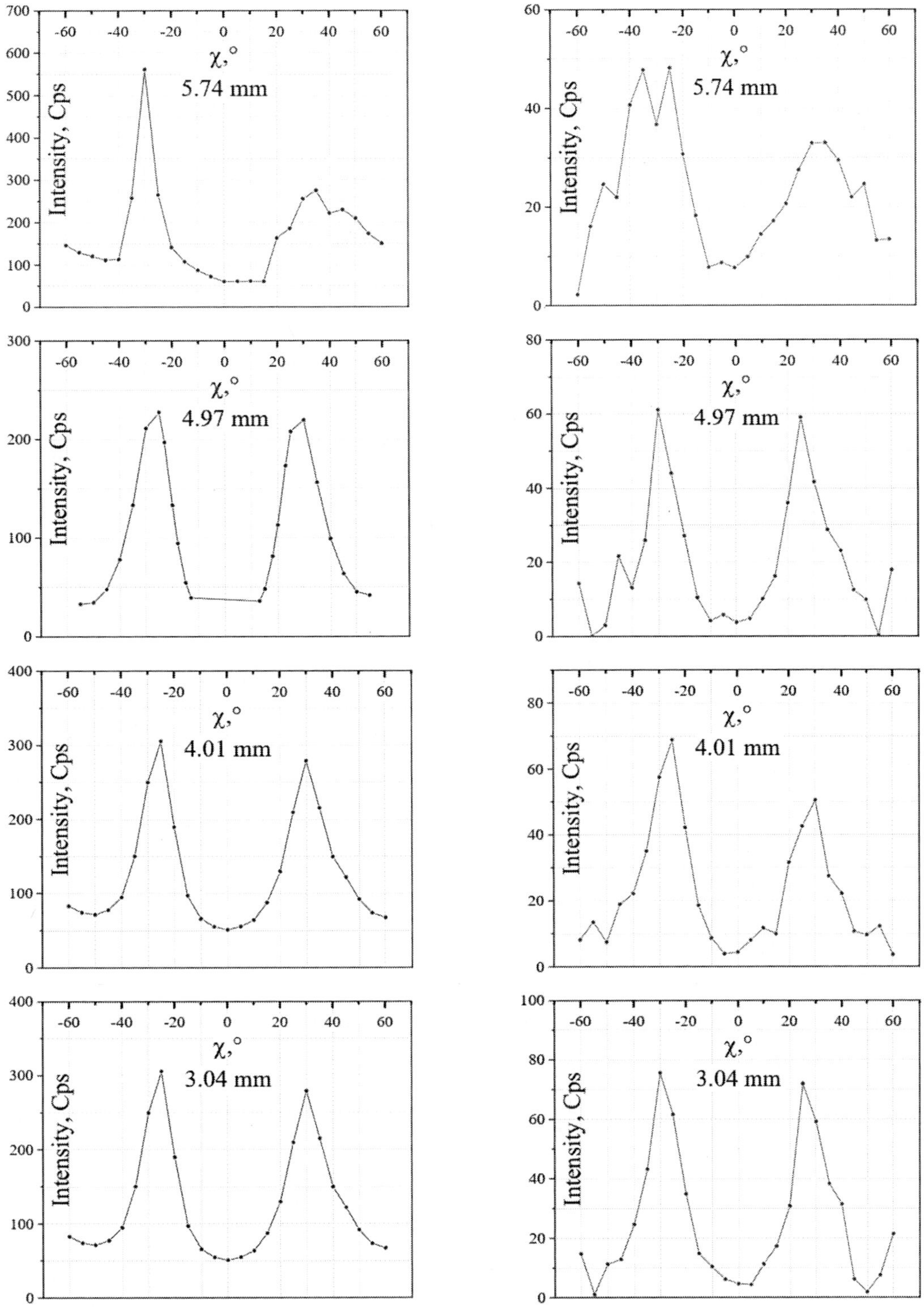

a b

Figure 3. χ cuts of aluminium {222} pole figures after different reductions (from 5.74 mm to 3.04 mm) in the rolling direction a) measured by centreless X-ray diffractometer (Figure 2) b) from the conventionally determined pole figures

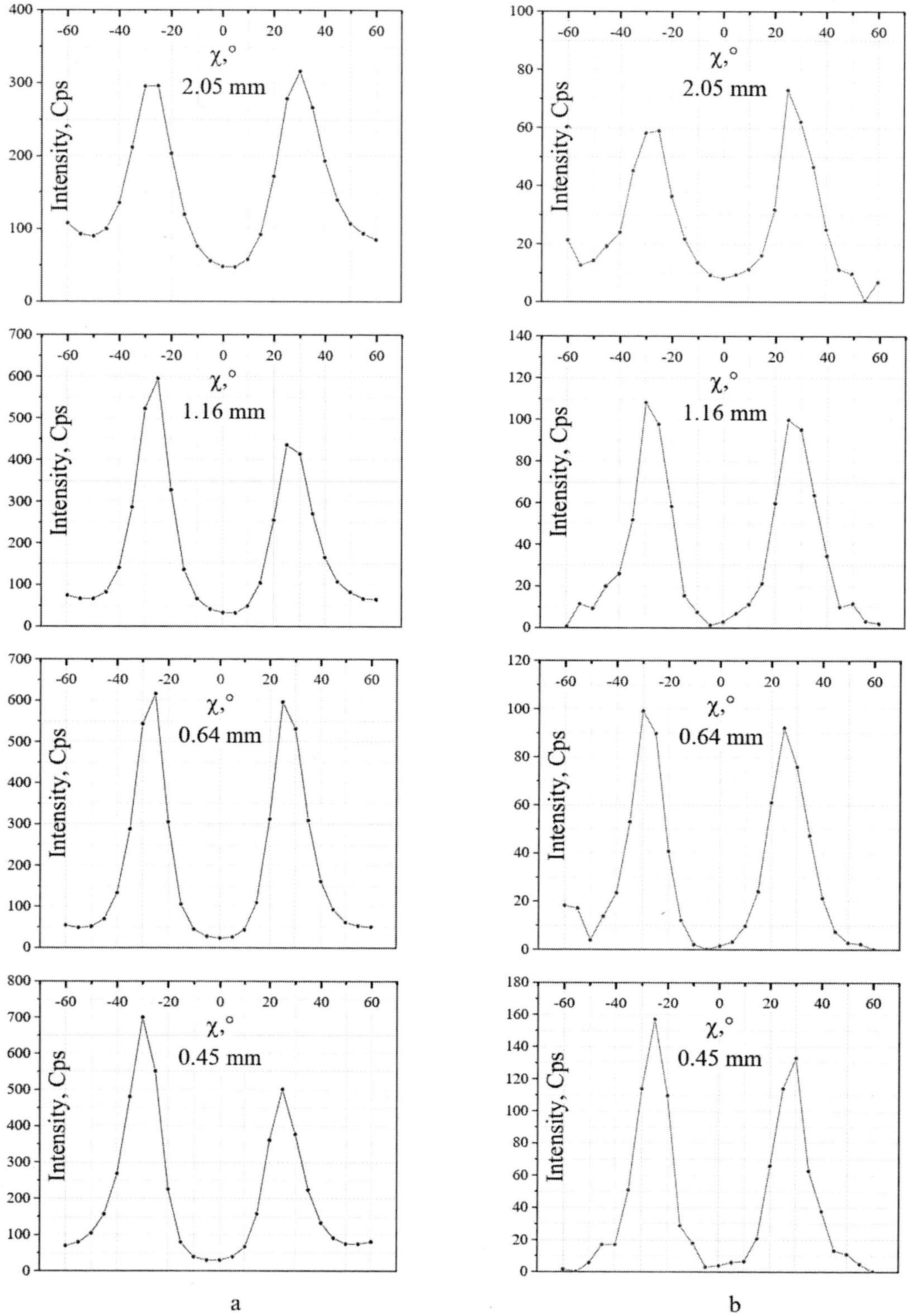

Figure 4. χ cuts of Aluminium {222} pole figures after different reduction (from 2.05 mm to 0.45 mm) in the rolling direction a) measured by centreless X-ray diffractometer (Figure 2) b) from the conventionally determined pole figures

4. Conclusions

The formation of anisotropy plays very important role the metal forming industry, and there are several methods to characterize it. Unfortunately, most of these methods are destructive, sample-cutting, and preparation always needed. In this paper, a unique non-destructive measurement technique was shown based on X-ray diffraction and compered with the conventional X-ray diffraction technique. The advantages of this method are obviously its non-destructive nature and the very short measuring time (~2 min.) what create the justification for implementation into the industry. Validation measurements were performed with a Stresstech G3R centreless X-ray diffractometer and a conventional, Bruker D8 Advance X-ray diffractometer equipped with an Eulerian cradle. The characterization of anisotropy by the pole figure is a well-established method, but it takes a long time. In some cases, especially industrial environment, it is enough to perform measurements in the specific directions (e.g. χ cuts in the rolling, cross and diagonal direction). With our centreless X-ray diffractometer, we managed to measure a χ cuts in the rolling direction under two minutes without sample cutting. The validation measurements were performed on a 3103 type aluminium alloy through 8 steps rolling. The results determined by the two methods were identical, even the effect of asymmetric tribological conditions was able to determine.

Acknowledgments

The described article was carried out as part of the GINOP-2.3.2-15-2016-00027 "Sustainable operation of the workshop of excellence for the research and development of crystalline and amorphous nanostructured materials" project implemented in the framework of the Szechenyi 2020 program. The realization of this project is supported by the European Union. The described research was also supported through the National Research, Development and Innovation Office – NKFIH K119566 project. M Sepsi's research was supported through the new national excellence program of the ministry of human capacities (Hungary).

References

[1] Engler O and Randle V 2010 *Introduction to texture analysis* (CRC Press).
[2] Suwas S and Ray K R 2014 *Crystallographic Texture of Materials* (London: Springer-Verlag)
[3] Engler O 2012 *Materials Science and Engineering A* **538** pp 69-80
[4] Kocks F, Tomé C and Wenk H 1998 *Texture and Anisotropy- Preferred Orientations in Polycrystals and their Effect on Materials Properties* (Cambridge: Cambridge University)
[5] Rollet A, Humphreys F, Hatherly M 2012 *Recrystallization and Related Annealing Phenomena* (Oxford:Elsevier) pp 256-274
[6] Vatne H, Engler O and Nes E 1997 *Materials Science and Technology* **13** pp 93-102
[7] ed Totten G, Howes M, and Inoue T 2008 *Handbook of Residual Stress and Deformation of Steel* (Materials Park: ASM International)

IOP Conf. Series: Materials Science and Engineering **426** (2018) 012043 doi:10.1088/1757-899X/426/1/012043

Structure and properties of closed-cell foam prepared from rPET

V A Szabó[1] and G Dogossy[1]

[1] Department of Material Science and Technology, Széchenyi István University

E-mail: szabo.veronika.anna@ga.sze.hu

Abstract. Nowadays, in the field of packaging technology, one-way polymer packaging materials are increasingly used. Food and packaging industry uses the most, the life cycle of which becomes short, they become waste very quick. Also included are light weight bottles of polyethylene terephthalate (PET). Large amounts of waste can be handled with multiple options. The worst is their dumping on dump sites, as these materials do not compost. Their thermal recycling is more favorable due to their high calorific value, but a real cycle can only be achieved by their physical recycling. However, during the multiple re-purposing, PET's mechanical and processing properties are reduced. In our research, chemical foaming has been investigated, resulting in a smaller weight, better specific property, and fully reusable product. During our investigations, the structure and mechanical properties of recycled PET (rPET) specimens manufactured with different composition, chain and foam additives were analyzed.

1. Introduction

From the 50's onwards, the use of fossil fuels and energy sources increased, which attracted the rapid development of the plastics industry. In 70's, polyethylene terephthalate (PET) appeared and soon began to dominate the liquid food packaging market. Consumer habits have changed. Due to its low production cost, low weight, good optical properties and gas tightness, PET has outperformed glass [1].

There are chemical and physical possibilities for recycling PET. During the chemical processes, various reactants are used to break the polyester into components which can then serve as a monomer to produce PET and other polymers [2]. By contrast, physical methods are more energy-saving, as they retain the energy invested previously in the material. However, the recycled product is degraded, and its properties deteriorate during use [3].

We want to introduce a property enhancement technology that does not prevent multiple processability. This technology is chemical foaming, this results in smaller weight, better specific properties and a fully reusable product. In our opinion, the chemically expanded structure of recycled polyethylene terephthalate is an appropriate raw material for industrial utilization.

Under the term polymer foam, we mean a two-phase system in which statically distributed variable gas bubbles are embedded in a polymer matrix. Almost all thermoplastics or many crosslinked polymers are suitable to produce a foamed product. Foaming technologies are basically divided into three groups: mechanical, physical and chemical. During the chemical foaming, gases can be formed by the addition of chemical blowing agents which break down when heated, while producing their type-dependent gas (pl. CO, CO_2, N_2, NH_3) [4].

Chemical blowing agents can be either organic or inorganic solids which decompose in the high temperature melt. Their solid decomposition products act as nodules. In addition, large quantities of gas are generated. The gas bubbles formed at the nodule expand as they cool down, thus form the cells. The formed foam structure is a closed cell, the cavities are well separated from each other, they are not connected, their surface is solid, the cell density is increased towards the core, i.e., structural foam can be adjusted [5, 6].

PET suffers from chain breakage while recycling, which also results in a decrease in the melts viscosity. This causes some difficulties during processing, partly during chemical foaming. Prior to chemical foaming, it is essential to restore the original structure of the raw material. For this purpose, various artificial additives were tested.

Molnár [7] investigated the effect of technological parameters with chain extenders and chemical blowing agents on the injection molded rPET foam. The results obtained during the experiments have shown that the pressure and the switching point significantly affect the foamed structure and its properties. If the pressure drops, the foaming ratio increases with the cell diameters being lowered. Increasing the switch point does not correlate with the total volume of cells, but their average size increases.

Numerous research deals with the effect of chain-extender additives. The effect of the PMDA (Piromellitic Dianhydride) chain extender on the recycled PET extrusion was studied by Raffa [8]. In terms of viscosity, the increasing number of chains increased. In contrast, in the range of lower shear rates, no such difference was seen in the case of higher shear rates.

Toth and his colleagues [9] tested the reaction of recycled PET with radioactive radiation. The base material was reinforced with glass fiber and added a reactive additive. PET scrap was prepared for the experiment and purified to reduce humidity at 540°C for 5 hours. The base material, epoxy additive, was added at 2 m/m%. The additive was Derakane D-410 with a molecular weight of 400 g/mol. Glass fiber was a Zoltek-type E type 10-15 mm long 8 μm diameter product. This was utilized in a 20% weight ratio. During the experiment a radioactive dose was used (10 kGy) to activate the additive. In the tests, bending, tensile, and impact resistance measurements were performed. Because of the bending test, it was found that the bending resistance increased slightly with 20% glass fiber added. In contrast, the additive has worsened the results. In the assay of shock resistance, it was observed that the additive results in a slight decrease again, while the irradiated dose can improve impact resistance. Because of the research, they received a quality-enhanced material. The original PET impact resistance was exceeded by the new reactive reinforced recycled PET. This could be a major milestone in the secondary use of PET.

Coccorullo and his colleagues [10] have tested for the molecular weight increase in the chemical blowing agent with PMDA and CT 534 (Hydrocerol) as excipients. PMDA is a by-product free, economically feasible additive. To avoid crosslinking, it is not advisable to use more than 0.2-0.3% of the additive in the mixture. A closed cell structure can be formed with the use of the appropriate amount. The primary purpose of the research is to produce high-density structural foams from low viscosity raw materials. The secondary aim is to create the right tool in which the foaming process can be reproduced, while changing the various parameters, the foam structure can also be modified.

Ronkay et al. [11] wish to increase the impact resistance with rubber-tight additives. Rubber-tight additives are filler particles which preferably provide a small film thickness with uniform dispersion. The functionalized end group of reactive additives improves its attachment to the PET matrix and provides a better distribution by changing the interface energy between the matrix and the additive.

2. Applied materials and methods

During the experiments, commercially available blue crystallized PET regranulate (rPET) was used. Our chemical blowing agent (CBA) was Tracell IM 7200, which was mixed with 4% of the raw material. Due to their high moisture absorption, hygroscopic plastics, such as PET, must be dried, especially before injection molding. The moisture content of the PET should not exceed 0.004% during processing. In this experiment, the base material was dewatered in a DEGA 2500 drying chamber to achieve this. Prior to extrusion, the recycled PET was dried for 24 hours in the drying chamber at 100°C. After extrusion the mixture was dried for another 12 hours at 140°C in the hot air-drying chamber. In the experiments, CESA Extend was used as a chain extender (CE) at a rate of 2%, while the Du-Pont Elovaloy PTW was used as an impact modifier (IM) additive at 10%. The exact composition and marking of the manufactured blends are given in Table 1.

Table 1. Composition of the injection molded products

	rPET [%]	CE [%]	IM [%]	CBA [%]
rPET (reference)	100	0	0	0
rPET4CBA	100	0	0	4
rPET2CE4CBA	100	2	0	4
rPET2CE10IM4CBA	100	2	10	4

The Arburg ALLROUNDER 420C Golden Edition injection molding machine was used to produce the specimens. To achieve the appropriate foam level, the pressure differences can be influenced by the breathing mold technique. During this process, the polymer mixed with the chemical blowing agent is injected into the mold cavity where the compact surface layer of the piece is formed during the cooling. Then, by minimally opening the tool, the space increases, the pressure decreases, hence the melt forms into the blown material inside as foaming takes place. During the cycle a 0.3 mm mold opening was used (Fig. 1) (Table 2).

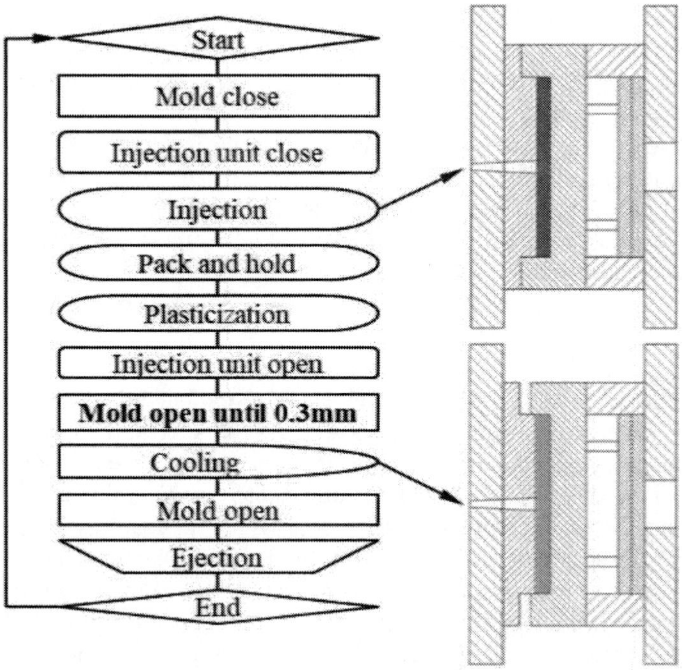

Figure 1. The injection molding cycle used with the breathing mold technique [12]

The specimens were produced with mold temperature of 35°C. The foam structure of the manufactured specimens was tested by YXLON Y.CT Modular Industrial Computer Tomography (CT). The impact resistance was measured on CEAST 65-45.000 impactor with 4 J hammer, 62 mm support span by the EN ISO 179 standard. The tensile test was done on the INSTRON 5582 universal tensile machine. The specimen were pulled at 1 mm/min tensile speed with 100 mm support span. To determine the modulus of elasticity, if the elongation reached the 1% value, it was switched to 5 mm/min. Measurements were carried out in accordance with EN ISO 527. Flexural tests (three-point bending tests) were also carried out on the INSTRON 5582 universal tensile machine according to EN ISO 178. During the test, the specimens were bent at 10 mm/min with 64 mm support span. Five assays were performed at room temperature with relative humidity of 50%; for each, averaging and standard deviation were calculated.

Table 2. Injection molding parameters

	Unit	Value
Clamping force	kN	150
Nozzle temperature	°C	260
Injection pressure	bar	650
Injection speed	cm^3/s	30
Pack pressure	bar	150-50-20
Pack pressure time	s	2-1
Mold opening	mm	0.3
Cooling	s	20
Mold temperature	°C	35

3. Results and discussions

In the case of injection molded specimens, we examined the characterization of the foam structure, the density of the products and the mechanical properties. The following subsections describe the results obtained.

3.1. Density

When determining the density of solid substances, the weight recorded in a known density liquid is compared to the mass measured in air. During the measurement, first determine the mass taken in air and then the mass dipped in the reference fluid. Density was determined in 98% ethanol using an Ohaus Explorer scale, based on Equation (1).

$$\rho_{calc} = \rho_{ethanol} \frac{m_{air}}{m_{air} - m_{ethanol}}, \tag{1}$$

where ρ_{calc} [g/cm³] is the density of the measured specimen, $\rho_{ethanol}$ [g/cm³] is the density of ethanol [g/cm³], m_{air} [g] is the weight of the specimen in air, $m_{ethanol}$ [g] is the weight of the specimen in ethanol.

One of the beneficial effects of foaming is weight loss. Because of the established cell structure, the density of the test pieces decreased. The average drop in density was 8.27%. In the case of a functional group mixed only with a chain extender, it can be said by increasingly adding chain extender to the mix it marginally effects the density (1.34%). In addition to the impact modifier, the density increased, but still did not reach the pre-foam density value (Fig. 2).

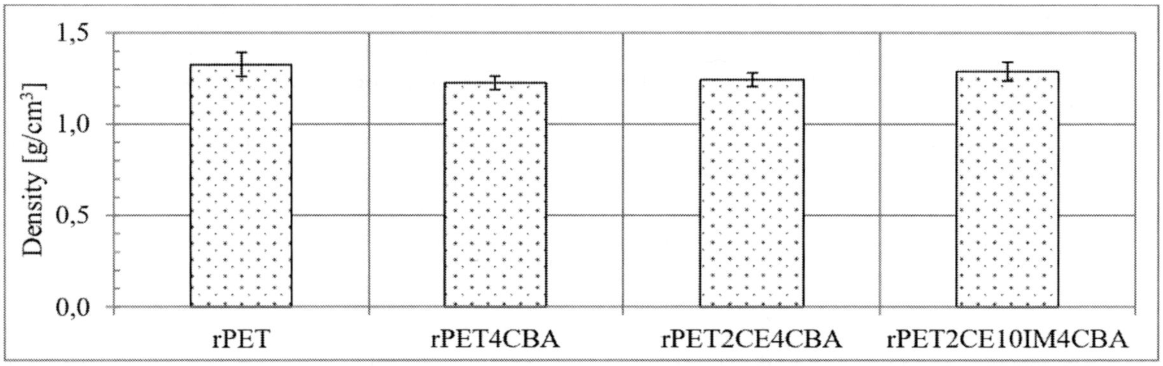

Figure 2. Density of the specimens

3.2. Morphology

During the test, we analyzed a 2 mm height segment of each specimem. On a test piece, several measurements gave the average of the results. During our testings, 1280 images/360° were taken.

When the 3D CT recordings of the test specimens are jointly examined, the effect of the chain extender on the cell structure is conspicuous (Fig. 3). In addition to the additive, a true foam structure is formed. The impact resistance increased the skin thickness, but there was no suction on the outer surface of the test piece.

Figure 3. 3D CT scan of specimens

Concerning porosity, it can be observed that the rate of foaming has fallen due to the chain extender, but the mixture reacted well to the impact modifier (Fig. 4).

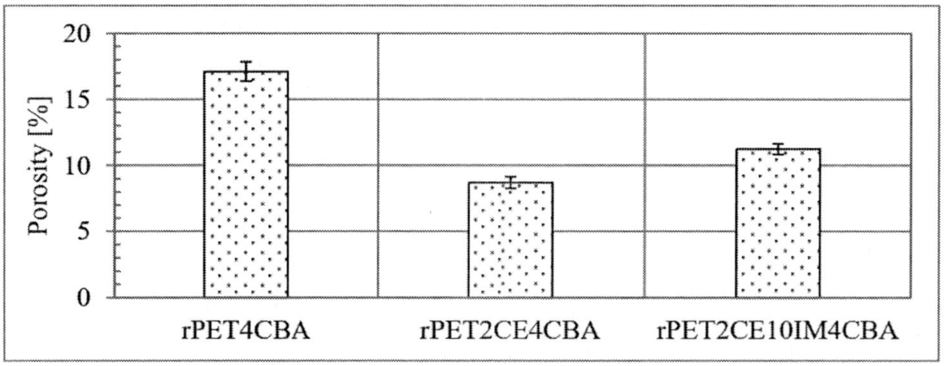

Figure 4. Porosity of the test specimens

3.3. Mechanical tests

We have investigated the effect of different blends on mechanical properties. The results are summarized in Table 3.

The tensile properties of the dumbbell specimens, using various additives were tested. The tensile modulus has increased due to the chain extender as an ordered cell structure has been achieved. Due to the impact modifier, the specimens lost their elasticity. The tensile strength improved by 19.28% due to the chain-extender masterbatch. Using the impact modifier, we achieved a 26.49% increase.

The flexural modulus of the specimens was reduced by the chain extender, but the impact resistance increased 14.55%, when compared to the results of the original foamed pieces. A 30.04% increase can be observed in the chain extended test pieces. The deviation values were high. Based on the results, it can be said: the flexural modulus can be increased using the impact modifier. During the tests, the flexural strength of the chain extended specimens was reduced. This is due to the increase in cell density. Due to the better cell distribution, the cell number increased while using the masterbatch. The flexural modulus has fallen to nearly a quarter of this. When adding impact modifier to the mix, we could see the improvement. A 62.16% increase was observed compared to the only chain extended test specimens. However, with the use of the two masterbatches, we did not achieve the results of the foam without additives.

The specimens lost their impact resistance by applying the chain extender agent. In addition to the impact modifier, Charpy's impact strength showed an improvement of 45.12%.

Table 3. Mechanical properties of the injection molded specimen

	Tensile test		Flexural test		Impact test
	E_T [GPa]	σ_T [MPa]	E_F [GPa]	$\sigma_{1,5h}$ [MPa]	a_{cN} [kJ/m^2]
rPET	4.58±0.03	50.03±1.48	7.47±0.61	67.14±0.65	3.45±0.38
rPET4CBA	1.32±0.24	19.92±3.92	3.64±0.57	40.39±1.26	1.50±0.10
rPET2CE4CBA	1.35±0.33	24.68±0.63	2.98±0.56	13.05±1.61	1.22±0.38
rPET2CE10IM4CBA	1.35±0.07	27.10±1.89	4.26±0.54	34.49±2.07	2.34±0.31

Generally, the resistance to wear has been reduced by a chain extender, however the cell structure has improved. The effectiveness of the impact modifier has been verified by several tests. With the use of the two additives, the resistance to mechanical stress increased and the cell structure improved.

4. Industrial application

The experiment presented is to reproduce the inner door handle of an automobile, from foamed rPET. For this we used the rPET2CE10IM4CBA blend we thought was best for this purpose.

We examined the width and height of the handle part of the injection molded internal handle using a caliper. Twenty samples were manufactured and parameterized, marking the number of work pieces tested. The tool is suitable to produce both the right and left side door handle. A total of 85 internal door handles have been created in automated production. Two 5-piece group of specimens were chosen for my examination. Approximately 23 minutes passed between the two groups. As a condition for reproducibility, a tolerance of 0.1 mm was determined for both the width and the height tests.

When measuring the height of left-hand handle, we got an average of 8.55 mm with a 0.019 mm standard deviation.

Thickness measurement resulted in an average value of 14.55 mm and a 0.013 mm standard deviation. All test pieces have performed within the tolerance set by us.

When analyzing the height of the test pieces on the right-side cavity, an average rating of 8.55 mm was obtained with a 0,019 mm standard deviation. The width test resulted in an average value of 14.54 mm and a 0.019 mm standard deviation. These values are within the tolerance range. Based on the measurement data, it was found that the reproducibility of the product is not affected by filling the right or left cavity because they are mirror-symmetric to each other. On the average width of the pieces 0.01% and 0.07% in length were measured. As a result of the investigation, it can be concluded that there is no difference between the right and the left door handle specimens during the production of the pieces. Using the tolerance field defined, we found that the piece is perfectly reproducible.

To analyze the reproducibility, the internal door handle was tested on an optical 3D coordinate measuring system (GOM) (Fig. 5). The resulting three-dimensional model was compared to the CAD model used in tool making. The joint was below 0.004. Due to this, it can be said the product can be manufactured using injection molding and a breathing mold technique.

Figure 5. GOM test result of the internal door handle

By analyzing CT scans, it can be concluded that higher cell diameters were generated during the production due to the increased volume (Fig. 6). Cell density and cell distribution resulted in a suitable foam structure.

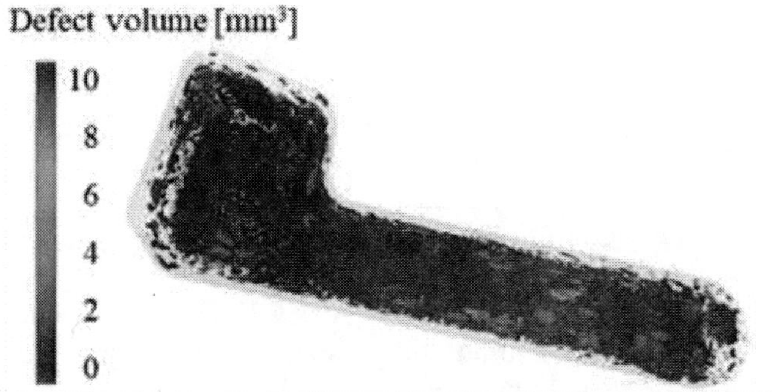

Figure 6. CT scan of the internal door handle

5. Conclusion

During our work, we aimed to recycle the ever-increasing amount of PET bottle waste by chemical foaming so as not to deteriorate its subsequent recycling properties. After the measurements and production, we found that the available crystallized blue bottle regranulate is suitable for producing a closed cell integral foam product. After testing the use of various additives, we can improve the resistance of the material against mechanical stresses. The new blend can be widely used in the automotive and construction industry.

IOP Conf. Series: Materials Science and Engineering **426** (2018) 012043 doi:10.1088/1757-899X/426/1/012043

Acknowledgement
The publishing of this paper was supported by the ÚNKP-17-1 New National Excellence Program of the Ministry of Human Capacities.

References
[1] Bach C *et al* 2011 *Water Res.* **3** 571-582
[2] Ragaert K *et al* 2017 *Waste Manage.* **69** 24-58
[3] Solarajan A K *et al* 2017 *J. Appl. Polym. Sci.* **32** 45181
[4] Mills N J 2007 *Polymer Foams Handbook: engineering and biomechanics applications and design guide* (Amsterdam: Butterworth-Heinemann) p 535
[5] Dogossy G 2015 *Porous structures* ed Zsoldos I (Győr: Universitas) chapter 3 pp 212-265
[6] Czvikovszky T *et al* 2000 *Fundamentals of polymer technology* (Budapest: Műegyetemi Kiadó) p 453
[7] Molnár B 2014 *Acta Tech. Jaur.* **1** 1-10
[8] Raffa P *et al* 2012 *React.Funct. Polym.* **1** 50-60
[9] Tóth K *et al* 2004 *Radiat. Phys. Chem.* **8** 143-147
[10] Coccorullo I *et al* 2009 *Express Polym. Lett.* **2** 84–96
[11] Ronkay F and Kasza D 2014 *Műanyag és Gumi* **3** 88-91
[12] Malnati P 2017 *Plast. Eng.* **5** 32-37

IOP Conf. Series: Materials Science and Engineering **426** (2018) 012044 doi:10.1088/1757-899X/426/1/012044

Effect of fibre sizing on the interlaminar properties of polyamide matrix composites

G Szebényi[1] and B Magyar[2]

[1]Assistant professor, BME, Faculty of Mechanical Engineering, Department of Polymer Engineering, Budapest, Hungary
[2]PhD student, BME, Faculty of Mechanical Engineering, Department of Polymer Engineering, Budapest, Hungary

E-mail: szebenyi@pt.bme.hu

Abstract. The aim of our study was to compare the interlaminar performance of PA based composites with different fiber sizing to investigate the importance of proper sizing. Composites were prepared by film-stacking from reinforcing materials with three available sizings compatible for vinylester, epoxy and polyamide. The composites were compared through standard static and self-developed dynamic and cyclic interlaminar shear test methods and scanning electron microscopy. The results showed that the proper sizing is crucial for the good mechanical performance of the composites, the PA sized fiber composites significantly better in all mechanical properties.

1. Introduction

Nowadays the application of fiber reinforced composites is spreading in the fields of both high performance materials [1, 2] and commercial applications [3, 4]. While in the previous decades long fiber reinforced composites were only applied in demanding low volume applications or mainly as covering materials for large volume production [5, 6], the tendency is shifting towards the use of long fiber reinforced composites in demanding large volume applications as load bearing structures [7, 8] and automotive structural parts [9, 10]. While the relatively high curing time of thermoset matrices was an obstacle, the emerging of high output technologies like HP-RTM (high pressure resin transfer moulding) [11] or T-RTM (thermoplastic resin transfer moulding) [12] can provide a low cycle time of some minutes compared to hours enabling the use of these materials in areas where thousands of parts have to be annually produced [13]. While these technologies open new ways for composites, some problems of these materials also become more emphasized. In standard thermoset materials proper sizing of the fibers gives adequate adhesion between the fibers and the matrix. Achieving such a level of adhesion [14, 15] is more challenging when using thermoplastic matrix materials, where in some cases chemical reactions are completely absent during production. In the application areas of most of these composites, the presence of constant vibration and dynamic loads can endanger the structural integrity of the composites by the formation of fiber-matrix debonding and delamination in the loaded areas. In all of the applications the interlaminar properties of the composites is a key factor for static and dynamic load bearing capability, failure method and fatigue cycle life. While the most reinforcing materials come with surface sizing developed for different kinds of thermoset matrices, where these are mostly used, some specific grades are already marketed with sizing especially selected for PA matrix applications.

IOP Conf. Series: Materials Science and Engineering **426** (2018) 012044 doi:10.1088/1757-899X/426/1/012044

The aim of our study was to compare the interlaminar performance of PA based composites with different fiber sizing to investigate the importance of proper sizing. The composites were compared through standard static and self-developed dynamic and cyclic interlaminar shear test methods and scanning electron microscopy.

2. Materials and Methods

2.1. Materials, composite production

The matrix material applied for the film-stacking method for the composite production was Schulamid 6 MV13 NATUR polyamide 6, similar to the caprolactam-based in situ systems used for the T-RTM production technology. The PA films were extruded by a Labtech Scientific film extruder line. The extruder zone temperatures were 230, 235, 245 and 250°C from the hopper to the die respectively. The screw rotational speed was 30 rpm. The three zone temperatures of the film die were set to 260°C. The pickup rolls in the production line were set to provide 1.1 m/minute and 1.7 m/minute pull speed at the rubber pickup rolls. The winding roll was set to 9 rpm. 155 mm wide and 150±10 μm thick film was extruded. The thickness of the film was measured by a Mitutoyo Digimatic digital micrometer.

Three reinforcement materials were selected for the specimen production with glass fibers with different types of fibre sizing prepared to be used for vinylester (VE), epoxy (EP) and polyamide (PA) by Saertex (Germany). All reinforcement materials had the same plain woven structure with a surface weight of 300 g/m^2 composed of 1200 tex glass rovings.

The composite sheets were produced by film-stacking method using a Schwabenthan Polystat 300S platen press. The platen temperature was set to 260°C on both sides according to the results of the preliminary tests for optimal impregnation. The produced composite layup consisted of 6 layers of reinforcement material at uniform 0/90 direction and 7 layers of matrix. The pressing was performed at 300 bars piston pressure, which resulted in 53.7 bar pressing pressure on the 310x310 mm composite plate surface. The cooling was started after the completion of the 3 minutes pressing cycle. The pressing was sustained during the whole process. The platens were opened at 60°C temperature.

Interlaminar shear strength (ILSS) test specimens were cut out of the laminates in the 0° direction according to ASTM D 3846 – 94 by a Mutronic Diadisc 4200 disc saw. The notches on the opposite sides of the specimens were also machined by the same device. The specimen geometry was defined according to ASTM D 3846 – 94, with notches on the opposite sides of the specimens 6.4 mm from eachother. The notch thickness was 1 mm (as a result of the diamond disc thickness), the notch depth was half of the laminate thickness to achieve a stable delamination in the laminate center plane.

2.2. Test methods

The interlaminar properties of the composites were investigated by a static, a dynamic and a cyclic method.

The static method used for the interlaminar characterization was the double notched (DNS) interlaminar shear test method according to ASTM D 3846 – 94 presented in Figure 1. The test was performed using a Zwick Z020 universal computer controlled tensile tester at 1.3 mm/minute crosshead speed.

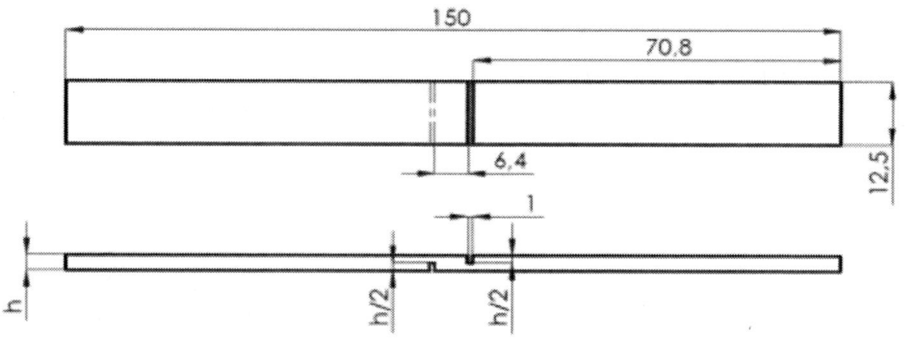

Figure 1. Double notched specimen

The dynamic method was a self-developed dynamic interlaminar shear test method [16] based on the static standard. The standard specimen was tested using a Ceast Resil Impactor Junior pendulum equipped with a 15 J tensile impact hammer. The specimen was fixed in the anvil of the tester and loaded through the 30 g weight crosshead attached to the other end of the specimen. The starting angle was 150° resulting in an impact velocity of 3.7 m/s. The energy required for the fracture was recorded by the machine. The energy required for the movement of the crosshead was deducted from the energy value. The resulting dynamic interlaminar shear strength was calculated from the corrected fracture energy divided by the shear cross-section between the notches.

The cyclic interlaminar shear tests were also carried out according to a self-developed method [16]. The standard specimens were investigated using an Instron 8872 computer controlled servo-hydraulic fatigue tester. The test was performed in displacement-controlled operation. The applied loading waveform was pulsating sinusoid with a frequency of 5 Hz with peak values set to the displacements corresponding to 50 and 10% of the failure displacements recorded during the static tests.

The scanning electron microscope (SEM) micrographs were taken using a JEOL JSM 6380LA microscope in high vacuum operation. The surfaces of the samples were gold spur coated prior to the investigations by a JEOL FC 1200 device to achieve the electrical conductivity of the surface to avoid local charge accumulation.

3. Results and Discussion

3.1. Static interlaminar properties

The results of the static interlaminar shear tests are presented in Figure 2. From the results the requirement of proper sizing can be clearly seen. The samples with VE and EP sizing showed poor interlaminar shear strength compared to the composite with proper PA sizing. The EP sizing slightly outperformed the VE sizing. Comparing the interlaminar shear strength values to the literature data [17, 18] it can be seen that even in case of the PA sized reinforcement, the interlaminar shear strength is inferior compared to the thermoset system results. This can be caused by the lack of chemical reactions during processing: no covalent bonds can be formed between the fibers and the matrix. Secondly even in case of thermoplastic matrices with relatively low melt viscosity the low shear speed of film-stacking can not force the melt into the reinforcing fiber rowings for proper impregnation. The T-RTM process, creating PA in situ from caprolactam can address both these problems, the monomer viscosity is less than the viscosity of most thermoset resins and because of the in situ process the matrix can covalently bond to the reinforcement sizing.

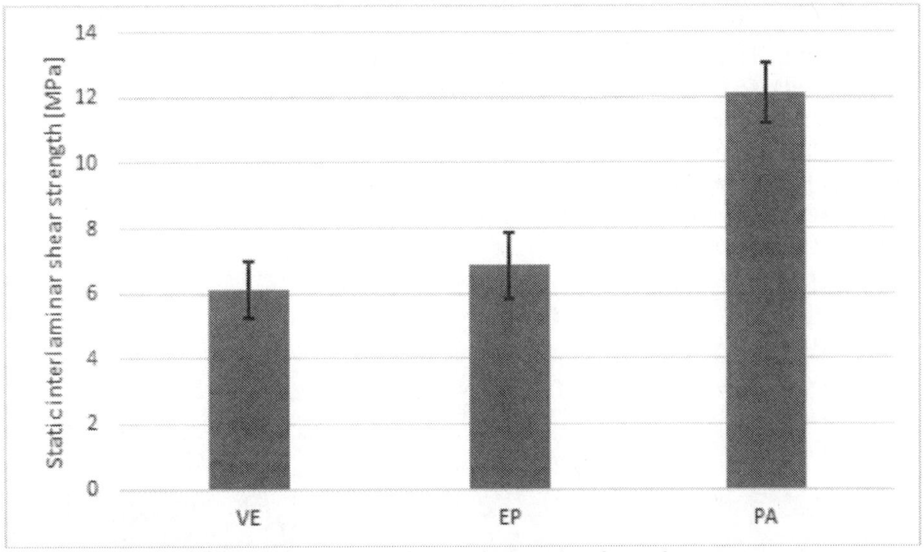

Figure 2. Results of the static interlaminar shear tests
(VE- fiber sizing for VE, EP – fiber sizing for EP, PA – fiber sizing for PA)

The results of the dynamic interlaminar shear tests are presented in Figure 3. Here, the impact produces a much higher strain ratio, the crack has to propagate much faster, the ductile matrix behaves much more brittle than in case of static loading. The targeted automotive sector requires high ductility of most of the composite parts, ductile behavior means more energy absorption in case of an accident, and less sharp fracture edges, which can cause injuries to passengers and pedestrians. In the results the clear superiority of the PA sizing is dominant. When comparing to standard Charpy strength results measured at similar neat PA materials [19], the dynamic interlaminar shear strength is much lower, especially with improper sizing. By constructing the T-RTM produced parts with neat PA edges, their performance can be probably greatly improved: the edges show much more ductile behavior and they can also hinder the crack to initiate at the free edges of the reinforced composite.

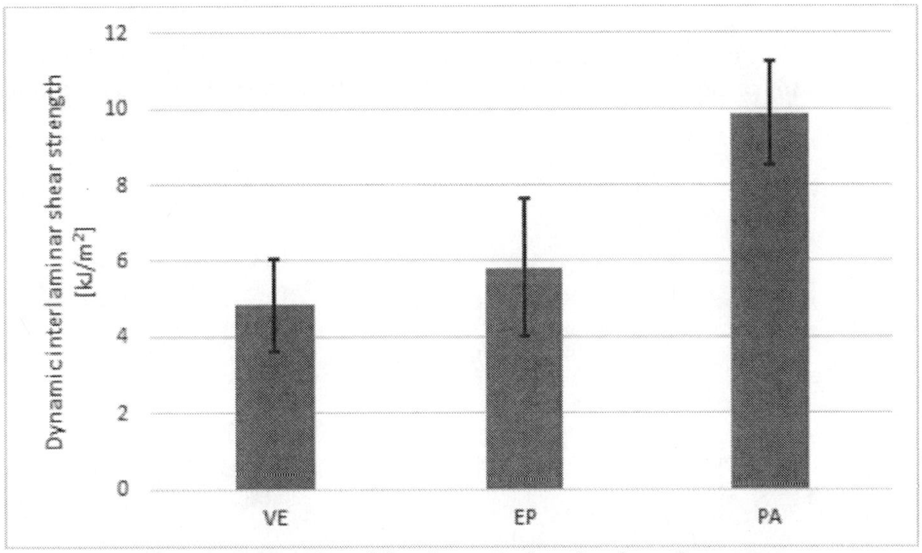

Figure 3. Results of the dynamic interlaminar shear tests
(VE- fiber sizing for VE, EP – fiber sizing for EP, PA – fiber sizing for PA)

The results of the cyclic interlaminar shear tests are presented in Figure 4. The PA sized fabrics performed much better. The load levels were defined based on the static interlaminar shear strength of the individual materials, so in case of the PA sized system the applied load was approximately two times higher compared to the other systems. Considering this, the performance of the PA sized system is even better. In case of fatigue loading of composites, the main failure process is not the propagation of an individual crack, but the formation of numerous micro-cracks in the system at the weakest points, which are most of the cases at the fiber-matrix interface. By improving the adhesion with proper sizing, the formation of these cracks can be avoided or at least postponed.

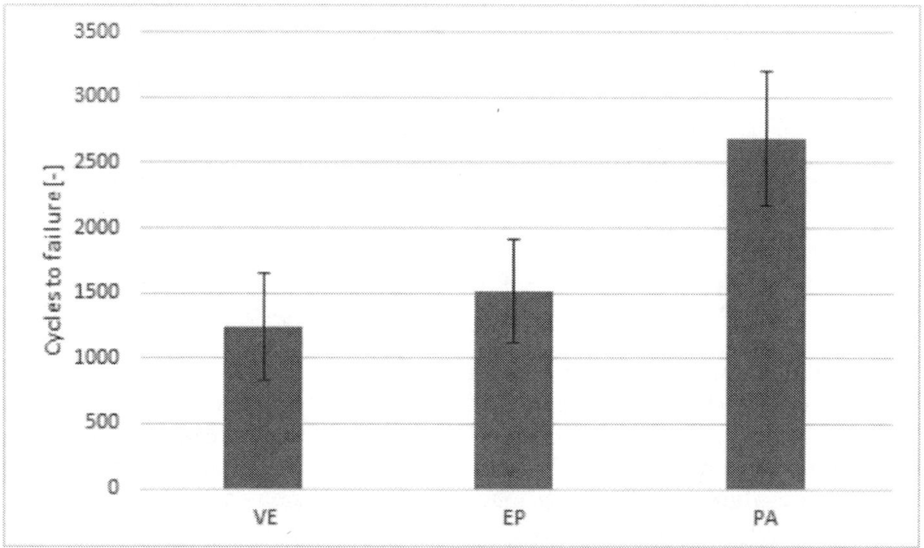

Figure 4. Results of the cyclic interlaminar shear tests
(VE- fiber sizing for VE, EP – fiber sizing for EP, PA – fiber sizing for PA)

The SEM micrographs of the different composites are presented in Figure 5. There is no significant difference in the structure of the composites. In all cases some trapped-in air voids and poorly impregnated rowings can be observed, which are common in composites produced by film-stacking. Although the mechanical performance showed clear difference in the adhesion between the fibers and the matrix, poor wetting or fiber matrix debonding at the interface are only present occasionally at all of the composites.

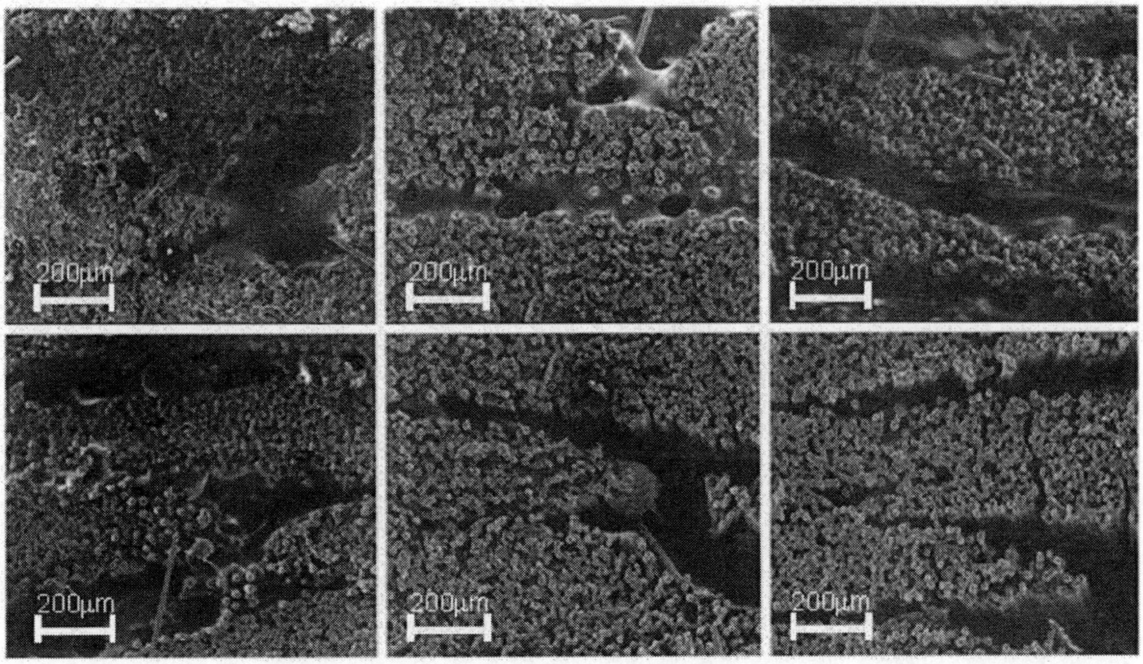

Figure 5. SEM micrographs of the composites containing reinforcement fibers with VE (left), EP (center) and PA (right) compatible sizing

4. Conclusions

According to the test results the proper fiber sizing is of key importance for the tested composite systems. In case of all, static, dynamic and cyclic interlaminar properties, the proper sizing showed significantly better mechanical performance and lifetime. Although the polar chemical structure of the binding groups of PA is similar to the structure of common thermoset resins used for reactive technologies in the composite industry, but the molecular chains, which are non-polar are much longer between the binding groups so the intermolecular forces will be significantly lower. Also the impregnation capability of the thermoplastic melt are much lower than in case of a thermoset resin, so more dry fiber bundles and trapped-in air voids have to be considered. With the rapid growth of the application of the reactive T-RTM technology more and more reinforcement manufacturers enter the market of PA compatible sized reinforcement materials, each with their own sizing composition. The methodology presented in our paper is capable for the determination of both static and cyclic interlaminar properties and the evaluation of new fiber sizing types.

Acknowledgements

This research was supported by the National Research, Development and Innovation Office (NVKP_16-1-2016-0046, OTKA FK 124352).

References

[1] Meola C, Boccardi S and Carlomagno G m 2017 *Infrared Thermography in the Evaluation of Aerospace Composite Materials,* ed C Meola, *et al.*: Woodhead Publishing) pp 1-24

[2] Holmes M 2017 Aerospace looks to composites for solutions *Reinforced Plastics* **61** 237-41

[3] GangaRao H 2017 *Applied Plastics Engineering Handbook (Second Edition),* ed M Kutz: William Andrew Publishing) pp 675-95

[4] Patil A, Patel A and Purohit R 2017 An overview of Polymeric Materials for Automotive Applications *Materials Today: Proceedings* **4** 3807-15
[5] Effing M 2017 Expert insights in Europe's booming composites market *Reinforced Plastics*
[6] Uribe B E B, Chiromito E M S, Carvalho A J F and Tarpani J R 2017 Low-cost, environmentally friendly route for producing CFRP laminates with microfibrillated cellulose interphase *Express Polymer Letters* **11** 47-59
[7] Subramani M, Arumugam A B and Ramamoorthy M 2017 Vibration Analysis of Carbon Fiber Reinforced Laminated Composite Skin with Glass Honeycomb Sandwich Beam Using HSDT *Periodica Polytechnica Mechanical Engineering; Vol 61 No 3 (2017)*
[8] Laurenzi S, Grilli A, Pinna M, De Nicola F, Cattaneo G and Marchetti M 2014 Process simulation for a large composite aeronautic beam by resin transfer molding *Composites Part B: Engineering* **57** 47-55
[9] 2015 New BMW 7 series features large CFRP parts *Reinforced Plastics* **59** 212
[10] 2017 Volvo adopts composite parts *Reinforced Plastics* **61** 116
[11] Fais C 2011 Lightweight automotive design with HP-RTM *Reinforced Plastics* **55** 29-31
[12] Maazouz A, Lamnawar K and Dkier M 2018 Chemorheological study and in-situ monitoring of PA6 anionic-ring polymerization for RTM processing control *Composites Part A: Applied Science and Manufacturing*
[13] Verrey J, Wakeman M D, Michaud V and Månson J A E 2006 Manufacturing cost comparison of thermoplastic and thermoset RTM for an automotive floor pan *Composites Part A: Applied Science and Manufacturing* **37** 9-22
[14] Yao T-T, Wu G-P and Song C 2017 Interfacial adhesion properties of carbon fiber/polycarbonate composites by using a single-filament fragmentation test *Composites Science and Technology* **149** 108-15
[15] Kishi H, Nakao N, Kuwashiro S and Matsuda S 2017 Carbon fiber reinforced thermoplastic composites from acrylic polymer matrices: Interfacial adhesion and physical properties *Express Polymer Letters* **11** 334-42
[16] Szebényi G, Magyar B and Iványicki T 2017 Comparison of static and fatigue interlaminar testing methods for continuous fiber reinforced polymer composites *Polymer Testing* **63** 307-13
[17] Liu Y, Yang J-P, Xiao H-M, Qu C-B, Feng Q-P, Fu S-Y and Shindo Y 2012 Role of matrix modification on interlaminar shear strength of glass fibre/epoxy composites *Composites Part B: Engineering* **43** 95-8
[18] Charkviani R V, Pavlov A A and Pavlova S A 2017 Interlaminar Strength and Stiffness of Layered Composite Materials *Procedia Engineering* **185** 168-72
[19] McKeen L W 2017 *Film Properties of Plastics and Elastomers (Fourth Edition)*: William Andrew Publishing) pp 187-227

Fatigue properties of EP/A356 aluminium matrix syntactic foams with different densities

A Szlancsik[1,2], B Katona[1,2], I N Orbulov[1,2], M Taherishargh[3] and T Fiedler[3]

[1]Department of Materials Science and Engineering, Budapest University of Technology and Economics, Budapest, Hungary
[2]MTA–BME Lendület Composite Metal Foams Research Group, Budapest, Hungary
[3]The University of Newcastle, School of Engineering, University Drive, Callaghan, Australia

E-mail: szlancsik@eik.bme.hu

Abstract. The fatigue properties of EP/A356 aluminium matrix syntactic foams were measured. Four different type of syntactic foam were produced in order to determine the relation between the density and the fatigue properties. The density range was set to be between 0.7 – 1.1 g/cm^3. The measured values indicated that, both the fatigue limit and the slope of the Wöhler curve increased with the density. However, the comparison of Globocer filled aluminium matrix syntactic foams with EP filled aluminium matrix syntactic foams suggested that this trend only applicable to foams with identical filler materials.

1. Introduction

Metal matrix syntactic foams (MMSFs) are materials with porous structure. This makes them a light weight material which is really important nowadays. Taking advantage of this property machine parts can be lighter what makes them more energy efficient, which is one of the main goals of the modern designers [1].

Based on the structure of this material it is mostly used as energy absorber. The compressive behaviour has been widely investigated in the literature [2-10]. It is well known that the compressive strength and the energy absorption of metal matrix syntactic foams (MMFs) are good. The wear and the thermal properties of these materials are also been studied [11-16].

Regarding the compressive properties of MMSFs only a few research contributions can be found in the databases. An interesting work on composite metallic foams (CMFs) that contain steel hollow spheres in aluminium alloy matrix (produced and manufactured by simple gravity casting) or in steel matrix (produced by a simple powder metallurgy method) was published by the research group of Vendra et al. Under repeated compression load, the CMFs proved high cyclic stability and the deformation of the CMF samples could be separated into three stages: (i) linear increment in strain with fatigue cycles, (ii) minimal deformation accumulation through relatively large number of cycles and (iii) fast deformation accumulation during a few cycles up to complete fracture. The deformation of the MMSFs occurred to be more uniform compared to regular metal foams, which deform by forming collapse bands at weaker sections [17].

On the other hand, Katona et al. [18] published on MMSFs, consisting of two grades of aluminium alloys (Al99.5 and AlSi12) and a set of specific oxide ceramic hollow spheres. The MMSFs were studied in the viewpoint of repeated loading. The final results of the repeated compressive load with the load asymmetry factor of R=0.1 provided full reliability design data for the studied materials in the whole lifetime part of the Wöhler curve, besides the fatigue limits were determined by the so called staircase method. Based on the experiments the Wöhler curves of the foams were plotted, considering the median curves, their given confidence statistical boundaries and the fatigue limit. Considering the matrix materials, the softer matrix proved higher load levels for the fatigue limit than the more rigid AlSi12 matrix. Regarding the average diameter of the filler ceramic hollow spheres, larger spheres proved to be better than the quite easily vulnerable smaller ones. Considering the failure modes, only one common fracture mode was separated for the studied foams: the samples were crashed along a single shear band, similar to the quasi-static case.

Other publications are available on the fatigue of 'conventional' metallic foams [19-21], from which a few generalities can be summarized: most of the investigations apply R=0.1 stress asymmetry factor, with pure compression load, however the test frequencies can differ significantly, however, its effect on the fatigue properties was not considered.

2. Materials and methods

2.1. Materials

A356 alloy was applied as matrix material, its chemical composition is listed in Table 1. As filler, expanded perlite was applied which was obtained from Australian Perlite Pty. and they had a size range of 3 – 4 mm. The chemical composition of the filler is also listed in Table 1.

Table 1. The chemical composition of A356 aluminium alloy and the expanded perlite in weight percent

	Chemical elements in weight percent							
Matrix material:	Si	Mg	Fe	Ti	Al			
A356	7.2	0.4	0.1	0.12	Balance			
Filler material:	SiO_2	Al_2O_3	Na_2O	K_2O	CaO	Fe_2O_3	MgO	TiO_2
Expanded perlite	75	14	3	4	1.3	1	0.3	0.2

The counter gravity infiltration process which was used to produce the EP/A356 syntactic foam is described in details in Taherishargh et al [22]. However, a modification was used: namely the compaction procedure, which is used to tailor the density values of the produced syntactic foams. This procedure is detailed in Broxtermann et al [23]. Four different density range were created using compaction masses. 5, 10 and 15 kg compaction masses were used, and also a reference type also created without compaction mass. The sample names refer to the producing process, samples with "1st batch" means that these samples were created with 15 kg compaction mass, samples with "2nd batch" designation refers to the 10 kg compaction mass, while "3rd batch" means 5 kg compaction mass, and finally "4th batch" refers to those samples which were created without compaction mass.

2.2. Fatigue test procedure

The produced syntactic foams were firstly investigated under quasi-static compression tests. These results are also detailed in Boxtermann et al [23]. This was necessary because the 1% compression strength is the base for the calculations of the load levels for the fatigue tests. Figure 1. depicts the scheme of the fatigue cycle with the most important test parameters.

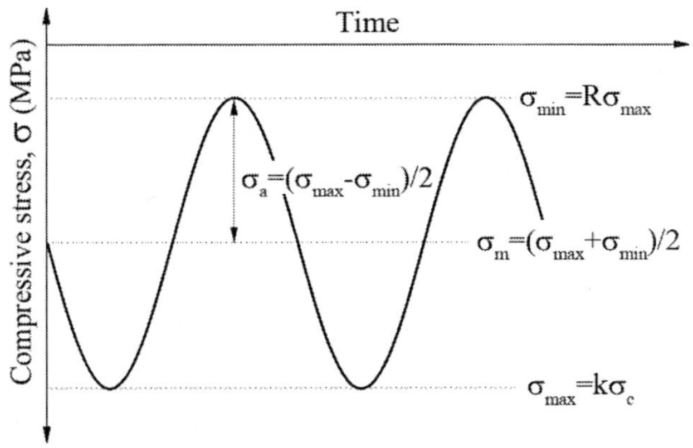

Figure 1. Parameters of the fatigue test

Figure 2. The measurement setup with the 10 kN load cell

The maximal stress was calculated from the 1% compressive stress (σ_c) – determined from the quasi-static tests –, by multiplying it with k, that is the load ratio which was varied between 70 – 100%. R is the asymmetry factor which was always set to R = 0.1. The frequency of the test was f = 10 Hz and the fatigue limit was taken as $2{\cdot}10^6$ cycle. The failure criterion was 2% engineering deformation according to the literature [24, 25]. The samples' diameter was D = 27 mm, and the height was H = 42 mm (the aspect ratio was H/D = 1.56). The fatigue tests were performed on a MTS 810 hydraulic testing machine, with a 10 kN load cell if the maximal load was under 8 kN. Otherwise a 250 kN load cell was used. The displacement was measured with the crosshead displacement due to the relatively large sample dimensions. Figure 2. shows the measurement setup.

3. Results and discussion

First, we had to determine the maximum stress for the fatigue tests. In order to get this data for every density a plot had to be created. Figure 3. depicts the density – compressive stress diagram, where the stress values are equal to the 1% compressive stress from the quasi-static measurement.

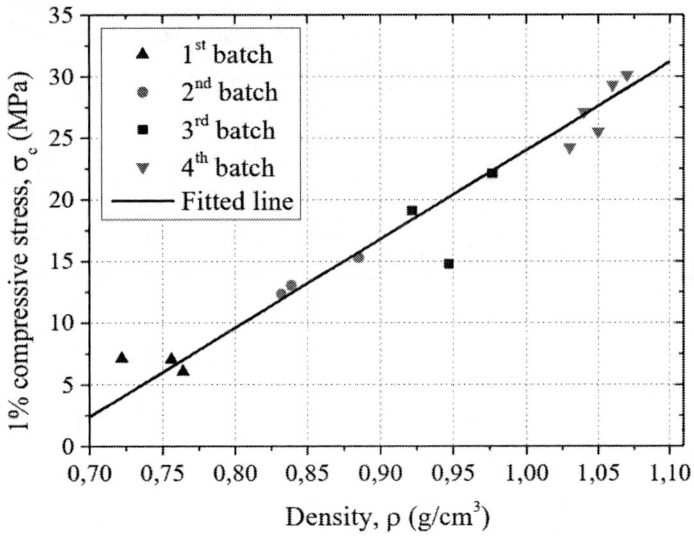

Figure 3. 1% compressive stress in function of the density

IOP Conf. Series: Materials Science and Engineering **426** (2018) 012045 doi:10.1088/1757-899X/426/1/012045

The 1% compressive strength showed linear relationship to the density, the fitted line's equation is

$$\rho = -48.28 + 72.86\sigma_c \tag{1}$$

With this equation the maximum stress was calculated individually, considering the load ratio too. From the maximal stress and R value, all of the other cyclic load parameters can be easily derived. Table 2. summarizes the properties of the specimens and the calculated parameters for the fatigue test.

Total duration of the measurements was ~ 700 h. Two types of failure mechanism were detected. These failure modes are similar to what Taherishargh et al. [26] found in their research. Type I failure mode means that a shear band appears across the cylindrical specimen with an angle of ~45° and the specimens usually completely fell apart along this shear band. Type II has a different method. The first part is similar to type I, namely a 45° shear band appears, but after a relatively small deformation the failure slows down because of the hardening of the material. This deceleration of the failure creates a plateau in the engineering deformation – number of cycles diagram. After this plateau a new shear band appears, which is perpendicular to the previous one. After the second shear band appears the samples completely fall apart. Figure 4. shows the mechanism of the failure types [26]. Type I typically occurred at those samples which survived at least 10.000 cycles, while type II occurred at lower lifetime. Type I only has one peak at the deformation speed diagram, while type II has a plateau zone after the first peak.

The Wöhler curves were plotted in a load ratio – failure cycle number graph (Figure 5.).

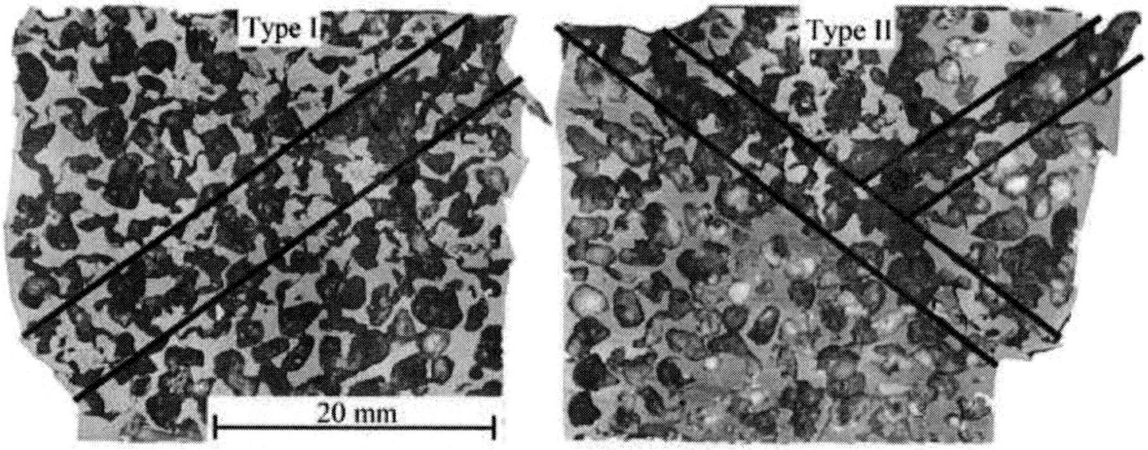

Figure 4. The cross-section of samples showing the deformation mechanism at low and high loading ratios (type I and type II) [26]

Table 2. Sample properties and the individual parameters for the fatigue test

Sample (batch num.-sample num.)	Density, ρ (g/cm^3)	Calculated compressive strength, σ_c (MPa)	Load ratio, k (-)	Minimal stress, σ_{min} (MPa)	Maximal stress, σ_{max} (MPa)
1-1	0.78	8.55	0.9	7.70	0.77
1-2	0.75	6.36	0.8	5.09	0.51
1-3	0.74	5.64	0.8	4.51	0.45
1-4	0.73	4.91	0.9	4.42	0.44
1-5	0.73	4.91	0.7	3.44	0.34
1-6	0.71	3.45	0.9	3.11	0.31
1-7	0.69	1.99	0.8	1.59	0.16

Table 2. Sample properties and the individual parameters for the fatigue test (continued)

Sample (batch num.-sample num.)	Density, ρ (g/cm^3)	Calculated compressive strength, σ_c (MPa)	Load ratio, k (-)	Minimal stress, σ_{min} (MPa)	Maximal stress, σ_{max} (MPa)
2-1	0.88	15.84	0.7	11.09	1.11
2-2	0.87	15.11	0.8	12.09	1.21
2-3	0.87	15.11	0.8	12.09	1.21
2-4	0.86	14.38	0.9	12.94	1.29
2-5	0.86	14.38	0.9	12.94	1.29
2-6	0.86	14.38	0.7	10.07	1.01
2-7	0.86	14.38	0.9	12.94	1.29
2-8	0.85	13.65	0.8	10.92	1.09
3-1	1.00	24.58	0.9	22.12	2.21
3-2	1.00	24.58	0.8	19.66	1.97
3-3	1.00	24.58	1.0	24.58	2.46
3-4	0.99	23.85	1.0	23.85	2.39
3-5	0.99	23.85	0.9	21.47	2.15
3-6	0.98	23.12	1.0	23.12	2.31
4-1	1.07	29.68	1.0	29.68	2.97
4-2	1.06	28.95	0.95	27.50	2.75
4-3	1.06	28.95	0.95	27.50	2.75
4-4	1.05	28.22	0.85	23.99	2.40
4-5	1.04	27.49	0.95	26.12	2.61
4-6	1.04	27.49	0.85	23.37	2.34

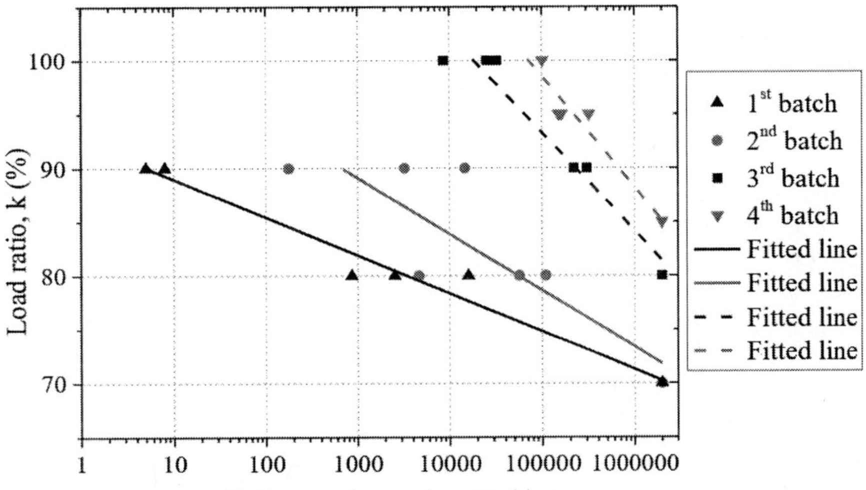

Figure 5. Load ratio – failure cycle number diagram

It can be seen that the slope of the Wöhler curves are increasing with the density, and the fatigue limit also increasing with the density. The Wöhler curves were compared to another type of syntactic foams, namely AlSi12 matrix filled with Globocer balloons [18]. The results of the two fatigue tests can be compared and proves that Wöhler curves dependency on the density only works with the same type of syntactic foam, because this type of syntactic foam has a density of 1.8 g/cm^3 (Figure 6.) and does not fit into the trends of the EP filled composites. This phenomenon indicates that the filler has significant effect on the fatigue properties of the MMSFs.

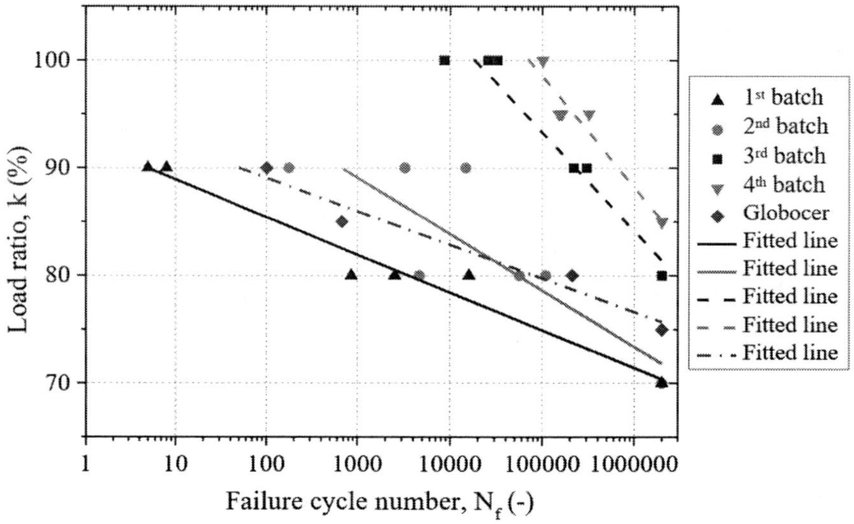

Figure 6. Load ratio – failure cycle number diagram supplemented with an AlSi12-Globocer syntactic foam's results

A large difference between the two type of syntactic foam is in the filler material, and in the shape of the quasi-static compressive curve [27-29]. The Globocer filled syntactic foams have a defined peak stress, which means if these type of materials go through that stage, it will collapse, but in the other hand, the EP filled syntactic foams do not have such a peak in their quasi-static curves, which means that if they exceed the 1% compression stress point they will only be stronger. This can cause, that the dependency from the density not apply for the Globocer filled syntactic foams.

4. Conclusions

From the investigation which was described above, the following conclusions can be drawn:

- The compaction method is a proper way to produce different density range syntactic foams from expanded perlite and A356 aluminium alloy.
- The slope of the Wöhler curves depends on the density of the syntactic foam.
- The fatigue limit depends on the density of the syntactic foams.
- The slope of the Wöhler curves and the fatigue limit dependency from the density only applies for the syntactic foams made from the same filler material.

Acknowledgment

This paper was supported by the János Bolyai Research Scholarship of the Hungarian Academy of Sciences and by the ÚNKP-17-3-I New National Excellence Program of the Ministry of Human Capacities.

References

[1] Gupta N and Rohatgi P K 2014 Metal Matrix Syntactic Foams *DEStech Publications*, Inc., Lancaster, Pennsylvania, USA

[2] Mondal D P, Goel M D and Das S 2009 Compressive deformation and energy absorption characteristics of closed cell aluminum-fly ash particle composite foam *Mat. Sci. Eng.* A **507** pp 102–109

[3] Katona B and Orbulov I N 2017 Structural Damages in Syntactic Metal Foams Caused by Monotone or Cyclic Compression *Periodica Polytechnica Mechanical Engineering* **61(2)** pp 146–152

[4] Mondal D P, Jha N, Badkul A, Das S and Khedle R 2012 High temperature compressive deformation behaviour of aluminum syntactic foam *Mater. Sci. Eng.* A **534** pp 521–529

[5] Zhang L P and Zhao Y Y 2007 Mechanical Response of Al Matrix Syntactic Foams Produced by Pressure *Journal of Composite Materials* **41 (17)** pp 2105–2117

[6] Taherishargh M *et al* 2014 On themechanical properties of heat-treated expanded perlite–aluminium syntactic foam *Mater Des.* **63** pp 375–383

[7] Luong D D, Shunmugasamy V C, Gupta N, Lehmhus D, Weise J and Baumeister J 2015 Quasi-static and high strain rates compressive response of iron and Invar matrix syntactic foams *Mater. Des.* **66** pp 516–531

[8] Peroni L, Scapin M, Avalle M, Weise J, Lehmhus D, Baumeister J and Busse M 2012 Syntactic iron foams – on deformation mechanisms and strain-rate dependence of compressive properties *Adv. Eng. Mater.* **14** pp 909–918

[9] Zou L C, Zhang Q, Pang B J, Wu G H, Jiang L T and Su H 2013 Dynamic compressive behavior of aluminum matrix syntactic foam and its multilayer structure *Mater. Des.* **45** pp 555–560

[10] Goel M D, Mondal D P, Yadav M S and Gupta S K 2014 Effect of strain rate and relative density on compressive deformation behavior of aluminum cenosphere syntactic foam *Mater. Sci. Eng.* A **590** pp 406–415

[11] Mondal D P, Jha N, Badkul A, Das S and Khedle R 2012 High temperature compressive deformation behaviour of aluminum syntactic foam *Mater. Sci. Eng.* A **534** pp 521–529

[12] Májlinger K, Bozóki B, Kalácska G, Keresztes R and Zsidai L 2016 Tribological properties of hybrid aluminum matrix syntactic foams *Tribology International* **99** pp 211–223

[13] Májlinger K 2015 Wear properties of hybrid AlSi12 matrix syntactic foams *Int. Jour. of Mat. Res.* **106 (11)** pp 1165–1173

[14] Szalóki I and Viharos Zs J 2018 Surface Structure Analysis of Syntactic Metal Foams Machined by Milling *Periodica Polytechnica Mechanical Engineering* **62 (1)** pp 26–32

[15] Fiedler T, Belova I V and Murch G E 2015 On the thermal properties of expanded perlite – Metallic syntactic foam *International Journal of Heat and Mass Transfer* **90** pp 1009–1014

[16] Bidar B, Shahraki F and Kalhori D M 2016 3D Numerical Modelling of Convective Heat Transfer through Two-sided Vertical Channel Symmetrically Filled with Metal Foams *Periodica Polytechnica Mechanical Engineering* **60 (4)** pp 193–202

[17] Vendra L, Neville B and Rabiei A 2009 Fatigue in aluminum–steel and steel–steel composite foams *Mater. Sci. Eng.* A **517** pp 146–153

[18] Katona B, Szebényi G and Orbulov I N 2017 Fatigue properties of ceramic hollow sphere filled aluminium matrix syntactic foams *Materials Science & Engineering* A **679** pp 350–357

[19] Kafka O L, Ingraham M D, Morrison D J *et al* 2014 Characterization of fatigue fractures in closed-cell aluminum foam using X-ray micro-computed tomography *J Mater Eng Perform* **23** pp 759–765

[20] Zhou J and Soboyejo W O 2004 Compression-compression fatigue of open cell aluminum foams: macro-/micromechanisms and the effects of heat treatment *Mater Sci Eng* A **369** pp 23–35

[21] Vendra L, Neville B and Rabiei A 2009 Fatigue in aluminumsteel and steel-steel composite foams *Mater Sci Eng* A **517** pp 146–153

[22] Taherishargh M B, Murch G E and Fiedler T 2014 Low-density expanded perlite–aluminium syntactic foam *Mater Sci Eng A* **604** pp 127–134

[23] Broxtermann S, Taherishargh M, Belova I V, Murch G E and Fiedler T 2017 On the compressive behaviour of high porosity expanded Perlite-Metal Syntactic Foam (P-MSF) *Journal of Alloys and Compounds* **691** pp 690-697

[24] Sugimura Y, Rabiei A, Evans A G *et al* 1999 Compression fatigue of a cellular Al alloy *Mater Sci Eng* A **269** pp 38–48

[25] Harte A M, Fleck N A and Ashby M F 1999 Fatigue failure of an open cell and a closed cell aluminum alloy foam. *Acta Mater* **47** pp 2511–2524

[26] Taherishargh M, Katona B, Fiedler T and Orbulov I N 2017 Fatigue properties of expanded perlite/aluminum syntactic foams *Journal of Composite Materials* **51 (6)** pp 773–781

[27] Taherishargh M, Sulong M A, Belova I V, Murch G E and Fiedler T 2015 On the particle size effect in expanded perlite aluminium syntactic foam *Materials and Design* **66** pp 294–303

[28] Orbulov I N 2012 Compressive properties of aluminium matrix syntactic foams *Materials Science and Engineering:* A **555** pp 52–56

[29] Májlinger K and Orbulov I N 2014 Characteristic compressive properties of hybrid metal matrix syntactic foams *Materials Science and Engineering:* A **606** pp 248–256

IOP Conf. Series: Materials Science and Engineering **426** (2018) 012046 doi:10.1088/1757-899X/426/1/012046

Strain induced martensitic transformations in FeMnCr TWIP steel

A Talgotra[1], E Nagy[2], M Sepsi[1], M Benke[1] and V Mertinger[1]

[1] University of Miskolc, Institute of Physical metallurgy, Metalforming and Nanotechnology, Miskolc-Egyetemváros, Hungary
[2] MTA-ME Material Science Research Group, Miskolc-Egyetemváros, Hungary

E-mail: femarjun@uni-miskolc.hu

Abstract. Martensitic transformations in FeMnCr steel were investigated by the thermomechanical treatments achieved by uniaxial tensile tests with different temperature and strain rate. During thermomechanical treatments, when a well-controlled combination of the temperature and the deformation rate is applied an extraordinary combination of microstructures for a strain induced α' and/or ε martensite and thermally induced ε martensite can be achieved in this steel. The volume fractions of the phases after tested at different temperatures were determined and the phase ratios of individual martensite were observed.

1. Introduction

Austenitic steels are best suited for automobile applications including press-formed parts because of their energy absorption or structural reinforcement. Austenitic FeMnCr steels have high strength, high toughness and formability because of the martensitic transformation. In these steels the temperature as well as strain-induced non thermoelastic martensitic phase transformation takes place. This is the so-called TRIP (transformation induced plasticity) and TWIP (twining induced plasticity) effect. TWIP steel can deform by both glide of individual dislocations and mechanical twinning [1, 2].

It has been known, that during plastic deformation of austenitic steels austenite can transform into α' martensite. Two transformation mechanisms have been observed. The first one describes the direct formation of α' martensite from austenite, while the second one involves a two-step reaction in which α' phase is formed from ε martensite. The austenite–martensite transformation depends on composition, deformation rate and temperature [3, 4]. During thermomechanical treatments, when a well-controlled combination of the temperature and the deformation rate is applied an extraordinary combination of microstructures can be achieved. Because of the ratio and quality of resulted phases determines the properties of the product an extraordinary mechanical property can be predicted after or during the treatments [5–8].

Although many articles/research papers have been written/published in the past on the Martensitic transformations in the TWIP/TRIP steels, the novelty of this project was to present the complex martensitic behaviour in the High Cr (Chromium) Austenitic FeMnCr TWIP steels, which is quite different from various other martensitic transformations which consists of two phases rather than the three phases system as observed in this work. The presence of Alpha and Epsilon martensites along with Austenite at the same time makes it complicated to determine the mechanical properties of such type of steels. In this work the main task was to determine the volume fraction (composition) of the martensites when the test temperature and percentage of deformation were varied.

2. Experimental approach

Alloy was produced at TU Bergakademie Freiberg. The composition of examined alloy can be seen in Table 1. The cast ingots were hot rolled to rods with diameter of 10 mm. Tensile test specimens and samples for transformation temperature examinations were machined from the rods. The specimens were solution treated at 1000 °C for 30 minutes under argon atmosphere and subsequent water quench was applied. The $\varepsilon \leftrightarrow \gamma$ transformation temperatures of the quenched alloy was determined by DSC (Differential Scanning Calorimetry) can be seen in Table 2.

Table 1. Composition of Steel, [wt %]

	C	Mn	Cr	Si	P	S
Steel 1	0.026	17.7	2.26	0.1	0.0051	0.029

Table 2. Martensite and Austenite start-finish temperatures [°C]

	A_s	A_f	M_s	M_f
Steel 1	194	224	149	121

The thermomechanical treatments were achieved by uniaxial tensile test in climate chamber of the 100 kN maximal load multipurpose Instron 5982 floor installed equipment in the Laboratory of Material Testing of Institute of Physical Metallurgy, Metal forming and Nanotechnology at the University of Miskolc. The samples in the climate chamber were heated up to 300 °C (in a separate furnace for achieving homogenization) before the tensile tests to reach pure austenitic state of the alloy. Then it was cooled down to the test temperature (Cooling substantially takes place inside the climate chamber). The test temperatures were selected as 200, 180, 160, 140, 125, 110 and 25°C. The samples were loaded up to the fracture, but the strain at 125 °C, 140 °C and 180 °C were also varied, the tests were interrupted at two smaller 0,25 and 0,35 true strain value.

After the tensile tests, specimens were cooled down to room temperature, and samples machined from the uniformly elongated part (gauge section) of the specimens for LM (Light Optical Microscopy) and SEM (Scanning Electron Microscopy) investigation. Grinding and mechanical polishing by Nital and Beraha etchants were used for sample preparation for microscopic investigation. The volume fraction of phases was determined on the cross section of sample. Full intensity profile fit XRD (X-Ray Diffractometry) methods was used.

3. Results and discussion

The true stress-true strain curves of fractured samples at different temperatures are shown on figure 1.
A sharp drop on the tensile strengths can be observed as the temperature transcends the 125 °C value while the elongation remains till 180 °C. Above this temperature a decrease happens in the value of elongation too. Whereas, at the lower temperatures i.e. between the room temperature and 125 °C, a slight increase in the elongation (decrease in cross-section area) and increase in the true stress without any change in the tensile strength is seen which also is an unusual manner of this type of steels.

To clarify of this attribute interrupted tensile tests were performed at 125 °C 140 °C and 180 °C. The true stress-true strain curves are shown on figure 2 - 4.

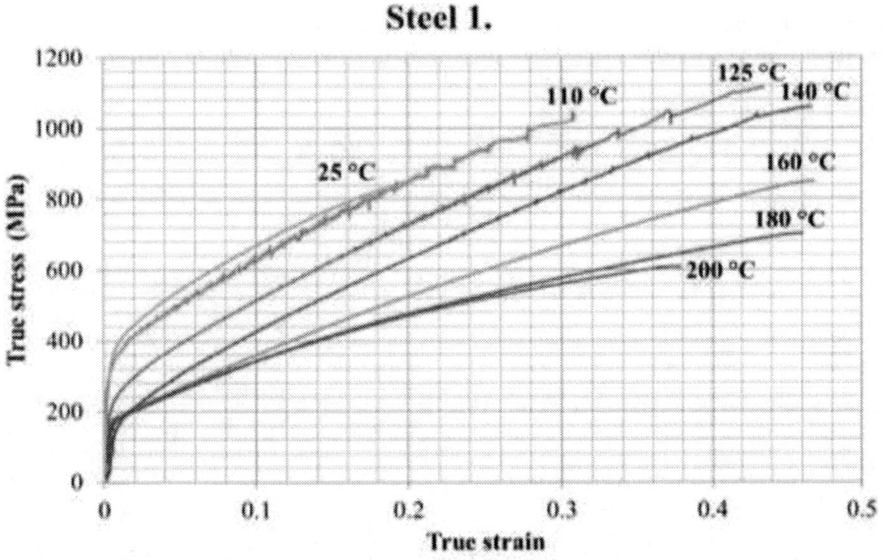

Figure 1 True stress- true strain curves of Steel 1 fractured at different temperature

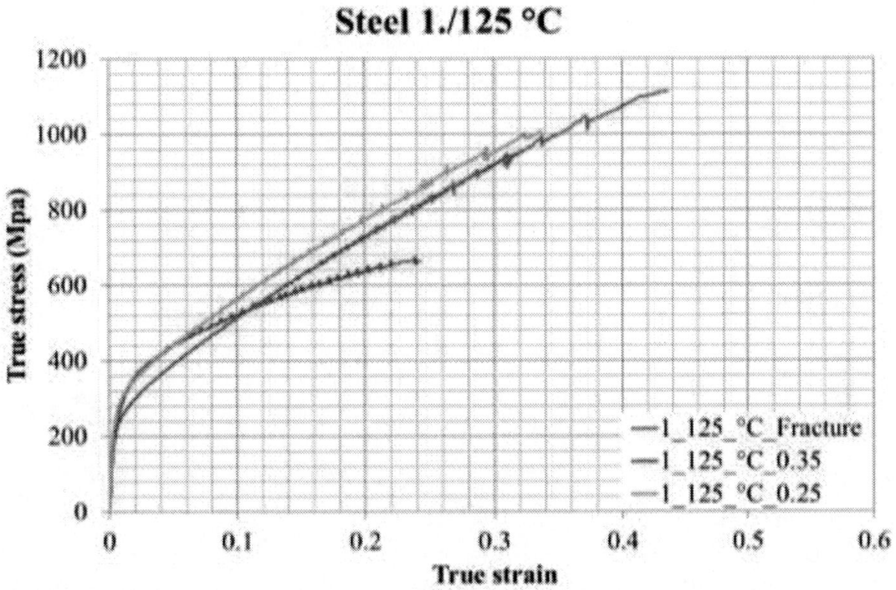

Figure 2. True stress-true strain curves of Steel 1 loaded at 125 °C, two interrupted and one up to fracture

In case of 180 °C the curves are identical. It is a remarkable effect that the fractured sample behaves very different from the two others at 140 °C. A very strong hardening effect can be observed, so a different sequence of the phase transformations can be predicted. If we see the transformation temperatures of this steel (Table 2.) we can conclude that the Ms temperature is very closed to the test temperature, so the thermally induced martensite transformation also can take place and interacts with the strain induced transformation. Just some thermal inhomogeneity of the climate chamber can produce such effect. The Ms temperature is much lower than 180 °C, so thermally induced martensite formation before the loading is not predictable in that case. The microscopic examination shows very fine microstructure (figure 5 and 6).

Zig-zag type needles and twins are also observable, the evaluation of the martensite formation is also clear in function of deformation rate, but to distinguish the volume fraction of the different martensites is impossible on based of microscopic examination.

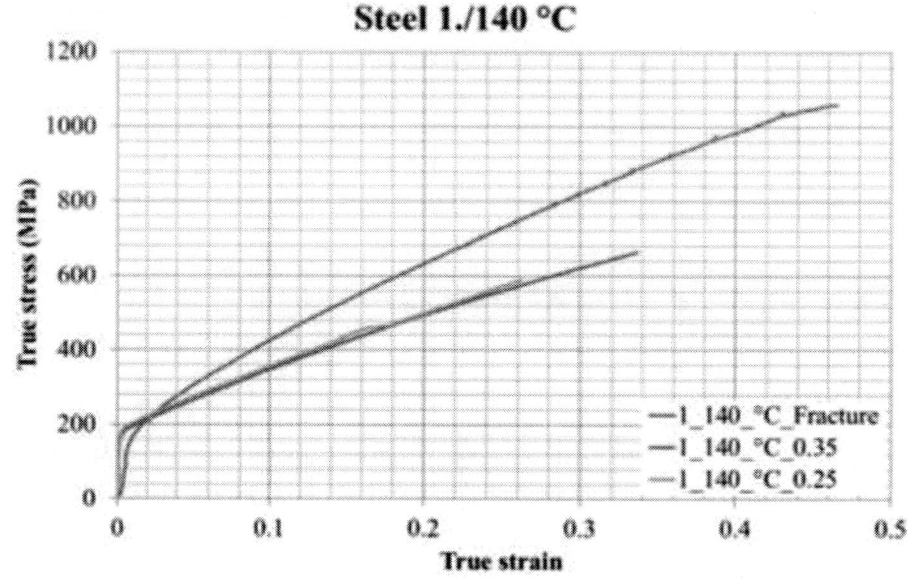

Figure 3. True stress-true strain curves of Steel 1 loaded at 140 °C, two interrupted and up to fracture

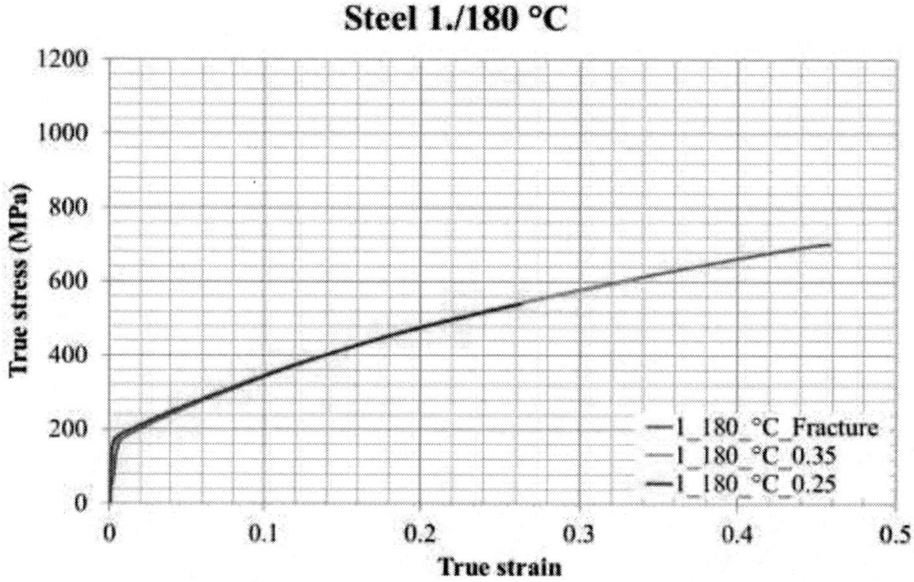

Figure 4. True stress-true strain curves of Steel 1 loaded at 180 °C, two interrupted and up to fracture

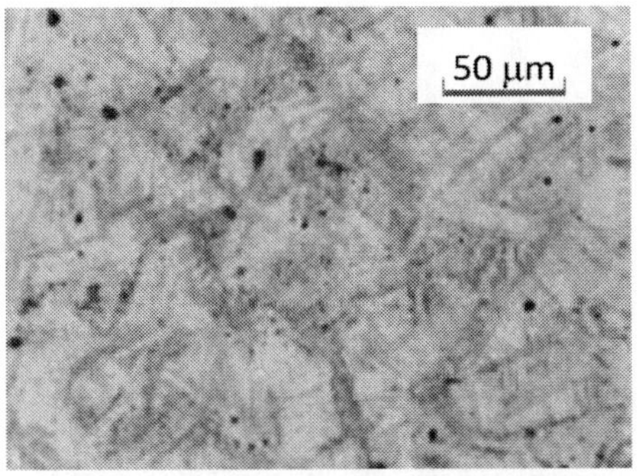

Steel 1: 125°C at 0,25

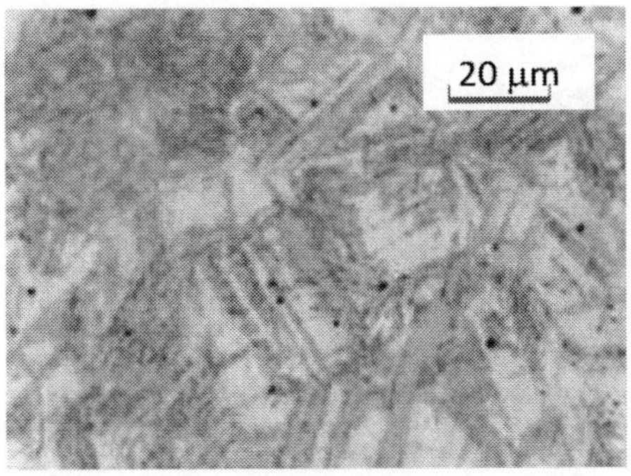

Steel 1: 125°C at 0,35

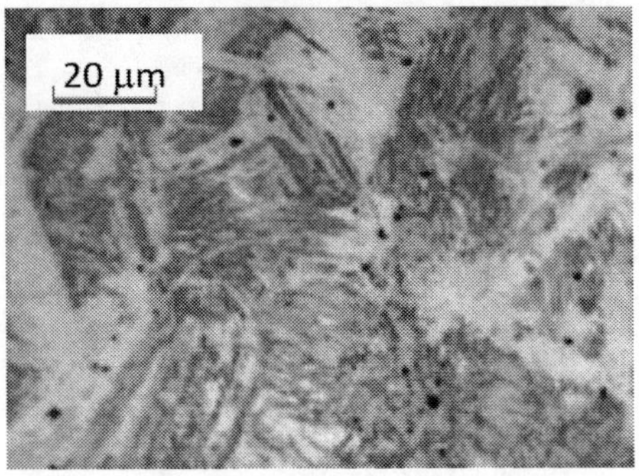

Steel 1: 140 °C at fracture

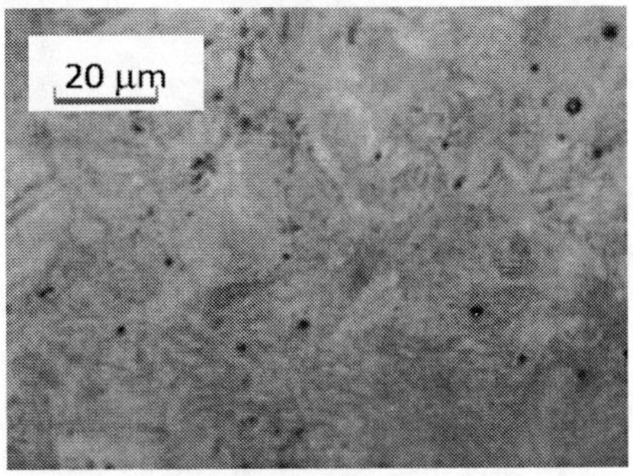

Steel 1: 180 °C at fracture

Figure 5. LM (Light optical microscopy) images of Steel loaded at 125 °C, 140 °C and 180 °C

As it was seen the steel tested at different temperature shows different behaviour in strength and strain value. This phenomenon can be related to the presence and the quantity of martensites. To identify the different phases XRD investigation was done. The figure 7 shows the XRD diffractograms of Steel 1 loaded at different temperatures and strain rate. The result of the volume fraction calculations is introduced at figure 8 and 9. It is shown the twining induced plasticity and the strain induced transformation coexists during deformation.

On the three highest temperature the volume fraction of the ε phase is lower after thermomechanical treatment than in the "as quenched" sample, so one can conclude in these samples the ε martensite forms during the subsequent cooling of tensile test and its volume is less because the plastic deformation stabilizes the austenite. This statement is confirmed by the figure 4 where the curves almost perfectly cover each other, so only the austenite phase suffer the transformation. On the lower isotherms significant differences can be observe on the tendency of curves what refer possible inhomogeneous phase constitutions. The volume fraction of ε martensite is higher at lower temperatures, so the formation of strain induced ε martensite is obvious. We have already shown in previous investigations the α' martensite can be only strain induced above the room temperature because Ms is below room temperature [9] [10]. It is in harmony with the results of partially loaded samples on the isotherm of 125 °C where the volume fraction is increasing by the strain rate. It can be declared that the volume fraction of the ε martensite is unchanged during the deformation and the increasing of the α' phase amount happens for the expense of austenite, what is not evident because α' may form also via ε phase.

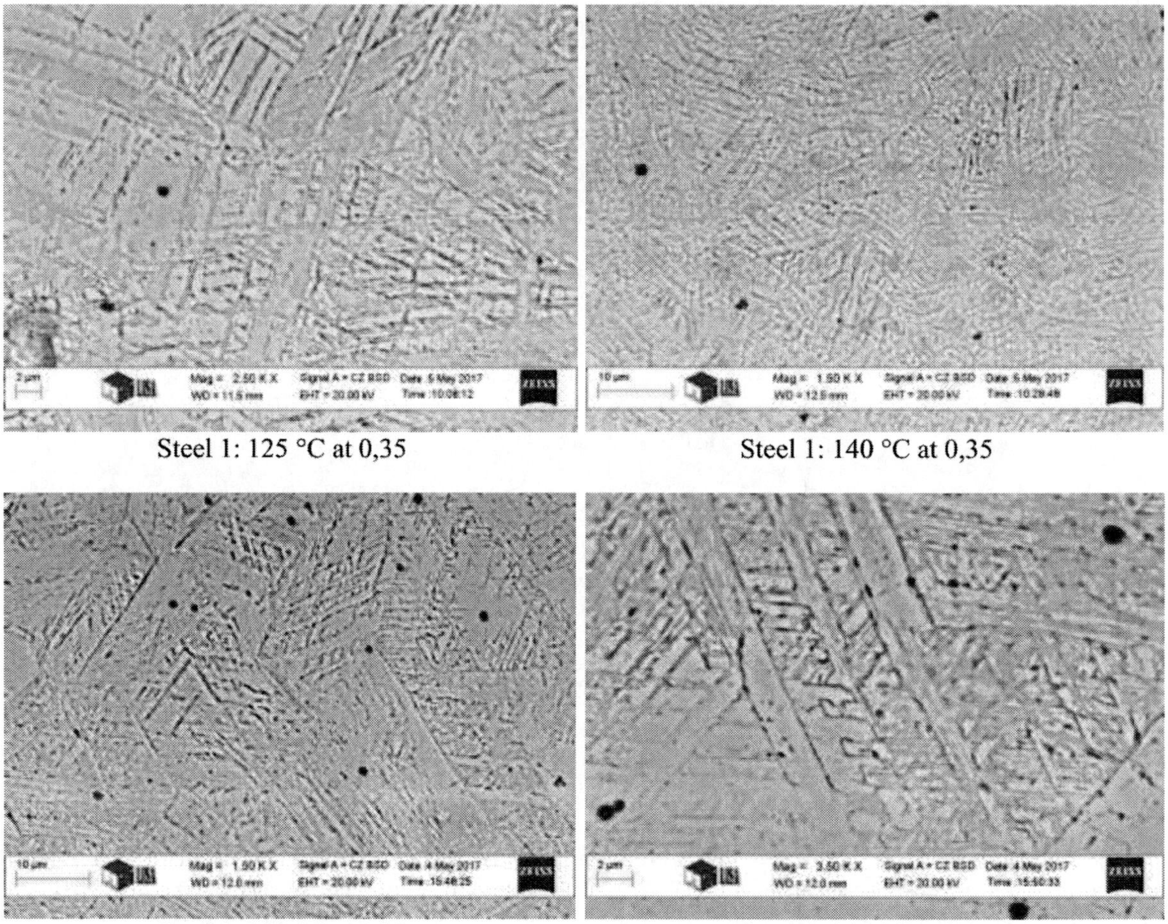

Steel 1: 125 °C at 0,35 | Steel 1: 140 °C at 0,35

Steel 1: 180 °C at 0,35 | Steel 1: 180 °C at 0,35

Figure 6. SEM (Scanning Electron Microscopy) images, deformation twins in Steel 1 loaded at: 125 °C, 140 °C and 180 °C

4. Summary

Solid state phase reactions in austenitic FeMnCr TWIP steel was studied induced by thermomechanical treatments. Uniaxial tensile load was performed at different temperatures (25 °C-180 °C) and different strain rate. XRD method was used to calculate the phase constitutions after the test. Based on the results the formation of different martensites (thermally or strain induced ε or strain induced α′) can be predicted. It was also concluded the deformation stabilizes the austenite.

Acknowledgement

The described article was carried out as part of the GINOP-2.3.2-15-2016-00027 “Sustainable operation of the workshop of excellence for the research and development of crystalline and amorphous nanostructured materials” project implemented in the framework of the Szechenyi 2020 program. The realization of this project is supported by the European Union. The described research was also supported through the National Research, Development and Innovation Office – NKFIH K119566 project

IOP Conf. Series: Materials Science and Engineering **426** (2018) 012046 doi:10.1088/1757-899X/426/1/012046

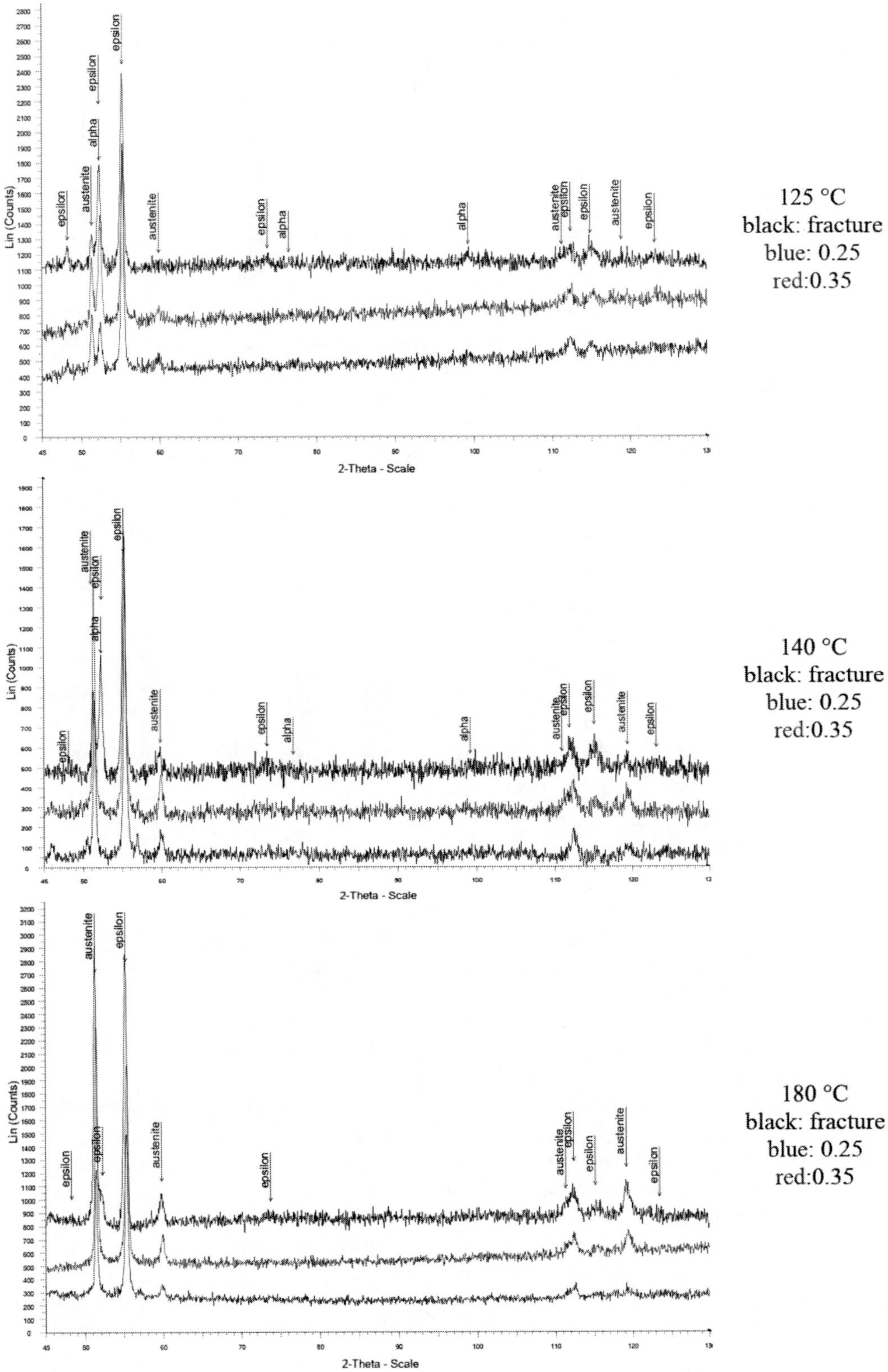

Figure 7. X-Ray diffractograms of Steel 1 loaded at different temperatures and strain rate

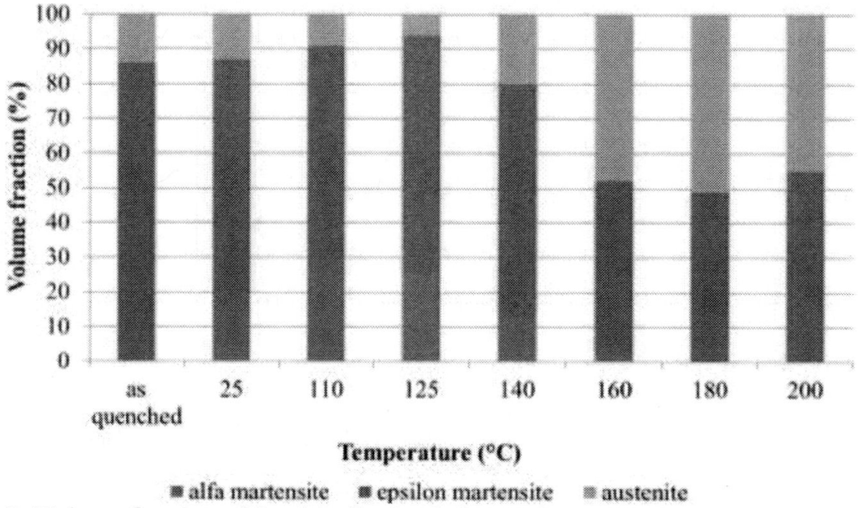

Figure 8. Volume fraction of Steel 1 after loaded at various test temperatures up to fracture

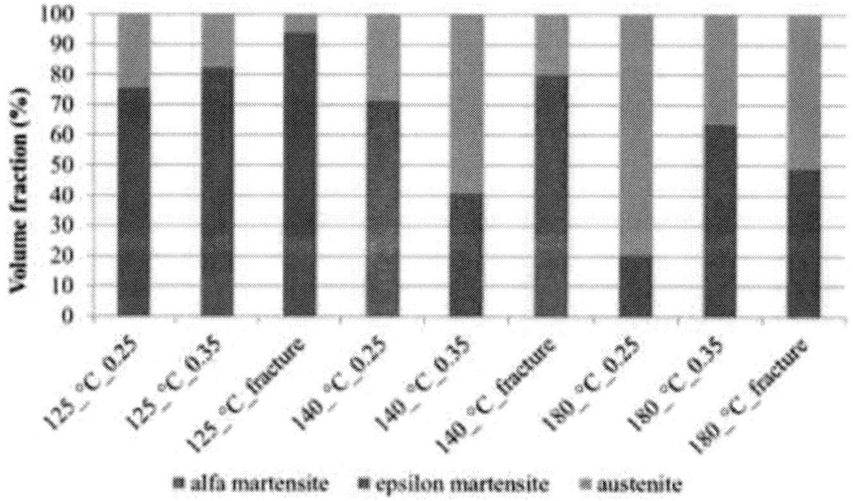

Figure 9. Volume fraction of Steel 1 at 125 °C, 140 °C and 180 °C test temperatures loaded up to different strain rate

References

[1] Bhadeshia H, Honeycombe R., 2000 *Steels, Microstructure and Properties* (Great Britain: Butterworth-Heinemann) 229
[2] Pla-Ferrand R., Sánches-Caballaero S., et al., 2011 TWIP/TRIP steels. Future trends in automotive industries *Fasc. of Manage. and Tech. Eng*. NR3
[3] Dash J., Otte H.M., 1963 *Acta Mater.* **11** 1169
[4] Han H.N., Lee G.G., et al. 2004 *Acta Mater.* **52** 5203
[5] Koyama M., Sawaguchi T., et al. 2011 Mat. Sci. Eng A **528** 7310
[6] Tang Z.Y., Misra R.D.K., et al., 2015 Mat. Sci. Eng A **624** 186
[7] Xu S., Ruan D., et al., 2013 Mat. Sci. Eng A **573** 132
[8] Shterner V., Timokhina I.B., Beladi H.: On the work-hardening behaviour of a high manganese TWIP steel at different deformation temperatures, Materials Science and Engineering A 669 (2016) 437-446
[9] Mertinger V., Erzsebet Nagy., 2008. *Materials Science and Engineering A* 718-722.
[10] Mertinger V., Erzsebet Nagy., 2015 *Materials Science Forum* **812**: 161-166.

IOP Conf. Series: Materials Science and Engineering **426** (2018) 012047 doi:10.1088/1757-899X/426/1/012047

Investigating the relationship between the theoretical layer thickness of the graphite lubricant film and the Kudo's friction number in the case of forged parts.

F Tancsics[1] and T Ibriksz[1]

[1] Széchenyi István University AHJK - ATT, Győr, Hungary

E-mail: tancsics@sze.hu , ibriksz@sze.hu

Abstract. In case of numerical analysis of forged parts, it is important to specify the friction coefficient characterizing the lubrication conditions. In our paper we present a combined method of a known SU (Simple Upsetting) deformation with EDS (Energy Dispersive Spectroscopy), which shows the relationship between the Kudo's friction number (this friction model is applicable for metalworking processes which produces high surface pressure) and the theoretical layer thickness of the graphite film covering the sink cavity. The tests were accomplished considering the solids content of non-synthetically produced lubricant concentrates diluted to varying degrees.

1. Graphite lubricants

One of the important technological step at hot impression die forging is removing the scale debrishes from cavity with compressed air and the uniform coverage of the cavity surfaces with layered crystalline solid lubricant that is mostly graphite. The operation has a significant impact on the life of the tools and the efficiency of the applied technology.

The characteristic feature of lubricant concentrate compositions of graphite content is that the so-called "solid lubricant" should be sprayed in a water-suspended state on the optimized temperature surface of cavity of the die [1-3] in technically acceptable dilution ratio. After vaporization of the carrier medium (water) the solid lubricant, which is a mixture of graphite and other different crystal structures, forms a continuous coating on the surfaces of the dies.

The most important constituent element of solid lubricant is the graphite, which is mostly derived from technologically processed mined flake graphite. According to mining forecasts, the demands of graphite flake amount in the world will increase to nearly fourfold in the next decade. The strong change is fueled by getting electric cars to the fore [4]. Taking this tendency into consideration, it is easy to recognize the importance of lubricants containing synthetic graphite, whose production costs are currently higher.

The goal of our work was to determine attributes for the qualification of mined flake graphite and synthetically produced graphite lubricants by linking the mined graphite content and average graphite particle size of lubricant to Kudo's friction number.

2. Analysis of test methods, suitable model

During deformation of the work piece the displacement of the work piece contour points, which are contacted to the tool surface by the solid lubricant of lubricant, are essentially influenced by the crystal structure of the graphite. Graphite is the main component of the heterogeneous media between the work piece and tool surfaces. This heterogeneous media, which contains large amount of graphite, is made up of mixtured thin fine-grained scale- and oxide particles, debris chipped off the tool surface and solid lubricant of lubricant.

It is assumed that the relationship of the graphite content of the lubricant, which is sprayed onto the tool surface, the graphite particle size and the Kudo's friction number, which affects the wear of the tools, can be detected in plant-operating conditions as well. The most important plant-operational markers (quality cost, specific lubricant usage, tool lifetime) can be determined based on this relationship, too. The amount of solid lubricant, which is sprayed onto the tool surface, can be calculated according to the following formula **Hiba! A hivatkozási forrás nem található.**:

$$m_F = c_{SZ} \cdot q_V \cdot \eta \int_{t_K}^{t_T} dt \qquad (1)$$

whereabout:

m_F	the mass of the solid lubricant sprayed onto the tool surface [mg]
c_{SZ}	% by weight of solid lubricant in the suspension [mg/cm^3]
q_V	volumetric flow rate of the sprayed suspension per time unit [cm^3/sec]
η	the efficiency of spraying (lubricant loss at lubricant deposition and at vaporization) [-]
t_K	the time to reach initial temperature of the adhesion [sec]
t_T	the total duration of spraying [sec]

Function (1) is applicable both to those tool surfaces on which the temperature is higher than the adhesion temperature of the lubricant and the ones that are in the temperature range beneficial to the adhesion of the lubricant. Lubrication cannot be explained under the adhesion temperature.

Adequate tool temperature for lubricant adhesion and the adequate adhesion temperature range of the tool surface can vary depending on the lubricant concentrate composition. Adequate tool temperature from adhesion point of view are well approachable both by mathematical way [3] and experimental way. It is advisable to carry out the control experiment during lubrication tests on pre-heated plates at different temperatures. The control lubrication tests should be carried out on pre-heated plates whose temperatures represent the typical operating conditions. During the lubrication test, the applicable lubrication temperatures are determined by spraying lubricant on test pieces of different temperatures and the lubricant adhesion is checked. The coating should form a coherent matte layer on the surface of the correspondent test piece. At the appropriate range of temperature the desired coating is formed. Under this temperature flow strips and balling appear, above it blisters appear on the coated surface.

We applied this method to examine the adhesion range of a mined graphite type lubricant by determining the typical adhesion temperature of the lubricant. The result of this examination was between 150-200°C. This result means that all surface elements of the die cavity have to be within this temperature range.

The lubrication systems used on forging machines are extremely varied. Generally, they are characterized with their spray nozzle characteristics (classically full cone) and the spray pattern (mostly circular or ellipse). At setup of spraying pattern the type of nozzles and the technological characteristics (volume delivered per unit time, lubrication time, atomised drop size, etc.), the movement conditions, the position sets relative to the parting line of the die, the lubrication mode (direct or indirect, i.e. lubricant mist above the surfaces) and the surface roughness of the die should be considered. The lubricant spraying can be accomplished by hand and robots as well. The advantage of manual spraying is its technological elasticity. The substance loss about lubricant dispersion can be minimized more effectively, that is virtually negligible because the spraying pattern can be directly changed through the spray distance (drop size) and the valve opening size (through-flow cross section).

Changing the drop size the surface temperature of die cavity can be controlled or changed by the kinematics of drop vaporization [3]. Determining the thickness of the lubricant adhering to the die cavity surface is more accurate if the surface of the tool is simpler, in other words, the number of the surface elements of the tool should be reduced as much as possible. Figure 1 shows a coating tape of an elongated ellipse-shaped spray pattern. The figure illustrates the theoretical distribution of the layer thickness and the dissipation band.

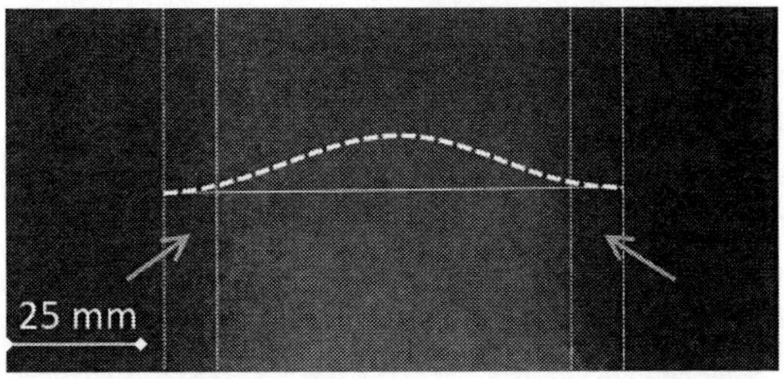

Figure 1. Distribution of lubricant layer thickness. The arrows show the interval of dissipation loss.

One of the most well-known tribometer method for layer thickness testing is the WHUST (Warm Hot Upsetting Sliding Test) test [9]. The method gives useful information on the optimum lubrication index (depending on the ratio of friction and normal forces) as well as on the appearance of scratches depending on the layer thickness of the lubricant applied on the surface.

In case of large surface pressures typical of forging, which induce equal or higher stresses than the yield stress of the tool steel, this method has not been proved yet. However, upsetting tests between parallel tool surfaces completely meet the forging conditions. During high-pressure deforming, the oxidation process becomes more intensive in the graphite crystals - due to the high pressure and temperature - which then accelerates above 600°C. During oxidation the layered crystal structure of graphite changes and its lubricity decreases [5]. Decrease in lubricity increases wear intensity.

The temperature range of the workpiece between 500-950°C (scale layer appearing) [6] and the temperature range of the tool surfaces above 450-500°C can be considered dangerous conditions for the intensity of tool cavity wear [1, 2, 8]. For the effect of high pressure on the surface partial sticking may occur as well. In this way the material flow will be realized inside the workpiece below the contacting surface [1, 7]. This should be avoided. The pressure distribution on the flat tool surface can be approximated by relation in Taylor's linearized form.

3. Operational experiments

The following conditions have been considered during the performing experiments.

Conditions to simplify:

- the number of surface elements of the tool was reduced to the minimum (two planes)
- adjusting the spray pattern with manual spraying results in negligible loss of material (dissipation, vaporization), therefore the spray efficiency: 1
- solid lubricant deposition during the test time, i.e. the change in concentration of the suspended lubricant, negligible
- the lubricant layer thickness is steady on the tool surface and the thickness value of this meet the average of the Gauss density function
- during upsetting the friction number on the contact surfaces is characterized by a Kudo's friction number whose value is constant
- during upsetting the material flow stress is the same in all elemental volumes of the workpiece

IOP Conf. Series: Materials Science and Engineering **426** (2018) 012047 doi:10.1088/1757-899X/426/1/012047

Boundary conditions:
- surface temperature of the upsetting tools may only change within the adhesion temperature range characteristic of the lubricant
- when spraying lubricant onto the tool cavity surface, the technological settings are unchanged
- the deformation of the upset workpiece is symmetrical, only samples corresponding to this condition can be tested
- the lubricant application is considered satisfactory when the lubricant forms matte and continuous dry coating on the surfaces of tool cavity.

The solid lubricant content is typical of the lubricant type and it essentially determines the thickness of the graphite layer to adhere onto the tool cavity surface. Figure 2 shows the evolution of solid lubricant content (orange dots) of an Aquanet type lubricant concentrate (blue dots), expressed as a percentage by weight, depending on dilution. On the horizontal axis, the weight ratio of water (X) and lubricant concentrate can be read directly and on the vertical axis the solid lubricant content of the suspension is expressed as weight percent. At a 1:5 dilution, a yellow square marker shows the solid lubricant content tested in the laboratory. As a result of the test in laboratory, the solid lubricant content was 4.2%. This value is practically equivalent to theoretical 4.25%, so the theoretical data of Figure 2 are acceptable.

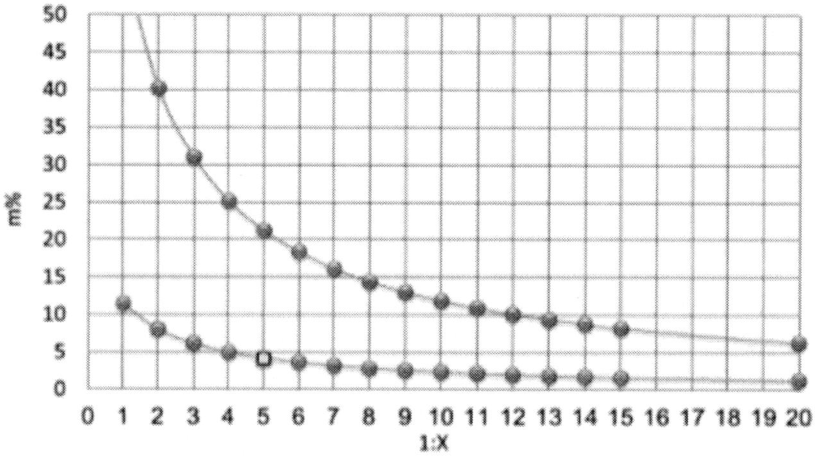

Figure 2. Aquanet type lubricant solid lubricant content at various dilutions

The upsetting tests were carried out in the forge-factory of *Rába Axle Ltd* on a maxima type presser of nominal force of 25 MN.

During the upsetting tests we recorded the following technological data: the dimension of lower tool flat surface is Ø200 mm, the dimension of upper tool flat surface is 320x300 mm, average surface roughness of tool surfaces Ra 1.6, the temperature of the tool surfaces measured in the axis line is between 150-160°C, Aquanet type lubricant usage in the following volume dilutions (lubricant concentrate / water) 1:4, 1:2, 1:1.

The temperature of the tool surfaces was measured by an INFRARED DT-882 optical pyrometer (50°C/+550°C). The lubricant was sprayed on the tool surfaces with manual spray gun in fine droplet (300-500 μm), with a flow rate of 0.43 liters per minute, with circular spreading pattern at 1 bar pressure.

The flow of volume was adjusted to the highest dilution (1:4) of lubricant, which was 50 cm^3, to ensure the layer thickness changing in dilution proportion. The technology settings of the manual spray gun remained unchanged during the test. Additional technological data: number of steel test pieces are 15, geometry data of the test pieces before induction heating Ø50×100 mm (1.54 kg), measured material loss after a single induction heating 0.91%. The average temperature of the induction heated test pieces is 1177+/-20°C, adjusted upsetting height is 60 mm. The surface temperatures of the test pieces were measured by an INFRATHERM (IMPAC) IGA 15 plus optical pyrometer (250°C/+1800°C).

The Kudo's friction number is independent of the material flow stress, so it can be determined directly from deformation data of the cold dimensions of upset test pieces. The measurement accuracy of deformed test pieces in our case also required the material loss taken into account as well.

4. Analysis of the test results

To evaluate the upset test pieces a MathCAD algorithm based on UBET (Upper-Bound Elemental Technique) method was used [1, 10]. By this method, the axial and radial velocity functions are drawn from such a kinematic permissioned velocity field which is derived from a third degree polynomial. By means of the material flow factor (k), which can be expressed from the components of the velocity field, the deformation of the test piece can be characterized by parabolic approximation. Using this method, the complete upsetting process can be modeled numerically by taking law of the volume constancy into account. The parabolic approximation of deformation is scientifically accepted [1, 11].

The relation between the material flow factor (k) and the Kudo's friction number (m) is expressed by relation (2) [1, 10]:

$$m = 1 - k . \tag{2}$$

Input data defined by measurement:
- geometry data of the test pieces after induction heating: Ø49.85x99.67 mm (1,526 kg),
- maximum diameter of upset test piece (barrelled): d_{max} [mm]
- height of upset test piece: h [mm]

Output data defined by calculation:
- Kudo's friction number: m
- upset profile curve of workpiece (image of the velocity field might be plotted as well)

Using the MathCAD program, we plotted the profile curves for different dilution ratios and determined Kudo’s friction numbers (Table 1).

Table 1. The relationship between Kudo’s friction numbers and different dilution ratios

parameters	dilution 1:4	dilution 1:2	dilution 1:1
d_{max}	67.62 mm	67.13 mm	66.35 mm
h	59.93 mm	60.07 mm	60.05 mm
m	0.445	0.387	0.275

The Kudo's friction number which is shown in the column of 1:4 dilution is practically equal to the result of the same diluted Aquanet lubricant that we got before by SU (Simple Upsetting) and RCT (Ring Compression Test) test methods [1]. The results were verified by the finite element solver of Simufact forming for the largest upset diameter based on the specified friction numbers and input geometric data.

Important settings for analysis:
- analysis: General/2D axis/FE (*Finite Element*)/Cold
- applied mesher types: Advancing Front Quad (2D/1 mm finite element size) and Quad Tree (2D/1,6 mm finite element size) whereabout the mesh maximum refinement and coarsening values are equally 5
- mesh element type: Quads

The maximum diameters depending on the measured height after upsetting (59.3 mm), in case of m=0.445 friction number are as follows: AFQ mesh: 67.656 mm and QT mesh: 67.508 mm.

We have stated that the results can be influenced by the type of mesher and the minor deviation of the profile curve from the classical parabolic curve. Following the FEA (Finite Element Analysis) inspection, the theoretical layer thickness (referring to simplification conditions) of the solid lubricant sprayed on the tool flat surface was plotted as a function of the Kudo's friction number (Figure 3).

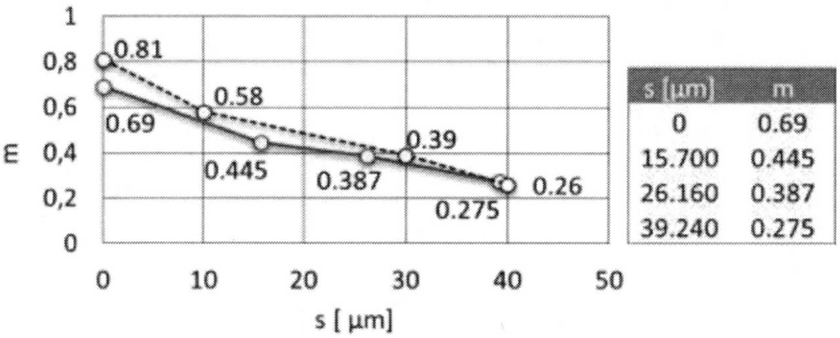

Figure 3. The change of the Kudo's friction number depending on the solid layer thickness of Aquanet type lubricant. Dashed line shows the result of the WHUST [9] test.

The layer thickness of solid lubricant sprayed on the tool surface was determined using Function (1) based on the atomization technology and geometry data of the tool. The lubricant required for full and proper covering of the tool surface was applied to the tool surface by hand spraying at a dilution ratio of 1:4 for 7 seconds. Later on, at each dilution ratio, the volume flown out during this time interval (50 cm^3) was applied. Thus, a different layer thickness per dilution came into being. It is noted here that constant 7 sec spraying time results in the same 15.70 µm layer thickness for all three dilutions (viscosity).

Figure 4. Graphite content typical of the solid substant referring to the Aquanet type lubricant.

The EDS (Energy Dispersive Spectroscopy) assay was used to determine the graphite content of the solid substant. Determination of the graphite content was made on the basis of an appropriate quantity of sample in compliance with the statistical rules. During the examinations we also found such sampling sites, which has shown extremely different results. Deviations were defined as an anomaly of local concentration and we excluded them from the evaluation. The chosen method requires very accurate sample preparation and measurement, therefore the tests were performed on sample groups with 10KV and 20KV excitation voltages. Figure 4 shows the test result typical of the samples.

Based on Figure 4 we can state that besides the graphite the oxides are also present in the solid lubricant, mostly in the form of $Na_2SiO_3/MgSiO_3$ and solid graphite oxide ($Na_2SiO_3/MgSiO_3$ is providing uniform distribution of lubricant).

The pure graphite content of solid substant is approaching 86%. Based on the EDS results, after correcting the relationship shown in Figure 3 we obtained the correlation shown in Figure 5. The graph describes connection between the Kudo's friction number and the theoretical layer thickness of graphite.

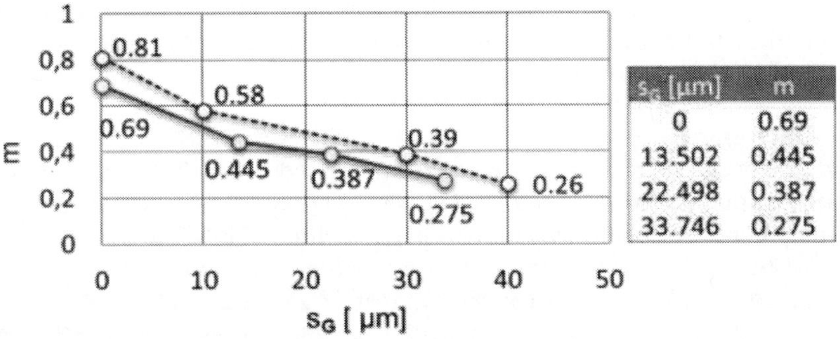

Figure 5. The change of the Kudo's friction number as a function of the graphite layer thickness of the Aquanet type lubricant. Dashed line shows the result of the WHUST [9] test.

Figure 5 shows that the results of completely independent test methods show similarity to the attributes which characterized the functions. However, the friction coefficients considering the same layer thickness differ in approximately the same proportions, and are less than as the WHUST method. This deviation can be explained by difference between the test methods [1] [3] [9] and by the different average size of the graphite particles typical of lubricant types as well. So, in our case, we also need to examine the average size of the graphite flakes.

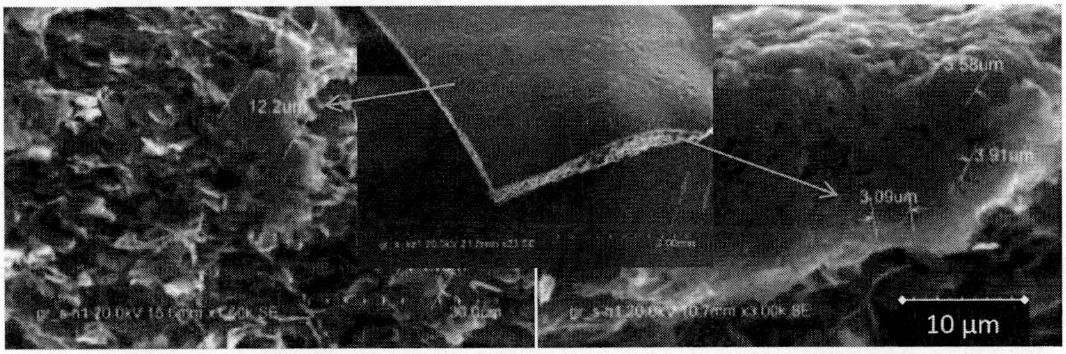

Figure 6. Graphite particle size changes depending on sampling location.

Figure 6 shows typical measurement results. Based on these results, the average particle size of the sample can be determined statistically by a good approximation. Here we note that only the well definable particle sizes were examined in 2D expansion.

It was found that the average typical particle size distribution is D50=6 μm, i.e. 6 microns is the median particle size distribution. The average size of flake fits in with the performance requirements of finely graphite lubricants applicable on finely machined tool surfaces (e.g. cavities of forging dies). Fine flakes create surface films by filling the micro-imperfections of the tool surface better (e.g. machining grooves). The average particle sizes of the WHUST test referred to above are bigger [9] than in our case but are still classified as fine particles in graphite based lubricants.

Comparing the observed differences of the particle size with the curves of Figure 5, further remarks can be taken:

– the difference in the dry lubrication state is caused by the different characteristics of test methods

– the differences in the lubrication state are generated by the average particle size difference of graphite.

The latter remark is particularly important because of finer graphite particles, as the graphite layer thickness may be further reduced to a certain extent (depending on the tool surface quality, lubricant type) without having the thickness of the solid lubricant layer changed. This is important because we can avoid the splitting of the solid substant layer. This also means that it is possible to use lubricants of differing composition in more favourable dilution.

5. Summary, conclusions

In our work we were searching the relationship between Kudo's friction number and the layer thicknesses of flake graphite based lubricants. Our tests were carried out under operating conditions and with forging equipment. Correlation was found between the Kudo's friction number and the graphite layer thickness of lubricant referring to the lubricant based on content of mined flake graphite. The significance of this relationship is shown by the fact that we also found a link between the layer thickness and the average particle size distribution, which allows the thickness of the graphite layer sprayed onto the cavity surface to be reduced. This can assist in making more flexible lubrication technologies and making economic decisions about different type of lubricants.

We would say thank you for the following project: „Dynamics and direction of autonomous vehicles in the synergy of the requirements of automated traffic systems". Code: EFOP-3.6.1-16-2016-00017.

References

[1] Tancsics, F.: Practical application of virtual forging technologies, GlobeEdit, Saarbrücken (2016), (written in Hungarian)

[2] Behrens, B.A., et al.: Numerical Analysis of Tool Failure in Hot Forging Processes, Obróbka Plastyczna Metalit, XIX nr 4, (2008), 11-17

[3] Mendez, P.F., et al.: Influence of heat transfer on the application of solid lubricant on hot dies, Elsevier Scripta Materialia, Vol. 59, No. 7, (2008), 784-787

[4] Spencer, R., Hill, L.: A flake's chance in cell: quantifying graphite demand, Specialty Minerals & Metals, Industry Overview Nov 20th (2016), 35

[5] Rudolph, S., Klein, F.: Untersuchungen zum Einsatz von Bornitrid in Gießereischlichten zum Vergießen von Aluminium- und Magnesium-Legierungen, Sonderdruck aus der Gießerei-Praxis, Heft 6 (1992), 81-84

[6] Barrau, O., et al.: Investigations of Friction and Wear Mechanisms of Hot Forging Tool Steels, Konferencia kiadvány: 6th International Tooling Conference, Karlstad, (2002), 95-111.

[7] Liu, C.: Modeling of Water and Lubricant Sprays in Hot Metal Working, PhD thesis, The Ohio State University, (2007).

[8] Sawamura, M., et al.: Estimation of Spray Lubrication and Die Temperature for Die Wear Life Prediction in Hot Forging, Research Report, R&D Review of Toyota CRDL, Vol. 40, No. 1, (2009), 50-56

[9] Daouben, E., et al.: Effects of Lubricant and Lubrication Parameters on Friction During Hot Steel Forging, International Journal of Material Forming, Vol. 1, No. 1, (2008), 1223-1226

[10] Tancsics, F., Halbritter, E.: Determination of Friction Coefficient During Upsetting Using a Kinematically Admissible Velocity Field, Strojnícky Časopis, Journal of Mechanical Engineering, Vol.63 No.4, (2012), 197-223

[11] Tancsics, F., Halbritter, E.: Inspection of barrelling of upset forgings based on digital photographs, Acta Technica Jaurinensis, Vol. 6, No 2 (2013), 107-117

IOP Conf. Series: Materials Science and Engineering **426** (2018) 012048 doi:10.1088/1757-899X/426/1/012048

The examination of the grain refining effects of titanium in case of AlSi7MgCu0.5 alloy

M Tokár[1], Gy Fegyverneki[1], M Alexa[1] and T Mende[2]

[1] University of Miskolc, Institute of Foundry, Miskolc-Egyetemváros, Hungary
[2] University of Miskolc, Institute of Physical Metallurgy, Metal Forming and Nanotechnology, Miskolc-Egyetemváros, Hungary

E-mail: monika.tokar@uni-miskolc.hu

Abstract. Due to the rapidly developing technologies and the various expectations of the customers, the continuous development of the products is necessary. In case of automotive industry, the increasingly more demanding expectations and requirements pose great challenge for the casting industry. One of the most significant requirements is the production of lightweight but high strength castings. Thus, aluminium became one of the most widely used base material for the automotive industry. However, to maintain a lasting presence on the market, continuous development and the improvement of the mechanical properties of the castings is necessary. During the experiments, the effect of grain refinement with AlTi5B1 master alloy on the mechanical properties, porosity, secondary dendrite arm spacings (SDAS) and grain size of AlSi7MgCu0.5 alloy was examined. The initial 1000 ppm titanium concentration was increased to 2000 ppm, with the addition of 200 ppm at each step.

1. Introduction

The crystallization of the Al-Si hypoeutectic alloy in the mould starts with the formation and growth of the nuclei of the aluminium solid solution. The nuclei continuously grow until reaching the eutectic temperature. The solidification ends with the crystallisation of the eutectic structure. The mechanical properties are highly affected by the number and the fineness of the grains in the solid solution: the finer the granular structure is, the better the mechanical properties are. Grain size can be decreased by accelerating the nucleation in the crystallisation process. Thus, the structure can be composed of more and smaller crystals. To achieve such structure, the so-called inoculation, grain refinement is applied [1].

Apart from wrought alloys, grain refinement has several benefits in cast alloys like improved mechanical properties that are uniform throughout the casting, distribution of second phase and microporosity on a fine scale, better feeding to eliminate shrinkage porosity, improved ability to achieve a uniform anodized surface, better strength and fatigue life. [2, 3]

The size and distribution of grains and their secondary dendrite arm spacing (SDAS), which is highly influenced by the cooling rate, are parameters to be tightly controlled [4]. The grain size of aluminium castings is usually between 0.1 and 10 mm. Castings have fine structure if the average diameter of the grains are less than 1 mm. Due to overheating, long holding time or slow solidification rate, the structure of aluminium alloys become coarse. Therefore, the mechanical properties are deteriorated. The aim of grain refinement is to eliminate the above-mentioned issues by adding an additive to the molten metal to increase the number of nuclei [5]. Elements which react with aluminium in a peritectic reaction during solidification are the most sufficient grain refiners. Thus, titanium is one of the most commonly used grain refiner. Grain refinement should always occur after degassing and cleaning (slag removal) on 730-750 °C, depending on the composition of the alloy [5].

2. Experimental part

The examined AlSi7MgCu0.5 foundry alloy melt was alloyed with AlTi5B1 master alloy under operating conditions to achieve finer grain structure and better mechanical properties. The master alloy was added to the melt right before degassing the melt with N_2. Portions of 1000 kg melts were alloyed in the holding furnace during the experiment. The measured density index values of the degassed melt were always below 2.0%, according to specification. The casting temperature of the melt was 733±3 °C. The holding time – the time until the casting of the test bars – was ~40 minutes.

The changes in the composition of the melt was continuously monitored by the composition analysis of the chill tests. Based on the results, the titanium master alloy was added to the melt. The sought and achieved Ti concentrations can be found in Table 1, while Table 2 contains information on the average composition of the examined alloy.

Table 1. The sought and achieved Ti concentrations

Portion	Sought Ti concentration	Achieved Ti concentration
Base material	1200 ppm= 0.12 m/m%	1227 ppm
1. alloying	1400 ppm= 0.14 m/m%	1453 ppm
2. alloying	1600 ppm= 0.16 m/m%	1623 ppm
3. alloying	1800 ppm= 0.18 m/m%	1879 ppm
4. alloying	2000 ppm= 0.20 m/m%	1970 ppm

Table 2. The average composition of the examined alloy

Elements	Al	Si	Ti	B	Fe	Cu	Mn	Mg	Zn	Sr
m/m%	91.80	6.95	0.10	0.00026	0.11	0.53	0.06	0.37	0.01	0.0238

In order to draw conclusions from the cooling curves of the melts about the degree if grain refinement, thermal analyses were carried out. Microstructure – grain number and secondary dendrite arm spacing – and strength tests – tensile tests, Brinell hardness test – were also carried out during the experiments. The tests were carried out at the specific sampling points of the test bars determined by standard.

2.1. *Thermal analysis*

The development of primary crystalline phase is accelerated by the addition of nucleating agents to homogenous melt. Thus, more grains and finer structure can be achieved. In case of the examined melts, the maximum temperature of the liquidus must increase, as less time is needed for the critical nucleus size, and consequently to start crystallisation, because of the addition of the proper grain refiner.

Based on the cooling curves of the degassed melts prepared under operating conditions, the maximum liquidus temperatures increased with the increased titanium concentration, as it can be observed in Figure 1. Because of data loosing we could not see the whole curve in case of 1400 ppm.

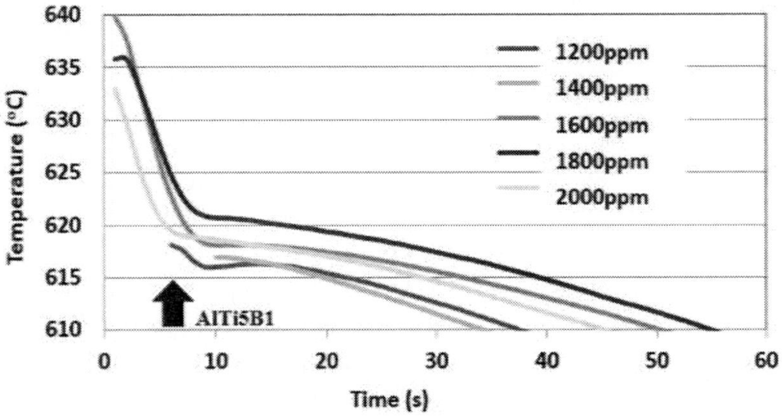

Figure 1. The liquidus temperature intervals of the cooling curves defined during thermal analysis

However, a decline can be observed in the cooling curve for the 2000 ppm titanium concentration, so the microstructure analyses of the test bars were also carried out. The results are discussed in detail in Chapter 2.2.

2.2. Microstructure analysis

The investigated areas of the experimental castings and thermal analysis samples can be seen in Figure 2.

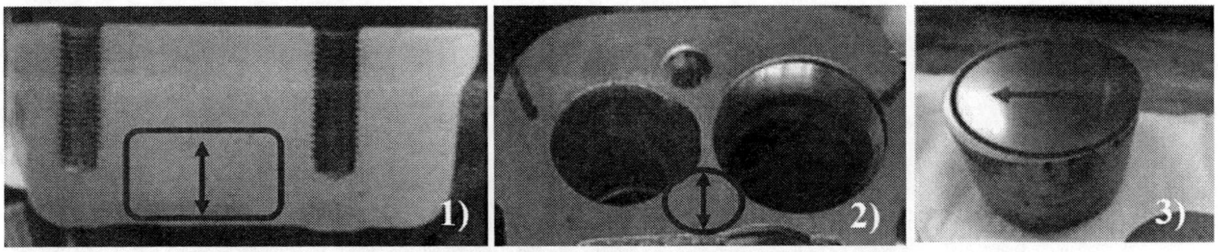

Figure 2. The investigated areas of the experimental castings (1, 2) and the thermal analysis sample (3)

2.2.1. Determination of grain numbers with colour etching

Barker [8] etching was carried out to examine the grain structure with a Struers LectroPol-5 instrument. The samples were examined with an optical microscope (M=25x). The approximate number of particles were determined on the surface of the samples with the help of their different colours. Five microstructure images of each sample have been made and calculations carried out based on them.

The average grain numbers determined at the 1st and 2nd sampling points can be seen in Figure 3.

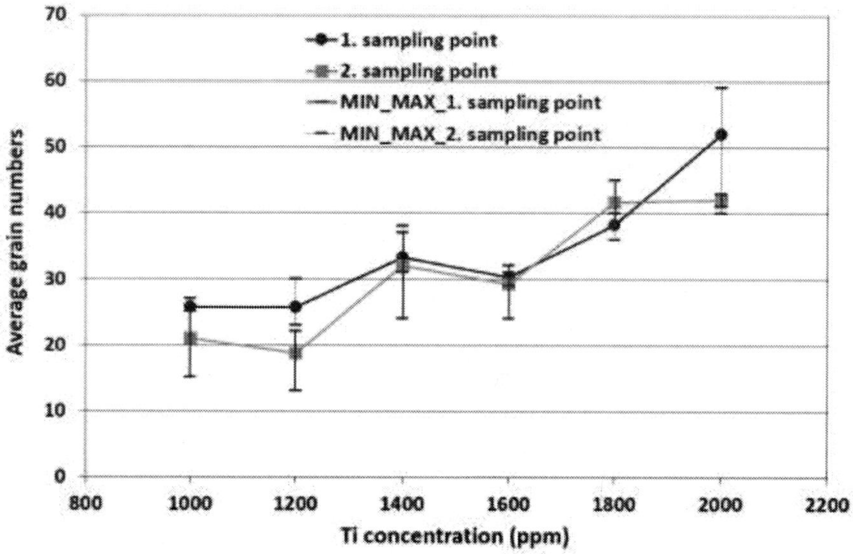

Figure 3. The average grain numbers at the 1st and 2nd sampling points of the test bars

The grain refining ability of titanium is confirmed by the number of grains, as the grain number increased throughout the experiments. The average grain numbers at the 1st and 2nd sampling points can be seen in Table 3., including the characteristic images of the Barker-etched microstructures as well.

Table 3. The number of grains at the 1st and 2nd sampling points of the test bars (magnification 25x)

Samples	Ti concentration					
	1000ppm	1200ppm	1400ppm	1600ppm	1800ppm	2000ppm
1. sampling point						
Average grain numbers	25,70	25,70	33,30	30,30	38,30	52,00
2. sampling point		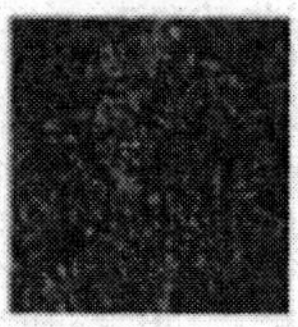				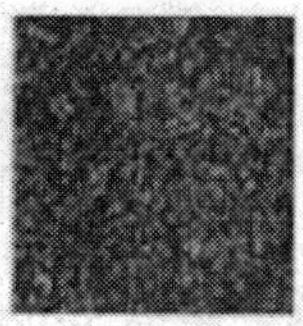
Average grain numbers	21,00	18,70	32,00	29,30	41,70	42,00
Casting temperature (°C)	730	732	736	734	734	735

As it can be observed in Table 4, the increased titanium concentration resulted in finer grain structure. The average grain number also increased with the titanium addition. Similar tendencies could be observed in case of both sampling points and the test bars for thermal analysis (Figure 4.). Unfortunately because of an industrial problem we could not take samples in case of 1000 ppm casting.

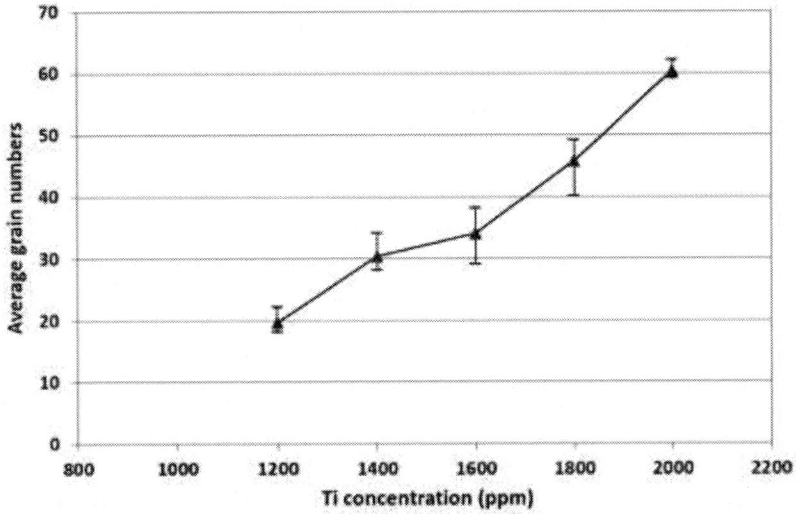

Figure 4. Average grain numbers of the thermal analysis test bars

The average grain numbers at the thermal test bars sampling points can be seen in Table 4, including the characteristic images of the Barker-etched microstructures as well.

Table 4. The number of grains at the sampling points of the thermal test bars (magnification 25x)

Samples	Ti concentration				
	1200ppm	1400ppm	1600ppm	1800ppm	2000ppm
2. sampling point	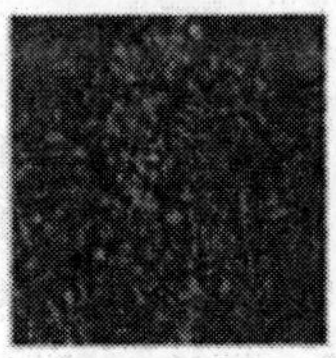	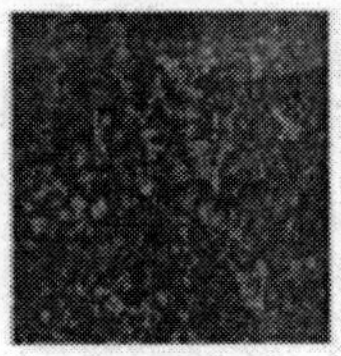	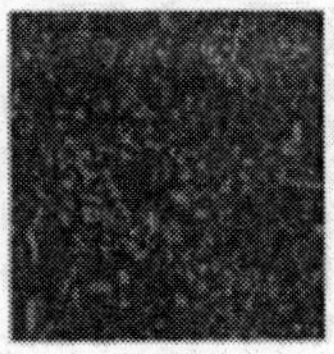	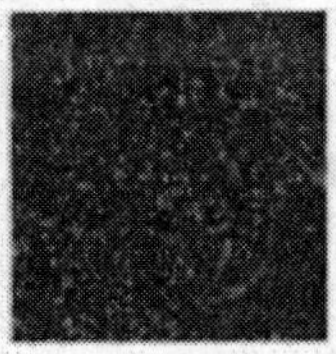	
Average grain numbers	19,70	30,30	34,00	45,70	60,30
Casting temperature (°C)	732	736	734	734	735

2.2.2. Determination of secondary dendrite arm spacing

The average secondary dendrite arm spacings (SDAS) were determined at the 1st and 2nd sampling points of the test bars. For the measurement of SDAS, first, 10 structural images per cross sections were prepared with the help of an optical microscope at 100x magnification. Three SDAS values were measured in each image and the results were averaged. The measured average secondary dendrite arm spacings are summarised in Figure 5. As it can be seen in Figure 5, the secondary dendrite arm spacings decreased in the experimental castings with the increased amount of grain refiner addition.

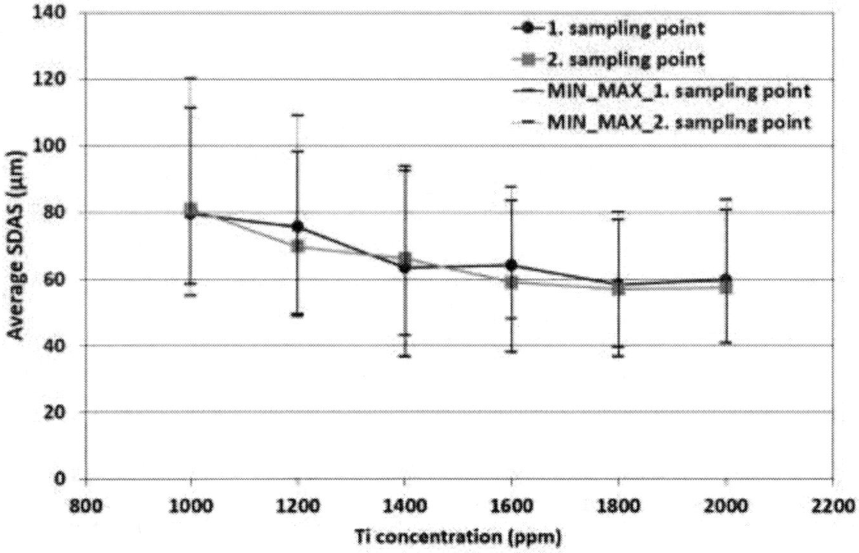

Figure 5. The average secondary dendrite arm spacings at the 1st and 2nd sampling points of the test bars

2.3. Mechanical properties

Mechanical tests were carried out on test bars prepared from the castings (3 castings /experiment) to examine if further titanium addition improves the mechanical properties. The mechanical tests were carried out after the T5 heat treatment of the castings. 3 castings were examined in case of each experiment, Brinell hardness was measured at 3 points and 3 tensile test specimens were used for tensile tests. The Brinell hardness tests were carried out with a 5-mm ball and 250 kp – 2451,66 N load. The test bars were cylindrical, 5 mm in diameter and 50 mm long.

Figure 6. shows the average Brinell hardness values. The Brinell hardness values were between 70 – 90 HB during alloying. The hardness was slightly improved by the increased titanium concentration, mainly in the middle range.

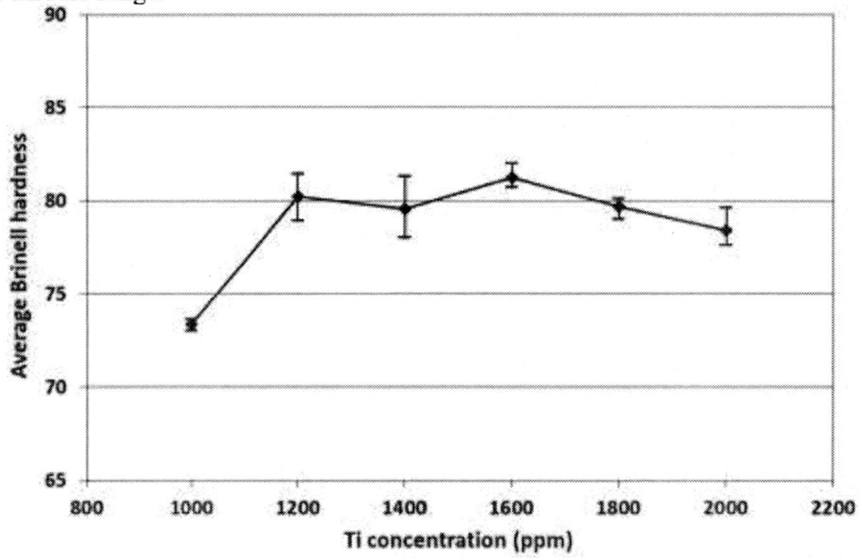

Figure 6. Average Brinell hardness values in relation to the titanium concentration

During tensile tests, the tensile strength, elongation and yield point of the test bars were examined. It can be seen in Figure 7 that the tensile strength values followed a maximum curve as the titanium concentration was increased. Above 1200 ppm titanium concentration, the tensile strength values decreased.

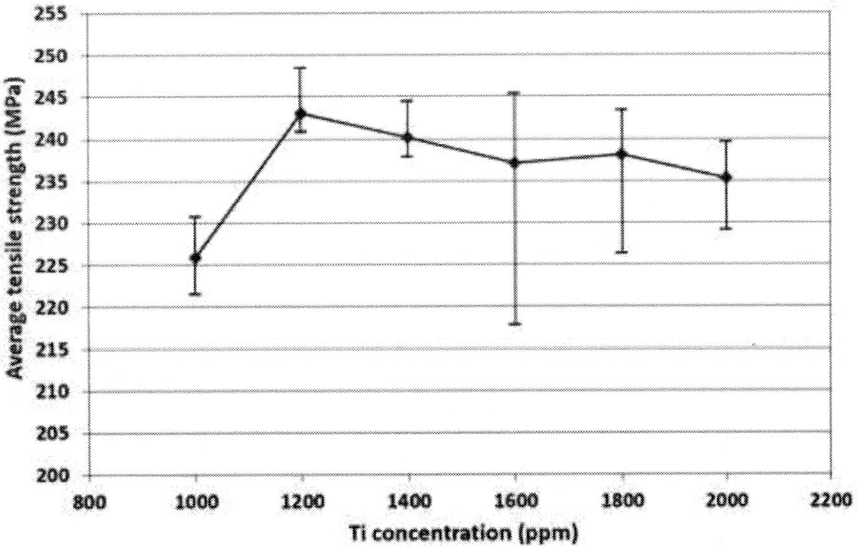

Figure 7. The changes in the average tensile strength in relation to the titanium concentration

The strength of the test bars can be improved by decreasing the secondary dendrite arm spacings, which was also confirmed by the experiments. It can be observed in Figure 8 that higher tensile strength values can be achieved than the average values of castings with 1000 ppm titanium concentration by decreasing the SDAS with alloying.

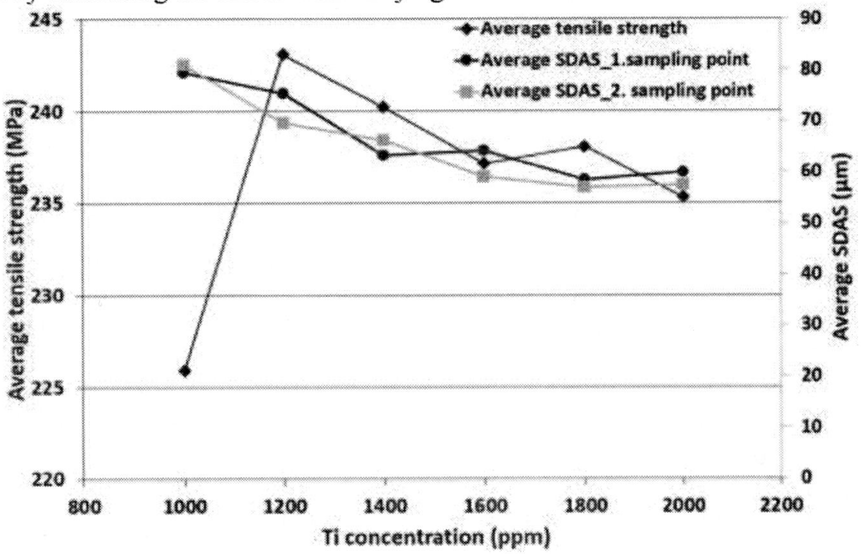

Figure 8. The changes in the average tensile strength and SDAS during alloying

The values of relative elongation were significantly higher than the required 1.5% *(Figure 9).*

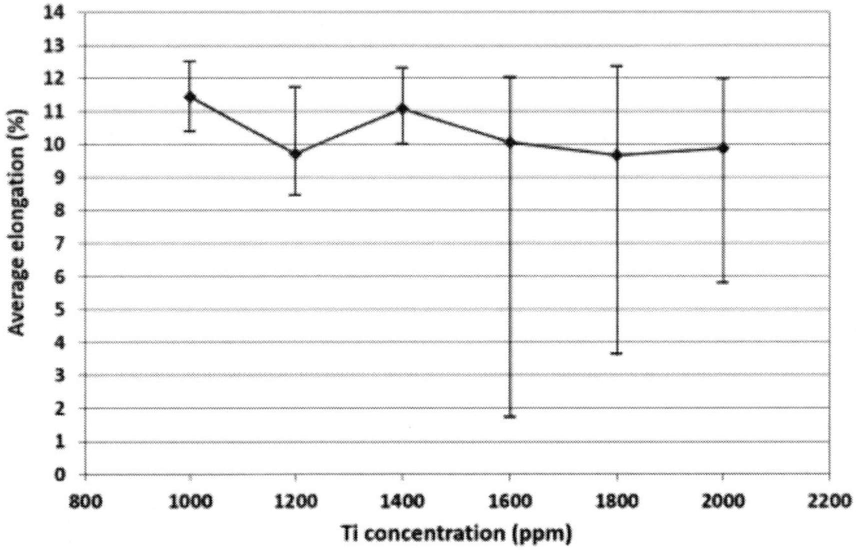

Figure 9. The average relative elongation in relation to titanium concentration

Conclusion

1. The increased titanium addition to the melt lead to higher liquidus maximum temperature, confirming that lower supercooling temperature is necessary for the crystallisation process to start in case of increased titanium concentration.
2. The number of grains increased and the secondary dendrite arm spacings decreased in both the experimental castings and the thermal analysis test bars when the titanium concentration was increased. 3. The strength tests revealed that the required mechanical properties can also be achieved with higher titanium concentrations. Based on the experiments, it can be concluded that the optimal titanium concentration is between 1000 and 1400 ppm. For the determination of a more precise concentration further experiments are needed.

Acknowledgement

The authors are grateful to Mária Ambrus for her work.

References

[1] Quested T E. 2004 Understanding mechanisms of grain refinement of aluminium alloys by inoculation *Mat. and Sci. Tech.* 20 pp 1357-1369
[2] D Apelian, G K Sigworth, K R Whaler 1984 Assessment of grain refinement and modification of Al-Si foundry alloys by thermal analysis *AFS Trans.* 92 pp 297-307
[3] K T KAshyap and T Chandrashekar 2001 Effects and mechanisms of grain refinement in aluminium alloys, *Bull. Mater. Sci.* Vol 24 No 4 August pp 345–353
[4] David Gloria Ibarra 1999 Control of Grain Refinement of Al-Si Alloys by Thermal Analysis *PhD dissertation* National Library of Canada
[5] Pál Jónás 2011 Könnyűfém öntészeti ismeretek *Nem. Tan.* Budapest

IOP Conf. Series: Materials Science and Engineering **426** (2018) 012049 doi:10.1088/1757-899X/426/1/012049

Optical properties of nano- and ultrananocrystalline diamond thin layers in the UV and visible spectral range

S Tóth[1,2], J Budai[2,3], M Veres[1] and M Koós[1]

[1]Wigner Research Centre for Physics of the Hungarian Academy of Sciences, Budapest, Hungary
[2]ELI-ALPS Research Institute, ELI-Hu Nonprofit Ltd, Szeged, Hungary
[3]Department of Optics and Quantum Electronics, University of Szeged, Szeged, Hungary

E-mail: toth.sara@wigner.mta.hu

Abstract The potential application of nano- and ultrananocrystalline diamond thin films in nanotechnology demands a detailed knowledge of the optical properties and their relation to the structure. By varying the atomic bonding structure and morphological properties of nanodiamond layers we have explored a relation between the structure and the spectral dependence of the refractive index and extinction coefficient in nanodiamond. For this purpose nanodiamond layers were prepared from CH_4, H_2 and Ar gas mixture in a chemical vapor deposition process with different Ar-content of the feed gas and self-bias of the substrate material. Optical parameters were determined by using a three-layer model in the evaluation of the spectral ellipsometric data. Surface topography was monitored and the average grain size was obtained from scanning electron microscopy images. Bonding characteristics of diamond and non-diamond phases were analyzed by Raman spectroscopy.

1. Introduction

Nanocrystalline (NCD) and ultrananocrystalline (UNCD) diamond receive increasing scientific attention due to their promising properties and broad application potential. Nanodiamond layer has already been used as coating material in microelectromechanical systems [1] to improve the wear and friction properties of components, as well as in implantable medical devices because of its biocompatibility and chemically inertness [2, 3]. Favorable optical properties of NCD films broaden the range of applications, like thin protective layers of optical components in the IR and visible spectral ranges.

Diamond film properties are strongly dependent on film morphology, especially when grains are on the nanometer scale. Microwave plasma assisted chemical vapor deposition (MW CVD) technique gives us the opportunity to change the morphology of diamond layers, as well as the grain size of diamond nanocrystals from μm to nm by fine tuning the deposition parameters. Main stages of MW CVD process has been established [4] as high density nucleation and plasma assisted film growth, however these are completely different in case of NCD and UNCD layers. UNCD films with grain size of 5-15 nm [5] are grown from Ar-rich plasma, where diamond crystals re-nucleate continuously during film growth, which results in approximately 10% grain boundary fraction in the total volume. In contrary, NCD films with grain size below 100 nm are grown in H-rich plasma, where the continuous etching of sp^2 carbon by atomic hydrogen during growth results lower portion of non-diamond phases in the diamond matrix.

The structure of diamond films deposited by MW CVD method has been intensively studied [6, 7] in last decades. Results showed that NCD and UNCD films possess smoother surface and lower friction coefficient than microcrystalline diamond, but the presence of sp^2 carbon containing grain boundary fraction in the diamond matrix influences their optical properties, such as refractive index and absorption coefficient. High precision applications require the sound knowledge of the optical behavior of NCD and UNCD thin layers in a wide spectral range.

The aim of this research was to analyze the relation between the optical properties and the structure on the atomic bonding scale, surface topography and crystallite size of UNCD thin films prepared by MW CVD method. Optical properties of UNCD layers were determined by spectral ellipsometry. Diamond and non-diamond phases were identified in the film structure by the assignment of scattering bands in the Raman spectra, while surface morphology and crystallite size distribution was discovered by scanning electron microscopy.

2. Experimental setup

NCD and UNCD layers were grown from a mixture of CH_4, H_2 and Ar gases as feed gas by using MW CVD (2.45 GHz) process in a bell jar type deposition reactor, where Si wafers served as substrates through the bias-enhanced nucleation and growth (BEN/BEG) process [8, 9]. The chamber pressure was kept at 40 mbar with 300 sccm total gas flow and the microwave power was 1200 W. The methane concentration in source gas was always kept at 1%, while the concentration of argon was varied from 10% to 95% by adjusting the flow rate of both hydrogen and argon. Bias voltage was set to 0, 50, 100 and 150 V, while substrate temperature was constant (650 °C) during the growth.

Ellipsometric measurements were performed using a Woollam M2000F ellipsometer. The ellipsometric spectra were recorded in the 1.24-5 eV spectral range at several angles of incidence (65°, 70°, 75°). A three-layer model was applied to evaluate the measured data, consisting of the silicon substrate, the NCD layer and the surface roughness. The optical data of the substrate were described using literature values [10]. Data were taken from ellipsometric determination of optical constants for silicon and thermally grown silicon dioxide via a multi-sample, multi-wavelength, multi-angle investigation [10], while for the effective medium approximation was used to characterize the surface roughness [11]. Cauchy dispersion relation extended with Urbach's absorption tail was used to describe the optical behavior of the NCD layer [12]. We also used Lorentz-, Gauss- and Harmonic-oscillator models to fit experimental data, but the mean squared error (MSE) values were the lowest in case of the Cauchy-model. From these measurements, the optical properties (refractive index, extinction coefficient) and thickness of the films were determined.

Renishaw 1000 B micro-Raman spectrometer attached to a Leica DM/LM microscope was used to examine Raman scattering at room temperature in the 800-2000 cm^{-1} wavenumber region with 488 nm laser excitation. The diameter of the excitation spot was 1 μm.

The morphology of the nanocrystalline diamond films was investigated by using a LEO 1540 XB field emission scanning electron microscope. The average grain size of diamond crystals was determined by the ImageJ software, which is able to count the number of grains and determine their size on the predetermined area of the analyzed picture and finally estimate the average grain size.

3. Results and discussion

Modification of film morphology in NCD films, especially the diamond grain size could be achieved by systematic variation of the deposition parameters, like Ar-concentration of the feed gas and self-bias voltage (U_{bias}) of the substrate, while the other parameters are fixed. One series of NCD layers were prepared at 10% Ar-content with variation of the self-bias in the 0-150 V range. The influence of self-bias on the morphology of these films can be seen on SEM images (Figure 1a.). It can be seen that U_{bias} has a remarkable influence on the surface topography and the grain size of the formed films. Diamond crystals with clearly distinguishable facets can be seen on the SEM images except for the film deposited at 150 V self-bias. The average grain size on the 0-50 V samples is around 100 nm. The first signs of the re-nucleation process are recognizable on the SEM image of the sample prepared at 100 V. The shape of bigger crystals is not well defined and smaller grains can be seen on their surface, which are the consequence of the re-nucleation caused by the increased growth speed. The film prepared at 150 V self-bias has uniform grain size distribution with 19 nm average size, characteristic for ultrananocrystalline diamonds.

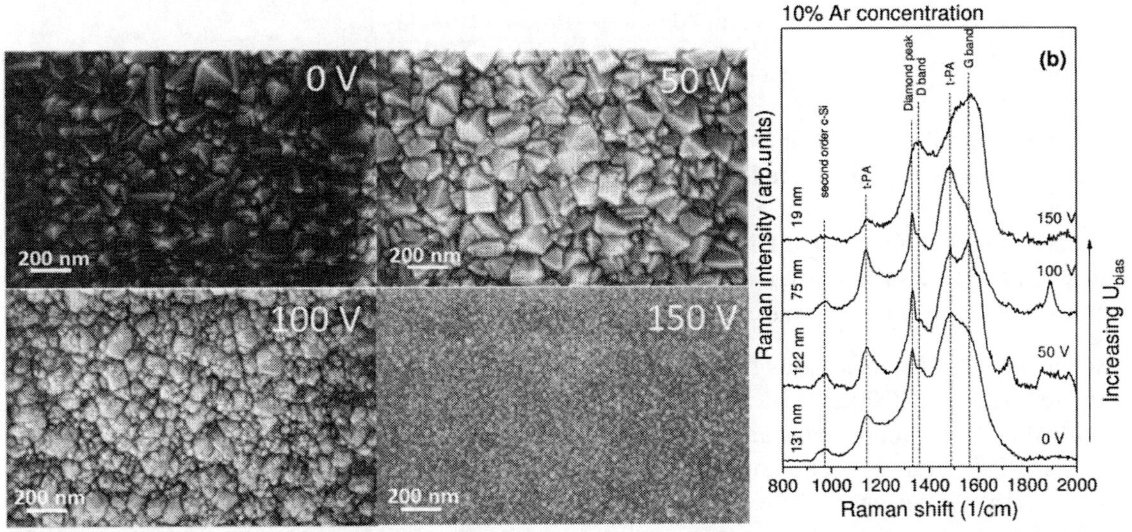

Figure 1. SEM images (a) and Raman spectra (b) of nanocrystalline diamond layers of different grain sizes deposited at 0, 50, 100 and 150 V self-biases from feed gas with 10% Ar-content.

Raman scattering spectra of these samples are shown on Figure 1b. in the 800-2000 cm^{-1} wavenumber region. Every spectrum was normalized to its maximum and shifted along the ordinate for better visibility. The presented spectra are typical for nanocrystalline diamond films and they exhibit different scattering peaks assigned to different phases composing the film structure. Nanodiamond films usually consist of diamond crystals, a non-diamond intergrain phase and the interface between these two, the so called grain boundary (Figure 2.). The characteristic Raman peak of bulk diamond crystals appears at 1332.3 cm^{-1} [13]. Raman spectra on Figure 1b. exhibit narrow peak at about 1333 cm^{-1} that can be assigned to diamond phase, however this band is hard to recognize on Raman spectrum of the 150 V sample. The peak position is shifted to higher wavenumbers (from 1332 to 1335 cm^{-1}) with increasing self-bias due to the increasing residual stress in diamond nanocrystals [14]. The well-known D and G bands near 1360 and 1560 cm^{-1}, respectively, are related to the presence of intergrain phases. The G-peak is related to the stretching vibration of the sp^2 C-C bonds (rings and chains), while the D-band is in connection with the breathing mode of the sp^2 carbon containing rings. Nanodiamond fingerprint peaks (1140 and 1480 cm^{-1}) usually appear in the Raman spectrum of nanosized diamond simultaneously and are related to the ν_1 and ν_3 modes of trans-polyacetylene [15], respectively. The increasing intensity of the fingerprint peaks from 0 to 100 V refers to the higher amounts of grain boundary phases, but in case of the 150 V sample the intensive G-band is caused by the high intergrain sp^2 carbon content. Additional relatively weak and broad band at 950-1000 cm^{-1} is related to the second-order Raman scattering of the crystalline Si substrate [16].

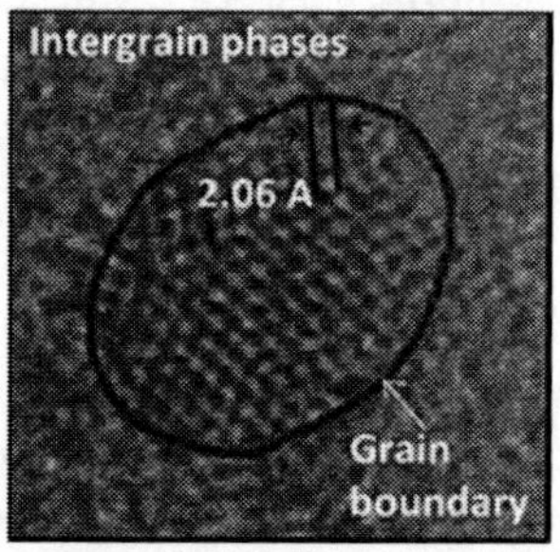

Figure 2. High-resolution transmission electron microscopy micrograph of a diamond nanocrystal. Reprinted with permission from [17].

The influence of the Ar-content of the feed gas on the film morphology can be seen on SEM images (Figure 3a.) taken on layers deposited at 0 V self-bias with increasing Ar-concentration (10-95 %). Between 10-50% Ar-content the layers consist of definite crystals with average grain size decreasing with the Ar-ratio in the feed gas. Smaller crystals appear beside big ones in layers deposited at 70% Ar-content, while the highest Ar-concentration results in films consisting of small nanocrystrals only. Formation of UNCD layer with columnar growth takes place at these deposition parameters. The film surface is not uniform and the structure differs from the high self-bias samples, even though their average grain size is similar.

Raman spectra of samples grown at 0 V self-bias with increasing Ar-concentration are shown on Figure 3b. in the 800-2000 cm^{-1} wavenumber region. All spectra were normalized to their maximum and shifted along the ordinate for better visibility. Characteristic Raman peak of diamond appears in the scattering spectrum of each sample, however its relative intensity decreases and the peak also broadens as the Ar-content of the feed gas increases. Diamond Raman peak can be found at 1334 cm^{-1} in the samples prepared at 10-70% Ar-concentration, while it is shifed to 1336 cm^{-1} in the case of 95% Ar ratio. The nanodiamond fingerprint peaks at 1140 and 1480 cm^{-1} characterize the bonding properties of grain boundarywhich change only slightly, as the peak positions do not vary with Ar-content. The changes of the G- and D-band parameters, however, indicate the modification of the bonding structure in the intergrain region. The estimation of the relative amount of diamond and non-diamond phases from the Raman spectrum produces unreliable results, because the scattering cross section is strongly dependent on the arrangement of the sp^2 bonds.

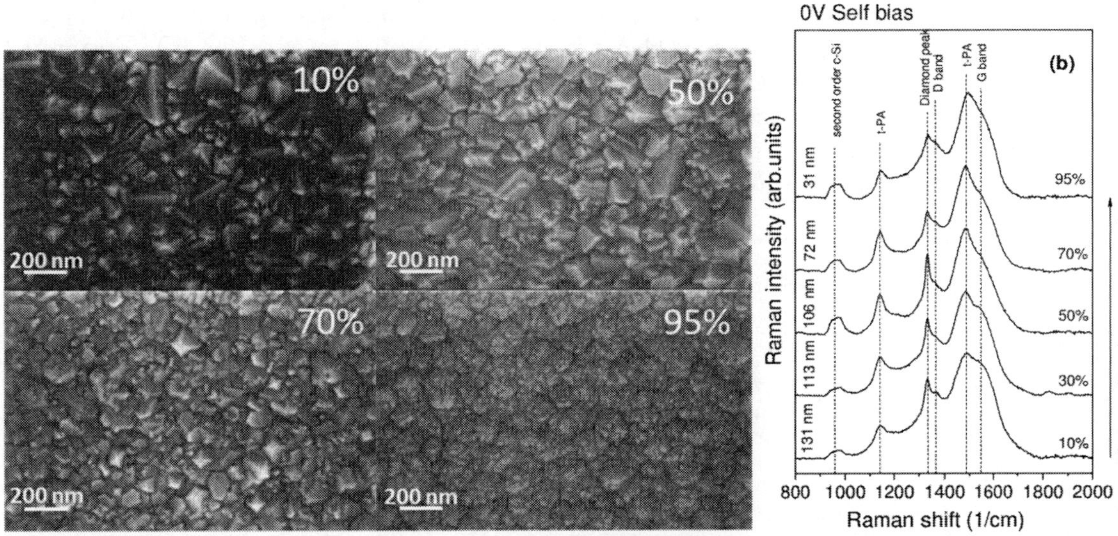

Figure 3. SEM images (a) and Raman spectra (b) of nanocrystalline diamond layers with different grain sizes deposited at 0 V self-bias and 10, 30, 50, 70 and 95% Ar-content of the feed gas.

From the analysis of the SEM images and the Raman scattering results it can be concluded that the deposition parameters (self-bias and Ar-content of the feed gas) have significant influence on the structural properties and the morphology of the deposited NCD films. Further, the effect of the morphological changes on the optical properties should be explored. Optical properties of the deposited layers can be monitored by spectral ellipsometry, and by using this method we can determine the spectral dependence of the extinction coefficient and the refractive index in a wide photon energy range. Results obtained on two selected nanocrystalline diamond layers (grain size: 106 and 78 nm) are presented on Figure 4a. and 4b. in the 1.24-5 eV photon energy range. Refractive index (n) and extinction coefficient (k) curves measured and calculated on bulk diamond sample [18] are shown on insert of Figure 4a and 4b, respectively. The refractive index of diamond crystal changes between 2.4 and 2.7 in the 1.24-5 eV photon energy range, while the extinction coefficient is zero here [18]. In the case of nanocrystalline diamond layers the dispersion curve of refractive index is similar to that of diamond, but the values are between 2.32 and 2.54. This is due to the lower optical density of the NCD film. The influence of the non-diamond phases on the refractive index can be clearly seen on Figure 4a. The NCD layer with larger average grain size (106 nm) has a lower ratio of non-diamond phase (grain boundary and intergrain regions), thus its refractive index is closer to that of bulk diamond. According to the Figure, the increase of the non-diamond phase contribution results in a smaller refractive index of the NCD layer. It is well known that diamond is a transparent material from the UV (225 nm) to the far infrared, thus the extinction coefficient is zero below 5.5 eV photon energy, which is the band gap of bulk diamond (see Figure 4b insert). At the same time, due to absorption loss NCD layers have non-zero extinction coefficient values in the same photon energy range. The dependence of the extinction coefficient on the photon energy is very similar in the weak absorption region (up to ~3.5 eV) in the NCD layers of the two different (106 and 78 nm) grain sizes. A more pronounced increase of the extinction coefficient can be observed from ~4 eV and the curve becomes steeper for the sample with smaller grain size (see Figure 4b.). This difference in the photon energy dependence indicates the increasing contribution of the non-diamond phases to the extinction coefficient. The absorption loss at photon energies below ~5.5 eV indicates the presence of localized states in the band gap of diamond. Non-diamond bonds, sp^2 C=C, sp^2 C=H, intrinsic and extrinsic defects can contribute to the formation of these localized states. Raman scattering measurements confirm the presence of sp^2 carbon phases in our NCD films and the change of their bonding structure and their spatial arrangement with the deposition parameters. Although the relative amount of these sp^2 phases cannot be deduced from the Raman spectra. However it seems to be plausible that the ratio of sp^2 phases increases with the decrease of the average grain size due to the higher surface area compared to the volume. Fluorescence measurements prove that NCD films exhibit relatively broad green (~2.3 eV) and blue (~3.2 eV) [19] luminescence bands attributed to intrinsic and in some cases extrinsic defects which also contributes to the localized states in the diamond band gap.

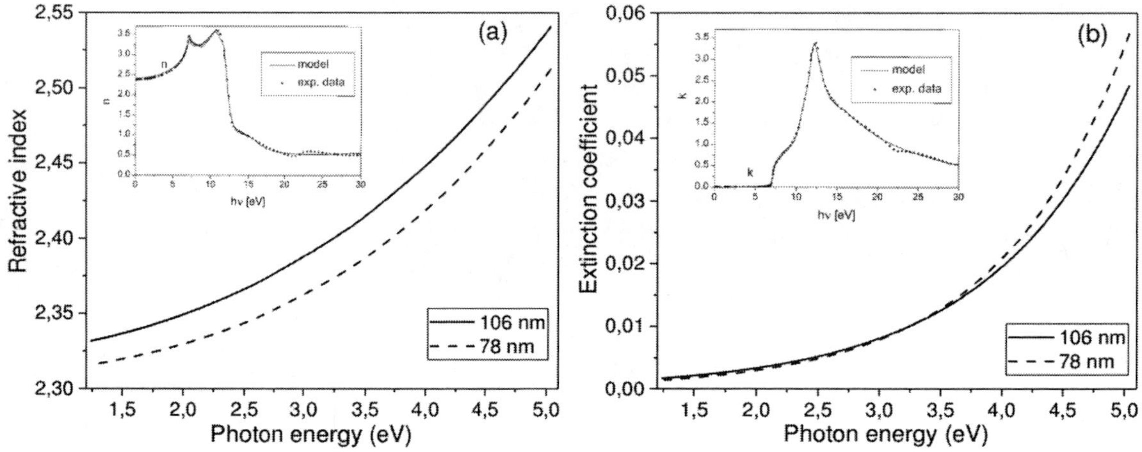

Figure 4. Spectral dependence of the refractive index (a) and the extinction coefficient (b) in nanocrystalline diamond films with average grain sizes of 106 and 78 nm. Spectral dependence of the refractive index (Figure 4a insert) and the extinction coefficient (Figure 4b insert) of bulk diamond [18]. Triangles - experimental data; solid curve - model.

Next we will examine diamond films of average grain sizes below 40 nm. The average grain size as a function of deposition parameters for these films are depicted on Figure 5a. It can clearly be seen that self-bias voltage has significant influence on the average grain size at a given gas mixture composition. In general, UNCD films with less than 40 nm are formed above 50 V self-bias values. With higher Ar-concentration even lower self-bias (0 and 50 V) is enough for the formation of films with grain sizes less than 40 nm (Figure 5a.). Series of NCD samples with average grain sizes close to those of so-called UNCD layers were chosen to represent changes in optical properties. Dispersion of the refractive index and extinction coefficient are plotted in Figure 5b. and 5c. for UNCD samples containing grains of 24-40 nm average size. Spectral dependence of the refractive index closely follows the characteristic behavior of the diamond crystal. However, in the case of 40 nm sample the values of the refractive index are around 90% of that of diamond crystals and with decreasing average grain size the refractive index shifts to smaller values. The increasing surface/volume ratio with smaller average grain size results in higher contribution of intergrain and grain boundary phases to the optical properties. The grain boundary and the intergrain phases contain different C-H bonds that cause lower average density of the UNCD layers, and thus the optical density also decreases.

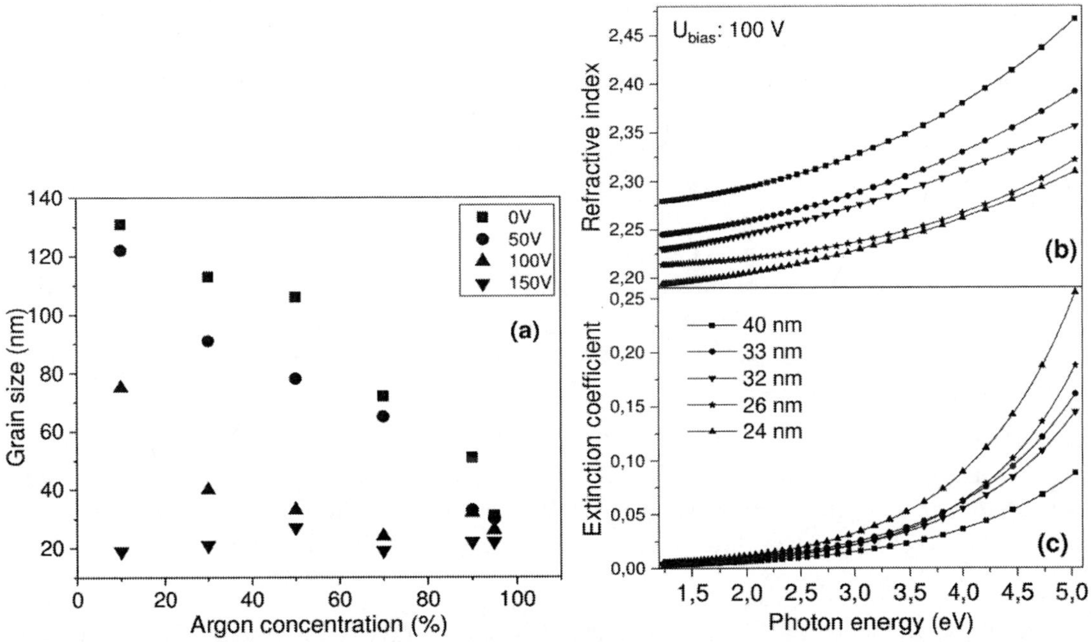

Figure 5. (a) Average grain size of analyzed NCD samples as a function of the deposition parameters. Spectral dependence of the refractive index (b) and the extinction coefficient (c) in UNCD films with different grain sizes.

Photon energy dependence of extinction coefficient for UNCD samples is presented on Figure 5c. Significantly higher extinction coefficient values can be observed in UNCD films (<40 nm) compared to those of the NCD layers. The absorption loss is nearly 5 times larger in the UNCD sample with grain size of 24 nm. The extinction coefficient is inversely proportional to the average grain size in the analyzed photon energy range. The difference between the extinction coefficient curves for samples of different grain sizes increases rapidly above ~3 eV. The observed dependence of the extinction coefficient on the average grain size is closely related to the fact that the bonding structure and the spatial arrangement of non-diamond phases is influenced by the grain size. The large values of the extinction coefficient for UNCD samples indicate extra contribution to the absorption loss in these layers. Raman scattering spectra on Figure 1b. and Figure 3b. show a meaningful decrease of the diamond Raman peak relative to the scattering contribution of other phases, and this can be related partly to the smaller grain sizes. The relative scattering intensity of the G band increases for the sample with 19 nm grain size (Figure 1b.), although similar change cannot be observed for the 31 nm sample (Figure 3b.). The appearance of a new scattering band which can be assigned to the formation of a new type of structure, cannot be observed in these spectra, however it is possible that this structure does not have Raman active modes. Thus we suggest that the extra contribution to the extinction coefficient in UNCD films can be attributed to the significant increase in the amount of non-diamond phase caused by the higher surface/volume ratio. Beside the optical absorption, which is related to the non-diamond bonding structure, contribution of different defect sites, both intrinsic and extrinsic ones, has to be considered. For example, the increase of the extinction coefficient at ~4 eV could be related to the presence of substitutional N defects [20]. Luminescence excitation spectroscopy performed on our UNCD samples confirmed these considerations.

4. Conclusions

Spectral dependence of the refractive index and the extinction coefficient were analyzed in nano- and ultrananocrystalline diamond layers. The average grain size was used as practical parameter for these layers, which is under 20 nm for the UNCD samples. The decrease of the average grain size indicates higher amount of non-diamond phases in the film structure. The evolution of the Raman scattering bands with the decreasing average grain size indicates changes in the bonding structure of the non-diamond phases. The difference between the optical parameters of nanodiamond layers and crystalline diamond in the whole studied spectral range becomes more pronounced as the average grain size of diamond layers decreases. This can be explained by the increasing ratio of non-diamond to diamond phases in the film. Since lower average grain size means larger grain boundary and intergrain region, the sp^2 C-C and sp^2 C-H bonds' contribution to the localized states in the band gap of diamond increases, together with the average density of layers.

Acknowledgement

Sára Tóth is grateful for the support of the János Bolyai Research Scholarship of the Hungarian Academy of Sciences. This work was supported by the Hungarian Science Foundation under contract number NKFI K-115805. The authors also wish to thank P. Csíkvári for technical support.

References

[1] Kriele A, Williams O A, Wolfer M, Brink D, Müller-Sebert W and Nebel C E 2009 *Appl. Phys. Lett.* **95** 031905
[2] Kopecek M, Bacakova L, Vacik J, Fendrych F, Vorlicek V, Kratochvilova I, Lisa V, Van Hove E, Mer Ch, Bergonzo P and Nesladek M 2008 *Phys. Status Solidi* A **205** 2146
[3] Auciello O, Gurman P, Guglielmotti M B, Olmedo D G, Berra A and Saravia M J 2014 *MRS Bulletin* **39** (Organic Spintronics) 621
[4] Williams O A and Nesladek M 2006 *Phys. Status Solidi A* **203** 3375
[5] Gruen D M 1999 *Annu. Rev. Mater. Sci.* **29** 211
[6] Hong B, Lee J, Collins R W, Kuang Y, Drawl W, Messier R, Tsong T T and Strausser Y E 1997 *Diam. Relat. Mater.* **6** 55
[7] Silva F, Benedic F, Bruno P and Gicquel A 2005 *Diam. Relat. Mater.* **14** 398
[8] Yugo S, Kanai T, Kimura T and Muto T 1991 *Appl. Phys. Lett.* **58** 1036
[9] Auciello O and Sumant A V 2010 *Diam. Relat. Mater.* **19** 699
[10] Herzinger C M, Johs B, McGahan W A, Woollam J A and Paulson W 1998 *J. Appl. Phys.* **83** 3323
[11] Fujiwara H 2007 Spectroscopic Ellipsometry, Principles and Applications, John Wiley Sons Inc., New York
[12] Liu Y C, Hsieh J H and Tung S K 2006 *Thin Solid Films* **510** 32
[13] Prawer S, Nugent K W, Jamieson D N, Orwa J O, Bursill L A and Peng J L 2000 *Chem. Phys Lett.* **332** 93
[14] Bergman L and Nemanich R J 1995 *J. Appl. Phys.* **78** 6709
[15] Brillante A, Hanfland M, Syassen K and Hocker J 1986 *Physica B+C* **139-140** 533
[16] Mishra P, Jain K P 2001 *Phys. Rev.* B **64** 073304
[17] Németh P, Garvie L A J and Buseck P R 2015 *Scientific Reports* **5** 18381
[18] Aleksandra B. Djurisic and E. Herbert Li 1998 *Appl. Optics* **37** 7273
[19] Himics L, Tóth S, Veres M, Czitrovszky A, Nagy A, Oszetzky D, Kerekes A, Kugler Sz, Rigó I, Tóth A, Koós M 2015 NATO Science For Peace and Security Series A: Chemistry And Biology **39** 93
[20] Tucker O D, Newton M E and Baker J M 1994 *Phys. Rev.* B **50** 15586

IOP Conf. Series: Materials Science and Engineering **426** (2018) 012050 doi:10.1088/1757-899X/426/1/012050

Comprehensive analysis of local corrosion degradation on austenitic pipeline

P Trampus[1], J Dobránszky[2], Z Kerner[3], J Knisz[4], F Oszvald[5], B Palotás[6], L Péter[7], M Réger[8] and B Verő[9]

[1] Research professor, University of Dunaújváros, Dunaújváros, Hungary
[2] Scientific advisor, Research Group for Composite Science and Technology, Budapest, Hungary
[3] Senior researcher, Centre for Energy Research, Hungarian Academy of Sciences, Budapest, Hungary
[4] Research scientist, VEIKI Energia+ Ltd, Budapest, Hungary
[5] Consultant, Oszvaldfer, Paks, Hungary
[6] Associate professor, University of Dunaújváros, Dunaújváros, Hungary
[7] Scientific advisor, Wigner Research Centre for Physics, Hungarian Academy of Sciences, Budapest, Hungary
[8] Full professor, Óbuda University, Budapest, Hungary
[9] Professor emeritus, University of Dunaújváros, Dunaújváros, Hungary

E-mail: trampus@uniduna.hu

Abstract. Corrosion degradation was observed in a nuclear power plant spent fuel cooling system. A systematic and comprehensive investigation program was developed to reveal the root cause of the degradation. Series of corrosion tests, mechanical, microstructural and microbiological investigations were carried out. Also, simulations of the operating conditions and welding parameters were performed. Based on the results, the major contributors to the degradation process were identified and a possible degradation mechanism was proposed.

1. Introduction

After 26 operating years, leakage was observed in the spent fuel pool cooling system at the Paks Nuclear Power Plant, Hungary. The cooling medium is boric acid with a concentration of > 14.5 $g \cdot dm^{-3}$; and average temperature of ~30°C. Two parallel loops operate alternately. Results of the immediate non-destructive examinations and subsequent comprehensive investigations made it clear that the leakage was caused by pitting corrosion on the inner surface of the cooling system pipelines. Some of the pits penetrated through the whole wall thickness. The structural material of the pipelines is Cr-Ni-alloyed, Ti-stabilized austenitic stainless steel (08Ch18N10T, nearly equivalent with steel 1.4541). The diameters vary from 100 to 300 mm; the wall thickness range is 6 to 10 mm. The pipes were laid down in concrete which made their investigations and subsequent replacements difficult and time consuming. *Figure 1* shows a through-wall pit detected by visual inspection and ultrasonic examination as well (left: corrosion products around a circumferential weld; middle: two pits after cleaning and decontamination; right: cross-section with the through-wall damage).

IOP Conf. Series: Materials Science and Engineering **426** (2018) 012050 doi:10.1088/1757-899X/426/1/012050

The plant operator asked technical support organizations to perform a root cause analysis of this service-induced, unexpected degradation phenomenon developed after quite a long operation period. The complexity of the question required knowledge from various technical areas such as material science, electrochemistry and corrosion science including microbiologically induced corrosion (MIC), welding technology and plant operation. The working team was composed of representatives of leading academic institutes and universities to complete the task. The description of the degradation phenomenon as well as the investigation process and analyses (Phase I – 2013-2014) were described previously [1]. The goal of the present paper is to summarize the results focusing primarily on those which were achieved following the first stage (Phase II of investigations – 2016-2017), and to suggest the most likely degradation mechanism.

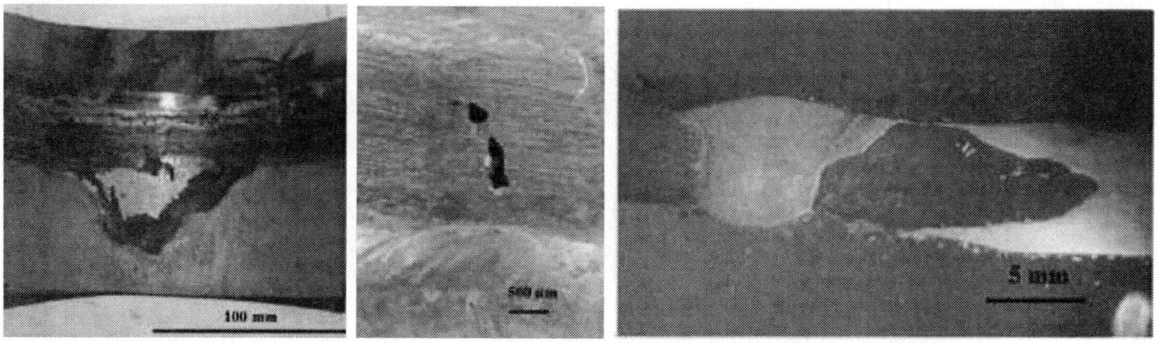

Figure 1. Pitting corrosion around the welded joint – pipe: Ø 219x10 mm (left: as is, middle: after cleaning, right: cross-section with the inside down)

2. Previous investigations

During Phase I samples were taken out from the leaking pipeline system, and the following characteristics were determined:

- long term corrosion potential, corrosion rate, electrochemical impedance, pitting potential (provoked by chloride ions and ferric chloride);
- bulk and local chemical composition (energy-dispersive X-ray spectroscopy – EDX, laser ablation inductively coupled plasma mass spectrometry – LA-ICP-MS), deposit composition;
- mechanical properties;
- microstructures (optical microscopy – OM, scanning electron microscopy – SEM, electron backscatter diffraction – EBSD);
- microbial presence (fluorescence microscopy, culture methods).

The electrochemical investigations were repeated on archive material and on identical new material as well.

The results [1] of these investigations showed that the corrosion resistance, chemical composition, mechanical properties and features of non-metallic inclusions were all in compliance with the relevant standard values. Most of the inner surface of the pipeline system was intact, conditions for the corrosion occurred locally. The majority of the corrosion was associated with heat tint covering the heat affected zone (HAZ) of site welds. Also, evidence of the presence of organic material on the corrosion surface was found, giving rise to suspicion of MIC process.

Obviously, more than one element of the entire system contributed to the corrosion. Major contributors include the oxide layer on the HAZ and associated corrosion sensitivity nearby welds, microbiological effects, and operating conditions such as flow velocity, stagnation and temperature ideal for microbial growth.

3. Current investigations and analyses

In Phase II complementary investigations, analyses, and modelling/simulation were performed. Also, the results of MIC monitoring started after completing Phase I were incorporated and evaluated.

3.1. Root cause investigations

The following investigations served for identifying the root cause of the localized corrosion degradation on the piping system of the spent fuel pool.

3.1.1. Effect of microbes on the corrosion phenomena. Over a two-year period, we were monitoring biomass changes in the bulk medium and found that most of the time the biomass was in the range of 10^2-10^3 ME·mL^{-1}, sometimes increasing up to 10^4 ME·mL^{-1} (ME = microbial equivalent, cells). Next generation sequencing of the bulk medium revealed that the majority of the microbes is *Ralstonia pickettii*, a mixotrophic species viable in ultrapure water [2, 3]. It can live in an autotrophic manner by using only H_2 as a sole source of energy and CO_2 as the sole carbon source. It is known to produce extracellular polymer material, a matrix needed for biofilm development. The sequencing revealed members of the *Desulfobacteriaceae* and *Desulfovibrionaceae* family, members of the *Burkholderia* family, the *Sediminbacterium* genus that have been implicated in iron oxidation [4], as well as the acid producing *Acinetobacter* genus [5] that has been shown to be a pioneer bacterium in biofilm development [6, 7]. Although the amount or the type of microbes present in the bulk fluid does not represent microbes on the surface, the results suggest that *Ralstonia pickettii* together with other *Ralstonia* species might initiate biofilm development in the presence of boric acid. Biological activity reaction test (BART) [8] performed on corrosion products collected from the bottom of the pipes revealed the presence of both iron oxidizing and iron reducing bacteria, heterotrophic micro-organisms, slime forming bacteria, as well as acid producing bacteria. Sulfate reducing bacteria were not detected. Corroded areas were analyzed by fluorescence staining to visualize live and dead cells inside the pit after cleaning (*Figure 2C*) and on corrosion deposit (*Figure 2A).*

IOP Conf. Series: Materials Science and Engineering **426** (2018) 012050 doi:10.1088/1757-899X/426/1/012050

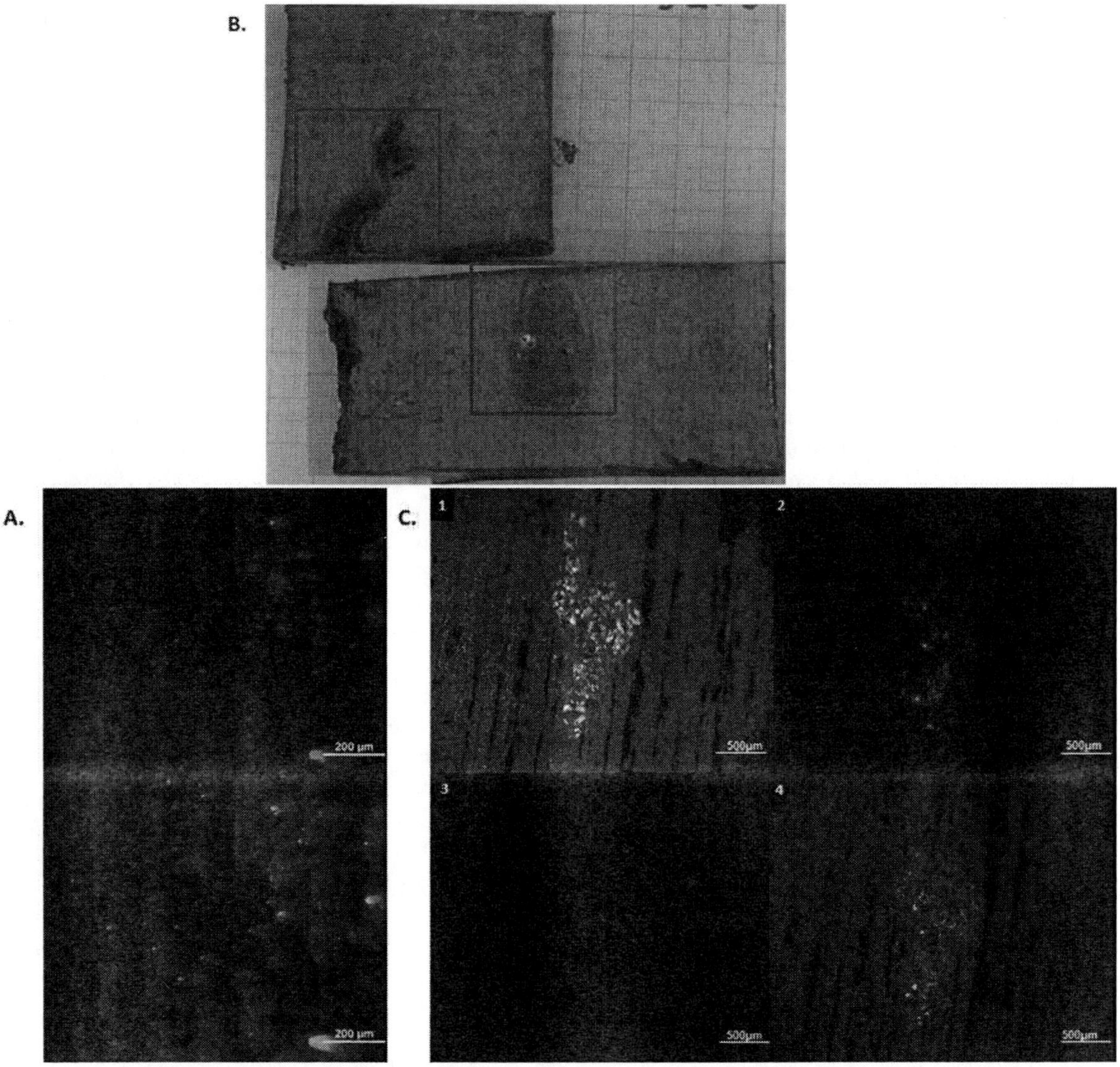

Figure 2. Fluorescence staining of corroded stainless steel.
A. Live cells (upper) and merged image (lower) of corrosion products; B. The corroded surface; C. (1) bright field image, (2) live cells, (3) dead cells, (4) merged image

The pit had a shiny metallic surface (*Figure 2B*: lower picture) suggesting an active corrosion mechanism, and live/dead stain revealed microbes inside the pit but not outside (*Figure 2C*).

A SEM/EDX analysis at another corrosion site revealed phosphorus, sulfur and carbon that may be indicative of microbial presence at the site of corrosion. Left picture (SEM) in *Figure 3* shows the pit and the porous corrosion product around it; on the right side the EDX spectrum can be seen.

IOP Conf. Series: Materials Science and Engineering **426** (2018) 012050 doi:10.1088/1757-899X/426/1/012050

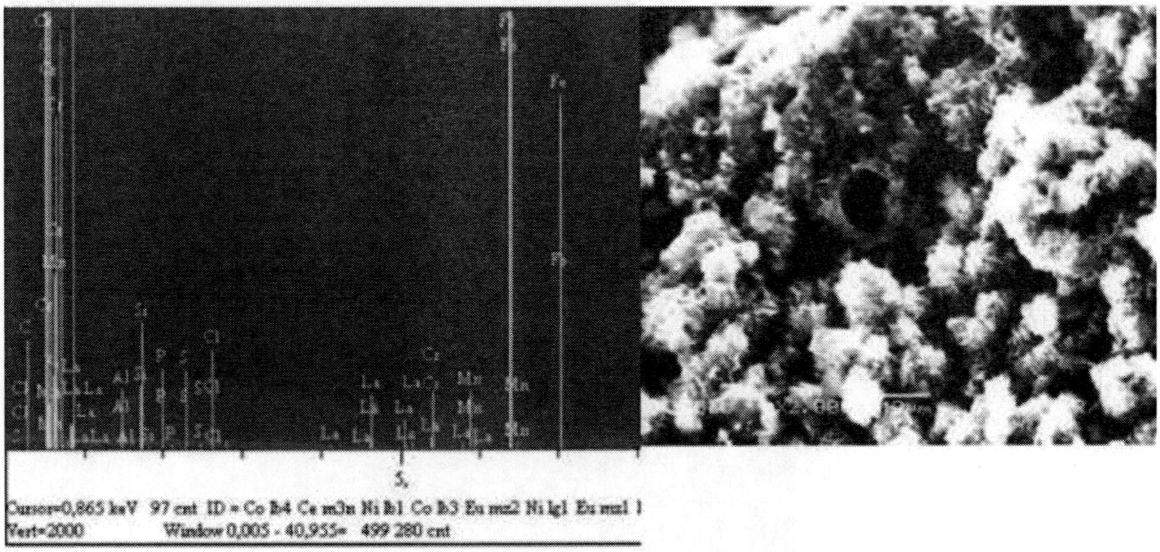

Figure 3. Porous corrosion product (inset) and its EDX-spectrum

3.1.2. Oxide layer. Both the literature and the results of our investigations suggested that the oxide layer appearing as heat tint, i.e. welding scale on some pipes plays leading role in the development of the corrosion process. In order to better understand as well as to validate the literature data with factual findings, we analyzed the main features of the oxide layer's composition and structure on samples prepared for modelling the oxide layer.

The glow discharge optical emission spectroscopy (GD-OES) is capable of measuring the layer's thickness and determining its tentative chemical composition. In *Figure 4* the GD-OES intensity measured on the transition to HAZ can be seen. First of all, the oxygen content and the distribution of Ni, which has a low inclination to migrate can indicate the boarder of the oxide layer with the thickness being approximately 10 nm. In the middle of the HAZ the thickness approaches 50 nm while on the normal surfaces the passive layer thickness does not exceed 3-4 nm. Intensity of Fe, Cr and Ni

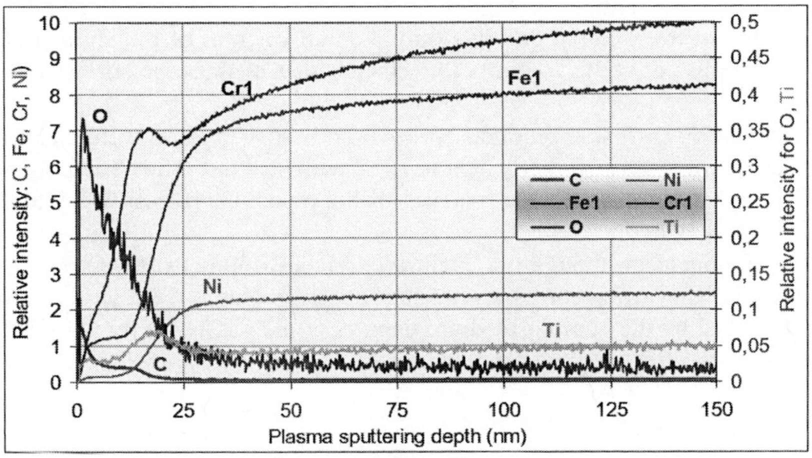

Figure 4. GD-OES intensity of various elements as a function of the sputtering depth, on clean surface

reaches the value characterizing the base material at this depth. The oxide layer's composition is detrimental from the corrosion's point of view because it contains less Cr than the normal passive layer and has a less beneficial Cr-oxide/Cr and Cr/Fe ratio. On the other hand, during the evolution of the thicker oxide layer more Cr will be detracted from the area beneath the scale than in the case of a normal passive layer.

Similar oxide layer thickness could be identified by transmission electron microscopy (TEM). On a segment perpendicular to the surface, the glue indicates the outer surface of the sample, see *Figure 5*, and immediately under it an amorphous oxide layer is visible with a thickness of 50–100 nm.

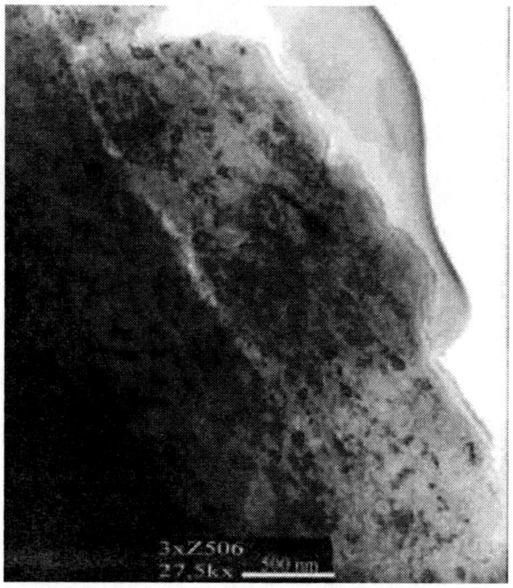

Figure 5. TEM picture of the welding scale covered thin foil

3.2. Records of visual inspections of the pipeline's inner surface

The video records made by professional inspection vendors between 2013 and 2017 were analyzed and evaluated. The following conclusions were drawn:

- Considering purely the number of degraded spots, ~60 % are located on or near welded joints. As the total surface of the welds represents only a small fraction of the entire surface of the piping system, vulnerability of welds to corrosion exceeds that of flat pipe surfaces by several orders of magnitude.
- The area covered by corrosion products spreads out in 80–90 % on welds, i.e. on root bead, HAZ and a band of parent material directly next to HAZ with heat tint. The remaining portion of corroded area could be identified on the base material of the pipes. Corrosion was mainly observed in the flow direction following the welds.
- There seems to be a clear, though not exclusive, relationship between corrosion of welds and that of other surfaces. The pitting corrosion can start in both welds and flat surfaces in areas covered by sediment delivered by the cooling medium degrading the passive layer, as well as in areas where the microbiological effect is more pronounced.

3.3. Simulations and analyses

3.3.1. Integrated modelling. A demonstration experiment to study the collective effect of each factor contributing to the corrosion phenomenon was performed. These factors are the following: differential aeration, welding without root protection, corrosion sediment, presence of microbes and chloride ions. Experimental parameters demonstrated extreme values for corrosion. The medium was much more aggressive than in the spent fuel pool resulting substantially accelerate the corrosion process. The experimental setup is shown on the left side of *Figure 6*, the equipment used for the experiment is shown on its right. The following were concluded:

- Long term operation (46 days) of differential aeration cells with large cathode/anode surface ratio confirmed the extreme corrosion sensitivity of weld produced without proper protection. These samples showed approximately ten times faster corrosion velocity than the samples without welds.
- Also, simulation of differential aeration in anaerobe environment revealed biofilm development on surfaces of solid materials. It occurred even in the presence of boric acid. A biofilm is considered as a potential initiation factor for microbiological corrosion degradation.
- On the weld root side of the specimens, the initiation stage of pitting corrosion could be observed, as well as indications for tuberculum development were found. It is to be emphasized that these degradation processes developed in a relatively short time.

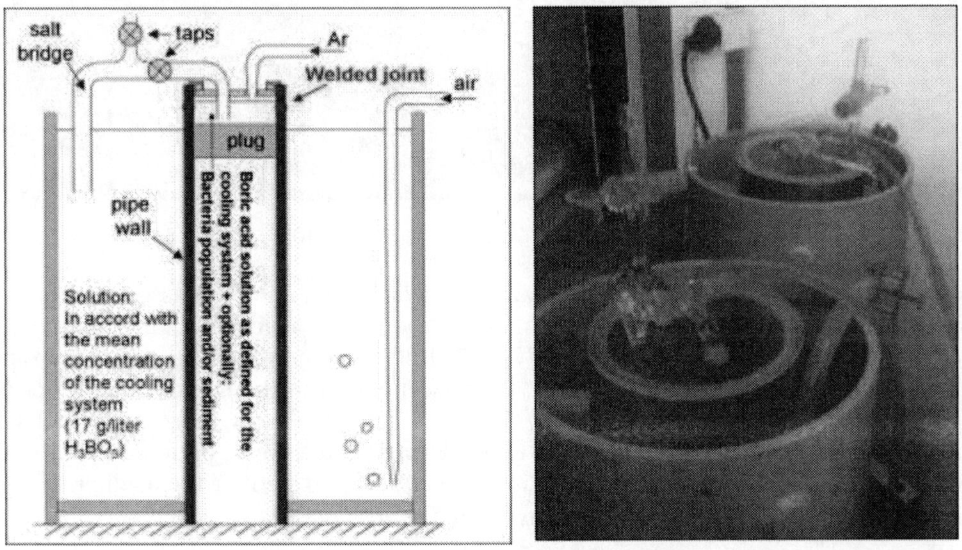

Figure 6. Setup of the differential aeration model (left) and the equipment (right)

3.3.2. Impact of welding parameters. Welding experiments were performed following the original WPS (Welding Procedure Specification); the following parameters were varied:

Deviation from WPS
Without root protection
With imperfect root protection
Parent materials were not prepared except edge preparation
Heat input limit was exceeded
Heat input limit was exceeded, and with imperfect root
With significantly enlarged root gap
Root protection only by root welding

The inner surfaces of the samples cut out from the welded joints were investigated by SEM; the chemical composition of the pipe in the vicinity of the joints was also analyzed. The pitting corrosion sensitivity was investigated using the ASTM G48 standard [9]. The tests were performed in two solutions with different concentration of $FeCl_3$ (3.6 % and 6 %) at two different temperatures (room-temperature and 50°C). The impact of the spent fuel cooling medium was simulated by mechanically as well as bacterially contaminated media. The fact that samples having no root protection showed the most intensive degradation supports the hypothesis that root protection has significant impact on the corrosion sensitivity of the welds. *Figure 7* shows the sample welded in accordance with WPS but without root protection. Pits can be seen both on the weld face and at the root in the sample tested at 50°C. The latter demonstrates the influence of the temperature. Significantly more corrosion products were formed when the solution contained MIC bacteria compared to a non-MIC environment.

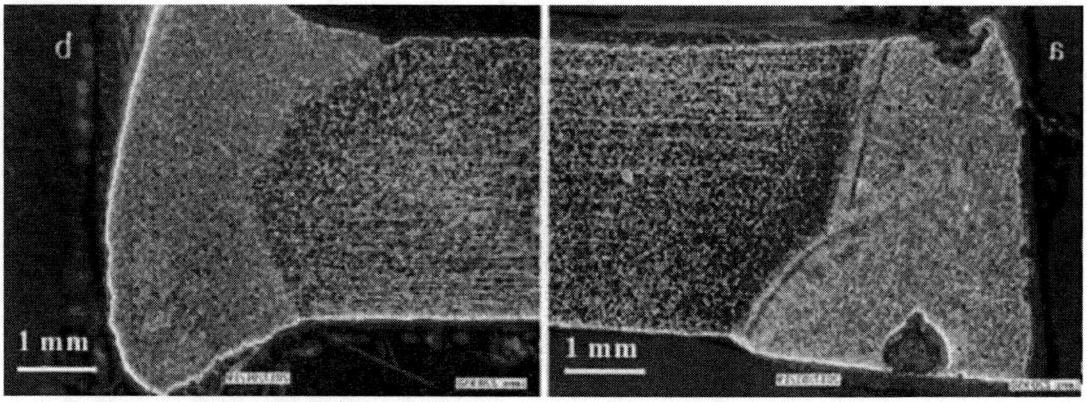

Figure 7. Test at 50°C, in 3.6 % $FeCl_3$ (a); at room-temperature, in 6 % $FeCl_3$ (b)

3.3.3. Gamma irradiation. The effect of the gamma irradiation on microbes associated with MIC was investigated to estimate their viability. A radiological test environment simulating the spent fuel pool was developed. Based on the results a dose of 100 kGy kills the majority of the microbes irradiated, but the solutions' sterility can only be achieved if the concentration of the microbes does not exceed an order of magnitude of 10^7 ME·mL^{-1}. It is likely that in the spent fuel pool, close to the fuel assemblies, the majority of microbes perishes within a few hours. However, due to the periodic flow of the cooling medium among the assemblies, gamma irradiation is insufficient to sterilize the medium, thus microbes can reach distant surfaces of the cooling system where the dose rate is low.

4. Evaluation of the results, conclusions

Evaluating all results including those achieved at the end of Phase I, the following uniform statements and conclusions can be given. From a corrosion's point of view, the spent fuel cooling system is considered as a complex and open system. Its corrosion behavior is determined by many contributing factors as well as by their interaction. The specific effect and the relative weight of the individual factors may vary and the observed localized pitting corrosion degradation cannot originate from a single, well defined factor or root cause.

The typical and average chloride ion concentration and boric acid content of the cooling medium applied either alone or in combination with other factors cannot initiate this corrosion process. The structural material of the pipes meets the standard requirements.

On significant number of the welds produced in the field, on the pipes' inside surface an undesired oxide layer (heat tint, weld scale) is present. The composition profile of these surfaces and the area beneath deviates from that providing the proper passive condition; thus, the corrosion resistance of the surface range is reduced as a consequence of the thermal process.

In the cooling medium with the typical boric acid content, various micro-organisms can grow; and several micro-organisms previously associated with MIC were identified. An optimal temperature for reproduction of micro-organisms is 15–40°C, which corresponds to the cooling system's operation temperature range. Depending on both the medium composition and operating conditions (periodic operation, stagnation) the thermal oxide layer around the welds, i.e. the weld scale fosters the attachment of micro-organisms leading to MIC. The development of biofilms induces a local reduction in oxygen supply of the area and lead to the creation of local aeration cells. Reactions within the biofilm are usually localized and influence electrochemical processes, typically accelerate them and subsequently result in localized corrosion.

5. A possible degradation process

Iron-oxidizing bacteria obtain their energy by oxidizing Fe^{2+} ion to Fe^{3+} ion. Due to this activity of the bacteria an orange-red color, cave-like iron-oxide sediment (tubercle) formed. The iron-oxide sediment blocks the oxygen and provides the right environment for iron reducing bacteria. Subsequently, this anoxic area will act as an anode compared to the surrounding oxygen-rich cathode. At the anode area, the oxidation of the metal takes place, which leads, through the hydrolysis of the newly produced metal ion, to the decrease in pH. These processes with the lack of oxygen as re-passivation agent subsequently lead to the acceleration of corrosion.

In the porous structure of the tubercles, due to local effects, the chloride-ion content of the cooling medium may locally increase. Taking into consideration the limited flow inside the biofilm, an anoxic and chloride-ion rich medium can develop strengthening the local cell effect quite dominantly. During medium flow, the corrosion product can be separated from the surface and be drifted towards the flow, some of these products may adhere to other surfaces in the system. The separated product can contain iron hydroxide as well as living and dead micro-organisms originating from the tubercles.

6. Summary

A local corrosion degradation phenomenon was identified after a long operation time in a nuclear power plant system. A comprehensive investigation program was developed to invent the root cause of the degradation. The program included various corrosion tests, mechanical, microstructural, and microbiological investigations and simulations of the operating conditions and welding parameters. Based on the evaluation of the results of the investigations and analyses, the major contributors to the corrosion degradation were identified. These were identified as being the oxide layer on the HAZ of the welds and associated corrosion sensitivity nearby, the effect of micro-organisms, and operating conditions such as flow velocity, stagnation and low temperature.

Acknowledgements

The authors would like to express their gratitude to G Radnóczy for the TEM investigations and to T Török for the GD-OES analyses.

References

[1] Trampus P *et al* 2017 Investigation of local corrosion degradation developed on a pipeline system in service period *Materials Science Forum* **885** 92

[2] Adley C C *et al* 2005 Ralstonia pickettii: biofilm formation in high purity water *The Biofilm Club* 151

[3] Gales G *et al* 2004 Action of an aerobic hydrogenotroph bacteria isolated from ultrapure water systems on AISI 304 stainless steel *Proc. Int. Conf. EUROCORR 2004 (Nice)* 11

[4] Kuever J 2014 *The Prokaryotes* Chapter: The family of Burkholderiaceae Springer 759

[5] Pavissich J P 2010 Culture dependent and independent analyses of bacterial communities involved in copper plumbing corrosion *Journal of Applied Microbiology* **109** 771

[6] Wang H B *et al* 2012 Effects of disinfectant and biofilm on the corrosion of cast iron pipes in a reclaimed water distribution system *Water Res* **46** 1070

[7] Percival S J 1999 The effect of molybdenum on biofilm development *Ind Microbiol Biotech* **23** 112

[8] http://www.dbi.ca/StandMeth/ {checked at December 12, 2017}

[9] ASTM G48 - 11 2015 Standard Test Methods for Pitting and Crevice Corrosion Resistance of Stainless Steels and Related Alloys by Use of Ferric Chloride Solution, ASTM International, West Conshohocken, PA.

IOP Conf. Series: Materials Science and Engineering **426** (2018) 012051 doi:10.1088/1757-899X/426/1/012051

Effects of different backing gases on 2404 duplex stainless steel welds

B Varbai[1, *], I Mészáros[1] and K Májlinger[1]

[1] Budapest University of Technology and Economics, Department of Materials Science and Engineering, Budapest, Hungary

E-mail: varbai@eik.bme.hu

Abstract. In our research butt welds were made on newly developed LDX 2404 lean duplex stainless steel grade with gas metal arc welding. For root protection three different backing gases were used: pure argon, 95 vol% nitrogen + 5 vol% hydrogen and pure nitrogen. In case of pure nitrogen backing the effect of different gas flow rates was also investigated. Based on our research, the nitrogen content in the backing gas resulted in increasing austenite content. The increasing nitrogen content in backing gas and the increasing austenite content resulted in increasing pitting corrosion resistance. The flow rate of the backing gas also has significant effect on the evolving microstructure and pitting corrosion resistance.

1. Introduction

High strength and novel corrosion resistant steels gain more attention in chemical, offshore and pulp industries [1–9]. Among stainless steels, duplex stainless steels (DSS) have excellent corrosion resistance due to the austenitic – ferritic microstructure, which is a result of austenite and ferrite forming alloying elements and heat treatment [10–15]. Because of the recent fluctuations in nickel and molybdenum prices of metal stock market, steel industries have developed new types of stainless steels with lower nickel and higher manganese and nitrogen alloying contents [16–18]. These special types of new DSS called "lean" duplex, indicating their lower nickel content [19–22]. Recently, lean DSS gaining more attention due to their high mechanical strength and corrosion resistance and lower price compared to the conventional DSS grades, such as EN 1.4462 (UNS S31803).

The welding of DSS is often challenging, because of their high alloying content and the metastable microstructure, which also depends on the weld thermal cycles [23–27]. For compensation of nitrogen loss during welding and to ensure optimal, 1:1 phase balance in duplex microstructure, nitrogen shielding gas is often used during arc welding [28–31]. However, nitrogen does not always solve the weldability issues of high nitrogen stainless steel as detrimental phases such as chromium-nitrides [32, 33] can precipitate and also porosity formation [34, 35] can occur.

The effects of nitrogen used as backing gas during welding is not well published yet in the professional open literature. Westin et al. [36–38] published their results on nitrogen, argon + nitrogen and nitrogen + hydrogen backing gas effects on the corrosion resistance and microstructure of conventional, lean and super DSS grades. According to their research nitrogen has an effect on the microstructure, even when used as a backing gas. As a result, in case of nitrogen gas root protection higher austenite content could be measured on weld root and the corrosion resistance improves.

In our research, butt welds were made with gas metal arc welding on a newly developed LDX 2404 lean duplex grade (UNS S82441, EN 1.4662). For root protection three different backing gases:

IOP Conf. Series: Materials Science and Engineering **426** (2018) 012051 doi:10.1088/1757-899X/426/1/012051

100 vol% argon, 95 vol% nitrogen + 5 vol% hydrogen and 100 vol% nitrogen was used. The effect of different backing gas flow rates were also investigated and found to have an effect on the evolving microstructure.

2. Materials and methods

2.1. Base and filler materials

The used base material was lean duplex LDX 2404 grade with lower nickel and higher nitrogen and manganese alloying content (Table 1.). The higher N and Mn alloying ensures the high yield strength ($R_{p0.2}$ = 640 MPa in case of cold rolled sheets) of LDX 2404 grade. Also, because of the high nitrogen and molybdenum (Mo = 1.6 wt%) content, the LDX 2404 grade offers good resistance to localized and uniform corrosion. For welding 3 mm thick sheets were used.

The used filler wire for MAG (metal active gas) welding was general duplex wire ISO 14343-B-SS2209 grade with 1.2 mm diameter. The filler wire contains higher nickel content (Ni = 8.5 wt%) in order to ensure adequate austenite formation during welding.

Table 1. Main alloying contents of the used LDX 2404 base material (BM) and the SS2209 welding wire (W). CrE and NiE values are calculated according to the DeLong diagram

Material designation	Measured alloying content in wt%									
	C	Si	Mn	Cr	Ni	Mo	N	Fe	CrE	NiE
LDX 2404 (BM)	0.02	0.7	3.0	24.0	3.6	1.6	0.27	bal.	26.75	13.80
SS 2209 (W)	0.02	0.5	1.6	22.8	8.5	3.1	0.17	bal.	26.65	15.00

2.2. Welding parameters

Robotized MAG welding machine was used to ensure constant welding parameters. The butt welds were made in flat position, without chamfering and keeping a constant 1 mm root gap. The applied heat input was 0.26 kJ · mm^{-1}, calculated with k = 0.8 thermal efficiency factor, according to EN 1011 standard.

For all welds on the face side the same shielding gas was used, 98 vol% Ar + 2 vol% O_2 – which mixture is recommended for DSS MAG welding – with 15 l · min^{-1} flow rate. For root protection three different backing gases with 9 l · min^{-1} flow rate were used: pure Ar, 95 vol% N_2 + 5 vol% H_2 (forming gas) and pure N_2 (with 4 l · min^{-1} flow rate too). One weld was done without root protection, exposed to ambient air. For the listed backing gas mixtures for specimen designation only; Ar, 95 N_2 + 5 H_2 and N_2 were used. After welding the samples were not pickled and re-passivated unlike industrial practices.

2.3. Microstructure investigations

Standard metallographic specimens were done from all welded joint cross-sections. The color etching was done, using Beraha mixture [39]: 85 ml H_2O + 15 ml HCl + 1 g $K_2S_2O_5$, using double etching method. On all metallographic images the austenite phases will appear light and the ferrite phases appear dark e.g. as seen on Figure 1.

The microstructure images were taken using Olympus PMG 3 type microscope. For taking the macro images of the welds face and root before and after corrosion tests Olympus SZX16 type stereomicroscope was used.

2.4. Austenite content measurement

For austenite content measurement two different methods were used. (I) on the as welded samples Feritscope measurements were done, using Fischer FMP30 instrument, which measures the volume of the ferromagnetic phases (ferrite in our case). And (II) on the images taken by microscope (from the etched cross-section of the welds) image analyzer software can be used to determine the austenite – ferrite phase ratio. The steps of the image analyzing are the following. The original image is loaded to an image analyzer software and the grayscale histogram is taken up on the 0 to 255 range (8 bit). The histogram has two peaks; one at a darker gray level and one at a lighter gray level. The difference between the two gray level peaks is equivalent to the level of contrast of the image. In order to count the darker and lighter number of pixels (which is in correlation to the austenite – ferrite ratio) the original image should be converted into a black and white image. To do this, the difference between the two peaks can be used as a threshold to determine if the pixel is to be converted to black or white. Whit this method the austenite-to-ferrite ratio can be calculated after correlated to the black and white pixel ratio.

The comparison of the two methods can be seen on Figure 2., where a 45° straight line is fitted on the points measured with both methods. A linear fit with $R^2 = 0.99$ is determined, meaning both of the two methods can be used for austenite content measurement with the same results. In this paper in the further diagrams Fischer type Feritscope was used to determine the austenite-to-ferrite ratio in the welds.

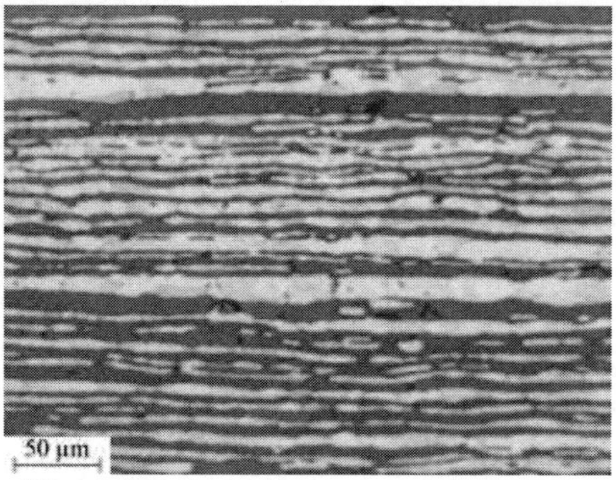

Figure 1. Etched microstructure of LDX 2404 base material (white is the austenite and dark is the ferrite phase)

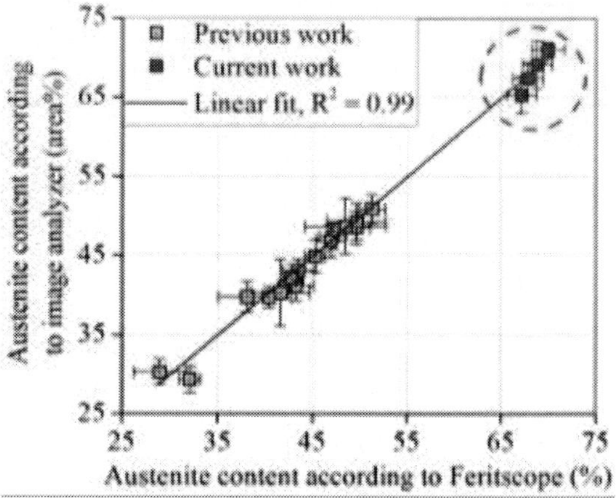

Figure 2. Comparison of the histogram based image analyzer and the electromagnetic based Feritscope austenite content measurements

2.5. Corrosion rate measurements

For the measurement of corrosion resistance ferric chloride pitting test was performed, according to code: ASTM G48 – Method A. For the test equal sections from all welded sheets were cut out using constant cooling, with the dimensions of 25 × 25 × 3 mm. The samples were immersed into 6 % $FeCl_3$ – H_2O solution for 72 hours at 50 ± 1 °C temperature. The weight of the samples was measured before and after the test with Denver Instrument SI-8007 instrument with 0.1 mg accuracy. Before and after the corrosion test ultrasonic cleaning was done in acetone in order to remove corrosion products and the samples were dried before the weight measurements.

3. Results and discussion

3.1. Macro- and microstructure of welds

With visual inspection no difference can be seen on the face side of the welds (Figure 3.). When no root side protection was applied, the root side oxidized and porous structure was visible. On the metallographic images the average measured oxidized layer depth is 20 μm.

Between the Ar and N_2 root protection no significant differences are visible on macroscale in the as-welded condition. When backing gas was not used for root protection an average of 100 × 100 μm oxidized areas are visible on the surface. These areas were later initiation sites for localized corrosion.

After the 72 hours corrosion test no differences were seen on face side of the welds as a function of the used backing gases, however significant differences were visible on the root side. When no root protection was applied the root side of the weld completely corroded, resulting the disappearance of the weld root after the corrosion test. In case of Ar root shielding localized corroded areas are visible with the average sizes of 500 × 500 μm. Also significant difference can be seen between the two samples welded with N_2 root protection, but different flow rates. In case of lower backing nitrogen gas flow rate ($4\ l \cdot min^{-1}$) shallow localized corroded areas are seen with average 100 × 100 μm size. With the higher

(9 l · min^{-1}) nitrogen backing flow rate much smaller (average 20 μm) localized corroded areas were seen.

As a summary it can be stated the higher flow rate pure nitrogen and argon root protection showed the best surface in terms of corrosion resistance.

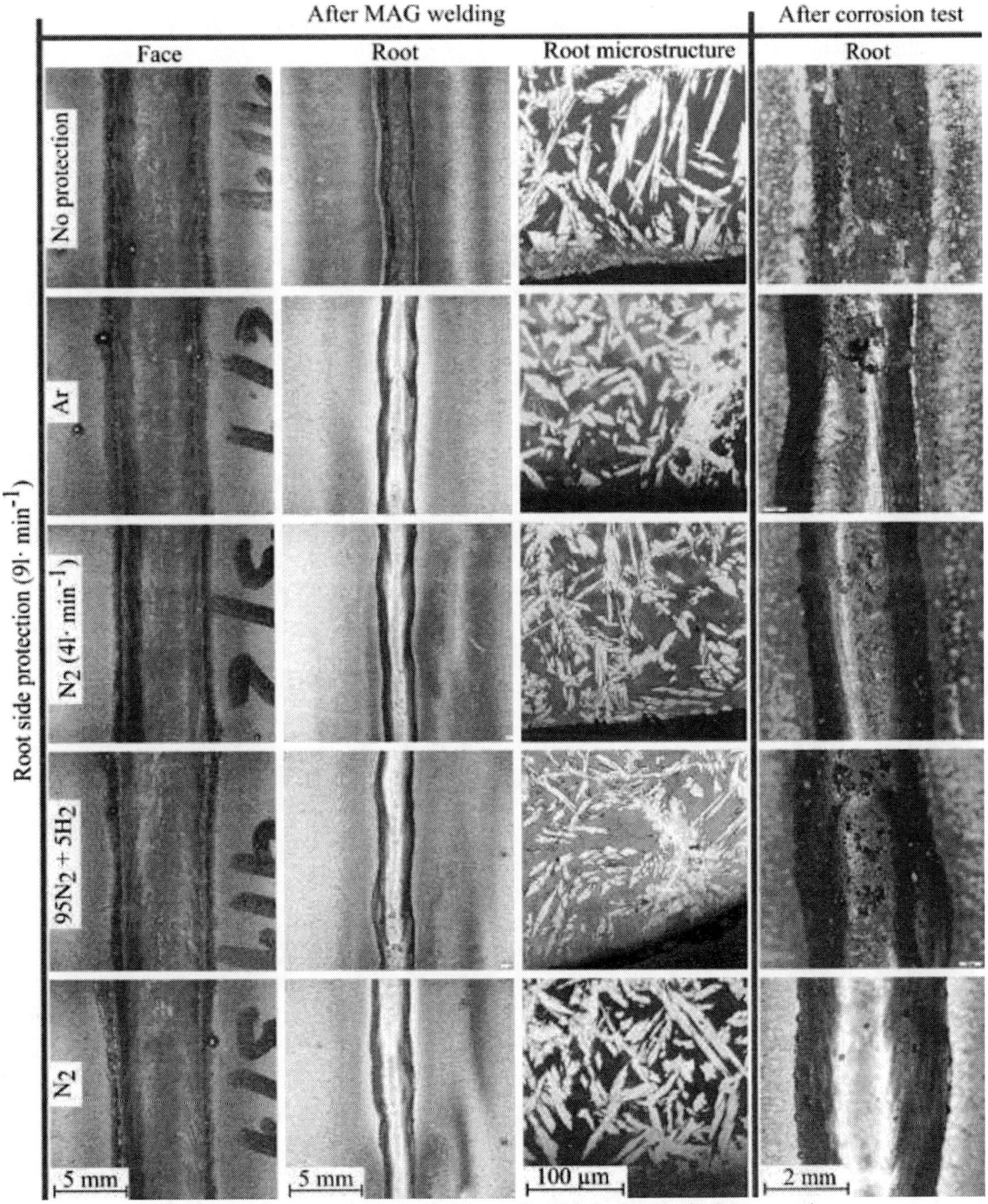

Figure 3. Macro- and microstructure of welds before and after corrosion test

3.2. Austenite ratio at the weld root

The average austenite ratio of the LDX 2404 base material is 59 vol% (Figure 4.). The austenite content of the weld face side did not change (65 ± 3 vol%) as a function of different backing gases used. However, on the root side of the welds different austenite ratios were measured.

Generally, the austenite content of the weld root increased with the increasing nitrogen content in the backing gas (Figure 4.). In case no root protection was used (the root side was exposed to air) the highest austenite content was measured, which is a measurement method error. The reason for this is the oxidation, which occurred on the root side and resulted in porous structure. The Feritscope measurement on this porous microstructure resulted in higher non-magnetic volume, which appears in lower ferrite content during the measurement. On the metallographic specimen of this weld, image analysis was done, which gave the result of 55.9 ±1.5 area% austenite content. It should be noted the image analysis cannot be done on the edge of the root side, only more inside the weld metal to have an adequate ~500 × 500 μm area for analysis.

On the edge of the weld root the lowest austenite content was measured in case of Ar root protection average 67 vol%, and the highest measured in case of N_2 was average 70 vol%. The lower flow rate of nitrogen resulted in lower austenite content on the root side; average 68 vol%. The 95 N_2 + 5 N_2 forming gas resulted an average 69 vol% austenite content.

3.3. Corrosion rates

The corrosion test were done without previous pickling and re-passivation of the samples. In case when no root side protection was used oxidation occurred on the weld root (Figure 3.). This oxidized layer has low corrosion resistance resulting a high corrosion rate of 55 mg · cm^{-2} (Figure 5.). When root protection was applied Ar backing gas resulted in the highest corrosion rate, which is 49.5 mg · cm^{-2}.

The lowest corrosion rate was measured in case of higher flow rate nitrogen, which is 45 mg · cm^{-2}. In case of the lower flow rate nitrogen backing 49 mg · cm^{-2} corrosion rate was measured. In case of forming gas the measured corrosion rate after 72 hours is 46 mg · cm^{-2} (Figure 5.). Overall, increasing nitrogen content in the backing gas resulted in increasing pitting corrosion resistance.

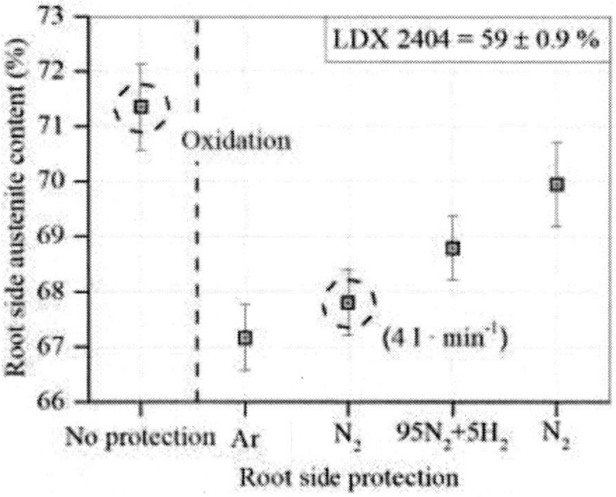

Figure 4. Austenite ratio of the weld root as a function of used root side protection gas

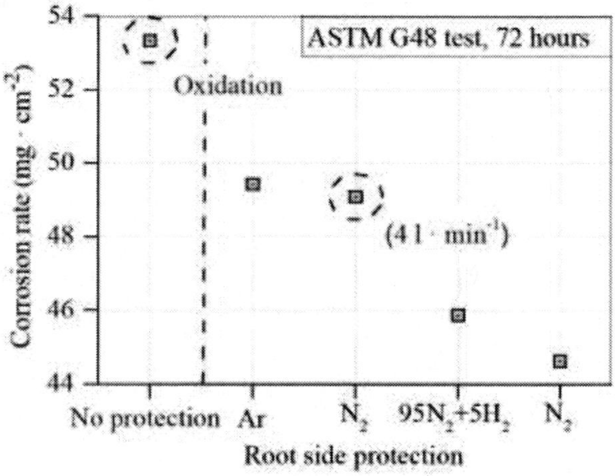

Figure 5. Corrosion rate after 72 hours as a function of used root side protection gas

4. Conclusions

Based on our performed research on MAG welding of LDX 2404 lean duplex stainless steel, using different root side protection, the following statements can be done:

- For austenite ratio determination both histogram based image analyser software and electromagnetic measurement based Feritscope can be used with the same quantitative mathematical approximation.
- The used backing gases had no effect on the weld face macro or microstructure.
- The increasing nitrogen content in the backing gas increased the austenite content on the root side of the weld metal.
- With increasing nitrogen content of the backing gas, decreased the pitting corrosion rate.
- The austenite content increased with the nitrogen backing gas flow rate.
- In case of no root protection, the root side of the welds oxidized (~ 20 μm deep) and the corrosion resistance decreased.

Acknowledgements

This paper has been supported by the János Bolyai Research Scholarship of the Hungarian Academy of Sciences grant number: BO/00196/16/6 and by the National Research, Development and Innovation Office - NKFIH, OTKA PD 120865 (K. Májlinger).

The research reported in this paper was supported by the Higher Education Excellence Program of the Ministry of Human Capacities in the frame of Nanotechnologyresearch area of Budapest University of Technology and Economics (BME FIKP-NANO).

References

[1] Lukács J 2010 *Procedia Eng.* **2** 1201–10
[2] Dobosy Á and Lukács J 2015 *Mater. Sci. Forum* **812** 29–34
[3] Dobosy Á and Lukács J 2017 T *Mater. Sci. Forum* **885** 111–6
[4] Boillot P and Peultier J 2014 *Procedia Eng.* **83** 309–21
[5] Balogh A et al 2012 *Prod. Process. Syst.* **5** 79–90
[6] Trampus P et al 2017 *Mater. Sci. Forum* **885** 92–7
[7] Haraszti F and Kovács T 2017 *IOP Conf. Series: Materials Science and Engineering* **175** 1-4
[8] Oyedemi K et al 2017 *Period. Polytech. Mech. Eng.* **61** 296
[9] Varbai B et al 2017 *Period. Polytech. Mech. Eng.* **61** 68–73
[10] Alcantara A S et al 2017 *Mater. Sci. Forum* **885** 190–5
[11] Dobránszky J et al 2004 *Spectrochimica Acta - Part B Atomic Spectroscopy* **59** 1781–8
[12] Fábián E R et al *Mater. Sci. Forum* **885** 245–50
[13] Mészáros I and Bögre B 2017 *Mater. Sci. Forum* **885** 184–9
[14] Nilsson J and Chai G 1997 *International Conference & Expo Duplex*
[15] Kim S-T et al 2011 *Corros. Sci.* **53** 1939–47
[16] Kalácska E et al 2017 *YPIC 2017: 3rd Young Welding Professionals International Conference* 20–6
[17] Charles J 2007 *Rev. Métallurgie* **104** 308–17
[18] Charles J et al 2012 *Mater. Appl. Ser.* **12** 1–17
[19] Cazottes S et al 2010 *8th Duplex Stainl. Steels Conf.*
[20] Johansson E and Pettersson R 2011 *Outokumpu magazine* 1–23
[21] Westin E M and Hertzman S 2014 *Weld. World* **58** 143–60
[22] Renaudot N et al 2012 *Metall. Ital.* **104** 29–35
[23] Damian J. K 2010 *Soldag. Inspeção* **15** 336–43
[24] Hosseini V A et al 2017 *Mater. Corros.* **68** 405–15
[25] Ciuffini A et al 2017 *Metals (Basel).* **7** 368
[26] Llorca-Isern N et al 2016 *Mater. Charact.* **112** 20–9
[27] Pérez A F M et al 2016 *Soldag. Inspeção* **21** 165–71

IOP Conf. Series: Materials Science and Engineering **426** (2018) 012051 doi:10.1088/1757-899X/426/1/012051

[28] Meyer A M and du Toit M 2001 *Weld. J.* 275–80
[29] Hertzman S and Charles J 2011 *Rev. Métallurgie* **108** 413–25
[30] Hertzman S et al 1996 *ISIJ Int.* **36** 968–76
[31] Nilsson J-O 2016 *Stainl. Steel World* 26–7
[32] Liao J 2001 *ISIJ Int.* **41** 460–7
[33] Ramirez A J et al 2003 *Metall. Mater. Trans. A* **34** 1575–97
[34] Zhao L et al 2008 *Sci. Technol. Weld. Join.* **14** 87–93
[35] Elmer J W 2015 *Weld. J.* **94** 313–25
[36] Westin E M et al 2013 *Weld. World* **57** 467–76
[37] Sales A M et al 2016 *Weld. World* **60** 877–82
[38] Westin E M et al 2014 *Weld. World* **58** 347–54
[39] Vander Voort G F and Manilova E P 2005 *Adv. Mater. Processes* 32-37

IOP Conf. Series: Materials Science and Engineering **426** (2018) 012052 doi:10.1088/1757-899X/426/1/012052

Efficiency of different kinds of aluminium during deoxidation at the Steelworks of ISD Dunaferr Co. Ltd.

K Wizner[1] and A Kovari[2]

[1] ISD DunaferrCo. Dunaujvaros, Hungary,
[2] University of Dunaújvaros, Dunaujvaros, Hungary

E-mail: wizner.krisztian@isd-dunaferr.hu, kovari@uniduna.hu

Abstract. In all areas of life, cost-effectiveness is increasingly emphasized. It is true for materials used in metallurgical processes too. In many cases, however, not only the composition of the materials used affects the efficiency, but also their consistency. The use of various Al carrier alloys used for deoxidation during converter steel production differs due to different degrees of burns. In this article we will compare the utilization of cast (solid) "mokka" aluminium and the loose waste "shredder" aluminium used in ISD DunaferrZrt. Steel Works during the deoxidation process and the economics of the use of two deoxidation aluminiums.
The experiment involved a total of 67 heats. Taking into account the results of the statistical analysis, 3 heats were excluded from the heats deoxidized with "mokka" aluminium. The calculations related to the experiment were performed on the basis of 35 heats deoxidized with "mokka" aluminium and 29 heats pre-deoxidized with "shredder" aluminium. In the calculation of aluminium utilization it can be observed that the "mokka" aluminium is ~ 2.3% better utilized in comparison to the „shredder" aluminium and it can be observed also based on the total amount of aluminium delivered during the whole production process, that the total aluminium utilization is ~ 3.7% better for heats where pre-deoxidation was made with "mokka" aluminium.
On the basis of economical calculations it can be stated that the use of "shredder" aluminium is only economical, if its price per tonne does not exceed ~ 80% of the price of "mokka" aluminium.

1. Introduction

In a BOF process [1, 2], a mixture of solid steel scrap and liquid hot metal (Dunaferr practice is 20-25% solid steel scrap and 75-80% liquid hot metal) is blown by high purity oxygen, in consequence of which hot metal components (Si, Mn, C, Fe, etc.) burn out (oxidize) in a thermodynamic order [3–5] and develop heat. The resulting heat melts the solid scrap.

As a result of oxygen blowing, the steel bath is highly over-saturated with oxygen, which can reach 400-1200 ppm (work experiment). For the rest of the steel production, the presence of oxygen is harmful, therefore the active oxygen level required for steel after ladle treatment is 2-5 ppm (Dunaferrpractice, depending on their quality). There are basically two methods for removing active oxygen in steel: diffusion and precipitation [1, 2, 4].

The primary requirement for diffusion oxygen removal is to have enough time to process, but this is not feasible due to the production rate, as the average tap to tap time is ~ 45 minutes, of which ~ 20 minutes is the duration of oxygen blowing.

Therefore, under the operating conditions, the precipitation method is applied to the steel which has a higher oxygen potential than iron and the reaction with oxygen forms a separate (removable) phase at the steel production temperature [1, 2, 4].

The most commonly used material in precipitation deoxidation for Si free steels is aluminium [1, 2, 4], the oxide of which forms a solid phase at the steel production temperature. During steel tapping, the aluminium-oxygen reaction takes place rapidly.

$$2[Al] + 3[FeO] = (Al_2O_3) + 3[Fe]$$

Aluminium oxide inclusions remained in steel are detrimental to the purity of steel [6–9] because they cause casting problems, surface defects, and subsequent mechanical anisotropy in the rolled product. There is limited scope for removal, so it is important to minimize its generation [10–12].

The consistency of aluminium used for deoxidation greatly influences the place of oxidation.

The smallest aluminium pieces with high surface area burn in the atmosphere already prior to entering the ladle due to radiant heat and ambient oxygen. Medium-sized aluminium pieces with medium surface cannot sink deep enough into the steel bath and therefore they cannot utilize in steel and discards into the secondary slag covering the steel bath, which is not harmful but ineffective.

Only larger aluminium pieces with small surface can effectively perform their deoxidation tasks in the steel bath.

Since the amount of aluminium used to deoxidize the steel bath significantly influences the quality of the steels produced and the cost of steel production, we thought it is worth considering the extent to which they should be used in steel.

In the steel mill of ISD DunaferrZrt. cast (solid) "mokka" aluminium and loose waste "shredder" aluminium is used for deoxidation.

Figure 1. "Mokka" (left) and "shredder" (right) deoxidizing aluminium

IOP Conf. Series: Materials Science and Engineering **426** (2018) 012052 doi:10.1088/1757-899X/426/1/012052

2. Implementation of the experiment

Analysis of the samples taken from both types of aluminiums showed that the content of Al is in the range of 95-99%, but because of its consistency shredder contains also non-Al-based waste (e.g. copper-based faucet, foundry magnesium block, etc.), therefore the minimum Al-contents indicated by distributors have been taken into account for calculations.

Table 1. Minimum Al-content of deoxidation aluminium

Al carrier	Min. Al
Mokka	99%
Shredder	97%
Wire	99%

Aluminium was used for pre-deoxidation in the form of "mokka" for 38 heats and in the form of shredder for 29 heats, of the 67 heats of the experiment. Heats with outliers were excluded based on the Boxplot analysis. Taking into account the results of the analysis 3 heats, which were deoxidized with "mokka", were excluded from the experiment, two of them because of the Al content of the final sample (Figure 2.) and one of them because of the quantity of Al used (Figure 3.).

The following studies were based on the data of 35 "mokka" pre-deoxidized and 29 "shredder" pre-deoxidized heats.

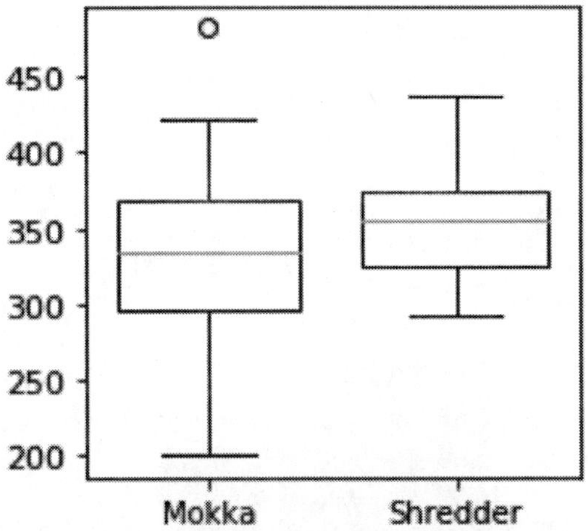

Figure 2. Al consumption analysis [kg]

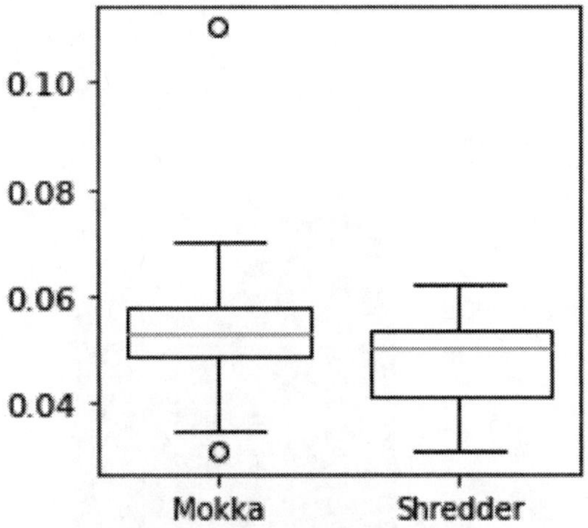

Figure 3. Final Al content analysis [%]

3. Determination of deoxidation Al utilization

In calculating the utilization, the aluminium burns resulting from the difference in the tapping temperature of heats produced using the two types of pre-deoxidizing agents and from the difference in the degree of active oxygen content are ignored because the aluminium content of steel does not change significantly due to these differences (only the 4th decimal is influenced during the measurement with the accurancy of 4 decimals). The tapping temperature and the measured active oxygen concentration are shown in Table 2.

Table 2. Steel temperature and active oxygen measurement results before tapping

	Temperature (°C)	O (ppm)
Mokka	1683	662
Shredder	1678	689

It is apparent from Table 3 that aluminium utilization of "mokka" aluminium is about 2.3% higher than that of the "shredder" aluminium during pre-deoxidation.

Table 3. Al-utilization of pre-deoxidizing Al

	Mokka	Shredder	
Pre-deoxidizing Al	277	298	kg
Al dissolved in steel	79,1	78,7	kg
Al utilization	28,6	26,4	%

Aluminium utilization throughout the process can only be observed by taking into account also the amount of correction Al-wire supplied at the ladle treatment station. In Table 4, the utilization of the Al amount of all deoxidation aluminium ("mokka" or "shredder" + wire) shows that total aluminium utilization is ~ 3.7% better when "mokka" aluminium is charged. Figure 4.and Figure 5. illustrate the utilization during BOF process of aluminium calculated based on the aluminium content of the sample taken after pre-deoxidation and that of the final sample.

Table 4. Utilization of all aluminium until final sampling

	Mokka	Shredder	
Total deoxidizing Al	325	344	kg
Al dissolved in steel	74,4	65,8	kg
Al utilization	22,9	19,1	%

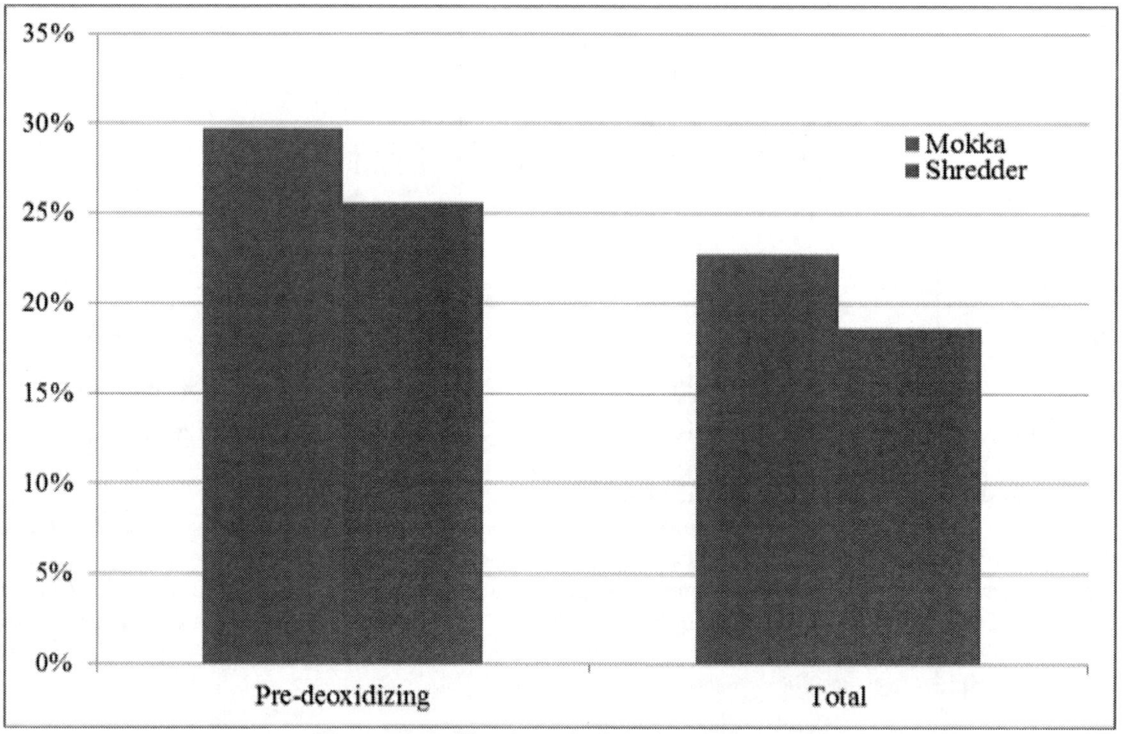

Figure 4. Utilization of Al during BOF process

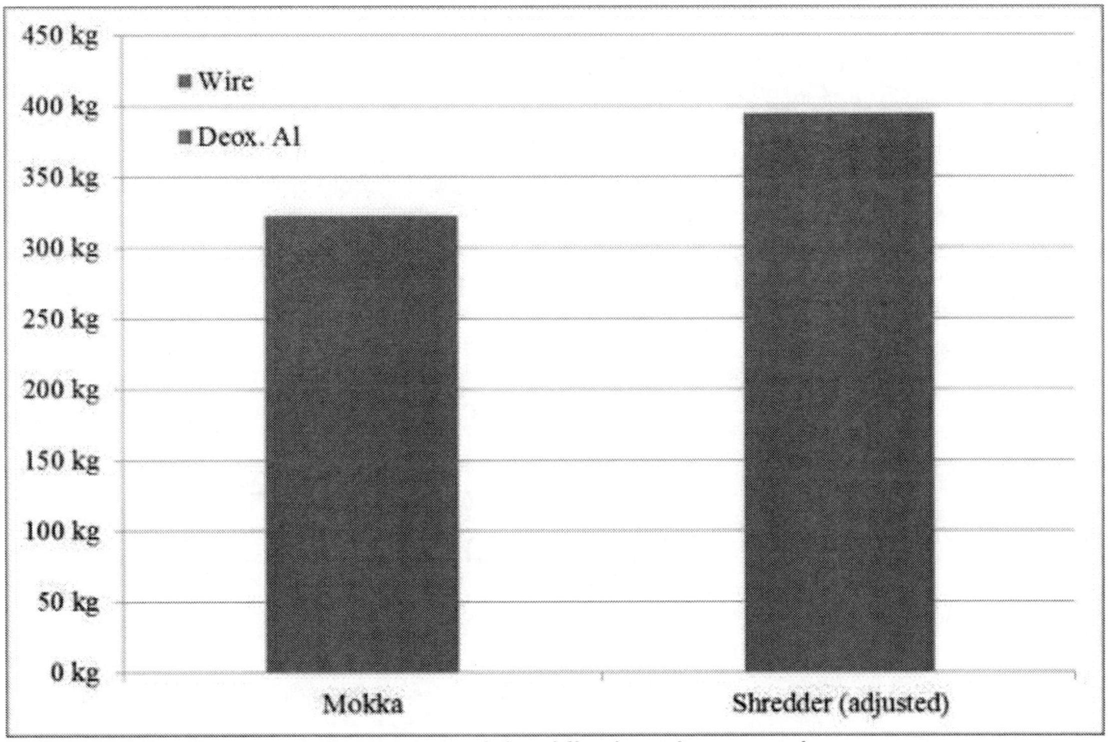

Figure 5. Total deoxidization Al consumption

4. Economic calculations

To perform the economic calculation, the amount of Al-wire used in the heats produced with the two types of pre-deoxidationaluminiums (Table 5) and the Al content of the final sample (Table 6) were tested by a statistical two-sample t-test.

Table 5. Results of two-sample t-test for the used Al-wire amount

	equal variance		non equal variance	
	Mokka	Shredder	Mokka	Shredder
Expected value	49,23	46,41	49,23	46,41
t value	0,4016		0,4269	
P(T<=t) double edged	0,6893		0,6709	
t critical double edged	19971		1,9983	

The results show that the difference between the Al-wire amounts used in addition to the two types of pre-deoxidizing agents does not significantly affect the result of the economic calculation and can be regarded as statistically identical. It is clear that the economy is influenced only by the price of the two different types of pre-deoxidizing agents and that the cost of Al-wire is negligible in comparison.

Table 6. Two-sample t-test for the final Al content

	equal variance		non equal variance	
	Mokka	Shredder	Mokka	Shredder
Expected value	0,0531	0,0477	0,0531	0,0477
t value	2,7188		2,6940	
P(T<=t) double edged	0,0085		0,0093	
t critical double edged	1,9990		2,0025	

IOP Conf. Series: Materials Science and Engineering **426** (2018) 012052 doi:10.1088/1757-899X/426/1/012052

The results show that the difference in the Al content in the final sample of heats produced with the two types of pre-deoxidizing agents is statistically not negligible, for comparability correction is required.

In the calculation of the correction, the amount of "shredder" aluminium (including burns and purity) was determined, with which the aluminium content of the final sample would equal to the aluminium content of the "mokka" aluminium at pre-deoxidized heats in the final sample. Based on the expected value of the two-sample t-test, an additional 40.54 kg of Al „shredder" is required to increase the Al content by ~ 0.005%. (Wire data are only for information)

The corrected results are shown in Table 7, and the specific results are shown in Table 8.

Table 7. The amount of deoxidation Al added during steel production

	Pre-deoxidizing Al	Al wire	ΣAl	
Mokka	279	49,2	329	kg
Shredder (adjusted)	348	46,4	395	kg

Table 8. Specific amount of deoxidation Al added during steel production

	Pre-deoxidizing Al	Al wire	ΣAl	
Mokka	2,02	0,36	2,38	kg/t
Shredder (adjusted)	2,52	0,34	2,86	kg/t

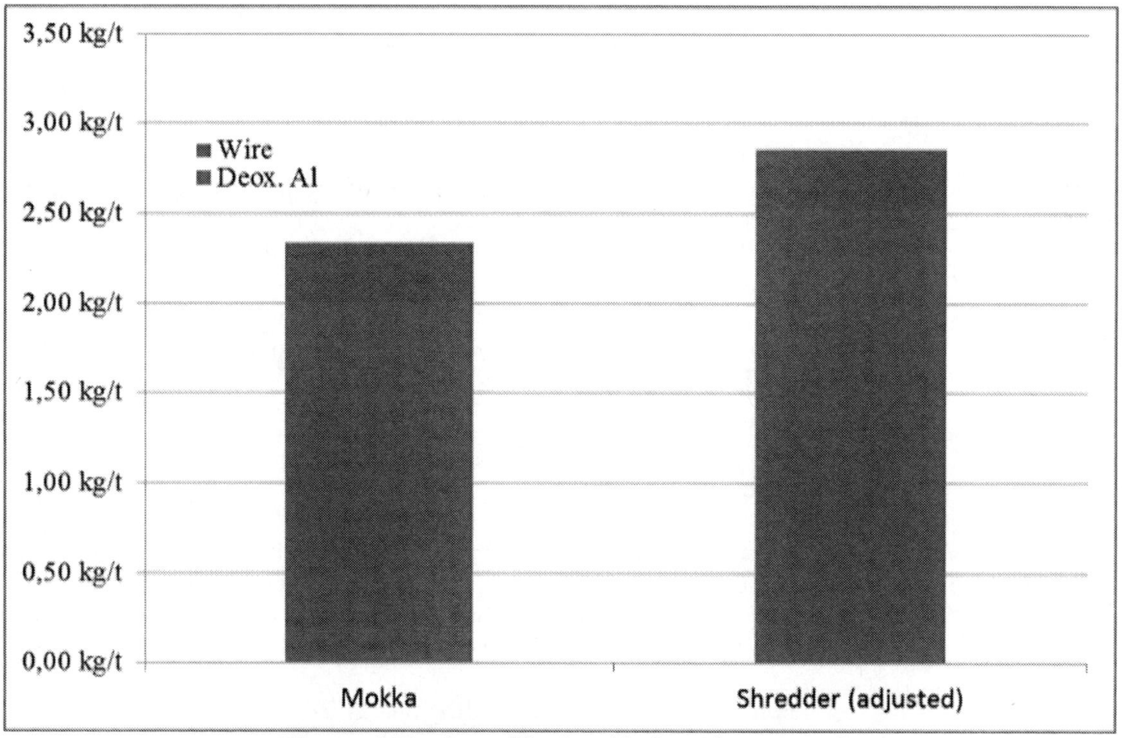

Figure 6. Total specific deoxidation Al consumption

The calculations made show that if the price of "shredder" per tonne exceeds ~ 80% of the price per tonne of "mokka", its use is no longer economical.

5. Summary

The experiment involved a total of 67 heats.

The experiment was based on the data of 35 "mokka" pre-deoxidized and 29 „shredder" pre-deoxidized heats.

In the calculation of aluminium utilization it can be observed that the "mokka" aluminium is ~2.3% better utilized in comparison to the "shredder" aluminium during pre-deoxidization and ~3.7% during the whole production process.

On the basis of economical calculations it can be stated that the use of shredder aluminium is only economical, if its price per tonne does not exceed ~ 80% of the price of "mokka" aluminium.

Acknowledgement

This publication is supported by the EFOP-3.6.1-16-2016-00003 project. The project is co-financed by the European Union.

References

[1] Simon S, Károly Gy 1983 *Acélkohászattan I.* (Budapest: Tankönyvkiadó) p. 215
[2] Horváth Z, Sziklavári K, Mihalik Á 1974 *Elméleti kohászattan* (Budapest: Tankönyvkiadó) p. 434
[3] Szabó Z 2006 *Alacsony zárványtartalmú acél gyártásának feltételei* (Dunaferr Műszaki Gazdasági Közlemények Vol 4.) pp. 203-208
[4] Károly Gy 2012/2013 Az *acélmetallurgia alapjai* (University of Miskolc Digital book) p. 127
[5] Simon S, Sziklavári K, SzőkeL 1974*Újabb technológiai megoldások az acélgyártásban* (Budapest: Műszaki Könyvkiadó) p. 402
[6] Szabó A 2009 *Szilícium-szegény, alumíniummal dezoxidált acél tisztaságának javítása argonos átöblítés finomításával* (Miskolc: University of Miskolc PhD dissertation) p. 143
[7] Károly Gy, Szélig Á, Takácsné Szabó A, Kiss L 2007 *Az Argonos átöblítés hatékonyságának szerepe az acélok tisztaságának javításában* (Miskolc: Mechatronika, Anyagtudomány Vol. 1, No. 3) pp. 73-86
[8] Zhang L, Pluschkell W, ThomasB. G. 2002 *Nucleation and growth of alumina inclusion during steel deoxidation* (Nashville, TN: 85th Steelmaking Conference, Vol.85) (Warrendale, PA: ISS), pp. 463-476
[9] Jin Y, Liu Z, TakataR 2010*Nucleation and Growth of Alumina Inclusion in Early Stages of Deoxidation: Numerical Modeling* (ISIJ International, Vol. 50, No. 3) pp. 371–379
[10] Wakoh M, Sano N 2007 *Behavior of Alumina Inclusions Just after Deoxidation* (ISIJ International, Vol. 47, No. 5) pp. 627–632
[11] Deng Z, Zhu M 2014 *Deoxidation Mechanism of Al-Killed Steel during Industrial Refining Process* (ISIJ International, Vol. 54, No. 7) pp. 1498–1506
[12] Bessho N, Yamasaki H, Fujii T, Nozaki T, Hiwasa S 1992 *Removal of Inclusion from Molten Steel in Continuous Casting Tundish* (ISIJ International, Vol. 32, No. 1) pp. 157-163

Institute of Physics
Dirac House, Temple Back
Bristol BS1 6BE UK

ISSN: 1757-8981
ISBN 978-1-5108-7365-0

CURRAN ASSOCIATES INC.
proceedings.com